Springer Proceedings in Mathematics & Statistics

Volume 155

More information about this series at http://www.springer.com/series/10533

Springer Proceedings in Mathematics & Statistics

This book series features volumes composed of select contributions from workshops and conferences in all areas of current research in mathematics and statistics, including OR and optimization. In addition to an overall evaluation of the interest, scientific quality, and timeliness of each proposal at the hands of the publisher, individual contributions are all refereed to the high quality standards of leading journals in the field. Thus, this series provides the research community with well-edited, authoritative reports on developments in the most exciting areas of mathematical and statistical research today.

George A. Anastassiou • Oktay Duman
Editors

Computational Analysis

AMAT, Ankara, May 2015 Selected Contributions

Springer

Editors
George A. Anastassiou
Department of Mathematical Sciences
University of Memphis
Memphis, TN, USA

Oktay Duman
Department of Mathematics
TOBB University of Economics and Technology
Ankara, Turkey

ISSN 2194-1009 ISSN 2194-1017 (electronic)
Springer Proceedings in Mathematics & Statistics
ISBN 978-3-319-28441-5 ISBN 978-3-319-28443-9 (eBook)
DOI 10.1007/978-3-319-28443-9

Library of Congress Control Number: 2016937456

Printed on acid-free paper

This Springer imprint is published by SpringerNature
The registered company is Springer International Publishing AG Switzerland

*Dedicated to **World Peace**!*

Attendees of AMAT 2015, TOBB University of Economics and Technology,
Ankara, Turkey, May 28–31

George A. Anastassiou and Oktay Duman
Ankara, Turkey, May 29, 2015

Preface

This special volume consists of selected papers of more theoretical nature presented in AMAT 2015 Conference—3rd International Conference on Applied Mathematics and Approximation Theory—which was held during May 28–31, 2015, in Ankara, Turkey, at TOBB University of Economics and Technology.

The AMAT 2015 Conference brought together researchers from all areas of applied mathematics and approximation theory, such as ODEs, PDEs, difference equations, applied analysis, computational analysis, harmonic analysis, signal theory, positive operators, statistical approximation, fuzzy approximation, fractional analysis, semigroups, inequalities, special functions, and summability. Previous conferences which had a similar approach to such diverse inclusiveness were held at the University of Memphis in 2008 and at TOBB Economics and Technology University in 2012.

We are particularly indebted to the Organizing Committee and the Scientific Committee for their great efforts. We also appreciate the plenary speakers: George A. Anastassiou (University of Memphis, USA), Martin Bohner (Missouri University of Science and Technology, USA), Alexander Goncharov (Bilkent University, Turkey), Varga Kalantarov (Koç University, Turkey), Gitta Kutyniok (Technische Universitt Berlin, Germany), Choonkil Park (Hanyang University, South Korea), Mircea Sofonea (University of Perpignan, France), and Tamaz Vashakmadze (Tbilisi State University, Georgia).

We would like also to thank the anonymous reviewers who helped us select the best articles for inclusion in this proceedings volume and also to the authors for their valuable contributions.

Finally, we are grateful to TOBB University of Economics and Technology, which hosted this conference and provided all of its facilities and also to the Central Bank of Turkey for financial support.

Memphis, TN, USA
Ankara, Turkey
November 1, 2015

George A. Anastassiou
Oktay Duman

Contents

List of Contributors

Hande Günay Akdemir Department of Mathematics, Giresun University, Turkey

Suad A. Alhihi Department of Mathematics, Al-Balqa Applied University, Alsalt, Jordan

Gökalp Alpan Department of Mathematics, Bilkent University, Ankara, Turkey

Birol Altın Department of Mathematics, Faculty of Science, Gazi University, Ankara, Turkey

Osman Altıntaş Departments of Mathematics Education & Secondary Education, Institute of Education Sciences, University of Başkent, Ankara, Turkey

Aisha Ahmed Amer Department of Mathematics, University of Al-Mergeb, Al-Khoms, Libya

Fariha Jumaa Amer Department of Mathematics, University of Al-Mergeb, Al-Khoms, Libya

George A. Anastassiou Department of Mathematical Sciences, University of Memphis, TN, USA

Yelda Aygar Department of Mathematics, Faculty of Science, Ankara University, Turkey

Nurgül Okur Bekar Departments of Statistics and Mathematics, Giresun University, Turkey

Tahira Nasreen Buttar Department of Mathematics, Lahore College for Women University, Pakistan

Olga Danilkina Department of Mathematics, The University of Dodoma, Tanzania

Ülkü Dinlemez Department of Mathematics, Faculty of Science, Gazi University, Ankara, Turkey

Oktay Duman Department of Mathematics, TOBB Economics and Technology University, Ankara, Turkey

Türkan Yeliz Gökçer Department of Mathematics, TOBB Economics and Technology University, Ankara, Turkey

Alexander Goncharov Department of Mathematics, Bilkent University, Ankara, Turkey

Burak Hatinoğlu Texas A&M University, College Station, TX, USA

Pembe Ipek Department of Mathematics, Karadeniz Technical University, Trabzon, Turkey

İmdat İşcan Department of Mathematics, Giresun University, Turkey

Zameddin I. Ismailov Department of Mathematics, Karadeniz Technical University, Trabzon, Turkey

Jamila Kalantarova Department of Mathematics, Yeditepe University, Istanbul, Turkey

Merve Kester Department of Mathematical Sciences, University of Memphis, TN, USA

Cemaliye Kürt Department of Mathematics, Eastern Mediterranean University, Famagusta, North Cyprus

Jung Rye Lee Department of Mathematics, Daejin University, Pocheon, South Korea

Alina Alb Lupaş Department of Mathematics and Computer Science, University of Oradea, Romania

Igor Neygebauer Department of Mathematics, University of Dodoma, Tanzania

Murat Olgun Department of Mathematics, Faculty of Science, Ankara University, Turkey

Mehmet Ali Özarslan Department of Mathematics, Eastern Mediterranean University, Famagusta, North Cyprus

Choonkil Park Department of Mathematics, Hanyang University, Seoul, South Korea

Won-Gil Park Mokwon University, Daejeon, South Korea

Mustafa Polat Department of Mathematics, Yeditepe University, Istanbul, Turkey

Themistocles M. Rassias National Technical University of Athens, Greece

Reza Saadati Iran University of Science and Technology, Tehran, Iran

Naila Sajid Department of Mathematics, Lahore College for Women University, Pakistan

Dong Yun Shin Department of Mathematics, University of Seoul, South Korea

Abdalla Tallafha Department of Mathematics, The University of Jordan, Amman, Jordan

Emre Taş Department of Mathematics, Ahi Evran University, Kırşehir, Turkey

Kemal Uslu Department of Mathematics, Selcuk University, Konya, Turkey

Şükran Uygun Department of Mathematics, Gaziantep University, Turkey

Tuba Vedi Department of Mathematics, Eastern Mediterranean University, Famagusta, North Cyprus

İlker Vural Domaniç Vocational School, Dumlupınar University, Kütahya, Turkey

İsmet Yüksel Department of Mathematics, Faculty of Science, Gazi University, Ankara, Turkey

Tuğba Yurdakadim Department of Mathematics, Hitit University, Çorum, Turkey

Slimane Zaim Département des Sciences de la Mtière, Faculté des Sciences, Université Hadj Lakhdar - Batna, Algeria

Chapter 1
Bivariate Left Fractional Pseudo-Polynomial Monotone Approximation

George A. Anastassiou

Abstract In this article we deal with the following general two-dimensional problem: Let f be a two variable continuously differentiable real-valued function of a given order, let L^* be a linear left fractional mixed partial differential operator and suppose that $L^*(f) \geq 0$ on a critical region. Then for sufficiently large $n, m \in \mathbb{N}$, we can find a sequence of pseudo-polynomials $Q^*_{n,m}$ in two variables with the property $L^*\left(Q^*_{n,m}\right) \geq 0$ on this critical region such that f is approximated with rates fractionally and simultaneously by $Q^*_{n,m}$ in the uniform norm on the whole domain of f. This restricted approximation is given via inequalities involving the mixed modulus of smoothness $\omega_{s,q}$, $s, q \in \mathbb{N}$, of highest order integer partial derivative of f.

1.1 Introduction

The topic of monotone approximation started in [10] has become a major trend in approximation theory. A typical problem in this subject is: given a positive integer k, approximate a given function whose kth derivative is ≥ 0 by polynomials having this property.

In [3] the authors replaced the kth derivative with a linear differential operator of order k. We mention this motivating result.

Theorem 1.1. *Let h, k, p be integers, $0 \leq h \leq k \leq p$ and let f be a real function, $f^{(p)}$ continuous in $[-1, 1]$ with modulus of continuity $\omega_1\left(f^{(p)}, x\right)$ there. Let $a_j(x)$, $j = h, h+1, \ldots, k$ be real functions, defined and bounded on $[-1, 1]$ and assume $a_h(x)$ is either $\geq$ some number $\alpha > 0$ or $\leq$ some number $\beta < 0$ throughout $[-1, 1]$. Consider the operator*

$$L = \sum_{j=h}^{k} a_j(x) \left[\frac{d^j}{dx^j}\right] \tag{1.1}$$

G.A. Anastassiou (✉)
Department of Mathematical Sciences, University of Memphis, TN 38152, USA
e-mail: ganastss@memphis.edu

G.A. Anastassiou, O. Duman (eds.), *Computational Analysis*, Springer Proceedings in Mathematics & Statistics 155, DOI 10.1007/978-3-319-28443-9_1

and suppose, throughout $[-1, 1]$,

$$L(f) \geq 0. \tag{1.2}$$

Then, for every integer $n \geq 1$, *there is a real polynomial* $Q_n(x)$ *of degree* $\leq n$ *such that*

$$L(Q_n) \geq 0 \text{ throughout } [-1, 1] \tag{1.3}$$

and

$$\max_{-1 \leq x \leq 1} |f(x) - Q_n(x)| \leq Cn^{k-p}\omega_1\left(f^{(p)}, \frac{1}{n}\right), \tag{1.4}$$

where C *is independent of* n *or* f.

Next let $n, m \in \mathbb{Z}_+$, P_θ denote the space of algebraic polynomials of degree $\leq \theta$. Consider the tensor product spaces $P_n \otimes C([-1, 1])$, $C([-1, 1]) \otimes P_m$ and their sum $P_n \otimes C([-1, 1]) + C([-1, 1]) \otimes P_m$, that is

$$\begin{aligned} &P_n \otimes C([-1, 1]) + C([-1, 1]) \otimes P_m \\ &= \left\{ \sum_{i=0}^{n} x^i A_i(y) + \sum_{j=0}^{m} B_j(x) y^j;\ A_i, B_j \in C([-1, 1]),\ x, y \in [-1, 1] \right\}. \end{aligned} \tag{1.5}$$

This is the space of pseudo-polynomials of degree $\leq (n, m)$, first introduced by A. Marchaud in 1924–1927 (see [7, 8]). Here $f^{(k,l)}$ denotes $\frac{\partial^{k+l} f}{\partial x^k \partial y^l}$, the (k, l)-partial derivative of f.

In this section we consider the space $C^{r,p}\left([-1, 1]^2\right) = \{f : [-1, 1]^2 \to \mathbb{R}; f^{(k,l)}$ is continuous for $0 \leq k \leq r, 0 \leq l \leq p\}$. Let $f \in C\left([-1, 1]^2\right)$; for $\delta_1, \delta_2 \geq 0$, define the mixed modulus of smoothness of order (s, q), $s, q \in \mathbb{N}$ (see [9, pp. 516–517]) by

$$\begin{aligned} \omega_{s,q}(f; \delta_1, \delta_2) \equiv \sup \Big\{ &\left| {}_x\Delta^s_{h_1} \circ_y \Delta^q_{h_2} f(x, y) \right| : (x, y), \\ &(x + sh_1, y + qh_2) \in [-1, 1]^2, |h_i| \leq \delta_i, i = 1, 2 \Big\}. \end{aligned} \tag{1.6}$$

Here

$$\begin{aligned} &{}_x\Delta^s_{h_1} \circ_y \Delta^q_{h_2} f(x, y) \\ &\equiv \sum_{\sigma=0}^{s} \sum_{\mu=0}^{q} (-1)^{s+q-\sigma-\mu} \binom{s}{\sigma} \binom{q}{\mu} f(x + \sigma h_1, y + \mu h_2) \end{aligned} \tag{1.7}$$

is a mixed difference of order (s, q).

We mention

Theorem 1.2 (Gonska [4]). *Let* $r, p \in \mathbb{Z}_+$, $s, q \in \mathbb{N}$, *and* $f \in C^{r,p}\left([-1,1]^2\right)$. *Let* $n, m \in \mathbb{N}$ *with* $n \geq \max\{4(r+1), r+s\}$ *and* $m \geq \max\{4(p+1), p+q\}$. *Then there exists a linear operator* $Q_{n,m}$ *from* $C^{r,p}\left([-1,1]^2\right)$ *into the space of pseudo-polynomials* $(P_n \otimes C([-1,1]) + C([-1,1]) \otimes P_m)$ *such that*

$$\left|(f - Q_{n,m}(f))^{(k,l)}(x,y)\right|$$
$$\leq M_{r,s} \cdot M_{p,q} (\Delta_n(x))^{r-k} \cdot (\Delta_m(y))^{p-l} \cdot \omega_{s,q}\left(f^{(r,p)}; \Delta_n(x), \Delta_m(y)\right), \quad (1.8)$$

for all $(0,0) \leq (k,l) \leq (r,p)$, $x, y \in [-1,1]$, *where*

$$\Delta_\theta(z) = \frac{\sqrt{1-z^2}}{\theta} + \frac{1}{\theta^2}, \quad \theta = n, m;\ z = x, y \in [-1,1]. \quad (1.9)$$

The constants $M_{r,s}$, $M_{p,q}$ *are independent of* f, (x,y) *and* (n,m); *they depend only on* (r,s), (p,q), *respectively.*

See also [5], saying that $Q_{n,m}^{(r,p)}(f)$ is continuous on $[-1,1]^2$.

We need the following result which is an easy consequence of the last theorem (see [9, p. 517]).

Corollary 1.3. *Let* $r, p \in \mathbb{Z}_+$, $s, q \in \mathbb{N}$, *and* $f \in C^{r,p}\left([-1,1]^2\right)$. *Let* $n, m \in \mathbb{N}$ *with* $n \geq \max\{4(r+1), r+s\}$ *and* $m \geq \max\{4(p+1), p+q\}$. *Then there exists a pseudo-polynomial* $Q_{n,m} \equiv Q_{n,m}(f) \in (P_n \otimes C([-1,1]) + C([-1,1]) \otimes P_m)$ *such that*

$$\left\| f^{(k,l)} - Q_{n,m}^{(k,l)} \right\|_\infty \leq \frac{\dot{C}}{n^{r-k} m^{p-l}} \cdot \omega_{s,q}\left(f^{(r,p)}; \frac{1}{n}, \frac{1}{m}\right), \quad (1.10)$$

for all $(0,0) \leq (k,l) \leq (r,p)$. *Here the constant* $\dot{C}$ *depends only on* r, p, s, q.

Corollary 1.3 was used in the proof of the main motivational result that follows.

Theorem 1.4 ([1]). *Let* h_1, h_2, v_1, v_2, r, p *be integers,* $0 \leq h_1 \leq v_1 \leq r$, $0 \leq h_2 \leq v_2 \leq p$ *and let* $f \in C^{r,p}\left([-1,1]^2\right)$, *with* $f^{(r,p)}$ *having a mixed modulus of smoothness* $\omega_{s,q}\left(f^{(r,p)}; x, y\right)$ *there,* $s, q \in \mathbb{N}$. *Let* $\alpha_{i,j}(x,y)$, $i = h_1, h_1+1, \ldots, v_1$; $j = h_2, h_2+1, \ldots, v_2$ *be real-valued functions, defined and bounded in* $[-1,1]^2$ *and suppose* $\alpha_{h_1 h_2}$ *is either* $\geq \alpha > 0$ *or* $\leq \beta < 0$ *throughout* $[-1,1]^2$. *Take the operator*

$$L = \sum_{i=h_1}^{v_1} \sum_{j=h_2}^{v_2} \alpha_{ij}(x,y) \frac{\partial^{i+j}}{\partial x^i \partial y^j} \quad (1.11)$$

and assume, throughout $[-1,1]^2$, *that*

$$L(f) \geq 0. \quad (1.12)$$

Then for any integers n, m with $n \geq \max\{4(r+1), r+s\}$, $m \geq \max\{4(p+1), p+q\}$, there exists a pseudo-polynomial

$$Q_{n,m} \in (P_n \otimes C([-1,1]) + C([-1,1]) \otimes P_m)$$

such that $L(Q_{m,n}) \geq 0$ throughout $[-1,1]^2$ and

$$\left\| f^{(k,l)} - Q_{n,m}^{(k,l)} \right\|_\infty \leq \frac{C}{n^{r-v_1} m^{p-v_2}} \cdot \omega_{s,q} \left(f^{(r,p)}; \frac{1}{n}, \frac{1}{m} \right), \tag{1.13}$$

for all $(0,0) \leq (k,l) \leq (h_1, h_2)$. Moreover we get

$$\left\| f^{(k,l)} - Q_{n,m}^{(k,l)} \right\|_\infty \leq \frac{C}{n^{r-k} m^{p-l}} \cdot \omega_{s,q} \left(f^{(r,p)}; \frac{1}{n}, \frac{1}{m} \right),$$

for all $(h_1+1, h_2+1) \leq (k,l) \leq (r,p)$. Also (1.13) is valid whenever $0 \leq k \leq h_1$, $h_2 + 1 \leq l \leq p$ or $h_1 + 1 \leq k \leq r$, $0 \leq l \leq h_2$. Here C is a constant independent of f and n, m. It depends only on r, p, s, q, L.

We are also motivated by Anastassiou [2].

We need

Definition 1.5 (See [6]). Let $[-1,1]^2$; $\alpha_1, \alpha_2 > 0$; $\alpha = (\alpha_1, \alpha_2)$, $f \in C\left([-1,1]^2\right)$, $x = (x_1, x_2)$, $t = (t_1, t_2) \in [-1,1]^2$. We define the left mixed Riemann–Liouville fractional two dimensional integral of order α

$$\left(I_{-1+}^{\alpha} f\right)(x) := \frac{1}{\Gamma(\alpha_1)\Gamma(\alpha_2)} \int_{-1}^{x_1} \int_{-1}^{x_2} (x_1 - t_1)^{\alpha_1 - 1} (x_2 - t_2)^{\alpha_2 - 1} f(t_1, t_2)\, dt_1 dt_2, \tag{1.14}$$

with $x_1, x_2 > -1$.

Notice here that $I_{-1+}^{\alpha}(|f|) < \infty$.

Definition 1.6. Let $\alpha_1, \alpha_2 > 0$ with $\lceil \alpha_1 \rceil = m_1$, $\lceil \alpha_2 \rceil = m_2$, ($\lceil \cdot \rceil$ ceiling of the number). Let here $f \in C^{m_1, m_2}\left([-1,1]^2\right)$. We consider the left Caputo type fractional partial derivative:

$$D_{*(-1)}^{(\alpha_1, \alpha_2)} f(x) := \frac{1}{\Gamma(m_1 - \alpha_1)\Gamma(m_2 - \alpha_2)} \times \int_{-1}^{x_1} \int_{-1}^{x_2} (x_1 - t_1)^{m_1 - \alpha_1 - 1} (x_2 - t_2)^{m_2 - \alpha_2 - 1} \frac{\partial^{m_1 + m_2} f(t_1, t_2)}{\partial t_1^{m_1} \partial t_2^{m_2}} dt_1 dt_2, \tag{1.15}$$

$\forall\, x = (x_1, x_2) \in [-1, 1]^2$, where Γ is the gamma function

$$\Gamma(\nu) = \int_0^\infty e^{-t} t^{\nu-1} dt, \quad \nu > 0. \tag{1.16}$$

We set

$$D_{*(-1)}^{(0,0)} f(x) := f(x), \quad \forall\, x \in [-1, 1]^2; \tag{1.17}$$

$$D_{*(-1)}^{(m_1,m_2)} f(x) := \frac{\partial^{m_1+m_2} f(x)}{\partial x_1^{m_1} \partial x_2^{m_2}}, \quad \forall\, x \in [-1, 1]^2. \tag{1.18}$$

Definition 1.7. We also set

$$D_{*(-1)}^{(0,\alpha_2)} f(x) := \frac{1}{\Gamma(m_2 - \alpha_2)} \int_{-1}^{x_2} (x_2 - t_2)^{m_2-\alpha_2-1} \frac{\partial^{m_2} f(x_1, t_2)}{\partial t_2^{m_2}} dt_2, \tag{1.19}$$

$$D_{*(-1)}^{(\alpha_1,0)} f(x) := \frac{1}{\Gamma(m_1 - \alpha_1)} \int_{-1}^{x_1} (x_1 - t_1)^{m_1-\alpha_1-1} \frac{\partial^{m_1} f(t_1, x_2)}{\partial t_1^{m_1}} dt_1, \tag{1.20}$$

and

$$D_{*(-1)}^{(m,\alpha_2)} f(x) := \frac{1}{\Gamma(m_2 - \alpha_2)} \int_{-1}^{x_2} (x_2 - t_2)^{m_2-\alpha_2-1} \frac{\partial^{m_1+m_2} f(x_1, t_2)}{\partial x_1^{m_1} \partial t_2^{m_2}} dt_2, \tag{1.21}$$

$$D_{*(-1)}^{(\alpha_1,m_2)} f(x) := \frac{1}{\Gamma(m_1 - \alpha_1)} \int_{-1}^{x_1} (x_1 - t_1)^{m_1-\alpha_1-1} \frac{\partial^{m_1+m_2} f(t_1, x_2)}{\partial t_1^{m_1} \partial x_2^{m_2}} dt_1. \tag{1.22}$$

In this article we extend Theorem 1.4 to the fractional level. Indeed here L is replaced by L^*, a linear left Caputo fractional mixed partial differential operator. Now the monotonicity property holds true only on the critical square of $[0, 1]^2$. Simultaneously fractional convergence remains true on all of $[-1, 1]^2$.

1.2 Main Result

We present

Theorem 1.8. *Let h_1, h_2, v_1, v_2, r, p be integers, $0 \leq h_1 \leq v_1 \leq r$, $0 \leq h_2 \leq v_2 \leq p$ and let $f \in C^{r,p}\left([-1,1]^2\right)$, with $f^{(r,p)}$ having a mixed modulus of smoothness $\omega_{s,q}\left(f^{(r,p)}; x, y\right)$ there, $s, q \in \mathbb{N}$. Let $\alpha_{ij}(x,y)$, $i = h_1, h_1+1, \ldots, v_1$; $j = h_2, h_2+1, \ldots, v_2$ be real-valued functions, defined and bounded in $[-1,1]^2$ and suppose $\alpha_{h_1 h_2}$ is either $\geq \alpha > 0$ or $\leq \beta < 0$ throughout $[0,1]^2$. Here $n, m \in \mathbb{N} : n \geq \max\{4(r+1), r+s\}$, $m \geq \max\{4(p+1), p+q\}$. Set*

$$l_{ij} := \sup_{(x,y)\in[-1,1]^2} \left|\alpha_{h_1 h_2}^{-1}(x,y)\, \alpha_{ij}(x,y)\right| < \infty \tag{1.23}$$

for all $h_1 \leq i \leq v_1$, $h_2 \leq j \leq v_2$. Let $\alpha_{1i}, \alpha_{2j} \geq 0$ with $\lceil \alpha_{1i} \rceil = i$, $\lceil \alpha_{2j} \rceil = j$, $i = 0, 1, \ldots, r$; $j = 0, 1, \ldots, p$, ($\lceil \cdot \rceil$ ceiling of the number), $\alpha_{10} = 0$, $\alpha_{20} = 0$. Consider the left fractional bivariate differential operator

$$L^* := \sum_{i=h_1}^{v_1} \sum_{j=h_2}^{v_2} \alpha_{ij}(x,y)\, D_{*(-1)}^{(\alpha_{1i},\alpha_{2j})}. \tag{1.24}$$

Assume $L^* f(x,y) \geq 0$, on $[0,1]^2$. Then there exists

$$Q_{n,m}^* \equiv Q_{n,m}^*(f) \in \left(P_n \otimes C([-1,1]) + C([-1,1]) \otimes P_m\right)$$

such that $L^* Q_{n,m}^*(x,y) \geq 0$, on $[0,1]^2$. Furthermore it holds:

(1)

$$\left\| D_{*(-1)}^{(\alpha_{1i},\alpha_{2j})}(f) - D_{*(-1)}^{(\alpha_{11},\alpha_{2j})} Q_{n,m}^* \right\|_{\infty,[-1,1]^2} \leq \frac{\dot{C} 2^{(i+j)-(\alpha_{1i}+\alpha_{2j})}}{\Gamma(i-\alpha_{1i}+1)\,\Gamma\left(j-\alpha_{2j}+1\right) n^{r-i} m^{p-j}} \cdot \omega_{s,q}\left(f^{(r,p)}; \frac{1}{n}, \frac{1}{m}\right), \tag{1.25}$$

where $\dot{C}$ is a constant that depends only on r, p, s, q; $(h_1+1, h_2+1) \leq (i,j) \leq (r,p)$, or $0 \leq i \leq h_1$, $h_2+1 \leq j \leq p$, or $h_1+1 \leq i \leq r$, $0 \leq j \leq h_2$,

(2)

$$\left\| D_{*(-1)}^{(\alpha_{1i},\alpha_{2j})}(f) - D_{*(-1)}^{(\alpha_{1i},\alpha_{2j})} Q_{n,m}^* \right\|_{\infty,[-1,1]^2} \leq \frac{c_{ij}}{n^{r-v_1} m^{p-v_2}} \cdot \omega_{s,q}\left(f^{(r,p)}; \frac{1}{n}, \frac{1}{m}\right) \tag{1.26}$$

for $(1,1) \leq (i,j) \leq (h_1, h_2)$, *where* $c_{ij} = \dot{C}A_{ij}$, *with*

$$A_{ij} := \left\{ \left[\sum_{\tau=h_1}^{v_1} \sum_{\mu=h_2}^{v_2} \frac{l_{\tau\mu} 2^{(\tau+\mu)-(\alpha_{1\tau}+\alpha_{2\mu})}}{\Gamma(\tau - a_{1\tau} + 1)\, \Gamma(\mu - \alpha_{2\mu} + 1)} \right] \right.$$

$$\times \left(\sum_{k=0}^{h_1-i} \frac{2^{h_1-\alpha_{1i}-k}}{k!\,\Gamma(h_1 - \alpha_{1i} - k + 1)} \right) \left(\sum_{\lambda=0}^{h_2-j} \frac{2^{h_2-\alpha_{2j}-\lambda}}{\lambda!\,\Gamma(h_2 - \alpha_{2j} - \lambda + 1)} \right)$$

$$\left. + \frac{2^{(i+j)-(\alpha_{1i}+\alpha_{2j})}}{\Gamma(i - \alpha_{1i} + 1)\, \Gamma(j - \alpha_{2j} + 1)} \right\} \tag{1.27}$$

(3)

$$\left\| f - Q^*_{n,m} \right\|_{\infty,[-1,1]^2} \leq \frac{c_{00}}{n^{r-v_1} m^{p-v_2}} \cdot \omega_{s,q}\left(f^{(r,p)}; \frac{1}{n}, \frac{1}{m} \right), \tag{1.28}$$

where $c_{00} := \dot{C}A_{00}$, *with*

$$A_{00} := \frac{1}{h_1! h_2!} \left(\sum_{\tau=h_1}^{v_1} \sum_{\mu=h_2}^{v_2} l_{\tau\mu} \frac{2^{(\tau+\mu)-(\alpha_{1\tau}+\alpha_{2\mu})}}{\Gamma(\tau - a_{1\tau} + 1)\, \Gamma(\mu - \alpha_{2\mu} + 1)} \right) + 1,$$

(4)

$$\left\| D^{(0,\alpha_{2j})}_{*(-1)}(f) - D^{(0,\alpha_{2j})}_{*(-1)} Q^*_{n,m} \right\|_{\infty,[-1,1]^2}$$

$$\leq \frac{c_{0j}}{n^{r-v_1} m^{p-v_2}} \cdot \omega_{s,q}\left(f^{(r,p)}; \frac{1}{n}, \frac{1}{m} \right), \tag{1.29}$$

where $c_{0j} = \dot{C}A_{0j}$, $j = 1, \ldots, h_2$, *with*

$$A_{0j} := \left[\frac{1}{h_1!} \left(\sum_{\tau=h_1}^{v_1} \sum_{\mu=h_2}^{v_2} l_{\tau\mu} \frac{2^{(\tau+\mu)-(\alpha_{1\tau}+\alpha_{2\mu})}}{\Gamma(\tau - a_{1\tau} + 1)\, \Gamma(\mu - \alpha_{2\mu} + 1)} \right) \right.$$

$$\left. \times \left(\sum_{\lambda=0}^{h_2-j} \frac{2^{h_2-\alpha_{2j}-\lambda}}{\lambda!\,\Gamma(h_2 - \alpha_{2j} - \lambda + 1)} \right) + \frac{2^{j-\alpha_{2j}}}{\Gamma(j - \alpha_{2j} + 1)} \right], \tag{1.30}$$

(5)

$$\left\| D_{*(-1)}^{(\alpha_{1i},0)}(f) - D_{*(-1)}^{(\alpha_{1i},0)} Q_{n,m}^{*} \right\|_{\infty,[-1,1]^2} \leq \frac{c_{i0}}{n^{r-v_1} m^{p-v_2}} \cdot \omega_{s,q}\left(f^{(r,p)}; \frac{1}{n}, \frac{1}{m}\right), \tag{1.31}$$

where $c_{i0} = \dot{C} A_{i0}$, $i = i, \ldots, h_1$, *with*

$$A_{i0} := \left[\frac{1}{h_2!} \left(\sum_{\tau=h_1}^{v_1} \sum_{\mu=h_2}^{v_2} l_{\tau\mu} \frac{2^{(\tau+\mu)-(\alpha_{1\tau}+\alpha_{2\mu})}}{\Gamma(\tau - a_{1\tau} + 1)\, \Gamma(\mu - \alpha_{2\mu} + 1)} \right) \times \left(\sum_{k=0}^{h_1-i} \frac{2^{h_1-\alpha_{1i}-k}}{k!\,\Gamma(h_1 - \alpha_{1i} - k + 1)} \right) + \frac{2^{i-\alpha_{1i}}}{\Gamma(i - \alpha_{1i} + 1)} \right]. \tag{1.32}$$

Proof. By Corollary 1.3 there exists

$$Q_{n,m} \equiv Q_{n,m}(f) \in (P_n \otimes C([-1,1]) + C([-1,1]) \otimes P_m)$$

such that

$$\left\| f^{(i,j)} - Q_{n,m}^{(i,j)} \right\|_\infty \leq \frac{\dot{C}}{n^{r-i} m^{p-j}} \cdot \omega_{s,q}\left(f^{(r,p)}; \frac{1}{n}, \frac{1}{m}\right), \tag{1.33}$$

for all $(0,0) \leq (i,j) \leq (r,p)$, while $Q_{n,m} \in C^{r,p}([-1,1])^2$. Here $\dot{C}$ depends only on r, p, s, q, where $n \geq \max\{4(r+1), r+s\}$ and $m \geq \max\{4(p+1), p+q\}$, with $r, p \in \mathbb{Z}_+$, $s, q \in \mathbb{N}$, $f \in C^{r,p}\left([-1,1]^2\right)$.

Indeed by [5] we have that $Q_{n,m}^{(r,p)}$ is continuous on $[-1,1]^2$.

We observe the following ($i = 0, 1, \ldots, r$; $j = 0, 1, \ldots, p$)

$$\left| D_{*(-1)}^{(\alpha_{1i},\alpha_{2j})} f(x_1,x_2) - D_{*(-1)}^{(\alpha_{1i},\alpha_{2j})} Q_{n,m}(x_1,x_2) \right| = \frac{1}{\Gamma(i-\alpha_{1i})\,\Gamma(j-\alpha_{2j})} \left| \int_{-1}^{x_1} \int_{-1}^{x_2} (x_1-t_1)^{i-\alpha_{1i}-1} (x_2-t_2)^{j-\alpha_{2j}-1} \times \left(\frac{\partial^{i+j} f(t_1,t_2)}{\partial t_1^i \partial t_2^j} - \frac{\partial^{i+j} Q_{n,m}(t_1,t_2)}{\partial t_1^i \partial t_2^j} \right) dt_1 dt_2 \right| \tag{1.34}$$

$$\leq \frac{1}{\Gamma(i-\alpha_{1i})\,\Gamma(j-\alpha_{2j})}\int_{-1}^{x_1}\int_{-1}^{x_2}(x_1-t_1)^{i-\alpha_{1i}-1}(x_2-t_2)^{j-\alpha_{2j}-1}$$

$$\times\left|\frac{\partial^{i+j}f(t_1,t_2)}{\partial t_1^i\partial t_2^j}-\frac{\partial^{i+j}Q_{n,m}(t_1,t_2)}{\partial t_1^i\partial t_2^j}\right|dt_1dt_2 \tag{1.35}$$

$$\leq \frac{1}{\Gamma(i-\alpha_{1i})\,\Gamma(j-\alpha_{2j})}$$

$$\times\left(\int_{-1}^{x_1}\int_{-1}^{x_2}(x_1-t_1)^{i-\alpha_{1i}-1}(x_2-t_2)^{j-\alpha_{2j}-1}dt_1dt_2\right) \tag{1.36}$$

$$\times\frac{\dot{C}}{n^{r-i}m^{p-j}}\cdot\omega_{s,q}\left(f^{(r,p)};\frac{1}{n},\frac{1}{m}\right)$$

$$=\frac{1}{\Gamma(i-\alpha_{1i})\,\Gamma(j-\alpha_{2j})}\frac{(x_1+1)^{i-\alpha_{1i}}}{i-\alpha_{1i}}\frac{(x_2+1)^{j-\alpha_{2j}}}{j-\alpha_{2j}}\frac{\dot{C}}{n^{r-i}m^{p-j}}$$

$$\times\,\omega_{s,q}\left(f^{(r,p)};\frac{1}{n},\frac{1}{m}\right) \tag{1.37}$$

$$=\frac{(x_1+1)^{i-\alpha_{1i}}}{\Gamma(i-\alpha_{1i}+1)}\frac{(x_2+1)^{j-\alpha_{2j}}}{\Gamma(j-\alpha_{2j}+1)}\frac{\dot{C}}{n^{r-i}m^{p-j}}\omega_{s,q}\left(f^{(r,p)};\frac{1}{n},\frac{1}{m}\right)$$

That is, there exists $Q_{n,m}$:

$$\left|D_{*(-1)}^{(\alpha_{1i},\alpha_{2j})}f(x_1,x_2)-D_{*(-1)}^{(\alpha_{1i},\alpha_{2j})}Q_{n,m}(x_1,x_2)\right| \tag{1.38}$$

$$\leq\frac{(x_1+1)^{i-\alpha_{1i}}(x_2+1)^{j-\alpha_{2j}}}{\Gamma(i-\alpha_{1i}+1)\,\Gamma(j-\alpha_{2j}+1)}\frac{\dot{C}}{n^{r-i}m^{p-j}}\cdot\omega_{s,q}\left(f^{(r,p)};\frac{1}{n},\frac{1}{m}\right),$$

$i=0,1,\ldots,r, j=0,1,\ldots,p$, $\forall\,(x_1,x_2)\in[-1,1]^2$.

We proved there exists $Q_{n,m}$ such that

$$\left\|D_{*(-1)}^{(\alpha_{1i},\alpha_{2j})}(f)-D_{*(-1)}^{(\alpha_{1i},\alpha_{2j})}Q_{n,m}^{*}\right\|_{\infty}$$

$$\leq\frac{2^{(i+j)-(\alpha_{1i}+\alpha_{2j})}\dot{C}}{\Gamma(i-\alpha_{1i}+1)\,\Gamma(j-\alpha_{2j}+1)\,n^{r-i}m^{p-j}}\cdot\omega_{s,q}\left(f^{(r,p)};\frac{1}{n},\frac{1}{m}\right), \tag{1.39}$$

$i=0,1,\ldots,r, j=0,1,\ldots,p$.

Define

$$\rho_{n,m} \equiv \dot{C}\omega_{s,q}\left(f^{(r,p)}; \frac{1}{n}, \frac{1}{m}\right)$$
$$\cdot \left[\sum_{i=h_1}^{v_1}\sum_{j=h_2}^{v_2}\left(l_{ij}\frac{2^{(i+j)-(\alpha_{1i}+\alpha_{2j})}}{\Gamma(i-a_{1i}+1)\,\Gamma\left(j-\alpha_{2j}+1\right)}n^{i-r}m^{j-p}\right)\right] \quad (1.40)$$

I. Suppose, throughout $[0,1]^2$, $\alpha_{h_1h_2}(x,y) \geq \alpha > 0$. Let $Q^*_{n,m}(x,y)$, $(x,y) \in [-1,1]^2$, as in (1.39), so that

$$\left\|D^{(\alpha_{1i},\alpha_{2j})}_{*(-1)}\left(f(x,y)+\rho_{n,m}\frac{x^{h_1}}{h_1!}\frac{y^{h_2}}{h_2!}\right)-D^{(\alpha_{1i},\alpha_{2j})}_{*(-1)}Q^*_{n,m}(x,y)\right\|_\infty$$
$$\leq \frac{2^{(i+j)-(\alpha_{1i}+\alpha_{2j})}\dot{C}}{\Gamma(i-\alpha_{1i}+1)\,\Gamma\left(j-\alpha_{2j}+1\right)n^{r-i}m^{p-j}}\omega_{s,q}\left(f^{(r,p)}; \frac{1}{n}, \frac{1}{m}\right) =: T_{ij}, \quad (1.41)$$

$i = 0, 1, \ldots, r; j = 0, 1, \ldots, p$.

If $(h_1+1, h_2+1) \leq (i,j) \leq (r,p)$, or $0 \leq i \leq h_1$, $h_2+1 \leq j \leq p$, or $h_1+1 \leq i \leq r$, $0 \leq j \leq h_2$, we get from the last

$$\left\|D^{(\alpha_{1i},\alpha_{2j})}_{*(-1)}(f) - D^{(\alpha_{1i},\alpha_{2j})}_{*(-1)}Q^*_{n,m}\right\|_\infty$$
$$\leq \frac{2^{(i+j)-(\alpha_{1i}+\alpha_{2j})}\dot{C}}{\Gamma(i-\alpha_{1i}+1)\,\Gamma\left(j-\alpha_{2j}+1\right)n^{r-i}m^{p-j}}\cdot\omega_{s,q}\left(f^{(r,p)}; \frac{1}{n}, \frac{1}{m}\right), \quad (1.42)$$

proving (1.25).

If $(0,0) \leq (i,j) \leq (h_1,h_2)$, we get

$$\left\|D^{(\alpha_{1i},\alpha_{2j})}_{*(-1)}f(x,y) + \rho_{n,m}D^{\alpha_{1i}}_{*(-1)}\left(\frac{x^{h_1}}{h_1!}\right)D^{\alpha_{2j}}_{*(-1)}\left(\frac{y^{h_2}}{h_2!}\right)\right.$$
$$\left.-D^{(\alpha_{1i},\alpha_{2j})}_{*(-1)}Q^*_{n,m}(x,y)\right\|_\infty \leq T_{ij}. \quad (1.43)$$

That is, for $i = 1, \ldots, h_1; j = 1, \ldots, h_2$, we obtain

$$\left\|D^{(\alpha_{1i},\alpha_{2j})}_{*(-1)}f(x,y) + \rho_{n,m}\left(\sum_{k=0}^{h_1-i}\frac{(-1)^k(x+1)^{h_1-\alpha_{1i}-k}}{k!\Gamma(h_1-\alpha_{1i}-k+1)}\right)\right.$$

$$\cdot\left(\sum_{\lambda=0}^{h_2-j}\frac{(-1)^{\lambda}(y+1)^{h_2-\alpha_{2j}-\lambda}}{\lambda!\Gamma\left(h_2-\alpha_{2j}-\lambda+1\right)}\right)-D_{*(-1)}^{\left(\alpha_{1i},\alpha_{2j}\right)}Q_{n,m}^{*}(x,y)\Bigg\|_{\infty}$$
$$\leq T_{ij}. \tag{1.44}$$

Hence for $(1,1)\leq(i,j)\leq(h_1,h_2)$, we have

$$\left\|D_{*(-1)}^{\left(\alpha_{1i},\alpha_{2j}\right)}f-D_{*(-1)}^{\left(\alpha_{1i},\alpha_{2j}\right)}Q_{n,m}^{*}\right\|_{\infty}$$
$$\leq\rho_{n,m}\left(\sum_{k=0}^{h_1-i}\frac{2^{h_1-\alpha_{1i}-k}}{k!\Gamma\left(h_1-\alpha_{1i}-k+1\right)}\right)\left(\sum_{\lambda=0}^{h_2-j}\frac{2^{h_2-\alpha_{2j}-\lambda}}{\lambda!\Gamma\left(h_2-\alpha_{2j}-\lambda+1\right)}\right)+T_{ij} \tag{1.45}$$

$$=\dot{C}\omega_{s,q}\left(f^{(r,p)};\frac{1}{n},\frac{1}{m}\right)$$
$$\times\left[\sum_{\bar{i}=h_1}^{v_1}\sum_{\bar{j}=h_2}^{v_2}l_{\bar{i}\bar{j}}\frac{2^{\left(\bar{i}+\bar{j}\right)-\left(\alpha_{1\bar{i}}+\alpha_{2\bar{j}}\right)}}{\Gamma\left(\bar{i}-\alpha_{1\bar{i}}+1\right)\Gamma\left(\bar{j}-\alpha_{2\bar{j}}+1\right)}\frac{1}{n^{r-\bar{i}}}\frac{1}{m^{p-\bar{j}}}\right]$$
$$\times\left(\sum_{k=0}^{h_1-i}\frac{2^{h_1-\alpha_{1i}-k}}{k!\Gamma\left(h_1-\alpha_{1i}-k+1\right)}\right)\left(\sum_{\lambda=0}^{h_2-j}\frac{2^{h_2-\alpha_{2j}-\lambda}}{\lambda!\Gamma\left(h_2-\alpha_{2j}-\lambda+1\right)}\right) \tag{1.46}$$
$$+\frac{2^{(i+j)-\left(\alpha_{1i}+\alpha_{2j}\right)}\dot{C}\omega_{s,q}\left(f^{(r,p)};\frac{1}{n},\frac{1}{m}\right)}{\Gamma\left(i-\alpha_{1i}+1\right)\Gamma\left(j-\alpha_{2j}+1\right)n^{r-i}m^{p-j}}$$
$$\leq\dot{C}\omega_{s,q}\left(f^{(r,p)};\frac{1}{n},\frac{1}{m}\right)\frac{1}{n^{r-v_1}m^{p-v_2}}A_{ij}, \tag{1.47}$$

where

$$A_{ij}:=\left\{\left[\sum_{\bar{i}=h_1}^{v_1}\sum_{\bar{j}=h_2}^{v_2}l_{\bar{i}\bar{j}}\frac{2^{\left(\bar{i}+\bar{j}\right)-\left(\alpha_{1\bar{i}}+\alpha_{2\bar{j}}\right)}}{\Gamma\left(\bar{i}-\alpha_{1\bar{i}}+1\right)\Gamma\left(\bar{j}-\alpha_{2\bar{j}}+1\right)}\right]\right.$$
$$\cdot\left(\sum_{k=0}^{h_1-i}\frac{2^{h_1-\alpha_{1i}-k}}{k!\Gamma\left(h_1-\alpha_{1i}-k+1\right)}\right)\left(\sum_{\lambda=0}^{h_2-j}\frac{2^{h_2-\alpha_{2j}-\lambda}}{\lambda!\Gamma\left(h_2-\alpha_{2j}-\lambda+1\right)}\right)$$
$$\left.+\frac{2^{(i+j)-\left(\alpha_{1i}+\alpha_{2j}\right)}}{\Gamma\left(i-\alpha_{1i}+1\right)\Gamma\left(j-\alpha_{2j}+1\right)}\right\}. \tag{1.48}$$

(Set $c_{ij}:=\dot{C}A_{ij}$)

We proved, for $(1,1) \leq (i,j) \leq (h_1,h_2)$, that

$$\left\| D_{*(-1)}^{(\alpha_{1i},\alpha_{2j})} f - D_{*(-1)}^{(\alpha_{1i},\alpha_{2j})} Q_{n,m}^* \right\|_\infty \leq \frac{c_{ij}}{n^{r-v_1} m^{p-v_2}} \omega_{s,q}\left(f^{(r,p)}; \frac{1}{n}, \frac{1}{m} \right). \tag{1.49}$$

So that (1.26) is established.

When $i = j = 0$ from (1.41) we obtain

$$\left\| f(x,y) + \rho_{n,m} \frac{x^{h_1}}{h_1!} \frac{y^{h_2}}{h_2!} - Q_{n,m}^*(x,y) \right\|_\infty \leq \frac{\dot{C}}{n^r m^p} \omega_{s,q}\left(f^{(r,p)}; \frac{1}{n}, \frac{1}{m} \right). \tag{1.50}$$

Hence

$$\begin{aligned} \left\| f - Q_{n,m}^* \right\|_\infty &\leq \frac{\rho_{n,m}}{h_1! h_2!} + \frac{\dot{C}}{n^r m^p} \omega_{s,q}\left(f^{(r,p)}; \frac{1}{n}, \frac{1}{m} \right) \qquad (1.51) \\ &= \frac{\dot{C}}{h_1! h_2!} \omega_{s,q}\left(f^{(r,p)}; \frac{1}{n}, \frac{1}{m} \right) \\ &\quad \cdot \left[\sum_{\bar{i}=h_1}^{v_1} \sum_{\bar{j}=h_2}^{v_2} l_{\bar{i}\bar{j}} \frac{2^{(\bar{i}+\bar{j})-(\alpha_{1\bar{i}}+\alpha_{2\bar{j}})}}{\Gamma(\bar{i}-\alpha_{1\bar{i}}+1)\Gamma(\bar{j}-\alpha_{2\bar{j}}+1)} \frac{1}{n^{r-\bar{i}}} \frac{1}{m^{p-\bar{j}}} \right] \\ &\quad + \frac{\dot{C}}{n^r m^p} \omega_{s,q}\left(f^{(r,p)}; \frac{1}{n}, \frac{1}{m} \right) \qquad (1.52) \\ &\leq \frac{\dot{C} \omega_{s,q}\left(f^{(r,p)}; \frac{1}{n}, \frac{1}{m} \right)}{n^{r-v_1} m^{p-v_2}} A_{00}, \qquad (1.53) \end{aligned}$$

where

$$A_{00} := \left[\frac{1}{h_1! h_2!} \sum_{\bar{i}=h_1}^{v_1} \sum_{\bar{j}=h_2}^{v_2} \frac{l_{\bar{i}\bar{j}} 2^{(\bar{i}+\bar{j})-(\alpha_{1\bar{i}}+\alpha_{2\bar{j}})}}{\Gamma(\bar{i}-\alpha_{1\bar{i}}+1)\Gamma(\bar{j}-\alpha_{2\bar{j}}+1)} + 1 \right]. \tag{1.54}$$

(Set $c_{00} = \dot{C} A_{00}$).

Then

$$\left\| f - Q_{n,m}^* \right\|_\infty \leq \frac{c_{00}}{n^{r-v_1} m^{p-v_2}} \omega_{s,q}\left(f^{(r,p)}; \frac{1}{n}, \frac{1}{m} \right). \tag{1.55}$$

So that (1.28) is established.

Next case of $i = 0, j = 1, \ldots, h_2$, from (1.41) we get

$$\left\| D_{*(-1)}^{(0,\alpha_{2j})} f(x,y) + \rho_{n,m} \frac{x^{h_1}}{h_1!} \left(\sum_{\lambda=0}^{h_2-j} \frac{(-1)^{\lambda} (y+1)^{h_2-\alpha_{2j}-\lambda}}{\lambda! \Gamma(h_2 - \alpha_{2j} - \lambda + 1)} \right) - D_{*(-1)}^{(0,\alpha_{2j})} Q_{n,m}^{*}(x,y) \right\|_{\infty} \leq T_{0j}. \tag{1.56}$$

Then

$$\left\| D_{*(-1)}^{(0,\alpha_{2j})} f - D_{*(-1)}^{(0,\alpha_{2j})} Q_{n,m}^{*} \right\|_{\infty} \leq \frac{\rho_{n,m}}{h_1!} \left(\sum_{\lambda=0}^{h_2-j} \frac{2^{h_2-\alpha_{2j}-\lambda}}{\lambda! \Gamma(h_2 - \alpha_{2j} - \lambda + 1)} \right) + T_{0j} \tag{1.57}$$

$$= \frac{\dot{C}}{h_1!} \omega_{s,q}\left(f^{(r,p)}; \frac{1}{n}, \frac{1}{m} \right) \cdot \left[\sum_{\bar{i}=h_1}^{v_1} \sum_{\bar{j}=h_2}^{v_2} l_{\bar{i}\bar{j}} \frac{2^{(\bar{i}+\bar{j})-(\alpha_{1\bar{i}}+\alpha_{2\bar{j}})}}{\Gamma(\bar{i} - \alpha_{1\bar{i}} + 1) \Gamma(\bar{j} - \alpha_{2\bar{j}} + 1)} \frac{1}{n^{r-\bar{i}}} \frac{1}{m^{p-\bar{j}}} \right] \tag{1.58}$$

$$\cdot \left(\sum_{\lambda=0}^{h_2-j} \frac{2^{h_2-\alpha_{2j}-\lambda}}{\lambda! \Gamma(h_2 - \alpha_{2j} - \lambda + 1)} \right) + \frac{2^{j-\alpha_{2j}} \dot{C}}{\Gamma(j - \alpha_{2j} + 1) n^r m^{p-j}} \omega_{s,q}\left(f^{(r,p)}; \frac{1}{n}, \frac{1}{m} \right)$$

$$\leq \frac{\dot{C} \omega_{s,q}\left(f^{(r,p)}; \frac{1}{n}, \frac{1}{m} \right)}{n^{r-v_1} m^{p-v_2}} A_{0j}, \tag{1.59}$$

where

$$A_{0j} := \left[\frac{1}{h_1!} \left(\sum_{\bar{i}=h_1}^{v_1} \sum_{\bar{j}=h_2}^{v_2} l_{\bar{i}\bar{j}} \frac{2^{(\bar{i}+\bar{j})-(\alpha_{1\bar{i}}+\alpha_{2\bar{j}})}}{\Gamma(\bar{i} - \alpha_{1\bar{i}} + 1) \Gamma(\bar{j} - \alpha_{2\bar{j}} + 1)} \right) \cdot \left(\sum_{\lambda=0}^{h_2-j} \frac{2^{h_2-\alpha_{2j}-\lambda}}{\lambda! \Gamma(h_2 - \alpha_{2j} - \lambda + 1)} \right) + \frac{2^{j-\alpha_{2j}}}{\Gamma(j - \alpha_{2j} + 1)} \right]. \tag{1.60}$$

(Set $c_{0j} := \dot{C} A_{0j}$)

We proved that (case of $i=0, j=1,\ldots,h_2$)

$$\left\| D_{*(-1)}^{(0,\alpha_{2j})} f - D_{*(-1)}^{(0,\alpha_{2j})} Q_{n,m}^{*} \right\|_{\infty} \leq \frac{c_{0j}}{n^{r-v_1} m^{p-v_2}} \omega_{s,q}\left(f^{(r,p)}; \frac{1}{n}, \frac{1}{m}\right). \tag{1.61}$$

establishing (1.29).

Similarly we get for $i=1,\ldots,h_1, j=0$, that

$$\left\| D_{*(-1)}^{(\alpha_{1i},0)} f - D_{*(-1)}^{(\alpha_{1i},0)} Q_{n,m}^{*} \right\|_{\infty} \leq \frac{c_{i0}}{n^{r-v_1} m^{p-v_2}} \omega_{s,q}\left(f^{(r,p)}; \frac{1}{n}, \frac{1}{m}\right), \tag{1.62}$$

where $c_{i0} := \dot{C} A_{i0}$, with

$$A_{i0} := \left[\frac{1}{h_2!} \left(\sum_{\bar{i}=h_1}^{v_1} \sum_{\bar{j}=h_2}^{v_2} l_{\bar{i}\bar{j}} \frac{2^{(\bar{i}+\bar{j})-(\alpha_{1\bar{i}}+\alpha_{2\bar{j}})}}{\Gamma(\bar{i}-\alpha_{1\bar{i}}+1)\,\Gamma(\bar{j}-\alpha_{2\bar{j}}+1)} \right) \right.$$
$$\left. \cdot \left(\sum_{k=0}^{h_1-i} \frac{2^{h_1-\alpha_{1i}-k}}{k!\,\Gamma(h_1-\alpha_{1i}-k+1)} \right) + \frac{2^{i-\alpha_{1i}}}{\Gamma(i-\alpha_{1i}+1)} \right\} \Bigg] \tag{1.63}$$

deriving (1.31).

So if $(x,y) \in [0,1]^2$, then

$$\alpha_{h_1h_2}^{-1}(x,y)\, L^*\left(Q_{n,m}^{*}(x,y)\right)$$
$$= \alpha_{h_1h_2}^{-1}(x,y)\, L^*\left(f(x,y)\right) + \rho_{n,m} \frac{(x+1)^{h_1-\alpha_{1i}}}{\Gamma(h_1-\alpha_{1i}+1)} \frac{(y+1)^{h_2-\alpha_{2j}}}{\Gamma(h_2-\alpha_{2j}+1)}$$
$$+ \sum_{i=h_1}^{v_1} \sum_{j=h_2}^{v_2} \alpha_{h_1h_2}^{-1}(x,y)\, \alpha_{ij}(x,y)$$
$$\left[D_{*(-1)}^{(\alpha_{1i},\alpha_{2j})} Q_{n,m}^{*}(x,y) - D_{*(-1)}^{(\alpha_{1i},\alpha_{2j})} f(x,y) - \rho_{n,m} D_{*(-1)}^{(\alpha_{1i},\alpha_{2j})} \left(\frac{x^{h_1}}{h_1!} \frac{y^{h_2}}{h_2!} \right) \right] \tag{1.64}$$
$$\overset{(1.41)}{\geq} \rho_{n,m} \frac{(x+1)^{h_1-\alpha_{1i}}}{\Gamma(h_1-\alpha_{1i}+1)} \frac{(y+1)^{h_2-\alpha_{2j}}}{\Gamma(h_2-\alpha_{2j}+1)} \tag{1.65}$$
$$- \sum_{i=h_1}^{v_1} \sum_{j=h_2}^{v_2} l_{ij} \frac{2^{(i+j)-(\alpha_{1i}+\alpha_{2j})}}{\Gamma(i-\alpha_{1i}+1)\,\Gamma(j-\alpha_{2j}+1)} \frac{\dot{C}}{n^{r-i} m^{p-j}} \omega_{s,q}\left(f^{(r,p)}; \frac{1}{n}, \frac{1}{m}\right)$$
$$= \rho_{n,m} \left[\frac{(x+1)^{h_1-\alpha_{1i}}}{\Gamma(h_1-\alpha_{1i}+1)} \frac{(y+1)^{h_2-\alpha_{2j}}}{\Gamma(h_2-\alpha_{2j}+1)} - 1 \right] \tag{1.66}$$

$$\geq \rho_{n,m}\left[\frac{1}{\Gamma(h_1-\alpha_{1i}+1)\,\Gamma(h_2-\alpha_{2j}+1)}-1\right]$$

$$=\rho_{n,m}\left[\frac{1-\Gamma(h_1-\alpha_{1i}+1)\,\Gamma(h_2-\alpha_{2j}+1)}{\Gamma(h_1-\alpha_{1i}+1)\,\Gamma(h_2-\alpha_{2j}+1)}\right]\geq 0. \quad (1.67)$$

Explanation: we have that $\Gamma(1)=1$, $\Gamma(2)=1$, and Γ is convex on $(0,\infty)$ and positive there, here $0\leq h_1-\alpha_{1h_1}, h_2-\alpha_{2h_2}<1$ and $1\leq h_1-\alpha_{1h_1}+1, h_2-\alpha_{2h_2}+1<2$. Thus $0<\Gamma(h_1-\alpha_{1h_1}+1)$, $\Gamma(h_2-\alpha_{2h_2}+1)\leq 1$, and

$$0\leq\Gamma(h_1-\alpha_{1h_1}+1)\,\Gamma(h_2-\alpha_{2h_2}+1)\leq 1. \quad (1.68)$$

And

$$1-\Gamma(h_1-\alpha_{1h_1}+1)\,\Gamma(h_2-\alpha_{2h_2}+1)\geq 0. \quad (1.69)$$

Therefore it holds

$$L^*(Q_{n,m}(x,y))\geq 0,\ \forall\ (x,y)\in[0,1]^2. \quad (1.70)$$

II. Suppose, throughout $[0,1]^2$, $\alpha_{h_1h_2}(x,y)\leq\beta<0$. Let $Q^{**}_{n,m}(x,y)$, $(x,y)\in[-1,1]^2$, as in (1.39), so that

$$\left\|D_{*(-1)}^{(\alpha_{1i},\alpha_{2j})}\left(f(x,y)-\rho_{n,m}\frac{x^{h_1}}{h_1!}\frac{y^{h_2}}{h_2!}\right)-D_{*(-1)}^{(\alpha_{1i},\alpha_{2j})}Q^{**}_{n,m}(x,y)\right\|_\infty$$

$$\leq\frac{2^{(i+j)-(\alpha_{1i}+\alpha_{2j})}\dot{C}}{\Gamma(i-\alpha_{1i}+1)\,\Gamma(j-\alpha_{2j}+1)\,n^{r-i}m^{p-j}}\omega_{s,q}\left(f^{(r,p)};\frac{1}{n},\frac{1}{m}\right), \quad (1.71)$$

$i=0,1,\ldots,r$, $j=0,1,\ldots,p$.

As earlier we produce the same convergence inequalities (1.25), (1.26), (1.28), (1.29), and (1.31).

So for $(x,y)\in[0,1]^2$ we get

$$\begin{aligned}&\alpha_{h_1h_2}^{-1}(x,y)\,L^*\left(Q^{**}_{n,m}(x,y)\right)\\&=\alpha_{h_1h_2}^{-1}(x,y)\,L^*(f(x,y))\\&-\rho_{n,m}\frac{(x+1)^{h_1-\alpha_{1i}}}{\Gamma(h_1-\alpha_{1i}+1)}\frac{(y+1)^{h_2-\alpha_{2j}}}{\Gamma(h_2-\alpha_{2j}+1)}\\&+\sum_{i=h_1}^{v_1}\sum_{j=h_2}^{v_2}\alpha_{h_1h_2}^{-1}(x,y)\,\alpha_{ij}(x,y)\end{aligned} \quad (1.72)$$

$$\cdot\left[D_{*(-1)}^{(\alpha_{1i},\alpha_{2j})}Q_{n,m}^{**}(x,y)-D_{*(-1)}^{(\alpha_{1i},\alpha_{2j})}f(x,y)+\rho_{n,m}D_{*(-1)}^{(\alpha_{1i},\alpha_{2j})}\left(\frac{x^{h_1}}{h_1!}\frac{y^{h_2}}{h_2!}\right)\right]$$

$$\overset{(1.71)}{\leq}-\rho_{n,m}\frac{(x+1)^{h_1-\alpha_{1i}}}{\Gamma(h_1-\alpha_{1i}+1)}\frac{(y+1)^{h_2-\alpha_{2j}}}{\Gamma(h_2-\alpha_{2j}+1)}$$

$$\left[\sum_{i=h_1}^{v_1}\sum_{j=h_2}^{v_2}l_{ij}\frac{2^{(i+j)-(\alpha_{1i}+\alpha_{2j})}}{\Gamma(i-\alpha_{1i}+1)\Gamma(j-\alpha_{2j}+1)}\frac{\dot{C}}{n^{r-i}m^{p-j}}\omega_{s,q}\left(f^{(r,p)};\frac{1}{n},\frac{1}{m}\right)\right]$$

$$+\rho_{n,m}\left[1-\frac{(x+1)^{h_1-\alpha_{1i}}}{\Gamma(h_1-\alpha_{1i}+1)}\frac{(y+1)^{h_2-\alpha_{2j}}}{\Gamma(h_2-\alpha_{2j}+1)}\right] \tag{1.73}$$

$$=\rho_{n,m}\left[\frac{\Gamma(h_1-\alpha_{1i}+1)\Gamma(h_2-\alpha_{2j}+1)-(x+1)^{h_1-\alpha_{1i}}(y+1)^{h_2-\alpha_{2j}}}{\Gamma(h_1-\alpha_{1i}+1)\Gamma(h_2-\alpha_{2j}+1)}\right]$$

$$\leq\rho_{n,m}\left[\frac{1-(x+1)^{h_1-\alpha_{1i}}(y+1)^{h_2-\alpha_{2j}}}{\Gamma(h_1-\alpha_{1i}+1)\Gamma(h_2-\alpha_{2j}+1)}\right]\leq 0. \tag{1.74}$$

Explanation: for $x, y \in [0, 1]$ we get that $x + 1, y + 1 \geq 1$, and $0 \leq h_1 - \alpha_{1h_1}$, $h_2 - \alpha_{2h_2} < 1$. Hence $(x+1)^{h_1-\alpha_{1i}}, (y+1)^{h_2-\alpha_{2j}} \geq 1$, and then

$$(x+1)^{h_1-\alpha_{1i}}(y+1)^{h_2-\alpha_{2j}} \geq 1,$$

so that

$$1-(x+1)^{h_1-\alpha_{1i}}(y+1)^{h_2-\alpha_{2j}} \leq 0. \tag{1.75}$$

Hence again

$$L^*\left(Q_{n,m}^{**}(x,y)\right) \geq 0, \text{ for } (x,y) \in [0,1]^2. \tag{1.76}$$

References

1. G.A. Anastassiou, Monotone approximation by pseudopolynomials, in *Progress in Approximation Theory* (Academic, Boston, MA, 1991), pp. 5–11
2. G.A. Anastassiou, Bivariate monotone approximation. Proc. Am. Math. Soc. **112**(4), 959–963 (1991)
3. G.A. Anastassiou, O. Shisha, Monotone approximation with linear differential operators. J. Approx. Theory **44**, 391–393 (1985)
4. H.H. Gonska, Simultaneously approximation by algebraic blending functions, in *Alfred Haar Memorial Conference, Budapest 1985*. Coloquia Mathematica Societatis Janos Bolyai, vol. 49 (North-Holland, Amsterdam, 1985), pp. 363–382
5. H.H. Gonska, Personal communication with author, 2-24-2014
6. T. Mamatov, S. Samko, Mixed fractional integration operators in mixed weighted Hölder spaces. Fract. Calculus Appl. Anal. **13**(3), 245–259 (2010)

7. A. Marchaud, Differences et deerivees d'une fonction de deux variables. C. R. Acad. Sci. **178,** 1467–1470 (1924)
8. A. Marchaud, Sur les derivees et sur les differences des fonctions de variables reelles. J. Math. Pures Appl. **6**, 337–425 (1927)
9. L.L. Schumaker, *Spline Functions: Basic Theory* (Wiley, New York, 1981)
10. O. Shisha, Monotone approximation. Pac. J. Math. **15**, 667–671 (1965)

Chapter 2
Bivariate Right Fractional Polynomial Monotone Approximation

George A. Anastassiou

Abstract Let $f \in C^{r,p}\left([0,1]^2\right)$, $r,p \in \mathbb{N}$, and let $\overline{L}$ be a linear right fractional mixed partial differential operator such that $\overline{L}(f) \geq 0$, for all (x,y) in a critical region of $[0,1]^2$ that depends on $\overline{L}$. Then there exists a sequence of two-dimensional polynomials $Q_{\overline{m_1},\overline{m_2}}(x,y)$ with $\overline{L}\left(Q_{\overline{m_1},\overline{m_2}}(x,y)\right) \geq 0$ there, where $\overline{m_1},\overline{m_2} \in \mathbb{N}$ such that $\overline{m_1} > r$, $\overline{m_2} > p$, so that f is approximated right fractionally simultaneously and uniformly by $Q_{\overline{m_1},\overline{m_2}}$ on $[0,1]^2$. This restricted right fractional approximation is accomplished quantitatively by the use of a suitable integer partial derivatives two-dimensional first modulus of continuity.

2.1 Introduction

The topic of monotone approximation started in [5] has become a major trend in approximation theory. A typical problem in this subject is: given a positive integer k, approximate a given function whose kth derivative is ≥ 0 by polynomials having this property.

In [2] the authors replaced the kth derivative with a linear differential operator of order k. We mention this motivating result.

Theorem 2.1. *Let h,k,p be integers, $0 \leq h \leq k \leq p$ and let f be a real function, $f^{(p)}$ continuous in $[-1,1]$ with modulus of continuity $\omega\left(f^{(p)},x\right)$ there. Let $a_j(x)$, $j = h, h+1, \ldots, k$ be real functions, defined and bounded on $[-1,1]$ and assume $a_h(x)$ is either $\geq$ some number $\alpha > 0$ or $\leq$ some number $\beta < 0$ throughout $[-1,1]$. Consider the operator*

$$L = \sum_{j=h}^{k} a_j(x)\left[\frac{d^j}{dx^j}\right]$$

and suppose, throughout $[-1,1]$,

$$L(f) \geq 0. \tag{2.1}$$

G.A. Anastassiou (✉)
Department of Mathematical Sciences, University of Memphis, TN 38152, USA
e-mail: ganastss@memphis.edu

G.A. Anastassiou, O. Duman (eds.), *Computational Analysis*, Springer Proceedings in Mathematics & Statistics 155, DOI 10.1007/978-3-319-28443-9_2

Then, for every integer $n \geq 1$, there is a real polynomial $Q_n(x)$ of degree $\leq n$ such that

$$L(Q_n) \geq 0 \text{ throughout } [-1, 1]$$

and

$$\max_{-1 \leq x \leq 1} |f(x) - Q_n(x)| \leq Cn^{k-p} \omega \left(f^{(p)}, \frac{1}{n} \right),$$

where C is independent of n or f.

We need

Definition 2.2 (D.D. Stancu [6]). Let $f \in C\left([0,1]^2\right)$, $[0,1]^2 = [0,1] \times [0,1]$, where (x_1, y_1), $(x_2, y_2) \in [0,1]^2$ and $\delta_1, \delta_2 \geq 0$. The first modulus of continuity of f is defined as follows:

$$\omega_1(f, \delta_1, \delta_2) = \sup_{\substack{|x_1 - x_2| \leq \delta_1 \\ |y_1 - y_2| \leq \delta_2}} |f(x_1, y_1) - f(x_2, y_2)|.$$

Definition 2.3. Let f be a real-valued function defined on $[0,1]^2$ and let m, n be two positive integers. Let $B_{m,n}$ be the Bernstein (polynomial) operator of order (m, n) given by

$$B_{m,n}(f; x, y) = \sum_{i=0}^{m} \sum_{j=0}^{n} f\left(\frac{i}{m}, \frac{j}{n}\right) \cdot \binom{m}{i} \cdot \binom{n}{j} \cdot x^i \cdot (1-x)^{m-i} \cdot y^j \cdot (1-y)^{n-j}. \tag{2.2}$$

For integers $r, s \geq 0$, we denote by $f^{(r,s)}$ the differential operator of order (r, s), given by

$$f^{(r,s)}(x, y) = \frac{\partial^{r+s} f(x, y)}{\partial x^r \partial y^s}.$$

We use

Theorem 2.4 (Badea and Badea [3]). *It holds that*

$$\begin{aligned} &\left\| f^{(k,l)} - (B_{m,n} f)^{(k,l)} \right\|_\infty \\ &\leq t(k,l) \cdot \omega_1\left(f^{(k,l)}; \frac{1}{\sqrt{m-k}}, \frac{1}{\sqrt{n-l}}\right) \\ &\quad + \max\left\{\frac{k(k-1)}{m}, \frac{l(l-1)}{n}\right\} \cdot \left\| f^{(k,l)} \right\|_\infty, \end{aligned} \tag{2.3}$$

where $m > k \geq 0$, $n > l \geq 0$ *are integers,* f *is a real-valued function on* $[0,1]^2$ *such that* $f^{(k,l)}$ *is continuous, and* t *is a positive real-valued function on* $\mathbb{Z}_+ = \{0,1,2,\dots\}$. *Here* $\|\cdot\|_\infty$ *is the supremum norm on* $[0,1]^2$.

Denote $C^{r,p}\left([0,1]^2\right) := \{f : [0,1]^2 \to \mathbb{R}; f^{(k,l)}$ is continuous for $0 \leq k \leq r$, $0 \leq l \leq p\}$.

In [1] the author proved the following main motivational result.

Theorem 2.5. *Let* h_1, h_2, v_1, v_2, r, p *be integers,* $0 \leq h_1 \leq v_1 \leq r$, $0 \leq h_2 \leq v_2 \leq p$ *and let* $f \in C^{r,p}\left([0,1]^2\right)$. *Let* $\alpha_{i,j}(x,y)$, $i = h_1, h_1+1, \dots, v_1$; $j = h_2, h_2+1, \dots, v_2$ *be real-valued functions, defined and bounded in* $[0,1]^2$ *and assume* $\alpha_{h_1 h_2}$ *is either* $\geq \alpha > 0$ *or* $\leq \beta < 0$ *throughout* $[0,1]^2$. *Consider the operator*

$$L = \sum_{i=h_1}^{v_1} \sum_{j=h_2}^{v_2} \alpha_{ij}(x,y) \frac{\partial^{i+j}}{\partial x^i \partial y^j} \tag{2.4}$$

and suppose that throughout $[0,1]^2$,

$$L(f) \geq 0.$$

Then for integers m, n *with* $m > r$, $n > p$, *there exists a polynomial* $Q_{m,n}(x,y)$ *of degree* (m,n) *such that* $L(Q_{m,n}(x,y)) \geq 0$ *throughout* $[0,1]^2$ *and*

$$\left\| f^{(k,l)} - Q_{m,n}^{(k,l)} \right\|_\infty \leq \frac{P_{m,n}(L,f)}{(h_1-k)!\,(h_2-l)!} + M_{m,n}^{k,l}(f), \tag{2.5}$$

all $(0,0) \leq (k,l) \leq (h_1,h_2)$. *Furthermore we get*

$$\left\| f^{(k,l)} - Q_{m,n}^{(k,l)} \right\|_\infty \leq M_{m,n}^{k,l}(f), \tag{2.6}$$

for all $(h_1+1, h_2+1) \leq (k,l) \leq (r,p)$. *Also (2.6) is true whenever* $0 \leq k \leq h_1$, $h_2+1 \leq l \leq p$ *or* $h_1+1 \leq k \leq r$, $0 \leq l \leq h_2$. *Here*

$$M_{m,n}^{k,l} \equiv M_{m,n}^{k,l}(f) \equiv t(k,l) \cdot \omega_1\left(f^{(k,l)}; \frac{1}{\sqrt{m-k}}, \frac{1}{\sqrt{n-l}}\right) + \max\left\{\frac{k(k-1)}{m}, \frac{l(l-1)}{n}\right\} \cdot \left\| f^{(k,l)} \right\|_\infty \tag{2.7}$$

and

$$P_{m,n} \equiv P_{m,n}(L,f) \equiv \sum_{i=h_1}^{v_1} \sum_{j=h_2}^{v_2} l_{ij} \cdot M_{m,n}^{i,j}, \tag{2.8}$$

where t *is a positive real-valued function on* $\mathbb{Z}_+^2$ *and*

$$l_{ij} \equiv \sup_{(x,y)\in[0,1]^2} \left| \alpha_{h_1 h_2}^{-1}(x,y) \cdot \alpha_{ij}(x,y) \right| < \infty. \tag{2.9}$$

In this article we extend Theorem 2.5 to the right fractional level. Indeed here L is replaced by $\overline{L}$, a linear right Caputo fractional mixed partial differential operator. Now the monotonicity property is only true on a critical region of $[0,1]^2$ that depends on $\overline{L}$ parameters. Simultaneous right fractional convergence remains true on all of $[0,1]^2$.

We need

Definition 2.6 (See [4]). Let $\alpha_1, \alpha_2 > 0$; $\alpha = (\alpha_1, \alpha_2)$, $f \in C\left([0,1]^2\right)$ and let $x = (x_1, x_2)$, $t = (t_1, t_2) \in [0,1]^2$. We define the right mixed Riemann–Liouville fractional two-dimensional integral of order α:

$$\left(I_{1-}^{\alpha} f\right)(x) := \frac{1}{\Gamma(\alpha_1)\Gamma(\alpha_2)} \int_{x_1}^{1} \int_{x_2}^{1} (t_1 - x_1)^{\alpha_1 - 1} (t_2 - x_2)^{\alpha_2 - 1} f(t_1, t_2)\, dt_1 dt_2,$$

with $x_1, x_2 < 1$. Here Γ stands for the Gamma function.

Notice here

$$I_{1-}^{\alpha}(|f|) < \infty. \tag{2.10}$$

Definition 2.7. Let $\alpha_1, \alpha_2 > 0$ with $\lceil \alpha_1 \rceil = m_1$, $\lceil \alpha_2 \rceil = m_2$, ($\lceil \cdot \rceil$ ceiling of the number). Let here $f \in C^{m_1, m_2}\left([0,1]^2\right)$. We consider the right (Caputo type) fractional partial derivative:

$$D_{1-}^{(\alpha_1, \alpha_2)} f(x) := \frac{(-1)^{m_1 + m_2}}{\Gamma(m_1 - \alpha_1)\Gamma(m_2 - \alpha_2)} \cdot \int_{x_1}^{1} \int_{x_2}^{1} (J_1 - x_1)^{m_1 - \alpha_1 - 1} (J_2 - x_2)^{m_2 - \alpha_2 - 1} \frac{\partial^{m_1 + m_2} f(J_1, J_2)}{\partial J_1^{m_1} \partial J_2^{m_2}} dJ_1 dJ_2, \tag{2.11}$$

$\forall\, x = (x_1, x_2) \in [0,1]^2$.

We set

$$D_{1-}^{(0,0)} f(x) := f(x),$$

$$D_{1-}^{(m_1, m_2)} f(x) := (-1)^{m_1 + m_2} \frac{\partial^{m_1 + m_2} f(x)}{\partial x_1^{m_1} \partial x_2^{m_2}}, \quad \forall\, x \in [0,1]^2. \tag{2.12}$$

Definition 2.8. We also set

$$D_{1-}^{(0, \alpha_2)} f(x) := \frac{(-1)^{m_2}}{\Gamma(m_2 - \alpha_2)} \int_{x_2}^{1} (J_2 - x_2)^{m_2 - \alpha_2 - 1} \frac{\partial^{m_2} f(x_1, J_2)}{\partial J_2^{m_2}} dJ_2, \tag{2.13}$$

$$D_{1-}^{(\alpha_1, 0)} f(x) := \frac{(-1)^{m_1}}{\Gamma(m_1 - \alpha_1)} \int_{x_1}^{1} (J_1 - x_1)^{m_1 - \alpha_1 - 1} \frac{\partial^{m_1} f(J_1, x_2)}{\partial J_1^{m_1}} dJ_1, \tag{2.14}$$

and

$$D_{1-}^{(m_1,\alpha_2)} f(x) := \frac{(-1)^{m_2}}{\Gamma(m_2 - \alpha_2)} \int_{x_2}^{1} (J_2 - x_2)^{m_2-\alpha_2-1} \frac{\partial^{m_1+m_2} f(x_1, J_2)}{\partial x_1^{m_1} \partial J_2^{m_2}} dJ_2, \tag{2.15}$$

$$D_{1-}^{(\alpha_1,m_2)} f(x) := \frac{(-1)^{m_1}}{\Gamma(m_1 - \alpha_1)} \int_{x_1}^{1} (J_1 - x_1)^{m_1-\alpha_1-1} \frac{\partial^{m_1+m_2} f(J_1, x_2)}{\partial J_1^{m_1} \partial x_2^{m_2}} dJ_1. \tag{2.16}$$

2.2 Main Result

We present our main result

Theorem 2.9. *Let h_1, h_2, v_1, v_2, r, p be integers, $0 \leq h_1 \leq v_1 \leq r$, $0 \leq h_2 \leq v_2 \leq p$ and let $f \in C^{r,p}\left([0,1]^2\right)$. Let $\alpha_{ij}(x,y)$, $i = h_1, h_1+1, \ldots, v_1; j = h_2, h_2+1, \ldots, v_2$ be real-valued functions, defined and bounded in $[0,1]^2$ and assume $\alpha_{h_1h_2}$ is either $\geq \alpha > 0$ or $\leq \beta < 0$ throughout $[0,1]^2$.*

Let integers $\overline{m_1}, \overline{m_2}$ with $\overline{m_1} > r$, $\overline{m_2} > p$. Set

$$l_{ij} := \sup_{(x,y)\in[0,1]^2} \left|\alpha_{h_1h_2}^{-1}(x,y) \cdot \alpha_{ij}(x,y)\right| < \infty. \tag{2.17}$$

Also set ($\lceil \alpha_{1i} \rceil = i$, $\lceil \alpha_{2j} \rceil = j$, $\lceil \cdot \rceil$ ceiling of number)

$$M_{\overline{m_1},\overline{m_2}}^{i,j} := M_{\overline{m_1},\overline{m_2}}^{i,j}(f) := \frac{1}{\Gamma(i - \alpha_{1i} + 1)\,\Gamma(j - \alpha_{2j} + 1)} \times \left\{ t(i,j)\,\omega_1\left(f^{(i,j)}; \frac{1}{\sqrt{\overline{m_1} - i}}, \frac{1}{\sqrt{\overline{m_2} - j}}\right) + \max\left\{\frac{i(i-1)}{\overline{m_1}}, \frac{j(j-1)}{\overline{m_2}}\right\} \cdot \left\|f^{(i,j)}\right\|_\infty \right\}, \tag{2.18}$$

$i = h_1, \ldots, v_1; j = h_2, \ldots, v_2$.

Here t is a positive real-valued function on $\mathbb{Z}_+^2$, $\|\cdot\|_\infty$ is the supremum norm on $[0,1]^2$. Call

$$P_{\overline{m_1},\overline{m_2}} := P_{\overline{m_1},\overline{m_2}}(f) = \sum_{i=h_1}^{v_1} \sum_{j=h_2}^{v_2} l_{ij} \cdot M_{\overline{m_1},\overline{m_2}}^{i,j}. \tag{2.19}$$

Let

$$0 \leq \alpha_{1h_1} \leq h_1 < \alpha_{11} < h_1 + 1 < \alpha_{12} < h_1 + 2 < \alpha_{13} < h_1 + 3 < \cdots < \alpha_{1v_1} < v_1 < \cdots < \alpha_{1r} < r,$$

with $\lceil \alpha_{1h_1} \rceil = h_1$*;*

$$0 \leq \alpha_{2h_2} \leq h_2 < \alpha_{21} < h_2 + 1 < \alpha_{22} < h_2 + 2 < \alpha_{23}$$
$$< h_2 + 3 < \cdots < \alpha_{2v_2} < v_2 < \cdots < \alpha_{2p} < p,$$

with $\lceil \alpha_{2h_2} \rceil = h_2$. *Here* $h_1 + h_2 = 2m$, $m = 0, 1, 2, \ldots$.

Consider the right fractional bivariate operator

$$\overline{L} := \sum_{i=h_1}^{v_1} \sum_{j=h_2}^{v_2} \alpha_{ij}(x, y)\, D_{1-}^{(\alpha_{1i}, \alpha_{2j})}. \tag{2.20}$$

Then there exists a polynomial $Q_{\overline{m_1}, \overline{m_2}}(x, y)$ *of degree* $(\overline{m_1}, \overline{m_2})$ *on* $[0, 1]^2$ *such that*

$$\left\| D_{1-}^{(\alpha_{1k}, \alpha_{2l})}(f) - D_{1-}^{(\alpha_{1k}, \alpha_{2l})}(Q_{\overline{m_1}, \overline{m_2}}) \right\|_\infty$$
$$\leq \frac{P_{\overline{m_1}, \overline{m_2}}}{(h_1 - k)!\,(h_2 - l)!} \left\{ \left[\sum_{\theta=0}^{h_1-k} \binom{h_1 - k}{\theta} \frac{\Gamma(h_1 - k - \theta + 1)}{\Gamma(h_1 - \alpha_{1k} - \theta + 1)} \right] \right.$$
$$\left. \left[\sum_{\rho=0}^{h_2-l} \binom{h_2 - l}{\rho} \frac{\Gamma(h_2 - l - \rho + 1)}{\Gamma(h_2 - \alpha_{2l} - \rho + 1)} \right] \right\} + M_{\overline{m_1}, \overline{m_2}}^{k,l}, \tag{2.21}$$

for $(0, 0) \leq (k, l) \leq (h_1, h_2)$.

If $(h_1 + 1, h_2 + 1) \leq (k, l) \leq (r, p)$, *or* $0 \leq k \leq h_1$, $h_2 + 1 \leq l \leq p$, *or* $h_1 + 1 \leq k \leq r$, $0 \leq l \leq h_2$, *then*

$$\left\| D_{1-}^{(\alpha_{1k}, \alpha_{2l})}(f) - D_{1-}^{(\alpha_{1k}, \alpha_{2l})}(Q_{\overline{m_1}, \overline{m_2}}) \right\|_\infty \leq M_{\overline{m_1}, \overline{m_2}}^{k,l}. \tag{2.22}$$

If $\overline{L}(f(0, 0)) \geq 0$, *then* $\overline{L}(Q_{\overline{m_1}, \overline{m_2}}(0, 0)) \geq 0$.

Let $0 < x < 1$, $0 < y < 1$, *with* $\alpha_{1h_1} \neq h_1$ *and* $\alpha_{2h_2} \neq h_2$, *such that*

$$\begin{aligned} 1 - x &\geq \Gamma(h_1 - \alpha_{1h_1} + 1)^{\overline{\left(\frac{1}{h_1 - \alpha_{1h_1}}\right)}}, \\ 1 - y &\geq \Gamma(h_2 - \alpha_{2h_2} + 1)^{\overline{\left(\frac{1}{h_2 - \alpha_{2h_2}}\right)}}, \end{aligned} \tag{2.23}$$

equivalently,

$$\begin{aligned} x &\leq 1 - \Gamma(h_1 - \alpha_{1h_1} + 1)^{\overline{\left(\frac{1}{h_1 - \alpha_{1h_1}}\right)}}, \\ y &\leq 1 - \Gamma(h_2 - \alpha_{2h_2} + 1)^{\overline{\left(\frac{1}{h_2 - \alpha_{2h_2}}\right)}}), \end{aligned} \tag{2.24}$$

and

$$\overline{L}(f(x, y)) \geq 0.$$

Then

$$\overline{L}\left(Q_{\overline{m_1},\overline{m_2}}(x,y)\right) \geq 0.$$

To prove Theorem 2.9 it takes some preparation. We need

Definition 2.10. Let f be a real-valued function defined on $[0,1]^2$ and let $\overline{m_1},\overline{m_2} \in \mathbb{N}$. Let $B_{\overline{m_1},\overline{m_2}}$ be the Bernstein (polynomial) operator of order $(\overline{m_1},\overline{m_2})$ given by

$$\begin{aligned} B_{\overline{m_1},\overline{m_2}}(f;x_1,x_2) &:= \sum_{i_1=0}^{\overline{m_1}} \sum_{i_2=0}^{\overline{m_2}} f\left(\frac{i_1}{\overline{m_1}},\frac{i_2}{\overline{m_2}}\right)\binom{\overline{m_1}}{i_1}\binom{\overline{m_2}}{i_2} \\ & x_1^{i_1}(1-x_1)^{\overline{m_1}-i_1} x_2^{i_2}(1-x_2)^{\overline{m_2}-i_2}. \end{aligned} \tag{2.25}$$

We need the following simultaneous approximation result.

Theorem 2.11 (Badea and Badea [3]). *It holds that*

$$\begin{aligned} &\left\| f^{(k_1,k_2)} - \left(B_{\overline{m_1},\overline{m_2}} f\right)^{(k_1,k_2)} \right\|_\infty \\ &\leq t(k_1,k_2)\,\omega_1\left(f^{(k_1,k_2)};\frac{1}{\sqrt{\overline{m_1}-k_1}},\frac{1}{\sqrt{\overline{m_2}-k_2}}\right) \\ &+ \max\left\{\frac{k_1(k_1-1)}{\overline{m_1}},\frac{k_2(k_2-1)}{\overline{m_2}}\right\}\cdot\left\|f^{(k_1,k_2)}\right\|_\infty, \end{aligned} \tag{2.26}$$

where $\overline{m_1} > k_1 \geq 0$, $\overline{m_2} > k_2 \geq 0$ *are integers, f is a real-valued function on* $[0,1]^2$, *such that* $f^{(k_1,k_2)}$ *is continuous, and t is a positive real-valued function on* $\mathbb{Z}_+^2$. *Here* $\|\cdot\|_\infty$ *is the supremum norm on* $[0,1]^2$.

Remark 2.12. We assume that $\overline{m_1} > m_1 = \lceil \alpha_1 \rceil$, $\overline{m_2} > m_2 = \lceil \alpha_2 \rceil$, where $\alpha_1,\alpha_2 > 0$.

We consider also

$$\begin{aligned} D_{1-}^{(\alpha_1,\alpha_2)}\left(B_{\overline{m_1},\overline{m_2}} f\right)(x_1,x_2) &= \frac{(-1)^{m_1+m_2}}{\Gamma(m_1-\alpha_1)\,\Gamma(m_2-\alpha_2)} \\ &\cdot \int_{x_1}^{1}\int_{x_2}^{1} (t_1-x_1)^{m_1-\alpha_1-1}(t_2-x_2)^{m_2-\alpha_2-1} \\ &\frac{\partial^{m_1+m_2}\left(B_{\overline{m_1},\overline{m_2}} f\right)(t_1,t_2)}{\partial t_1^{m_1}\,\partial t_2^{m_2}}\,dt_1\,dt_2, \end{aligned} \tag{2.27}$$

$\forall\,(x_1,x_2) \in [0,1]^2$.

Proposition 2.13. *Let* $\alpha_1, \alpha_2 > 0$ *with* $\lceil \alpha_1 \rceil = m_1$, $\lceil \alpha_2 \rceil = m_2$, $f \in C^{m_1,m_2}\left([0,1]^2\right)$, *where* $\overline{m_1}, \overline{m_2} \in \mathbb{N} : \overline{m_1} > m_1$, $\overline{m_2} > m_2$. *Then*

$$\left\| D_{1-}^{(\alpha_1,\alpha_2)} f - D_{1-}^{(\alpha_1,\alpha_2)} \left(B_{\overline{m_1},\overline{m_2}} f\right) \right\|_\infty \leq \frac{1}{\Gamma(m_1 - \alpha_1 + 1)\,\Gamma(m_2 - \alpha_2 + 1)}$$
$$\cdot \left\{ t(m_1, m_2)\, \omega_1 \left(f^{(m_1,m_2)}; \frac{1}{\sqrt{\overline{m_1} - m_1}}, \frac{1}{\sqrt{\overline{m_2} - m_2}} \right) \right.$$
$$\left. + \max\left\{ \frac{m_1(m_1-1)}{\overline{m_1}}, \frac{m_2(m_2-1)}{\overline{m_2}} \right\} \cdot \left\| f^{(m_1,m_2)} \right\|_\infty \right\}, \tag{2.28}$$

Proof. We observe the following:

$$\left| D_{1-}^{(\alpha_1,\alpha_2)} f(x_1,x_2) - D_{1-}^{(\alpha_1,\alpha_2)} \left(B_{\overline{m_1},\overline{m_2}} f\right)(x_1,x_2) \right|$$
$$= \frac{1}{\Gamma(m_1-\alpha_1)\,\Gamma(m_2-\alpha_2)} \left| \int_{x_1}^{1} \int_{x_2}^{1} (t_1 - x_1)^{m_1-\alpha_1-1} (t_2 - x_2)^{m_2-\alpha_2-1} \right.$$
$$\left. \cdot \left(\frac{\partial^{m_1+m_2} f(t_1,t_2)}{\partial t_1^{m_1} \partial t_2^{m_2}} - \frac{\partial^{m_1+m_2} \left(B_{\overline{m_1},\overline{m_2}} f\right)(t_1,t_2)}{\partial t_1^{m_1} \partial t_2^{m_2}} \right) dt_1 dt_2 \right| \tag{2.29}$$

$$\leq \frac{1}{\Gamma(m_1-\alpha_1)\,\Gamma(m_2-\alpha_2)} \int_{x_1}^{1} \int_{x_2}^{1} (t_1 - x_1)^{m_1-\alpha_1-1} (t_2 - x_2)^{m_2-\alpha_2-1} \tag{2.30}$$
$$\cdot \left| \frac{\partial^{m_1+m_2} f(t_1,t_2)}{\partial t_1^{m_1} \partial t_2^{m_2}} - \frac{\partial^{m_1+m_2} \left(B_{\overline{m_1},\overline{m_2}} f\right)(t_1,t_2)}{\partial t_1^{m_1} \partial t_2^{m_2}} \right| dt_1 dt_2$$

$$\overset{(2.26)}{\leq} \left\{ t(m_1, m_2)\, \omega_1 \left(f^{(m_1,m_2)}; \frac{1}{\sqrt{\overline{m_1} - m_1}}, \frac{1}{\sqrt{\overline{m_2} - m_2}} \right) \right.$$
$$\left. + \max\left\{ \frac{m_1(m_1-1)}{\overline{m_1}}, \frac{m_2(m_2-1)}{\overline{m_2}} \right\} \cdot \left\| f^{(m_1,m_2)} \right\|_\infty \right\} \tag{2.31}$$

$$\cdot \frac{1}{\Gamma(m_1-\alpha_1)\,\Gamma(m_2-\alpha_2)} \int_{x_1}^{1} \int_{x_2}^{1} (t_1 - x_1)^{m_1-\alpha_1-1} (t_2 - x_2)^{m_2-\alpha_2-1} dt_1 dt_2$$

$$= \frac{(1-x_1)^{m_1-\alpha_1} (1-x_2)^{m_2-\alpha_2}}{\Gamma(m_1-\alpha_1+1)\,\Gamma(m_2-\alpha_2+1)}$$
$$\cdot \left\{ t(m_1, m_2)\, \omega_1 \left(f^{(m_1,m_2)}; \frac{1}{\sqrt{\overline{m_1} - m_1}}, \frac{1}{\sqrt{\overline{m_2} - m_2}} \right) \right.$$
$$\left. + \max\left\{ \frac{m_1(m_1-1)}{\overline{m_1}}, \frac{m_2(m_2-1)}{\overline{m_2}} \right\} \left\| f^{(m_1,m_2)} \right\|_\infty \right\}, \tag{2.32}$$

$\forall\ (x_1, x_2) \in [0,1]^2$.

Proof (Proof of Theorem 2.9).

We need for $(0,0) \le (k,l) \le (h_1,h_2)$ to calculate

$$D_{1-}^{(\alpha_{1k},\alpha_{2l})}\left(\frac{x^{h_1}}{h_1!}\frac{y^{h_2}}{h_2!}\right) = \frac{(-1)^{k+l}}{\Gamma(k-\alpha_{1k})\,\Gamma(l-\alpha_{2l})} \cdot \int_x^1\int_y^1 (J_1-x)^{k-\alpha_{1k}-1}(J_2-y)^{l-\alpha_{2l}-1}\frac{J_1^{h_1-k}}{(h_1-k)!}\frac{J_2^{h_2-l}}{(h_2-l)!}dJ_1dJ_2 \tag{2.33}$$

$$= \frac{(-1)^{k+l}}{(h_1-k)!\,(h_2-l)!}\left\{\left(\frac{1}{\Gamma(k-\alpha_{1k})}\int_x^1 (J_1-x)^{k-\alpha_{1k}-1}J_1^{h_1-k}dJ_1\right)\cdot\left(\frac{1}{\Gamma(l-\alpha_{2l})}\int_y^1 (J_2-y)^{l-\alpha_{2l}-1}J_2^{h_2-l}dJ_2\right)\right\}. \tag{2.34}$$

We find that

$$\int_x^1 J_1^{h_1-k}(J_1-x)^{k-\alpha_{1k}-1}dJ_1$$

$$= (-1)^{h_1-k}\int_x^1 (-J_1)^{h_1-k}(J_1-x)^{k-\alpha_{1k}-1}dJ_1$$

$$= (-1)^{h_1-k}\int_x^1 ((1-J_1)-1)^{h_1-k}(J_1-x)^{k-\alpha_{1k}-1}dJ_1$$

$$= (-1)^{h_1-k}\int_x^1\left(\sum_{\theta=0}^{h_1-k}\binom{h_1-k}{\theta}(1-J_1)^{h_1-k-\theta}(-1)^{\theta}\right)\times(J_1-x)^{k-\alpha_{1k}-1}dJ_1 \tag{2.35}$$

$$= \sum_{\theta=0}^{h_1-k}\binom{h_1-k}{\theta}(-1)^{h_1-k+\theta}\times\int_x^1 (1-J_1)^{(h_1-k-\theta+1)-1}(J_1-x)^{k-\alpha_{1k}-1}dJ_1$$

$$= \sum_{\theta=0}^{h_1-k}\binom{h_1-k}{\theta}(-1)^{h_1-k+\theta}\times\left\{\frac{\Gamma(h_1-k-\theta+1)\,\Gamma(k-\alpha_{1k})}{\Gamma(h_1-\alpha_{1k}-\theta+1)}(1-x)^{h_1-\alpha_{1k}-\theta}\right\}. \tag{2.36}$$

Similarly we find

$$\int_y^1 (J_2 - y)^{l-\alpha_{2l}-1} J_2^{h_2-l} dJ_2$$

$$= \sum_{\rho=0}^{h_2-l} \binom{h_2 - l}{\rho} (-1)^{h_2-l+\rho}$$

$$\times \left\{ \frac{\Gamma(h_2 - l - \rho + 1)\, \Gamma(l - \alpha_{2l})}{\Gamma(h_2 - \alpha_{2l} - \rho + 1)} (1-y)^{h_2-\alpha_{2l}-\rho} \right\}. \tag{2.37}$$

Therefore

$$D_{1-}^{(\alpha_{1k},\alpha_{2l})} \left(\frac{x^{h_1}}{h_1!} \frac{y^{h_2}}{h_2!} \right) = \frac{1}{(h_1-k)!\,(h_2-l)!}$$

$$\cdot \left\{ \left[\sum_{\theta=0}^{h_1-k} \binom{h_1-k}{\theta} (-1)^{h_1+\theta} \left\{ \frac{\Gamma(h_1-k-\theta+1)}{\Gamma(h_1-\alpha_{1k}-\theta+1)} (1-x)^{h_1-\alpha_{1k}-\theta} \right\} \right] \right.$$

$$\left. \cdot \left[\sum_{\rho=0}^{h_2-l} \binom{h_2-l}{\rho} (-1)^{h_2+\rho} \left\{ \frac{\Gamma(h_2-l-\rho+1)}{\Gamma(h_2-\alpha_{2l}-\rho+1)} (1-y)^{h_2-\alpha_{2l}-\rho} \right\} \right] \right\}. \tag{2.38}$$

Here we use a lot Proposition 2.13.

Case (i). Assume that on $[0,1]^2$, $\alpha_{h_1 h_2} \geq \alpha > 0$.
Call

$$Q_{\overline{m_1},\overline{m_2}}(x,y) := B_{\overline{m_1},\overline{m_2}}(f;x,y) + P_{\overline{m_1},\overline{m_2}} \frac{x^{h_1}}{h_1!} \frac{y^{h_2}}{h_2!}. \tag{2.39}$$

Then by (2.28) we get

$$\left\| D_{1-}^{(\alpha_{1k},\alpha_{2l})} \left(f + P_{\overline{m_1},\overline{m_2}} \frac{x^{h_1}}{h_1!} \frac{y^{h_2}}{h_2!} \right) - D_{1-}^{(\alpha_{1k},\alpha_{2l})} \left(Q_{\overline{m_1},\overline{m_2}}(x,y) \right) \right\|_\infty \leq M_{\overline{m_1},\overline{m_2}}^{k,l}, \tag{2.40}$$

for all $0 \leq k \leq r$, $0 \leq l \leq p$.

When $(0,0) \leq (k,l) \leq (h_1,h_2)$, using (2.38) inequality (2.40) becomes

$$\left\| D_{1-}^{(\alpha_{1k},\alpha_{2l})} (f) + P_{\overline{m_1},\overline{m_2}} \frac{1}{(h_1-k)!\,(h_2-l)!} \right.$$

$$\cdot \left\{ \left[\sum_{\theta=0}^{h_1-k} \binom{h_1-k}{\theta} (-1)^{h_1+\theta} \left\{ \frac{\Gamma(h_1-k-\theta+1)}{\Gamma(h_1-\alpha_{1k}-\theta+1)} (1-x)^{h_1-\alpha_{1k}-\theta} \right\} \right] \right.$$

$$\cdot\left[\sum_{\rho=0}^{h_2-l}\binom{h_2-l}{\rho}(-1)^{h_2+\rho}\left\{\frac{\Gamma(h_2-l-\rho+1)}{\Gamma(h_2-\alpha_{2l}-\rho+1)}(1-y)^{h_2-\alpha_{2l}-\rho}\right\}\right]\Bigg\}$$

$$-D_{1-}^{(\alpha_{1k},\alpha_{2l})}\left(Q_{\overline{m_1},\overline{m_2}}(x,y)\right)\Big\|_{\infty}\leq M_{\overline{m_1},\overline{m_2}}^{k,l}, \tag{2.41}$$

for all $(0,0)\leq(k,l)\leq(h_1,h_2)$, $x,y\in[0,1]$.

Using (2.41) and triangle inequality we obtain for $(0,0)\leq(k,l)\leq(h_1,h_2)$ that

$$\left\|D_{1-}^{(\alpha_{1k},\alpha_{2l})}(f)-D_{1-}^{(\alpha_{1k},\alpha_{2l})}\left(Q_{\overline{m_1},\overline{m_2}}\right)\right\|_{\infty}\leq\frac{P_{\overline{m_1},\overline{m_2}}}{(h_1-k)!\,(h_2-l)!}$$
$$\cdot\left\{\left[\sum_{\theta=0}^{h_1-k}\binom{h_1-k}{\theta}\left\{\frac{\Gamma(h_1-k-\theta+1)}{\Gamma(h_1-\alpha_{1k}-\theta+1)}\right\}\right]\right.$$
$$\left.\cdot\left[\sum_{\rho=0}^{h_2-l}\binom{h_2-l}{\rho}\left\{\frac{\Gamma(h_2-l-\rho+1)}{\Gamma(h_2-\alpha_{2l}-\rho+1)}\right\}\right]\right\}+M_{\overline{m_1},\overline{m_2}}^{k,l} \tag{2.42}$$

proving (2.21).

Next if $(h_1+1,h_2+1)\leq(k,l)\leq(r,p)$, or $0\leq k\leq h_1$, $h_2+1\leq l\leq p$, or $h_1+1\leq k\leq r$, $0\leq l\leq h_2$, we get by (2.40) that

$$\left\|D_{1-}^{(\alpha_{1k},\alpha_{2l})}(f)-D_{1-}^{(\alpha_{1k},\alpha_{2l})}\left(Q_{\overline{m_1},\overline{m_2}}\right)\right\|_{\infty}\leq M_{\overline{m_1},\overline{m_2}}^{k,l}, \tag{2.43}$$

proving (2.22).

Furthermore, if (x,y) in critical region, see (2.23), we get

$$\begin{aligned}\alpha_{h_1h_2}^{-1}(x,y)\overline{L}\left(Q_{\overline{m_1},\overline{m_2}}(x,y)\right) &= \alpha_{h_1h_2}^{-1}(x,y)\overline{L}(f(x,y))\\ &+P_{\overline{m_1},\overline{m_2}}\frac{(1-x)^{h_1-\alpha_{1h_1}}(1-y)^{h_2-\alpha_{2h_2}}}{\Gamma\left(h_1-\alpha_{1h_1}+1\right)\Gamma\left(h_2-\alpha_{2h_2}+1\right)}\\ &+\sum_{i=h_1}^{v_1}\sum_{j=h_2}^{v_2}\alpha_{h_1h_2}^{-1}(x,y)\,\alpha_{ij}(x,y)\\ &\cdot D_{1-}^{(\alpha_{1i},\alpha_{2j})}\left[Q_{\overline{m_1},\overline{m_2}}(x,y)-f(x,y)-P_{\overline{m_1},\overline{m_2}}\frac{x^{h_1}}{h_1!}\frac{y^{h_2}}{h_2!}\right]\\ &\overset{(2.40)}{\geq} P_{\overline{m_1},\overline{m_2}}\frac{(1-x)^{h_1-\alpha_{1h_1}}(1-y)^{h_2-\alpha_{2h_2}}}{\Gamma\left(h_1-\alpha_{1h_1}+1\right)\Gamma\left(h_2-\alpha_{2h_2}+1\right)}\\ &-\sum_{i=h_1}^{v_1}\sum_{j=h_2}^{v_2}l_{ij}M_{\overline{m_1},\overline{m_2}}^{i,j}\end{aligned} \tag{2.44}$$

$$= P_{\overline{m_1},\overline{m_2}} \frac{(1-x)^{h_1-\alpha_{1h_1}} (1-y)^{h_2-\alpha_{2h_2}}}{\Gamma\left(h_1-\alpha_{1h_1}+1\right)\Gamma\left(h_2-\alpha_{2h_2}+1\right)} - P_{\overline{m_1},\overline{m_2}}$$

$$= P_{\overline{m_1},\overline{m_2}} \left[\frac{(1-x)^{h_1-\alpha_{1h_1}} (1-y)^{h_2-\alpha_{2h_2}}}{\Gamma\left(h_1-\alpha_{1h_1}+1\right)\Gamma\left(h_2-\alpha_{2h_2}+1\right)} - 1\right] \quad (2.45)$$

$$= P_{\overline{m_1},\overline{m_2}} \cdot \left[\frac{(1-x)^{h_1-\alpha_{1h_1}} (1-y)^{h_2-\alpha_{2h_2}} - \Gamma\left(h_1-\alpha_{1h_1}+1\right)\Gamma\left(h_2-\alpha_{2h_2}+1\right)}{\Gamma\left(h_1-\alpha_{1h_1}+1\right)\Gamma\left(h_2-\alpha_{2h_2}+1\right)}\right]$$

$$=: (*). \quad (2.46)$$

We know $\Gamma(1) = 1$, $\Gamma(2) = 1$, and Γ is convex and positive on $(0,\infty)$. Here $0 \le h_1 - \alpha_{1h_1} < 1$ and $0 \le h_2 - \alpha_{2h_2} < 1$, hence $1 \le h_1 - \alpha_{1h_1} + 1 < 2$, $1 \le h_2 - \alpha_{2h_2} + 1 < 2$. Thus $0 < \Gamma(h_1 - \alpha_{1h_1} + 1)$, $\Gamma(h_2 - \alpha_{2h_2} + 1) \le 1$, and $1 - \Gamma(h_1 - \alpha_{1h_1} + 1)\Gamma(h_2 - \alpha_{2h_2} + 1) \ge 0$. Clearly acting as in (2.44)–(2.46), when $\overline{L}(f(0,0)) \ge 0$, we get $\overline{L}(Q_{\overline{m_1},\overline{m_2}}(0,0)) \ge 0$.

Also clearly here on the critical region (2.23) we have $(*) \ge 0$. That is, there $\overline{L}(Q_{\overline{m_1},\overline{m_2}}(x,y)) \ge 0$.

Case (ii). Assume that throughout $[0,1]^2$, $\alpha_{h_1h_2} \le \beta < 0$. Consider

$$Q^{-}_{\overline{m_1},\overline{m_2}}(x,y) :\equiv B_{\overline{m_1},\overline{m_2}}(f;x,y) - P_{\overline{m_1},\overline{m_2}} \frac{x^{h_1}}{h_1!}\frac{y^{h_2}}{h_2!}.$$

Then by (2.28) we get

$$\left\| D_{1-}^{(\alpha_{1k},\alpha_{2l})}\left(f - P_{\overline{m_1},\overline{m_2}} \frac{x^{h_1}}{h_1!}\frac{y^{h_2}}{h_2!}\right) - D_{1-}^{(\alpha_{1k},\alpha_{2l})}\left(Q^{-}_{\overline{m_1},\overline{m_2}}(x,y)\right)\right\|_{\infty} \le M^{k,l}_{\overline{m_1},\overline{m_2}}, \quad (2.47)$$

all $0 \le k \le r$, $0 \le l \le p$.

Working similarly as earlier in this proof we derive again (2.21), (2.22). Furthermore, if (x,y) in critical region, see (2.23), we get

$$\alpha^{-1}_{h_1h_2}(x,y)\,\overline{L}\left(Q^{-}_{\overline{m_1},\overline{m_2}}(x,y)\right)$$

$$= \alpha^{-1}_{h_1h_2}(x,y)\,\overline{L}(f(x,y)) - P_{\overline{m_1},\overline{m_2}} \frac{(1-x)^{h_1-\alpha_{1h_1}} (1-y)^{h_2-\alpha_{2h_2}}}{\Gamma(h_1-\alpha_{1h_1}+1)\Gamma(h_2-\alpha_{2h_2}+1)}$$

$$+ \sum_{i=h_1}^{v_1}\sum_{j=h_2}^{v_2} \alpha^{-1}_{h_1h_2}(x,y)\,\alpha_{ij}(x,y)$$

$$\cdot D_{1-}^{(\alpha_{1i},\alpha_{2j})}\left[Q^{-}_{\overline{m_1},\overline{m_2}}(x,y) - f(x,y) + P_{\overline{m_1},\overline{m_2}} \frac{x^{h_1}}{h_1!}\frac{y^{h_2}}{h_2!}\right] \quad (2.48)$$

$$\overset{(2.47)}{\leq} -P_{\overline{m_1},\overline{m_2}} \frac{(1-x)^{h_1-\alpha_{1h_1}} (1-y)^{h_2-\alpha_{2h_2}}}{\Gamma(h_1-\alpha_{1h_1}+1)\,\Gamma(h_2-\alpha_{2h_2}+1)}$$

$$+\sum_{i=h_1}^{v_1}\sum_{j=h_2}^{v_2} l_{ij} M_{\overline{m_1},\overline{m_2}}^{i,j} \tag{2.49}$$

$$= P_{\overline{m_1},\overline{m_2}} \left(1 - \frac{(1-x)^{h_1-\alpha_{1h_1}} (1-y)^{h_2-\alpha_{2h_2}}}{\Gamma(h_1-\alpha_{1h_1}+1)\,\Gamma(h_2-\alpha_{2h_2}+1)}\right) \tag{2.50}$$

$$= P_{\overline{m_1},\overline{m_2}}$$

$$\cdot \left(\frac{\Gamma(h_1-\alpha_{1h_1}+1)\,\Gamma(h_2-\alpha_{2h_2}+1) - (1-x)^{h_1-\alpha_{1h_1}} (1-y)^{h_2-\alpha_{2h_2}}}{\Gamma(h_1-\alpha_{1h_1}+1)\,\Gamma(h_2-\alpha_{2h_2}+1)}\right)$$

$$=: (**)\,. \tag{2.51}$$

We know $\Gamma(1) = 1$, $\Gamma(2) = 1$, and Γ is convex and positive on $(0,\infty)$. Here $0 \leq h_1 - \alpha_{1h_1} < 1$ and $0 \leq h_2 - \alpha_{2h_2} < 1$, hence $1 \leq h_1 - \alpha_{1h_1} + 1 < 2$, $1 \leq h_2 - \alpha_{2h_2} + 1 < 2$. Thus $0 < \Gamma(h_1 - \alpha_{1h_1} + 1)$, $\Gamma(h_2 - \alpha_{2h_2} + 1) \leq 1$, and

$$\Gamma(h_1-\alpha_{1h_1}+1)\,\Gamma(h_2-\alpha_{2h_2}+1) - 1 \leq 0.$$

Clearly acting as in (2.48)–(2.51), when $\overline{L}(f(0,0)) \geq 0$, we get

$$\overline{L}\left(Q_{\overline{m_1},\overline{m_2}}^{-}(0,0)\right) \geq 0.$$

Also clearly here on the critical region (2.23) we get $(**) \leq 0$. That is, there $\overline{L}\left(Q_{\overline{m_1},\overline{m_2}}^{-}(x,y)\right) \geq 0$.

References

1. G.A. Anastassiou, Bivariate monotone approximation. Proc. Am. Math. Soc. **112**(4), 959–963 (1991)
2. G.A. Anastassiou, O. Shisha, Monotone approximation with linear differential operators. J. Approx. Theory **44**, 391–393 (1985)
3. I. Badea, C. Badea, On the order of simultaneously approximation of bivariate functions by Bernstein operators. Anal. Numer. Theor. Approx. **16**, 11–17 (1987)
4. S. Iqbal, K. Krulic, J. Pecaric, On an inequality of H. G. Hardy. J. Inequal. Appl., 23 pp. (2010). Article ID 264347
5. O. Shisha, Monotone approximation. Pac. J. Math. **15**, 667–671 (1965)
6. D.D. Stancu, Studii Si Cercetari Stiintifice **XI** (2), 221–233 (1960)

Chapter 3
Uniform Approximation with Rates by Multivariate Generalized Discrete Singular Operators

George A. Anastassiou and Merve Kester

Abstract Here we establish the uniform approximation properties of multivariate generalized discrete versions of Picard, Gauss–Weierstrass, and Poisson–Cauchy singular operators over $\mathbb{R}^N$, $N \geq 1$. We treat both the unitary and non-unitary cases of the operators above. We give quantitatively the pointwise and uniform convergence of these operators to the unit operator by involving the multivariate higher order modulus of smoothness.

3.1 Introduction

This article is motivated mainly by Favard [4], where J. Favard in 1944 introduced the discrete version of Gauss–Weierstrass operator

$$(F_n f)(x) = \frac{1}{\sqrt{\pi n}} \sum_{\nu=-\infty}^{\infty} f\left(\frac{\nu}{n}\right) \exp\left(-n\left(\frac{\nu}{n} - x\right)^2\right), \tag{3.1}$$

$n \in \mathbb{N}$, which has the property that $(F_n f)(x)$ converges to $f(x)$ pointwise for each $x \in \mathbb{R}$, and uniformly on any compact subinterval of $\mathbb{R}$, for each continuous function f ($f \in C(\mathbb{R})$) that fulfills $|f(t)| \leq Ae^{Bt^2}$, $t \in \mathbb{R}$, where A, B are positive constants. We are also motivated by Anastassiou [1], Anastassiou and Kester [2] and Anastassiou and Mezei [3].

In this article, we define the multivariate generalized discrete versions of the Picard, Gauss–Weierstrass, and Poisson–Cauchy singular operators and we study their uniform approximation properties. We cover thoroughly the unitary and non-unitary cases and their interconnections.

G.A. Anastassiou • M. Kester (✉)
Department of Mathematical Sciences, University of Memphis, TN 38152, USA
e-mail: ganastss@memphis.edu; mkester@memphis.edu

G.A. Anastassiou, O. Duman (eds.), *Computational Analysis*, Springer Proceedings in Mathematics & Statistics 155, DOI 10.1007/978-3-319-28443-9_3

3.2 Background

In [1], for $r \in \mathbb{N}$, $m \in \mathbb{Z}_+$, the author defined

$$\alpha_{j,r}^{[m]} := \begin{cases} (-1)^{r-j} \binom{r}{j} j^{-m}, & \text{if } j = 1, 2, \ldots, r, \\ 1 - \sum\limits_{j=1}^{r} (-1)^{r-j} \binom{r}{j} j^{-m}, & \text{if } j = 0, \end{cases} \tag{3.2}$$

and

$$\delta_{k,r}^{[m]} := \sum_{j=1}^{r} \alpha_{j,r}^{[m]} j^{k}, \quad k = 1, 2, \ldots, m \in \mathbb{N}. \tag{3.3}$$

The author observed that

$$\sum_{j=0}^{r} \alpha_{j,r}^{[m]} = 1, \tag{3.4}$$

and

$$-\sum_{j=1}^{r} (-1)^{r-j} \binom{r}{j} = (-1)^{r} \binom{r}{0}. \tag{3.5}$$

In [1], the author let μ_{ξ_n} be a probability Borel measure on $\mathbb{R}^N$, $N \geq 1$, $\xi_n > 0$ for $n \in \mathbb{N}$. Then, the author defined the multiple smooth singular integral operators as

$$\theta_{r,n}^{[m]} (f; x_1, \ldots, x_N) := \sum_{j=0}^{r} \alpha_{j,r}^{[m]} \int_{\mathbb{R}^N} f (x_1 + s_1 j, x_2 + s_2 j, \ldots, x_N + s_N j) \, d\mu_{\xi_n} (s), \tag{3.6}$$

where $s := (s_1, \ldots, s_N)$, $x := (x_1, \ldots, x_N) \in \mathbb{R}^N$; $n, r \in \mathbb{Z}$, $m \in \mathbb{Z}_+$, $f : \mathbb{R}^N \to \mathbb{R}$ is a Borel measurable function, and also $(\xi_n)_{n\in\mathbb{N}}$ is a bounded sequence of positive real numbers.

In [1], the author stated

Remark 3.1. The operators $\theta_{r,n}^{[m]}$ are not in general positive.

Furthermore, the author observed

Lemma 3.2. *The operators $\theta_{r,n}^{[m]}$ preserve the constant functions in N variables.*

In [1], the author needed

Definition 3.3. Let $f \in C_B(\mathbb{R}^N)$, the space of all bounded and continuous functions on $\mathbb{R}^N$. Then, the rth multivariate modulus of smoothness of f is given by

$$\omega_r (f; h) := \sup_{\sqrt{u_1^2 + \ldots + u_N^2} \leq h} \left\| \Delta_{u_1, u_2, \ldots, u_N}^{r} (f) \right\|_{\infty} < \infty, \quad h > 0, \tag{3.7}$$

where $\|\cdot\|_\infty$ is the sup-norm and

$$\begin{aligned}\Delta_u^r f(x) &:= \Delta_{u_1,u_2,\ldots,u_N}^r f(x_1,\ldots,x_N)\\ &= \sum_{j=0}^{r} (-1)^{r-j}\binom{r}{j} f(x_1+ju_1, x_2+ju_2,\ldots,x_N+ju_N).\end{aligned} \tag{3.8}$$

Let $m \in \mathbb{N}$ and let $f \in C^m(\mathbb{R}^N)$.
Assume that all partial derivatives of f of order m are bounded, i.e.

$$\left\| \frac{\partial^m f(\cdot,\cdot,\ldots,\cdot)}{\partial x_1^{\alpha_1}\ldots\partial x_N^{\alpha_N}} \right\|_\infty < \infty, \tag{3.9}$$

for all $\alpha_j \in \mathbb{Z}^+, j = 1,\ldots,N; \sum_{j=1}^{N} \alpha_j = m$.

In [1], the author proved

Theorem 3.4. *Let $m \in \mathbb{N}$, $f \in C^m(\mathbb{R}^N)$, $N \geq 1$, $x \in \mathbb{R}^N$. Assume $\left\| \frac{\partial^m f(\cdot,\ldots,\cdot)}{\partial x_1^{\alpha_1}\ldots\partial x_N^{\alpha_N}} \right\|_\infty < \infty$, for all $\alpha_j \in \mathbb{Z}^+$, $j = 1,\ldots,N$: $|\alpha| := \sum_{j=1}^{N} \alpha_j = m$. Let μ_{ξ_n} be a Borel probability measure on $\mathbb{R}^N$, for $\xi_n > 0$, $(\xi_n)_{n\in\mathbb{N}}$ bounded sequence. Assume that for all $\alpha := (\alpha_1,\ldots,\alpha_N)$, $\alpha_i \in \mathbb{Z}^+$, $i = 1,\ldots,N$, $|\alpha| := \sum_{i=1}^{N} \alpha_i = m$, we have that*

$$u_{\xi_n} := \int_{\mathbb{R}^N} \left(\prod_{i=1}^{N} |s_i|^{\alpha_i}\right)\left(1+\frac{\|s\|_2}{\xi_n}\right)^r d\mu_{\xi_n}(s) < \infty. \tag{3.10}$$

For $\tilde{j} = 1,\ldots,m$, and $\alpha := (\alpha_1,\ldots,\alpha_N)$, $\alpha_i \in \mathbb{Z}^+$, $i = 1,\ldots,N$, $|\alpha| := \sum_{i=1}^{N} \alpha_i = \tilde{j}$, call

$$c_{\alpha,n} := c_{\alpha,n,\tilde{j}} := \int_{\mathbb{R}^N} \prod_{i=1}^{N} s_i^{\alpha_i} d\mu_{\xi_n}(s_1,\ldots,s_N). \tag{3.11}$$

Then

i) *For all $x \in \mathbb{R}^N$,*

$$\begin{aligned} E_{r,n}^{[m]}(x) &:= \left| \theta_{r,n}^{[m]}(f;x) - f(x) - \sum_{\tilde{j}=1}^{m} \delta_{\tilde{j},r}^{[m]} \left(\sum_{\substack{\alpha_1,\ldots,\alpha_N \geq 0:\\ |\alpha|=\tilde{j}}} \frac{c_{\alpha,n,\tilde{j}} f_\alpha(x)}{\prod_{i=1}^{N} \alpha_i!} \right) \right| \\ &\leq \sum_{\substack{\alpha_1,\ldots,\alpha_N\geq 0\\ |\alpha|=m}} \frac{(\omega_r(f_\alpha,\xi_n))}{\left(\prod_{i=1}^{N}\alpha_i!\right)} \left(\int_{\mathbb{R}^N} \left(\prod_{i=1}^{N}|s_i|^{\alpha_i}\right)\left(1+\frac{\|s\|_2}{\xi_n}\right)^r d\mu_{\xi_n}(s)\right). \end{aligned} \tag{3.12}$$

ii)

$$\left\| E_{r,n}^{[m]} \right\|_{\infty} \leq R.H.S.\ (3.12). \tag{3.13}$$

Given that $\xi_n \to 0$*, as* $n \to \infty$*, and* u_{ξ_n} *is uniformly bounded, then we obtain that* $\left\| E_{r,n}^{[m]} \right\|_{\infty} \to 0$ *with rates.*

iii) *It holds also that*

$$\left\| \theta_{r,n}^{[m]}(f) - f \right\|_{\infty} \leq \sum_{\tilde{j}=1}^{m} \left| \delta_{\tilde{j},r}^{[m]} \right| \left(\sum_{\substack{\alpha_1,\ldots,\alpha_N \geq 0: \\ |\alpha| = \tilde{j}}} \frac{\left| c_{\alpha,n,\tilde{j}} \right| \|f_\alpha\|_\infty}{\prod_{i=1}^{N} \alpha_i!} \right) + R.H.S.\ (3.12), \tag{3.14}$$

given that $\|f_\alpha\|_\infty < \infty$*, for all* $\alpha : |\alpha| = \tilde{j}$*,* $\tilde{j} = 1, \ldots, m$*. Furthermore, as* $\xi_n \to 0$ *when* $n \to \infty$*, assuming that* $c_{\alpha,n,\tilde{j}} \to 0$*, while* u_{ξ_n} *is uniformly bounded, we conclude that*

$$\left\| \theta_{r,n}^{[m]}(f) - f \right\|_{\infty} \to 0 \tag{3.15}$$

with rates.

In [1], for the case of $m = 0$, the author gave

Theorem 3.5. *Let* $f \in C_B\left(\mathbb{R}^N\right)$ *(the space of all bounded and continuous functions on* $\mathbb{R}^N$*),* $N \geq 1$*. Then*

$$\left\| \theta_{r,n}^{[0]} f - f \right\|_{\infty} \leq \left(\int_{\mathbb{R}^N} \left(1 + \frac{\|s\|_2}{\xi_n} \right)^r d\mu_{\xi_n}(s) \right) \omega_r(f, \xi_n), \tag{3.16}$$

under the assumption

$$\Phi_{\xi_n} := \int_{\mathbb{R}^N} \left(1 + \frac{\|s\|_2}{\xi_n} \right)^r d\mu_{\xi_n}(s) < \infty. \tag{3.17}$$

As $n \to \infty$ *and* $\xi_n \to 0$*, given that* Φ_{ξ_n} *are uniformly bounded, we obtain that*

$$\left\| \theta_{r,n}^{[0]} f - f \right\|_{\infty} \to 0 \tag{3.18}$$

with rates.

3.3 Main Results

Here, let μ_{ξ_n} be a Borel probability measure on $\mathbb{R}^N$, $N \geq 1$, $0 < \xi_n \leq 1$, $n \in \mathbb{N}$. Assume that $\nu := (\nu_1, \ldots, \nu_N)$, $x := (x_1, \ldots, x_N) \in \mathbb{R}^N$ and $f : \mathbb{R}^N \to \mathbb{R}$ is a Borel measurable function.

1. When

$$\mu_{\xi_n}(\nu) = \frac{e^{-\frac{1}{\xi_n}\sum_{i=1}^{N}|\nu_i|}}{\sum_{\nu_1=-\infty}^{\infty} \cdots \sum_{\nu_N=-\infty}^{\infty} e^{-\frac{\sum_{i=1}^{N}|\nu_i|}{\xi_n}}}, \tag{3.19}$$

we define generalized multiple discrete Picard operators as:

$$P_{r,n}^{*\,[m]}(f; x_1, \ldots, x_N) = \frac{\sum_{\nu_1=-\infty}^{\infty} \cdots \sum_{\nu_N=-\infty}^{\infty} \left(\sum_{j=0}^{r} \alpha_{j,r}^{[m]} f(x_1 + j\nu_1, \ldots, x_N + j\nu_N) \right) e^{-\frac{1}{\xi_n}\sum_{i=1}^{N}|\nu_i|}}{\sum_{\nu_1=-\infty}^{\infty} \cdots \sum_{\nu_N=-\infty}^{\infty} e^{-\frac{1}{\xi_n}\sum_{i=1}^{N}|\nu_i|}}. \tag{3.20}$$

2. When

$$\mu_{\xi_n}(\nu) = \frac{e^{-\frac{1}{\xi_n}\sum_{i=1}^{N}\nu_i^2}}{\sum_{\nu_1=-\infty}^{\infty} \cdots \sum_{\nu_N=-\infty}^{\infty} e^{-\frac{1}{\xi_n}\sum_{i=1}^{N}\nu_i^2}}, \tag{3.21}$$

we define generalized multiple discrete Gauss–Weierstrass operators as:

$$W_{r,n}^{*\,[m]}(f; x_1, \ldots, x_N) = \frac{\sum_{\nu_1=-\infty}^{\infty} \cdots \sum_{\nu_N=-\infty}^{\infty} \left(\sum_{j=0}^{r} \alpha_{j,r}^{[m]} f(x_1 + j\nu_1, \ldots, x_N + j\nu_N) \right) e^{-\frac{1}{\xi_n}\sum_{i=1}^{N}\nu_i^2}}{\sum_{\nu_1=-\infty}^{\infty} \cdots \sum_{\nu_N=-\infty}^{\infty} e^{-\frac{1}{\xi_n}\sum_{i=1}^{N}\nu_i^2}}. \tag{3.22}$$

3. Let $\hat{\alpha} \in \mathbb{N}$ and $\beta > \frac{1}{\hat{\alpha}}$. When

$$\mu_{\xi_n}(\nu) = \frac{\prod_{i=1}^{N} \left(\nu_i^{2\hat{\alpha}} + \xi_n^{2\hat{\alpha}} \right)^{-\beta}}{\sum_{\nu_1=-\infty}^{\infty} \cdots \sum_{\nu_N=-\infty}^{\infty} \prod_{i=1}^{N} \left(\nu_i^{2\hat{\alpha}} + \xi_n^{2\hat{\alpha}} \right)^{-\beta}}, \tag{3.23}$$

we define the generalized multiple discrete Poisson–Cauchy operators as:

$$Q_{r,n}^{*[m]}(f;x_1,\ldots,x_N) = \frac{\sum\limits_{\nu_1=-\infty}^{\infty}\cdots\sum\limits_{\nu_N=-\infty}^{\infty}\left(\sum\limits_{j=0}^{r}\alpha_{j,r}^{[m]}f(x_1+j\nu_1,\ldots,x_N+j\nu_N)\right)\prod\limits_{i=1}^{N}\left(\nu_i^{2\hat{\alpha}}+\xi_n^{2\hat{\alpha}}\right)^{-\beta}}{\sum\limits_{\nu_1=-\infty}^{\infty}\cdots\sum\limits_{\nu_N=-\infty}^{\infty}\left(\prod\limits_{i=1}^{N}\left(\nu_i^{2\hat{\alpha}}+\xi_n^{2\hat{\alpha}}\right)^{-\beta}\right)}. \tag{3.24}$$

4. When

$$\mu_{\xi_n}(\nu) = \frac{e^{-\frac{1}{\xi_n}\sum\limits_{i=1}^{N}|\nu_i|}}{\left(1+2\xi_n e^{-\frac{1}{\xi_n}}\right)^N}, \tag{3.25}$$

we define the generalized multiple discrete non-unitary Picard operators as:

$$P_{r,n}^{[m]}(f;x_1,\ldots,x_N) = \frac{\sum\limits_{\nu_1=-\infty}^{\infty}\cdots\sum\limits_{\nu_N=-\infty}^{\infty}\left(\sum\limits_{j=0}^{r}\alpha_{j,r}^{[m]}f(x_1+j\nu_1,\ldots,x_N+j\nu_N)\right)e^{-\frac{1}{\xi_n}\sum\limits_{i=1}^{N}|\nu_i|}}{\left(1+2\xi_n e^{-\frac{1}{\xi_n}}\right)^N}. \tag{3.26}$$

5. When

$$\mu_{\xi_n}(\nu) = \frac{e^{-\frac{1}{\xi_n}\sum\limits_{i=1}^{N}\nu_i^2}}{\left(\sqrt{\pi\xi_n}\left(1-\operatorname{erf}\left(\frac{1}{\sqrt{\xi_n}}\right)\right)+1\right)^N}, \tag{3.27}$$

we define the generalized multiple discrete non-unitary Gauss–Weierstrass operators as:

$$W_{r,n}^{[m]}(f;x_1,\ldots,x_N) = \frac{\sum\limits_{\nu_1=-\infty}^{\infty}\cdots\sum\limits_{\nu_N=-\infty}^{\infty}\left(\sum\limits_{j=0}^{r}\alpha_{j,r}^{[m]}f(x_1+j\nu_1,\ldots,x_N+j\nu_N)\right)e^{-\frac{1}{\xi_n}\sum\limits_{i=1}^{N}\nu_i^2}}{\left(\sqrt{\pi\xi_n}\left(1-\operatorname{erf}\left(\frac{1}{\sqrt{\xi_n}}\right)\right)+1\right)^N}, \tag{3.28}$$

where $\operatorname{erf}(x) = \frac{2}{\sqrt{\pi}}\int\limits_0^x e^{-t^2}dt$ with $\operatorname{erf}(\infty) = 1$.

Additionally, in this article we assume that $0^0 = 1$.

We observe

Proposition 3.6. *Let* $\alpha := (\alpha_1, \ldots, \alpha_N)$, $\alpha_i \in \mathbb{Z}^+$, $i = 1, \ldots, N \in \mathbb{N}$, $|\alpha| := \sum_{i=1}^{N} \alpha_i = m \in \mathbb{N}$. *Then, there exist* $K_1 > 0$ *such that*

$$u^*_{P,\xi_n} = \frac{\sum_{\nu_1=-\infty}^{\infty} \cdots \sum_{\nu_N=-\infty}^{\infty} \left(\prod_{i=1}^{N} |\nu_i|^{\alpha_i}\right) \left(1 + \frac{\|\nu\|_2}{\xi_n}\right)^r e^{-\frac{1}{\xi_n} \sum_{i=1}^{N} |\nu_i|}}{\sum_{\nu_1=-\infty}^{\infty} \cdots \sum_{\nu_N=-\infty}^{\infty} e^{-\frac{1}{\xi_n} \sum_{i=1}^{N} |\nu_i|}} \le K_1 < \infty, \tag{3.29}$$

for all $\xi_n \in (0, 1]$ *where* $n \in \mathbb{N}$ *and* $\nu = (\nu_1, \ldots, \nu_N)$.

Proof. First, we observe that

$$\sum_{\nu_1=-\infty}^{\infty} \cdots \sum_{\nu_N=-\infty}^{\infty} e^{-\frac{1}{\xi_n} \sum_{i=1}^{N} |\nu_i|} > 1. \tag{3.30}$$

Thus, we obtain

$$u^*_{P,\xi_n} \le \sum_{\nu_1=-\infty}^{\infty} \cdots \sum_{\nu_N=-\infty}^{\infty} \left(\prod_{i=1}^{N} |\nu_i|^{\alpha_i}\right) \left(1 + \frac{\|\nu\|_2}{\xi_n}\right)^r e^{-\frac{1}{\xi_n} \sum_{i=1}^{N} |\nu_i|} := R_{1,\alpha_i}. \tag{3.31}$$

Since

$$\|\nu\|_2 = \sqrt{\nu_1^2 + \cdots + \nu_N^2} \le \sum_{i=1}^{N} |\nu_i|, \tag{3.32}$$

we get

$$\begin{aligned} R_{1,\alpha_i} &\le \sum_{\nu_1=-\infty}^{\infty} \cdots \sum_{\nu_N=-\infty}^{\infty} \left(\prod_{i=1}^{N} |\nu_i|^{\alpha_i}\right) \left(1 + \frac{1}{\xi_n} \sum_{i=1}^{N} |\nu_i|\right)^r e^{-\frac{1}{\xi_n} \sum_{i=1}^{N} |\nu_i|} \\ &= 2^N \sum_{\nu_1=1}^{\infty} \cdots \sum_{\nu_N=1}^{\infty} \left(\prod_{i=1}^{N} \nu_i^{\alpha_i}\right) \left(1 + \frac{1}{\xi_n} \sum_{i=1}^{N} \nu_i\right)^r e^{-\frac{1}{\xi_n} \sum_{i=1}^{N} \nu_i}. \end{aligned} \tag{3.33}$$

We also observe that, for each $\nu_i \ge 1$, we have

$$\begin{aligned} \left(1 + \frac{1}{\xi_n} \sum_{i=1}^{N} \nu_i\right)^r &\le 2^r \left(\frac{1}{\xi_n} \sum_{i=1}^{N} \nu_i\right)^r = 2^r \xi_n^{-r} \left(\sum_{i=1}^{N} \nu_i\right)^r \\ &\le 2^r \xi_n^{-r} N^r \prod_{i=1}^{N} \nu_i^r \le 2^r \xi_n^{-Nr} N^r \prod_{i=1}^{N} \nu_i^r \\ &= 2^r N^r \prod_{i=1}^{N} \left(\frac{\nu_i}{\xi_n}\right)^r \le 2^r N^r \prod_{i=1}^{N} \left(1 + \frac{\nu_i}{\xi_n}\right)^r. \end{aligned} \tag{3.34}$$

Thus,

$$\begin{aligned} u^*_{P,\xi_n} &\le R_{1,\alpha_i} \le 2^{N+r} N^r \sum_{\nu_1=1}^{\infty} \cdots \sum_{\nu_N=1}^{\infty} \left(\prod_{i=1}^{N} \nu_i^{\alpha_i} \right) \left(\prod_{i=1}^{N} \left(1 + \frac{\nu_i}{\xi_n} \right)^r \right) e^{-\frac{1}{\xi_n} \sum_{i=1}^{N} \nu_i} \\ &= 2^{N+r} N^r \sum_{\nu_1=1}^{\infty} \cdots \sum_{\nu_N=1}^{\infty} \left(\prod_{i=1}^{N} \nu_i^{\alpha_i} \left(1 + \frac{\nu_i}{\xi_n} \right)^r e^{-\frac{\nu_i}{\xi_n}} \right) \\ &= 2^{N+r} N^r \prod_{k=1}^{N} \sum_{\nu_k=1}^{\infty} \nu_k^{\alpha_k} \left(1 + \frac{\nu_k}{\xi_n} \right)^r e^{-\frac{\nu_k}{\xi_n}}. \end{aligned} \tag{3.35}$$

In [2], for $\alpha_k \in \mathbb{N}$, the authors showed that

$$\begin{aligned} M_{1,\alpha_k} &:= \sum_{\nu_k=1}^{\infty} \nu_k^{\alpha_k} \left(1 + \frac{\nu_k}{\xi_n} \right)^r e^{-\frac{\nu_k}{\xi_n}} \\ &\le 2^{2r} r! \left(\lambda_{\alpha_k} + (2\alpha_k + 1)^{\alpha_k} e^{-\frac{(2\alpha_k+1)}{2}} + (\alpha_k)! 2^{\alpha_k+1} \right) < \infty, \end{aligned} \tag{3.36}$$

where

$$\lambda_{\alpha_k} := \sum_{\nu_k=1}^{2\alpha_k} \nu_k^{\alpha_k} e^{-\frac{\nu_k}{2}} < \infty. \tag{3.37}$$

So if $\alpha_k \in \mathbb{N}$, by (3.35)–(3.37), we obtain

$$u^*_{P,\xi_n} \le N^r 2^{2rN+N+r} (r!)^N \prod_{k=1}^{N} \left(\lambda_{\alpha_k} + (2\alpha_k + 1)^{\alpha_k} e^{-\frac{(2\alpha_k+1)}{2}} + (\alpha_k)! 2^{\alpha_k+1} \right) < \infty, \tag{3.38}$$

for all $\xi_n \in (0, 1]$. For $\alpha_k = 0$, we have that

$$M_{1,0} = \sum_{\nu_k=1}^{\infty} \left(1 + \frac{\nu_k}{\xi_n} \right)^r e^{-\frac{\nu_k}{\xi_n}} \le \sum_{\nu_k=1}^{\infty} \nu_k \left(1 + \frac{\nu_k}{\xi_n} \right)^r e^{-\frac{\nu_k}{\xi_n}} = M_{1,1} < \infty, \tag{3.39}$$

so that, by (3.36), we have

$$M_{1,0} < \infty \tag{3.40}$$

for all $\xi_n \in (0, 1]$.
Thus, by (3.35), (3.38), and (3.40), we get

$$u^*_{P,\xi_n} \le R_{1,\alpha_i} < \infty, \tag{3.41}$$

so that the proof is complete.

Next, we have

Proposition 3.7. *Let* $\alpha := (\alpha_1, \ldots, \alpha_N)$, $\alpha_i \in \mathbb{Z}^+$, $i = 1, \ldots, N \in \mathbb{N}$, $|\alpha| := \sum_{i=1}^{N} \alpha_i = m$. *Then, there exist* $K_2 > 0$ *such that*

$$u^*_{W,\xi_n} = \frac{\sum\limits_{\nu_1=-\infty}^{\infty} \cdots \sum\limits_{\nu_N=-\infty}^{\infty} \left(\prod\limits_{i=1}^{N} |\nu_i|^{\alpha_i} \right) \left(1 + \frac{\|\nu\|_2}{\xi_n}\right)^r e^{-\frac{1}{\xi_n} \sum\limits_{i=1}^{N} \nu_i^2}}{\sum\limits_{\nu_1=-\infty}^{\infty} \cdots \sum\limits_{\nu_N=-\infty}^{\infty} e^{-\frac{1}{\xi_n} \sum\limits_{i=1}^{N} \nu_i^2}} \leq K_2 < \infty, \tag{3.42}$$

for all $\xi_n \in (0, 1]$ *where* $n \in \mathbb{N}$ *and* $\nu = (\nu_1, \ldots, \nu_N)$.

Proof. We observe that

$$\sum_{\nu_1=-\infty}^{\infty} \cdots \sum_{\nu_N=-\infty}^{\infty} e^{-\frac{1}{\xi_n} \sum\limits_{i=1}^{N} \nu_i^2} > 1. \tag{3.43}$$

Thus, we obtain

$$u^*_{W,\xi_n} \leq \sum_{\nu_1=-\infty}^{\infty} \cdots \sum_{\nu_N=-\infty}^{\infty} \left(\prod_{i=1}^{N} |\nu_i|^{\alpha_i} \right) \left(1 + \frac{\|\nu\|_2}{\xi_n}\right)^r e^{-\frac{1}{\xi_n} \sum\limits_{i=1}^{N} \nu_i^2} := R_{2,\alpha_i}. \tag{3.44}$$

On the other hand, we have

$$\nu_i^2 \geq |\nu_i| \tag{3.45}$$

which yields that

$$\frac{\nu_i^2}{\xi_n} \geq \frac{|\nu_i|}{\xi_n} \geq 1 \tag{3.46}$$

and

$$e^{-\frac{\nu_i^2}{\xi_n}} \leq e^{-\frac{|\nu_i|}{\xi_n}}, \tag{3.47}$$

for all $i = 1, \ldots, N$. Thus

$$e^{-\frac{1}{\xi_n} \sum\limits_{i=1}^{N} \nu_i^2} = \prod_{i=1}^{N} e^{-\frac{\nu_i^2}{\xi_n}} \leq \prod_{i=1}^{N} e^{-\frac{|\nu_i|}{\xi_n}} = e^{-\frac{1}{\xi_n} \sum\limits_{i=1}^{N} |\nu_i|}. \tag{3.48}$$

Therefore, by (3.31), (3.44), and (3.48), we get

$$u^*_{W,\xi_n} \leq R_{2,\alpha_i} \leq R_{1,\alpha_i}. \tag{3.49}$$

Hence, by Proposition 3.6, the proof is done.

We get

Proposition 3.8. *Let* $\alpha := (\alpha_1, \ldots, \alpha_N)$, $\alpha_i \in \mathbb{Z}^+$, $i = 1, \ldots, N \in \mathbb{N}$, $|\alpha| := \sum_{i=1}^{N} \alpha_i = m$. *Then, there exist* $K_3 > 0$ *such that*

$$u^*_{Q,\xi_n} = \frac{\sum_{\nu_1=-\infty}^{\infty} \cdots \sum_{\nu_N=-\infty}^{\infty} \left(\prod_{i=1}^{N} |\nu_i|^{\alpha_i} \right) \left(1 + \frac{\|\nu\|_2}{\xi_n} \right)^r \left(\prod_{i=1}^{N} \left(\nu_i^{2\hat{\alpha}} + \xi_n^{2\hat{\alpha}} \right)^{-\beta} \right)}{\sum_{\nu_1=-\infty}^{\infty} \cdots \sum_{\nu_N=-\infty}^{\infty} \left(\prod_{i=1}^{N} \left(\nu_i^{2\hat{\alpha}} + \xi_n^{2\hat{\alpha}} \right)^{-\beta} \right)}$$
$$\leq K_3 < \infty, \tag{3.50}$$

for all $\xi_n \in (0, 1]$ *where* $\hat{\alpha}$, $n \in \mathbb{N}$, $\beta > \max\left\{ \frac{1+r+\alpha_i}{2\hat{\alpha}}, \frac{r+2}{2\hat{\alpha}} \right\}$ *for all* $i = 1, \ldots, N$, *and* $\nu = (\nu_1, \ldots, \nu_N)$.

Proof. First we observe that

$$\sum_{\nu_i=-\infty}^{\infty} \left(\nu_i^{2\hat{\alpha}} + \xi_n^{2\hat{\alpha}} \right)^{-\beta} \geq \xi_n^{-2\hat{\alpha}\beta} \tag{3.51}$$

for all $i = 1, \ldots, N$. Thus, we get

$$\begin{aligned}
&\sum_{\nu_1=-\infty}^{\infty} \cdots \sum_{\nu_N=-\infty}^{\infty} \left(\prod_{i=1}^{N} \left(\nu_i^{2\hat{\alpha}} + \xi_n^{2\hat{\alpha}} \right)^{-\beta} \right) \\
&= \prod_{i=1}^{N} \sum_{\nu_i=-\infty}^{\infty} \left(\nu_i^{2\hat{\alpha}} + \xi_n^{2\hat{\alpha}} \right)^{-\beta} \\
&\geq \prod_{i=1}^{N} \xi_n^{-2\hat{\alpha}\beta} = \xi_n^{-2N\hat{\alpha}\beta}.
\end{aligned} \tag{3.52}$$

Hence, by (3.32), (3.34), and (3.52), we get

$$\begin{aligned}
u^*_{Q,\xi_n} &\leq R_{3,\alpha_i} \\
&:= \xi_n^{2N\hat{\alpha}\beta} \sum_{\nu_1=-\infty}^{\infty} \cdots \sum_{\nu_N=-\infty}^{\infty} \left(\prod_{i=1}^{N} |\nu_i|^{\alpha_i} \right) \left(1 + \frac{\|\nu\|_2}{\xi_n} \right)^r \left(\prod_{i=1}^{N} \left(\nu_i^{2\hat{\alpha}} + \xi_n^{2\hat{\alpha}} \right)^{-\beta} \right) \\
&= \xi_n^{2N\hat{\alpha}\beta} 2^N \sum_{\nu_1=1}^{\infty} \cdots \sum_{\nu_N=1}^{\infty} \left(\prod_{i=1}^{N} \nu_i^{\alpha_i} \right) \left(1 + \frac{\|\nu\|_2}{\xi_n} \right)^r \left(\prod_{i=1}^{N} \left(\nu_i^{2\hat{\alpha}} + \xi_n^{2\hat{\alpha}} \right)^{-\beta} \right) \\
&\leq N^r \xi_n^{2N\hat{\alpha}\beta} 2^{N+r} \sum_{\nu_1=1}^{\infty} \cdots \sum_{\nu_N=1}^{\infty} \left(\prod_{i=1}^{N} \nu_i^{\alpha_i} \left(1 + \frac{\nu_i}{\xi_n} \right)^r \left(\nu_i^{2\hat{\alpha}} + \xi_n^{2\hat{\alpha}} \right)^{-\beta} \right) \\
&= N^r 2^r \prod_{i=1}^{N} \left(2\xi_n^{2\hat{\alpha}\beta} \sum_{\nu_i=1}^{\infty} \nu_i^{\alpha_i} \left(1 + \frac{\nu_i}{\xi_n} \right)^r \left(\nu_i^{2\hat{\alpha}} + \xi_n^{2\hat{\alpha}} \right)^{-\beta} \right).
\end{aligned} \tag{3.53}$$

In [1], for $\alpha_i \in \mathbb{N}$ and $\beta > \frac{1+r+\alpha_i}{2\hat{\alpha}}$ the authors proved

$$\begin{aligned}
M_{2,\alpha_i} &:= 2\xi_n^{2\hat{\alpha}\beta} \sum_{\nu_i=1}^{\infty} \nu_i^{\alpha_i} \left(1 + \frac{\nu_i}{\xi_n} \right)^r \left(\nu_i^{2\hat{\alpha}} + \xi_n^{2\hat{\alpha}} \right)^{-\beta} \\
&\leq 2^{r+1} \sum_{\nu_i=1}^{\infty} \left(\frac{1}{\nu_i} \right)^{2\hat{\alpha}\beta-\alpha_i-r} < \infty,
\end{aligned} \tag{3.54}$$

for all $\xi_n \in (0, 1]$. Clearly we also have that

$$R_{3,\alpha_i} < \infty. \tag{3.55}$$

Hence, by (3.53) and (3.54), we have

$$u^*_{Q,\xi_n} \leq N^r 2^r \prod_{i=1}^{N} M_{2,\alpha_i} < \infty, \tag{3.56}$$

for all $\xi_n \in (0, 1]$, $\alpha_i \in \mathbb{N}$, $\beta > \frac{1+r+\alpha_i}{2\hat{\alpha}}$, and $i = 1, \ldots, N$.
For $\alpha_i = 0$, we observe that

$$\begin{aligned} M_{2,0} &= 2\xi_n^{2\hat{\alpha}\beta} \sum_{\nu_i=1}^{\infty} \left(1 + \frac{\nu_i}{\xi_n}\right)^r \left(\nu_i^{2\hat{\alpha}} + \xi_n^{2\hat{\alpha}}\right)^{-\beta} \\ &\leq 2\xi_n^{2\hat{\alpha}\beta} \sum_{\nu_i=1}^{\infty} \nu_i \left(1 + \frac{\nu_i}{\xi_n}\right)^r \left(\nu_i^{2\hat{\alpha}} + \xi_n^{2\hat{\alpha}}\right)^{-\beta} \\ &= M_{2,1}. \end{aligned} \tag{3.57}$$

Thus, by (3.54), when $\alpha_i = 0$, we get

$$u^*_{Q,\xi_n} \leq N^r 2^r \prod_{i=1}^{N} M_{2,1} < \infty, \tag{3.58}$$

for all $\xi_n \in (0, 1]$, and $\beta > \frac{r+2}{2\hat{\alpha}}$. Therefore by (3.56) and (3.58), the proof is done.

We need

Definition 3.9. We define

$$\begin{aligned} c_{\alpha,n,\tilde{j}} &:= \frac{\sum\limits_{\nu_1=-\infty}^{\infty} \cdots \sum\limits_{\nu_N=-\infty}^{\infty} \left(\prod\limits_{i=1}^{N} \nu_i^{\alpha_i}\right) e^{-\frac{1}{\xi_n}\sum\limits_{i=1}^{N}|\nu_i|}}{\sum\limits_{\nu_1=-\infty}^{\infty} \cdots \sum\limits_{\nu_N=-\infty}^{\infty} e^{-\frac{1}{\xi_n}\sum\limits_{i=1}^{N}|\nu_i|}} \\ &= \frac{\sum\limits_{\nu_1=-\infty}^{\infty} \cdots \sum\limits_{\nu_N=-\infty}^{\infty} \left(\prod_{i=1}^{N} \nu_i^{\alpha_i} e^{-\frac{|\nu_i|}{\xi_n}}\right)}{\sum\limits_{\nu_1=-\infty}^{\infty} \cdots \sum\limits_{\nu_N=-\infty}^{\infty} \left(\prod\limits_{i=1}^{N} e^{-\frac{|\nu_i|}{\xi_n}}\right)} = \prod_{i=1}^{N} \left(\frac{\sum\limits_{\nu_i=-\infty}^{\infty} \nu_i^{\alpha_i} e^{-\frac{|\nu_i|}{\xi_n}}}{\sum\limits_{\nu_i=-\infty}^{\infty} e^{-\frac{|\nu_i|}{\xi_n}}} \right), \end{aligned} \tag{3.59}$$

$$p_{\alpha,n,\tilde{j}} := \frac{\sum_{\nu_1=-\infty}^{\infty} \cdots \sum_{\nu_N=-\infty}^{\infty} \left(\prod_{i=1}^{N} \nu_i^{\alpha_i} \right) e^{-\frac{1}{\xi_n} \sum_{i=1}^{N} \nu_i^2}}{\sum_{\nu_1=-\infty}^{\infty} \cdots \sum_{\nu_N=-\infty}^{\infty} e^{-\frac{1}{\xi_n} \sum_{i=1}^{N} \nu_i^2}}$$

$$= \frac{\sum_{\nu_1=-\infty}^{\infty} \cdots \sum_{\nu_N=-\infty}^{\infty} \left(\prod_{i=1}^{N} \nu_i^{\alpha_i} e^{-\frac{\nu_i^2}{\xi_n}} \right)}{\sum_{\nu_1=-\infty}^{\infty} \cdots \sum_{\nu_N=-\infty}^{\infty} \left(\prod_{i=1}^{N} e^{-\frac{\nu_i^2}{\xi_n}} \right)} = \prod_{i=1}^{N} \left(\frac{\sum_{\nu_i=-\infty}^{\infty} \nu_i^{\alpha_i} e^{-\frac{\nu_i^2}{\xi_n}}}{\sum_{\nu_i=-\infty}^{\infty} e^{-\frac{\nu_i^2}{\xi_n}}} \right), \tag{3.60}$$

and

$$q_{\alpha,n,\tilde{j}} := \frac{\sum_{\nu_1=-\infty}^{\infty} \cdots \sum_{\nu_N=-\infty}^{\infty} \left(\prod_{i=1}^{N} \nu_i^{\alpha_i} \left(\nu_i^{2\hat{\alpha}} + \xi_n^{2\hat{\alpha}} \right)^{-\beta} \right)}{\sum_{\nu_1=-\infty}^{\infty} \cdots \sum_{\nu_N=-\infty}^{\infty} \prod_{i=1}^{N} \left(\nu_i^{2\hat{\alpha}} + \xi_n^{2\hat{\alpha}} \right)^{-\beta}}$$

$$= \prod_{i=1}^{N} \left(\frac{\sum_{\nu_i=-\infty}^{\infty} \nu_i^{\alpha_i} \left(\nu_i^{2\hat{\alpha}} + \xi_n^{2\hat{\alpha}} \right)^{-\beta}}{\sum_{\nu_i=-\infty}^{\infty} \left(\nu_i^{2\hat{\alpha}} + \xi_n^{2\hat{\alpha}} \right)^{-\beta}} \right). \tag{3.61}$$

We will use

Lemma 3.10. *For* $\tilde{j} = 1, \ldots, m$, *and* $\alpha := (\alpha_1, \ldots, \alpha_N)$, $\alpha_i \in \mathbb{Z}^+$, $i = 1, \ldots, N$, $|\alpha| := \sum_{i=1}^{N} \alpha_i = \tilde{j}$, *we have that*

$$c_{\alpha,n} := c_{\alpha,n,\tilde{j}} := \frac{\sum_{\nu_1=-\infty}^{\infty} \cdots \sum_{\nu_N=-\infty}^{\infty} \left(\prod_{i=1}^{N} \nu_i^{\alpha_i} \right) e^{-\frac{1}{\xi_n} \sum_{i=1}^{N} |\nu_i|}}{\sum_{\nu_1=-\infty}^{\infty} \cdots \sum_{\nu_N=-\infty}^{\infty} e^{-\frac{1}{\xi_n} \sum_{i=1}^{N} |\nu_i|}} < \infty. \tag{3.62}$$

for all $\xi_n \in (0, 1]$. *Additionally, let* $\alpha_i \in \mathbb{N}$, *then as* $\xi_n \to 0$ *when* $n \to \infty$, *we get* $c_{\alpha,n,\tilde{j}} \to 0$.

Proof. Let $\alpha_i \in \mathbb{N}$. In [2], the authors showed that

$$\frac{\sum_{\nu_i=-\infty}^{\infty} \nu_i^{\alpha_i} e^{-\frac{|\nu_i|}{\xi_n}}}{\sum_{\nu_i=-\infty}^{\infty} e^{-\frac{|\nu_i|}{\xi_n}}} < \infty, \tag{3.63}$$

for each $i = 1, \ldots, N \in \mathbb{N}$, and for all $\xi_n \in (0, 1]$. Then by (3.59) and (3.63), we get

$$c_{\alpha,n,\tilde{j}} = \prod_{i=1}^{N} \left(\frac{\sum\limits_{\nu_i=-\infty}^{\infty} \nu_i^{\alpha_i} e^{-\frac{|\nu_i|}{\xi_n}}}{\sum\limits_{\nu_i=-\infty}^{\infty} e^{-\frac{|\nu_i|}{\xi_n}}} \right) < \infty, \tag{3.64}$$

for all $\xi_n \in (0, 1]$. Moreover, by (3.31) and (3.41), we observe that

$$0 \leq c_{\alpha,n,\tilde{j}} \leq \sum_{\nu_1=-\infty}^{\infty} \cdots \sum_{\nu_N=-\infty}^{\infty} \left(\prod_{i=1}^{N} \nu_i^{\alpha_i} \right) e^{-\frac{1}{\xi_n} \sum\limits_{i=1}^{N} |\nu_i|} \leq R_{1,\alpha_i} < \infty. \tag{3.65}$$

Additionally, we observe that

$$\sum_{\nu_1=-\infty}^{\infty} \cdots \sum_{\nu_N=-\infty}^{\infty} \left(\prod_{i=1}^{N} \nu_i^{\alpha_i} \right) e^{-\frac{1}{\xi_n} \sum\limits_{i=1}^{N} |\nu_i|}$$

$$= \begin{cases} 0 & \text{if } \alpha_i \text{ is odd,} \\ 2^N \sum\limits_{\nu_1=1}^{\infty} \cdots \sum\limits_{\nu_N=1}^{\infty} \left(\prod\limits_{i=1}^{N} \nu_i^{\alpha_i} \right) e^{-\frac{1}{\xi_n} \sum\limits_{i=1}^{N} \nu_i} & \text{if } \alpha_i \text{ is even,} \end{cases} \tag{3.66}$$

see also [2].

Assume that α_i is even. Therefore,

$$K^1_{\xi_n,\nu_i} := 2^N \sum_{\nu_1=1}^{\infty} \cdots \sum_{\nu_N=1}^{\infty} \left(\prod_{i=1}^{N} \nu_i^{\alpha_i} \right) e^{-\frac{1}{\xi_n} \sum\limits_{i=1}^{N} \nu_i} = 2^N \prod_{i=1}^{N} \left(\sum_{\nu_i=1}^{\infty} \nu_i^{\alpha_i} e^{-\frac{\nu_i}{\xi_n}} \right). \tag{3.67}$$

Observe that the function $f(x) = x^{\alpha_i} e^{-\frac{x}{\xi_n}}$ is positive, continuous, and decreasing on $[1, \infty)$. Hence, by Smarandache [5], we obtain

$$\begin{aligned} \sum_{\nu_i=1}^{\infty} \nu_i^{\alpha_i} e^{-\frac{\nu_i}{\xi_n}} &\leq f(1) + \int_1^{\infty} \nu_i^{\alpha_i} e^{-\frac{\nu_i}{\xi_n}} d\nu_i \\ &= e^{-\frac{1}{\xi_n}} + \int_1^{\infty} \nu_i^{\alpha_i} e^{-\frac{\nu_i}{\xi_n}} d\nu_i. \end{aligned} \tag{3.68}$$

By the integral calculation in [3, p. 86], it is easy to see that

$$\begin{aligned} \int_1^{\infty} \nu_i^{\alpha_i} e^{-\frac{\nu_i}{\xi_n}} d\nu_i &\leq \int_0^{\infty} \nu_i^{\alpha_i} e^{-\frac{\nu_i}{\xi_n}} d\nu_i \\ &= \alpha_i! \xi_n^{\alpha_i+1}. \end{aligned} \tag{3.69}$$

Hence, by (3.67)–(3.69), we get

$$K^1_{\xi_n,\nu_i} \le e^{-\frac{1}{\xi_n}} + \alpha_i!\xi_n^{\alpha_i+1} \to 0, \text{ as } \xi_n \to 0. \tag{3.70}$$

Thus, by (3.65), and (3.70) we have as $\xi_n \to 0$ when $n \to \infty$, $c_{\alpha,n,\tilde{j}} \to 0$.

Now, let $\alpha_i = 0$. For each ν_i, we have $\left(\sum_{n=1}^{\infty} \frac{1}{n^2} = \frac{\pi^2}{6}, \text{ Euler 1741} \right)$

$$\sum_{\nu_i=-\infty}^{\infty} e^{-\frac{|\nu_i|}{\xi_n}} = 1 + 2\sum_{\nu_i=1}^{\infty} e^{-\frac{\nu_i}{\xi_n}}$$
$$\le 1 + 2\sum_{i=1}^{\infty} \frac{1}{\nu_i^2} = 1 + \frac{\pi^2}{3} < \infty. \tag{3.71}$$

Thus,

$$c_{\alpha,n,\tilde{j}} := \frac{\sum_{\nu_1=-\infty}^{\infty} \cdots \sum_{\nu_N=-\infty}^{\infty} e^{-\frac{1}{\xi_n}\sum_{i=1}^{N}|\nu_i|}}{\sum_{\nu_1=-\infty}^{\infty} \cdots \sum_{\nu_N=-\infty}^{\infty} e^{-\frac{1}{\xi_n}\sum_{i=1}^{N}|\nu_i|}}$$
$$= \prod_{i=1}^{N} \left(\frac{\sum_{\nu_i=-\infty}^{\infty} e^{-\frac{|\nu_i|}{\xi_n}}}{\sum_{\nu_i=-\infty}^{\infty} e^{-\frac{|\nu_i|}{\xi_n}}} \right) = 1 < \infty. \tag{3.72}$$

Next, we give our results for $P^{*\,[m]}_{r,n}$

Theorem 3.11. *Let* $m \in \mathbb{N}$, $f \in C^m\left(\mathbb{R}^N\right)$, $N \ge 1$, $x \in \mathbb{R}^N$. *Assume* $\left\| \frac{\partial^m f(\cdot,\ldots,\cdot)}{\partial x_1^{\alpha_1} \ldots \partial x_N^{\alpha_N}} \right\|_\infty < \infty$, *for all* $\alpha_j \in \mathbb{Z}^+$, $j = 1, \ldots, N$: $|\alpha| := \sum_{j=1}^{N} \alpha_j = m$. *Let* μ_{ξ_n} *be a Borel probability measure on* $\mathbb{R}^N$ *defined as in* (3.19), *for* $\xi_n \in (0, 1]$. *Then for all* $x \in \mathbb{R}^N$, *we have*

i)

$$\left| P^{*\,[m]}_{r,n}(f;x) - f(x) - \sum_{\tilde{j}=1}^{m} \delta^{[m]}_{\tilde{j},r} \left(\sum_{\substack{\alpha_1,\ldots,\alpha_N \ge 0: \\ |\alpha|=\tilde{j}}} \frac{c_{\alpha,n,\tilde{j}} f_\alpha(x)}{\prod_{i=1}^{N} \alpha_i!} \right) \right| \le \sum_{\substack{\alpha_1,\ldots,\alpha_N \ge 0: \\ |\alpha|=\tilde{j}}} \frac{(\omega_r(f_\alpha, \xi_n))}{\left(\prod_{i=1}^{N} \alpha_i!\right)} u^*_{P,\xi_n}, \tag{3.73}$$

ii)

$$\left\| P^{*\,[m]}_{r,n}(f) - f - \sum_{\tilde{j}=1}^{m} \delta^{[m]}_{\tilde{j},r} \left(\sum_{\substack{\alpha_1,\ldots,\alpha_N \ge 0: \\ |\alpha|=\tilde{j}}} \frac{c_{\alpha,n,\tilde{j}} f_\alpha}{\prod_{i=1}^{N} \alpha_i!} \right) \right\|_\infty \le R.H.S. \ (3.73). \tag{3.74}$$

When $\xi_n \to 0$, as $n \to \infty$, we obtain that

$$\left\| P_{r,n}^{*\,[m]}(f) - f - \sum_{\tilde{j}=1}^{m} \delta_{\tilde{j},r}^{[m]} \left(\sum_{\substack{\alpha_1,\ldots,\alpha_N \geq 0: \\ |\alpha| = \tilde{j}}} \frac{c_{\alpha,n,\tilde{j}} f_\alpha}{\prod_{i=1}^{N} \alpha_i!} \right) \right\|_\infty \to 0$$

with rates,

iii) *it holds also that*

$$\left\| P_{r,n}^{*\,[m]}(f) - f \right\|_\infty \leq \sum_{\tilde{j}=1}^{m} \left| \delta_{\tilde{j},r}^{[m]} \right| \left(\sum_{\substack{\alpha_1,\ldots,\alpha_N \geq 0: \\ |\alpha| = \tilde{j}}} \frac{\left| c_{\alpha,n,\tilde{j}} \right| \|f_\alpha\|_\infty}{\prod_{i=1}^{N} \alpha_i!} \right) + R.H.S.\ (3.73), \tag{3.75}$$

given that $\|f_\alpha\|_\infty < \infty$, for all $\alpha : |\alpha| = \tilde{j}$, $\tilde{j} = 1, \ldots, m$. Furthermore, for $\alpha_j \in \mathbb{N}$, as $\xi_n \to 0$ when $n \to \infty$, $c_{\alpha,n,\tilde{j}} \to 0$, and u_{P,ξ_n}^* is uniformly bounded, we conclude that

$$\left\| P_{r,n}^{*\,[m]}(f) - f \right\|_\infty \to 0 \tag{3.76}$$

with rates.

Proof. By *Theorem* 3.4, *Proposition* 3.6, and *Lemma* 3.10.

For the case of $m = 0$, we have

Theorem 3.12. *Let $f \in C_B(\mathbb{R}^N)$, $N \geq 1$. Then*

$$\left\| P_{r,n}^{*\,[0]}(f) - f \right\|_\infty \leq \Phi_{P,\xi_n}^* \omega_r(f, \xi_n), \tag{3.77}$$

where

$$\Phi_{P,\xi_n}^* := \frac{\sum_{\nu_1=-\infty}^{\infty} \cdots \sum_{\nu_N=-\infty}^{\infty} \left(1 + \frac{\|\nu\|_2}{\xi_n}\right)^r e^{-\frac{1}{\xi_n} \sum_{i=1}^{N} |\nu_i|}}{\sum_{\nu_1=-\infty}^{\infty} \cdots \sum_{\nu_N=-\infty}^{\infty} e^{-\frac{1}{\xi_n} \sum_{i=1}^{N} |\nu_i|}} \tag{3.78}$$

is uniformly bounded for all $\xi_n \in (0, 1]$.

Proof. We observe

$$\sum_{\nu_1=-\infty}^{\infty} \cdots \sum_{\nu_N=-\infty}^{\infty} \left(1 + \frac{\|\nu\|_2}{\xi_n}\right)^r e^{-\frac{1}{\xi_n} \sum_{i=1}^{N} |\nu_i|} \leq R_{1,1} < \infty. \tag{3.79}$$

Thus, by Theorem 3.5 and (3.41), the proof is done.

Next we show

Lemma 3.13. *For* $\tilde{j} = 1, \ldots, m$*, and* $\alpha := (\alpha_1, \ldots, \alpha_N)$*,* $\alpha_i \in \mathbb{Z}^+$*,* $i = 1, \ldots, N$*,* $|\alpha| := \sum_{i=1}^{N} \alpha_i = \tilde{j}$*, we have that*

$$p_{\alpha,n} := p_{\alpha,n\tilde{j}} := \frac{\sum_{\nu_1=-\infty}^{\infty} \cdots \sum_{\nu_N=-\infty}^{\infty} \left(\prod_{i=1}^{N} \nu_i^{\alpha_i}\right) e^{-\frac{1}{\xi_n}\sum_{i=1}^{N}\nu_i^2}}{\sum_{\nu_1=-\infty}^{\infty} \cdots \sum_{\nu_N=-\infty}^{\infty} e^{-\frac{1}{\xi_n}\sum_{i=1}^{N}\nu_i^2}} < \infty. \tag{3.80}$$

for all $\xi_n \in (0, 1]$*. Additionally, let* $\alpha_i \in \mathbb{N}$*, then as* $\xi_n \to 0$ *when* $n \to \infty$*, we get* $p_{\alpha,n\tilde{j}} \to 0$.

Proof. Let $\alpha_i \in \mathbb{N}$ for $i = 1, 2, \ldots, N \in \mathbb{N}$. In [2], the authors showed that

$$\frac{\sum_{\nu=-\infty}^{\infty} \nu_i^{\alpha_i} e^{\frac{-\nu_i^2}{\xi_n}}}{\sum_{\nu=-\infty}^{\infty} e^{\frac{-\nu_i^2}{\xi_n}}} < \infty, \tag{3.81}$$

for all $\xi_n \in (0, 1]$. Therefore, by (3.60) and (3.81), we obtain

$$p_{\alpha,n\tilde{j}} < \infty, \tag{3.82}$$

for all $\xi_n \in (0, 1]$.

Additionally, when $\alpha_i \in \mathbb{N}$, by (3.48), (3.67), and (3.70), we have

$$0 \leq p_{\alpha,n\tilde{j}} \leq \sum_{\nu_1=-\infty}^{\infty} \cdots \sum_{\nu_N=-\infty}^{\infty} \left(\prod_{i=1}^{N} \nu_i^{\alpha_i}\right) e^{-\frac{1}{\xi_n}\sum_{i=1}^{N}\nu_i^2} := K^2_{\xi_n,\nu_i} \leq K^1_{\xi_n,\nu_i}. \tag{3.83}$$

see also [2].

Thus, we have as $\xi_n \to 0$ when $n \to \infty$, $p_{\alpha,n\tilde{j}} \to 0$.

Now, let $\alpha_i = 0$. For each ν_i, we have $\left(\sum_{n=1}^{\infty} \frac{1}{n^2} = \frac{\pi^2}{6}, \text{Euler 1741}\right)$

$$\sum_{\nu_i=-\infty}^{\infty} e^{-\frac{\nu_i^2}{\xi_n}} = 1 + 2\sum_{\nu_i=1}^{\infty} e^{-\frac{\nu_i^2}{\xi_n}} \leq 1 + 2\sum_{i=1}^{\infty} \frac{1}{\nu_i^2} = 1 + \frac{\pi^2}{3} < \infty. \tag{3.84}$$

Thus,

$$p_{\alpha,n\tilde{j}} := \frac{\sum_{\nu_1=-\infty}^{\infty} \cdots \sum_{\nu_N=-\infty}^{\infty} e^{-\frac{1}{\xi_n}\sum_{i=1}^{N}\nu_i^2}}{\sum_{\nu_1=-\infty}^{\infty} \cdots \sum_{\nu_N=-\infty}^{\infty} e^{-\frac{1}{\xi_n}\sum_{i=1}^{N}\nu_i^2}} = \prod_{i=1}^{N} \left(\frac{\sum_{\nu_i=-\infty}^{\infty} e^{-\frac{\nu_i^2}{\xi_n}}}{\sum_{\nu_i=-\infty}^{\infty} e^{-\frac{\nu_i^2}{\xi_n}}} \right) = 1 < \infty. \tag{3.85}$$

Next, we state our results for the operators $W^{*[m]}_{r,n}$

Theorem 3.14. *Let* $m \in \mathbb{N}, f \in C^m(\mathbb{R}^N)$, $N \geq 1$, $x \in \mathbb{R}^N$. *Assume* $\left\| \frac{\partial^m f(\cdot,\ldots,\cdot)}{\partial x_1^{\alpha_1} \ldots \partial x_N^{\alpha_N}} \right\|_\infty < \infty$, *for all* $\alpha_j \in \mathbb{Z}^+$, $j = 1, \ldots, N : |\alpha| := \sum_{j=1}^{N} \alpha_j = m$. *Let* μ_{ξ_n} *be a Borel probability measure on* $\mathbb{R}^N$ *defined as in* (3.21), *for* $\xi_n \in (0, 1]$. *Then for all* $x \in \mathbb{R}^N$, *we have*

i)

$$\left| W_{r,n}^{*[m]}(f;x) - f(x) - \sum_{\tilde{j}=1}^{m} \delta_{\tilde{j},r}^{[m]} \left(\sum_{\substack{\alpha_1,\ldots,\alpha_N \geq 0: \\ |\alpha| = \tilde{j}}} \frac{p_{\alpha,n,\tilde{j}} f_\alpha(x)}{\prod_{i=1}^{N} \alpha_i!} \right) \right| \leq \sum_{\substack{\alpha_1,\ldots,\alpha_N \geq 0 \\ |\alpha| = m}} \frac{(\omega_r(f_\alpha, \xi_n))}{\left(\prod_{i=1}^{N} \alpha_i!\right)} u_{W,\xi_n}^*, \tag{3.86}$$

ii)

$$\left\| W_{r,n}^{*[m]}(f) - f - \sum_{\tilde{j}=1}^{m} \delta_{\tilde{j},r}^{[m]} \left(\sum_{\substack{\alpha_1,\ldots,\alpha_N \geq 0: \\ |\alpha| = \tilde{j}}} \frac{p_{\alpha,n,\tilde{j}} f_\alpha}{\prod_{i=1}^{N} \alpha_i!} \right) \right\|_\infty \leq R.H.S.\ (3.86). \tag{3.87}$$

When $\xi_n \to 0$, *as* $n \to \infty$, *we obtain that*

$$\left\| W_{r,n}^{*[m]}(f) - f - \sum_{\tilde{j}=1}^{m} \delta_{\tilde{j},r}^{[m]} \left(\sum_{\substack{\alpha_1,\ldots,\alpha_N \geq 0: \\ |\alpha| = \tilde{j}}} \frac{p_{\alpha,n,\tilde{j}} f_\alpha}{\prod_{i=1}^{N} \alpha_i!} \right) \right\|_\infty \to 0$$

with rates,

iii) *it holds also that*

$$\left\| W_{r,n}^{*[m]}(f) - f \right\|_\infty \leq \sum_{\tilde{j}=1}^{m} \left| \delta_{\tilde{j},r}^{[m]} \right| \left(\sum_{\substack{\alpha_1,\ldots,\alpha_N \geq 0: \\ |\alpha| = \tilde{j}}} \frac{\left| p_{\alpha,n,\tilde{j}} \right| \|f_\alpha\|_\infty}{\prod_{i=1}^{N} \alpha_i!} \right) + R.H.S.\ (3.86), \tag{3.88}$$

given that $\|f_\alpha\|_\infty < \infty$, *for all* $\alpha : |\alpha| = \tilde{j}, \tilde{j} = 1, \ldots, m$. *Furthermore, for* $\alpha_j \in \mathbb{N}$, *as* $\xi_n \to 0$ *when* $n \to \infty$, $p_{\alpha,n\tilde{j}} \to 0$, *and* u_{W,ξ_n}^* *is uniformly bounded, we conclude that*

$$\left\| W_{r,n}^{*[m]}(f) - f \right\|_\infty \to 0 \tag{3.89}$$

with rates.

Proof. By *Theorem* 3.4, *Proposition* 3.7, and *Lemma* 3.13.

For the case of $m = 0$, we have

Theorem 3.15. *Let $f \in C_B\left(\mathbb{R}^N\right)$, $N \geq 1$. Then*

$$\left\| W_{r,n}^{*\,[0]}(f) - f \right\|_{\infty} \leq \Phi_{W,\xi_n}^{*} \omega_r(f, \xi_n), \tag{3.90}$$

where

$$\Phi_{W,\xi_n}^{*} := \frac{\sum_{\nu_1=-\infty}^{\infty} \cdots \sum_{\nu_N=-\infty}^{\infty} \left(1 + \frac{\|\nu\|_2}{\xi_n}\right)^r e^{-\frac{1}{\xi_n} \sum_{i=1}^{N} \nu_i^2}}{\sum_{\nu_1=-\infty}^{\infty} \cdots \sum_{\nu_N=-\infty}^{\infty} e^{-\frac{1}{\xi_n} \sum_{i=1}^{N} \nu_i^2}} \tag{3.91}$$

is uniformly bounded for all $\xi_n \in (0, 1]$.

Proof. We observe

$$\sum_{\nu_1=-\infty}^{\infty} \cdots \sum_{\nu_N=-\infty}^{\infty} \left(1 + \frac{\|\nu\|_2}{\xi_n}\right)^r e^{-\frac{1}{\xi_n} \sum_{i=1}^{N} \nu_i^2} \leq R_{2,1} < \infty. \tag{3.92}$$

Thus, by Theorem 3.5 and (3.49), the proof is done.

We also need

Lemma 3.16. *For $\tilde{j} = 1, \ldots, m$, and $\alpha := (\alpha_1, \ldots, \alpha_N)$, $\alpha_i \in \mathbb{Z}^+$, $i = 1, \ldots, N$, $|\alpha| := \sum_{i=1}^{N} \alpha_i = \tilde{j}$, we have that*

$$q_{\alpha,n} := q_{\alpha,n\tilde{j}} := \frac{\sum_{\nu_1=-\infty}^{\infty} \cdots \sum_{\nu_N=-\infty}^{\infty} \left(\prod_{i=1}^{N} \nu_i^{\alpha_i} \left(\nu_i^{2\hat{\alpha}} + \xi_n^{2\hat{\alpha}}\right)^{-\beta}\right)}{\sum_{\nu_1=-\infty}^{\infty} \cdots \sum_{\nu_N=-\infty}^{\infty} \prod_{i=1}^{N} \left(\nu_i^{2\hat{\alpha}} + \xi_n^{2\hat{\alpha}}\right)^{-\beta}} < \infty. \tag{3.93}$$

for all $\xi_n \in (0, 1]$ where $\hat{\alpha} \in \mathbb{N}$ and $\beta > \frac{\alpha_i + r + 1}{2\hat{\alpha}}$. Additionally, let $\alpha_i \in \mathbb{N}$, then as $\xi_n \to 0$ when $n \to \infty$, we get $q_{\alpha,n\tilde{j}} \to 0$.

Proof. Let $\alpha_i \in \mathbb{N}$ for $i = 1, 2, \ldots, N \in \mathbb{N}$. In [2], the authors showed that

$$\frac{\sum_{\nu_i=-\infty}^{\infty} \nu_i^{\alpha_i} \left(\nu_i^{2\hat{\alpha}} + \xi_n^{2\hat{\alpha}}\right)^{-\beta}}{\sum_{\nu_i=-\infty}^{\infty} \left(\nu_i^{2\hat{\alpha}} + \xi_n^{2\hat{\alpha}}\right)^{-\beta}} < \infty, \tag{3.94}$$

for all $\xi_n \in (0, 1]$ where $\hat{\alpha} \in \mathbb{N}$ and $\beta > \frac{\alpha_i+r+1}{2\hat{\alpha}}$. Therefore, by (3.61) and (3.94), we obtain

$$q_{\alpha,n\tilde{j}} < \infty, \tag{3.95}$$

for all $\xi_n \in (0, 1]$ where $\hat{\alpha} \in \mathbb{N}$ and $\beta > \frac{\alpha_i+r+1}{2\hat{\alpha}}$.
Moreover, we observe that when $\alpha_i \in \mathbb{N}$, by the proof of Proposition 3.8, we have

$$0 \leq q_{\alpha,n\tilde{j}} \leq \xi_n^{2N\hat{\alpha}\beta} \sum_{\nu_1=-\infty}^{\infty} \cdots \sum_{\nu_N=-\infty}^{\infty} \left(\prod_{i=1}^{N} \nu_i^{\alpha_i} \left(\nu_i^{2\hat{\alpha}} + \xi_n^{2\hat{\alpha}} \right)^{-\beta} \right) \tag{3.96}$$

We define

$$K^3_{\xi_n,\nu_i} := \sum_{\nu_1=-\infty}^{\infty} \cdots \sum_{\nu_N=-\infty}^{\infty} \left(\prod_{i=1}^{N} \nu_i^{\alpha_i} \left(\nu_i^{2\hat{\alpha}} + \xi_n^{2\hat{\alpha}} \right)^{-\beta} \right). \tag{3.97}$$

We observe that

$$K^3_{\xi_n,\nu_i} = \begin{cases} 0, & \text{if } \alpha_i \text{ is odd,} \\ 2^N \prod_{i=1}^{N} \left(\sum_{\nu_i=1}^{\infty} \nu_i^{\alpha_i} \left(\nu_i^{2\hat{\alpha}} + \xi_n^{2\hat{\alpha}} \right)^{-\beta} \right), & \text{if } \alpha_i \text{ is even.} \end{cases}$$

see also [2].
Assume that α_i is even. We observe that

$$\left(\nu_i^{2\hat{\alpha}} + \xi_n^{2\hat{\alpha}} \right)^{-\beta} \leq \nu_i^{-2\hat{\alpha}\beta}. \tag{3.98}$$

Then

$$\sum_{\nu_i=1}^{\infty} \nu_i^{\alpha_i} \left(\nu_i^{2\hat{\alpha}} + \xi_n^{2\hat{\alpha}} \right)^{-\beta} \leq \sum_{\nu_i=1}^{\infty} \nu_i^{\alpha_i - 2\hat{\alpha}\beta} = \sum_{\nu_i=1}^{\infty} \left(\frac{1}{\nu_i} \right)^{2\hat{\alpha}\beta - \alpha_i} < \infty,$$

for all ν_i when $\beta > \frac{\alpha_i+1}{2\hat{\alpha}}$. Thus, we have

$$K^3_{\xi_n,\nu_i} < \infty. \tag{3.99}$$

Hence, by (3.97)–(3.99), we get

$$\xi_n^{2N\hat{\alpha}\beta} K^3_{\xi_n,\nu_i} \to 0, \text{ as } \xi_n \to 0. \tag{3.100}$$

Thus, by (3.96) and (3.100), for $\hat{\alpha} \in \mathbb{N}$ and $\beta > \frac{\alpha_i+1}{2\hat{\alpha}}$, we have as $\xi_n \to 0$ when $n \to \infty$, $q_{\alpha,n\tilde{j}} \to 0$.

Now, let $\alpha_i = 0$. We observe that

$$\sum_{\nu_i=-\infty}^{\infty}\left(\nu_i^{2\hat{\alpha}}+\xi_n^{2\hat{\alpha}}\right)^{-\beta} \leq \sum_{\nu_i=1}^{\infty}\nu_i^{-2\hat{\alpha}\beta} = 2\sum_{\nu_i=1}^{\infty}\left(\frac{1}{\nu_i}\right)^{2\hat{\alpha}\beta} < \infty,$$

when $\beta > \frac{1}{2\hat{\alpha}}$.
Thus,

$$q_{\alpha,n,\tilde{j}} := \frac{\sum_{\nu_1=-\infty}^{\infty}\cdots\sum_{\nu_N=-\infty}^{\infty}\prod_{i=1}^{N}\left(\nu_i^{2\hat{\alpha}}+\xi_n^{2\hat{\alpha}}\right)^{-\beta}}{\sum_{\nu_1=-\infty}^{\infty}\cdots\sum_{\nu_N=-\infty}^{\infty}\prod_{i=1}^{N}\left(\nu_i^{2\hat{\alpha}}+\xi_n^{2\hat{\alpha}}\right)^{-\beta}}$$

$$= \prod_{i=1}^{N}\left(\frac{\sum_{\nu_i=-\infty}^{\infty}\left(\nu_i^{2\hat{\alpha}}+\xi_n^{2\hat{\alpha}}\right)^{-\beta}}{\sum_{\nu_i=-\infty}^{\infty}\left(\nu_i^{2\hat{\alpha}}+\xi_n^{2\hat{\alpha}}\right)^{-\beta}}\right) = 1 < \infty. \tag{3.101}$$

Next, we state our results for the operators $Q_{r,n}^{*[m]}$

Theorem 3.17. *Let $m \in \mathbb{N}$, $f \in C^m\left(\mathbb{R}^N\right)$, $N \geq 1$, where $\hat{\alpha}$, $n \in \mathbb{N}$, $\beta > \max\left\{\frac{1+r+\alpha_i}{2\hat{\alpha}}, \frac{r+2}{2\hat{\alpha}}\right\}$, and $x \in \mathbb{R}^N$. Assume $\left\|\frac{\partial^m f(\cdot,\ldots,\cdot)}{\partial x_1^{\alpha_1}\ldots\partial x_N^{\alpha_N}}\right\|_\infty < \infty$, for all $\alpha_j \in \mathbb{Z}^+$, $j = 1,\ldots,N : |\alpha| := \sum_{j=1}^{N}\alpha_j = m$. Let μ_{ξ_n} be a Borel probability measure on $\mathbb{R}^N$ defined as* (3.23)*, for $\xi_n \in (0,1]$. Then for all $x \in \mathbb{R}^N$, we have*

i)

$$\left|Q_{r,n}^{*\,[m]}(f;x) - f(x) - \sum_{\tilde{j}=1}^{m}\delta_{\tilde{j},r}^{[m]}\left(\sum_{\substack{\alpha_1,\ldots,\alpha_N\geq 0:\\ |\alpha|=\tilde{j}}}\frac{q_{\alpha,n,\tilde{j}}f_\alpha(x)}{\prod_{i=1}^{N}\alpha_i!}\right)\right|$$

$$\leq \sum_{\substack{\alpha_1,\ldots,\alpha_N\geq 0\\ |\alpha|=m}}\frac{\left(\omega_r\left(f_\alpha,\xi_n\right)\right)}{\left(\prod_{i=1}^{N}\alpha_i!\right)}u_{Q,\xi_n}^{*}, \tag{3.102}$$

ii)

$$\left\|Q_{r,n}^{*\,[m]}(f) - f - \sum_{\tilde{j}=1}^{m}\delta_{\tilde{j},r}^{[m]}\left(\sum_{\substack{\alpha_1,\ldots,\alpha_N\geq 0:\\ |\alpha|=\tilde{j}}}\frac{q_{\alpha,n,\tilde{j}}f_\alpha}{\prod_{i=1}^{N}\alpha_i!}\right)\right\|_\infty \leq R.H.S.\ (3.102). \tag{3.103}$$

When $\xi_n \to 0$, as $n \to \infty$, we obtain that

$$\left\| Q_{r,n}^{*\,[m]}(f) - f - \sum_{\tilde{j}=1}^{m} \delta_{\tilde{j},r}^{[m]} \left(\sum_{\substack{\alpha_1,\ldots,\alpha_N \geq 0: \\ |\alpha|=\tilde{j}}} \frac{q_{\alpha,n\tilde{j}} f_\alpha}{\prod_{i=1}^{N} \alpha_i!} \right) \right\|_\infty \to 0$$

with rates,

iii) *it holds also that*

$$\left\| Q_{r,n}^{*\,[m]}(f) - f \right\|_\infty \leq \sum_{\tilde{j}=1}^{m} \left| \delta_{\tilde{j},r}^{[m]} \right| \left(\sum_{\substack{\alpha_1,\ldots,\alpha_N \geq 0: \\ |\alpha|=\tilde{j}}} \frac{\left| q_{\alpha,n\tilde{j}} \right| \|f_\alpha\|_\infty}{\prod_{i=1}^{N} \alpha_i!} \right) + R.H.S.\ (3.102), \tag{3.104}$$

given that $\|f_\alpha\|_\infty < \infty$, for all $\alpha : |\alpha| = \tilde{j}$, $\tilde{j} = 1, \ldots, m$. Furthermore, for $\alpha_j \in \mathbb{N}$, as $\xi_n \to 0$ when $n \to \infty$, $q_{\alpha,n\tilde{j}} \to 0$, and u_{Q,ξ_n}^{*} is uniformly bounded, we conclude that

$$\left\| Q_{r,n}^{*\,[m]}(f) - f \right\|_\infty \to 0 \tag{3.105}$$

with rates.

Proof. By *Theorem* 3.4, *Proposition* 3.8 ,and *Lemma* 3.16.

For the case of $m = 0$, we have

Theorem 3.18. *Let $f \in C_B(\mathbb{R}^N)$, $\hat{\alpha}, N \in \mathbb{N}$, $\beta > \frac{r+2}{2\hat{\alpha}}$. Then*

$$\left\| Q_{r,n}^{*\,[0]}(f) - f \right\|_\infty \leq \Phi_{Q,\xi_n}^{*} \omega_r(f, \xi_n), \tag{3.106}$$

where

$$\Phi_{Q,\xi_n}^{*} := \frac{\sum_{\nu_1=-\infty}^{\infty} \cdots \sum_{\nu_N=-\infty}^{\infty} \left(1 + \frac{\|\nu\|_2}{\xi_n}\right)^r \prod_{i=1}^{N} \left(\nu_i^{2\hat{\alpha}} + \xi_n^{2\hat{\alpha}}\right)^{-\beta}}{\sum_{\nu_1=-\infty}^{\infty} \cdots \sum_{\nu_N=-\infty}^{\infty} \prod_{i=1}^{N} \left(\nu_i^{2\hat{\alpha}} + \xi_n^{2\hat{\alpha}}\right)^{-\beta}} \tag{3.107}$$

is uniformly bounded for all $\xi_n \in (0, 1]$.

Proof. We observe that

$$\Phi_{Q,\xi_n}^{*} \leq R_{3,1} < \infty, \tag{3.108}$$

when $\beta > \frac{r+2}{2\hat{\alpha}}$. Thus, by Theorem 3.5, the proof is done.

Our final result for unitary case is the following:

Remark 3.19. The operators $P_{r,n}^{*\,[m]}$, $W_{r,n}^{*\,[m]}$, and $Q_{r,n}^{*\,[m]}$ are not in general positive. For instance, consider the function $g(x_1, \ldots, x_N) = \sum_{i=1}^{N} x_i^2$ and also take $r = 2$, $m = 3$. Observe that $g \geq 0$, however

$$P_{r,n}^{*\,[m]}(g; 0, \ldots, 0) = \frac{\sum_{\nu_1=-\infty}^{\infty} \cdots \sum_{\nu_N=-\infty}^{\infty} \left(\sum_{j=0}^{2} j^2 \alpha_{j,2}^{[3]} \left(\sum_{i=1}^{N} \nu_i^2 \right) \right) e^{-\frac{1}{\xi_n} \sum_{i=1}^{N} |\nu_i|}}{\sum_{\nu_1=-\infty}^{\infty} \cdots \sum_{\nu_N=-\infty}^{\infty} e^{-\frac{1}{\xi_n} \sum_{i=1}^{N} |\nu_i|}}. \tag{3.109}$$

We have that

$$\sum_{j=0}^{2} \alpha_{j,2}^{[3]} j^2 = \alpha_{1,2}^{[3]} + 4\alpha_{2,2}^{[3]} = -2 + \frac{1}{2} = -\frac{3}{2}. \tag{3.110}$$

Therefore, we obtain

$$P_{r,n}^{*\,[m]}(g; 0, \ldots, 0) = -\tfrac{3}{2} \cdot \left(\frac{\sum_{\nu_1=-\infty}^{\infty} \cdots \sum_{\nu_N=-\infty}^{\infty} \left(\sum_{i=1}^{N} \nu_i^2 \right) e^{-\frac{1}{\xi_n} \sum_{i=1}^{N} |\nu_i|}}{\sum_{\nu_1=-\infty}^{\infty} \cdots \sum_{\nu_N=-\infty}^{\infty} e^{-\frac{1}{\xi_n} \sum_{i=1}^{N} |\nu_i|}} \right) < 0. \tag{3.111}$$

Similarly, we can show the same result for $W_{r,n}^{*\,[m]}$ and $Q_{r,n}^{*\,[m]}$.

Now, we give our results for the non-unitary operators $P_{r,n}^{[m]}$ and $W_{r,n}^{[m]}$. We start with

Definition 3.20. We define the following error quantities:

$$E_{n,P}^{[0]}(f; x) := P_{r,n}^{[0]}(f; x) - f(x) \tag{3.112}$$

and

$$E_{n,W}^{[0]}(f; x) := W_{r,n}^{[0]}(f; x) - f(x). \tag{3.113}$$

Additionally we introduce the errors

$$E_{n,P}^{[m]}(f; x) := P_{r,n}^{[m]}(f; x) - f(x) - \sum_{\tilde{j}=1}^{m} \delta_{\tilde{j},r}^{[m]} \left(\sum_{\substack{\alpha_1, \ldots, \alpha_N \geq 0: \\ |\alpha| = \tilde{j}}} \frac{\tilde{c}_{\alpha,n,\tilde{j}} f_\alpha(x)}{\prod_{i=1}^{N} \alpha_i!} \right) \tag{3.114}$$

and

$$E_{n,W}^{[m]}(f;x) := W_{r,n}^{[m]}(f;x) - f(x) - \sum_{\tilde{j}=1}^{m} \delta_{\tilde{j},r}^{[m]} \left(\sum_{\substack{\alpha_1,\ldots,\alpha_N \geq 0: \\ |\alpha|=\tilde{j}}} \frac{\tilde{p}_{\alpha,n,\tilde{j}} f_\alpha(x)}{\prod_{i=1}^{N} \alpha_i!} \right), \tag{3.115}$$

where

$$\tilde{c}_{\alpha,n,\tilde{j}} := \frac{\sum_{\nu_1=-\infty}^{\infty} \cdots \sum_{\nu_N=-\infty}^{\infty} \left(\prod_{i=1}^{N} \nu_i^{\alpha_i} \right) e^{-\frac{1}{\xi_n} \sum_{i=1}^{N} |\nu_i|}}{\left(1 + 2\xi_n e^{-\frac{1}{\xi_n}} \right)^N} \tag{3.116}$$

and

$$\tilde{p}_{\alpha,n,\tilde{j}} := \frac{\sum_{\nu_1=-\infty}^{\infty} \cdots \sum_{\nu_N=-\infty}^{\infty} \left(\prod_{i=1}^{N} \nu_i^{\alpha_i} \right) e^{-\frac{1}{\xi_n} \sum_{i=1}^{N} \nu_i^2}}{\left(\sqrt{\pi \xi_n} \left(1 - \operatorname{erf}\left(\frac{1}{\sqrt{\xi_n}} \right) \right) + 1 \right)^N}. \tag{3.117}$$

We observe

Remark 3.21. Let

$$m_{\xi_n,P} := \frac{\sum_{\nu_1=-\infty}^{\infty} \cdots \sum_{\nu_N=-\infty}^{\infty} e^{-\frac{\sum_{i=1}^{N} |\nu_i|}{\xi_n}}}{\left(1 + 2\xi_n e^{-\frac{1}{\xi_n}} \right)^N} \tag{3.118}$$

and

$$m_{\xi_n,W} := \frac{\sum_{\nu_1=-\infty}^{\infty} \cdots \sum_{\nu_N=-\infty}^{\infty} e^{-\frac{1}{\xi_n} \sum_{i=1}^{N} \nu_i^2}}{\left(\sqrt{\pi \xi_n} \left(1 - \operatorname{erf}\left(\frac{1}{\sqrt{\xi_n}} \right) \right) + 1 \right)^N}. \tag{3.119}$$

In [2], the authors showed that

$$\frac{\sum_{\nu_i=-\infty}^{\infty} e^{-\frac{|\nu_i|}{\xi_n}}}{1 + 2\xi_n e^{-\frac{1}{\xi_n}}} \to 1 \text{ as } \xi_n \to 0^+, \tag{3.120}$$

and

$$\frac{\sum_{\nu_i=-\infty}^{\infty} e^{-\frac{\nu_i^2}{\xi}}}{1 + \sqrt{\pi \xi_n} \left(1 - \operatorname{erf}\left(\frac{1}{\sqrt{\xi_n}} \right) \right)} \to 1 \text{ as } \xi_n \to 0^+.$$

Therefore, we have

$$m_{\xi_n,P} = \prod_{i=1}^{N} \left(\frac{\sum_{\nu_i=-\infty}^{\infty} e^{-\frac{|\nu_i|}{\xi_n}}}{1 + 2\xi_n e^{-\frac{1}{\xi_n}}} \right) \to 1 \text{ as } \xi_n \to 0^+, \tag{3.121}$$

and

$$m_{\xi_n,W} = \prod_{i=1}^{N} \left(\frac{\sum\limits_{\nu_i=-\infty}^{\infty} e^{-\frac{\nu_i^2}{\xi}}}{1 + \sqrt{\pi \xi_n}\left(1 - \operatorname{erf}\left(\frac{1}{\sqrt{\xi_n}}\right)\right)} \right) \to 1 \text{ as } \xi_n \to 0^+. \tag{3.122}$$

Furthermore, we observe that

$$\begin{aligned} E_{n,P}^{[0]}(f;x) &:= P_{r,n}^{[0]}(f;x) - f(x) - f(x)m_{\xi_n,P} + f(x)m_{\xi_n,P} \\ &= m_{\xi_n,P}\left(\frac{P_{r,n}^{[0]}(f;x)}{m_{\xi_n,P}} - f(x) \right) + f(x)(m_{\xi_n,P} - 1). \end{aligned} \tag{3.123}$$

Since

$$\frac{P_{r,n}^{[0]}(f;x)}{m_{\xi_n,P}} = P_{r,n}^{*[0]}(f;x), \tag{3.124}$$

we get

$$\left| E_{n,P}^{[0]}(f;x) \right| \le m_{\xi_n,P} \left| P_{r,n}^{*[0]}(f;x) - f(x) \right| + |f(x)| \left| m_{\xi_n,P} - 1 \right|. \tag{3.125}$$

Similarly we obtain the inequalities

$$\left| E_{n,W}^{[0]}(f;x) \right| \le m_{\xi_n,W} \left| W_{r,n}^{*[0]}(f;x) - f(x) \right| + |f(x)| \left| m_{\xi_n,W} - 1 \right|, \tag{3.126}$$

$$\begin{aligned} \left| E_{n,P}^{[m]}(f;x) \right| \le m_{\xi_n,P} \Bigg| P_{r,n}^{*\,[m]}(f;x) - f(x) - \sum_{\tilde{j}=1}^{m} \delta_{\tilde{j},r}^{[m]} \\ \times \left(\sum_{\substack{\alpha_1,\ldots,\alpha_N \ge 0: \\ |\alpha|=\tilde{j}}} \frac{c_{\alpha,n,\tilde{j}} f_\alpha(x)}{\prod\limits_{i=1}^{N} \alpha_i!} \right) \Bigg| + |f(x)| \left| m_{\xi_n,P} - 1 \right|, \end{aligned} \tag{3.127}$$

and

$$\left|E_{n,W}^{[m]}(f;x)\right| \leq m_{\xi_n,W}\left|W_{r,n}^{*[m]}(f;x) - f(x) - \sum_{\tilde{j}=1}^{m}\delta_{\tilde{j},r}^{[m]}\right.$$

$$\left.\times\left(\sum_{\substack{\alpha_1,\ldots,\alpha_N\geq 0:\\ |\alpha|=\tilde{j}}}\frac{p_{\alpha,n,\tilde{j}}f_\alpha(x)}{\prod_{i=1}^{N}\alpha_i!}\right)\right| + |f(x)|\left|m_{\xi_n,W}-1\right|. \quad (3.128)$$

Thus, we derive

Theorem 3.22. *Let $m \in \mathbb{N}$, $f \in C^m(\mathbb{R}^N)$, $N \geq 1$, $x \in \mathbb{R}^N$. Assume $\left\|\frac{\partial^m f(\cdot,\ldots,\cdot)}{\partial x_1^{\alpha_1}\ldots\partial x_N^{\alpha_N}}\right\|_\infty < \infty$, for all $\alpha_j \in \mathbb{Z}^+$, $j = 1,\ldots,N : |\alpha| := \sum_{j=1}^{N}\alpha_j = m$. Let μ_{ξ_n} be a Borel measure on $\mathbb{R}^N$ defined as in* (3.25)*, for $\xi_n \in (0,1]$. Then for all $x \in \mathbb{R}^N$, we have*

i)

$$\left|E_{n,P}^{[m]}(f;x)\right| \leq m_{\xi_n,P}\sum_{\substack{\alpha_1,\ldots,\alpha_N\geq 0\\ |\alpha|=m}}\frac{\omega_r(f_\alpha,\xi_n)}{\left(\prod_{i=1}^{N}\alpha_i!\right)}u_{P,\xi_n}^{*} + |f(x)|\left|m_{\xi_n,P}-1\right|. \quad (3.129)$$

ii)

$$\left\|E_{n,P}^{[m]}(f)\right\|_\infty \leq m_{\xi_n,P}\sum_{\substack{\alpha_1,\ldots,\alpha_N\geq 0\\ |\alpha|=m}}\frac{\omega_r(f_\alpha,\xi_n)}{\left(\prod_{i=1}^{N}\alpha_i!\right)}u_{P,\xi_n}^{*} + \|f\|_\infty\left|m_{\xi_n,P}-1\right|. \quad (3.130)$$

When $\xi_n \to 0$, as $n \to \infty$, we obtain that $\left\|E_{n,P}^{[m]}(f;x)\right\|_\infty \to 0$ with rates.

Proof. By *Theorem* 3.11 *and* (3.127).

For the case of $m = 0$, we have

Theorem 3.23. *Let $f \in C_B(\mathbb{R}^N)$, $N \geq 1$. Then*

$$\left\|E_{n,P}^{[0]}(f)\right\|_\infty \leq m_{\xi_n,P}\Phi_{P,\xi_n}^{*}\omega_r(f,\xi_n) + \|f\|_\infty\left|m_{\xi_n,P}-1\right|. \quad (3.131)$$

Proof. By *Theorem* 3.12 *and* (3.125).

We demonstrate also

Theorem 3.24. *Let $m \in \mathbb{N}, f \in C^m\left(\mathbb{R}^N\right)$, $N \geq 1$, $x \in \mathbb{R}^N$. Assume* $\left\| \frac{\partial^m f(\cdot,\ldots,\cdot)}{\partial x_1^{\alpha_1} \ldots \partial x_N^{\alpha_N}} \right\|_\infty$ $< \infty$, *for all* $\alpha_j \in \mathbb{Z}^+, j = 1, \ldots, N : |\alpha| := \sum_{j=1}^{N} \alpha_j = m$. *Let* μ_{ξ_n} *be a Borel measure on* $\mathbb{R}^N$ *defined as in* (3.27), *for* $\xi_n \in (0, 1]$. *Then for all* $x \in \mathbb{R}^N$, *we have*

i)

$$\left| E_{n,W}^{[m]}(f;x) \right| \leq m_{\xi_n,W} \sum_{\substack{\alpha_1,\ldots,\alpha_N \geq 0 \\ |\alpha|=m}} \frac{\omega_r\left(f_\alpha, \xi_n\right)}{\left(\prod_{i=1}^{N} \alpha_i!\right)} u_{W,\xi_n}^* + |f(x)| \left| m_{\xi_n,W} - 1 \right|. \tag{3.132}$$

ii)

$$\left\| E_{n,W}^{[m]}(f) \right\|_\infty \leq m_{\xi_n,W} \sum_{\substack{\alpha_1,\ldots,\alpha_N \geq 0 \\ |\alpha|=m}} \frac{\omega_r\left(f_\alpha, \xi_n\right)}{\left(\prod_{i=1}^{N} \alpha_i!\right)} u_{W,\xi_n}^* + \|f\|_\infty \left| m_{\xi_n,W} - 1 \right|. \tag{3.133}$$

When $\xi_n \to 0$, *as* $n \to \infty$, *we obtain that* $\left\| E_{n,W}^{[m]}(f) \right\|_\infty \to 0$ *with rates.*

Proof. By *Theorem* 3.14 *and* (3.128).

Our final result is for the case of $m = 0$

Theorem 3.25. *Let* $f \in C_B\left(\mathbb{R}^N\right)$, $N \geq 1$. *Then*

$$\left\| E_{n,W}^{[0]}(f;x) \right\|_\infty \leq m_{\xi_n,W} \Phi_{W,\xi_n}^* \omega_r(f, \xi_n) + \|f\|_\infty \left| m_{\xi_n,W} - 1 \right|.$$

Proof. By *Theorem* 3.15 *and* (3.126).

References

1. G.A. Anastassiou, General uniform approximation theory by multivariate singular integral operators. Annales Polonici Mathematici **103**(1), 15–25 (2012)
2. G.A. Anastassiou, M. Kester, Quantitative uniform approximation by generalized discrete singular operators. Studia Mathematica Babes-Bolyai **60**(1), 39–60 (2015)
3. G.A. Anastassiou, R.A. Mezei, *Approximation by Singular Integrals* (Cambridge Scientific Publishers, Cambridge, 2012)
4. J. Favard, Sur les multiplicateurs d'interpolation. J. Math. Pures Appl. **23**(9), 219–247 (1944)
5. F. Smarandache, A triple inequality with series and improper integrals (2006), arxiv.org/ftp/mat/papers/0605/0605027.pdf

Chapter 4
Summation Process by Max-Product Operators

Türkan Yeliz Gökçer and Oktay Duman

Abstract In this study, we focus on the approximation to continuous functions by max-product operators in the sense of summation process. We also study error estimation corresponding to this approximation. At the end, we present an application to max-product Bernstein operators.

4.1 Introduction

In the 1950s, the approximation to continuous function by positive linear operators has been examined by Korovkin [16] (see also [1, 3]). Under a weaker linearity condition, Bede et.al. studied some approximating operators, the so-called max-product operators (see [4–12]). Later this non-linear approximation process has been studied by Duman [14] via the concept of statistical convergence in order to overcome the lack of the classical convergence. More detailed results on the statistical approximation theory may be found in the monograph book [2]. In the present paper, motivating the results in [14] and also using a general summability process given by Bell [13] we obtain some new approximation results for these max-product operators. We also compute its error estimation by using the classical modulus of continuity. We should remark that our results are not only different from the ones in [14] but also include many approximation process, such as ordinary convergence, arithmetic mean convergence, and almost convergence. An application presented at the end of this paper clearly explains why we really need such a summability process in the approximation.

We first remind the concept of summation process.

Let $\mathscr{A} = \{A^v\} = \{[a_{jn}^v]\}$ $(j, n, v \in \mathbb{N})$ be the sequence of infinite matrices, i.e., for each fixed $v \in \mathbb{N}$, we have an infinite matrix A^v. Then, we say that a sequence $x := (x_n)$ is $\mathscr{A}$-summable to a number L if

$$\lim_{j\to\infty} \sum_{n=1}^{\infty} a_{jn}^v x_n, \text{ uniformly in } v,$$

T.Y. Gökçer (✉) • O. Duman
Department of Mathematics, TOBB Economics and Technology University, Ankara, Turkey
e-mail: y.gokcer@etu.edu.tr; oduman@etu.edu.tr; okitayduman@gmail.com

G.A. Anastassiou, O. Duman (eds.), *Computational Analysis*, Springer Proceedings in Mathematics & Statistics 155, DOI 10.1007/978-3-319-28443-9_4

where we assume that the series $\sum_{n=1}^{\infty} a_{jn}^{\upsilon} x_n$ is convergent for each $n, \upsilon \in \mathbb{N}$. This limit is denoted by

$$\mathscr{A} - \lim_{j \to \infty} x_j = L.$$

Also we say that $\mathscr{A}$ is regular if it preserves both convergence and the limit value. For the regularity of $\mathscr{A}$, Bell [13] proved the following useful characterization which is similar to the well-known Silverman–Toeplitz conditions:

- $\forall n \in \mathbb{N}$, $\lim\limits_{j \to \infty} a_{jn}^{\upsilon} = 0$, uniformly in υ;
- $\lim\limits_{j \to \infty} \sum\limits_{n=1}^{\infty} a_{jn}^{\upsilon} = 1$ uniformly in υ;
- there are positive integers N, M such that $\forall j, \upsilon \in \mathbb{N}$, $\sum\limits_{n=1}^{\infty} \left| a_{jn}^{\upsilon} \right| < \infty$, and for $j \geq N$ and $\forall \upsilon \in \mathbb{N}$, $\sum\limits_{n=1}^{\infty} \left| a_{jn}^{\upsilon} \right| \leq M$.

This general summation process includes many well-known (regular) summability methods as follows:

- If, for each $\upsilon \in \mathbb{N}$, we take $A^{\upsilon} = I$, the identity matrix, then $\mathscr{A}$-summability method reduces to classical convergence;
- For each $\upsilon \in \mathbb{N}$, taking $A^{\upsilon} = C_1 = [c_{jn}]$, the Cesàro matrix given by $c_{jn} = 1/j$ if $n = 1, 2, \ldots, j$; and $c_{jn} = 0$ otherwise, then we immediately get Cesàro (arithmetic) mean convergence.
- Assume that $A^{\upsilon} = F^{\upsilon} = [c_{jn}^{\upsilon}]$, where

$$c_{jn}^{\upsilon} := \begin{cases} 1/j, & \text{if } \upsilon \leq j \leq j + \upsilon - 1 \\ 0, & \text{otherwise.} \end{cases}$$

  In this case, $\mathscr{A}$-summability method coincides with the almost convergence introduced by Lorentz [17].

Finally, we should remark that a summability process and the concept of statistical convergence introduced by Fast [15] cannot be comparable with each other.

4.2 Approximation in Sense of Summation Process

In this section, with the help of a general summability process given in the first section, we obtain approximation results by means of max-product operators. Now we consider the following operators:

Let (X, d) be an arbitrary compact metric space; and let $C(X, [0, \infty))$ denote the space of all non-negative continuous functions on X. Then we consider the following max-product operators which are defined by

$$L_n(f; x) := \bigvee_{k=0}^{n} K_{n,k}(x) \cdot f(x_{n,k}), \tag{4.1}$$

where $x \in X$, $n \in \mathbb{N}$, $x_{n,k} \in X$ $(k = 0, 1, \ldots, n)$, $f \in C(X, [0, \infty))$, and $K_{n,k}(.)$ is a non-negative continuous function on X. Here the symbol $\bigvee$ represents the maximum operation. The max-product operators in (4.1) are first introduced in the paper [5]. The operators are positive but do not to be linear. Actually they obey the property of the pseudo-linearity as follows:

$$L_n(\alpha \cdot f \bigvee \beta \cdot g) = \alpha L_n(f) \bigvee \beta L_n(g)$$

holds for all $f, g \in C(X, [0, \infty))$ and for any non-negative numbers α, β (see [5]).

The following lemma is useful for us to prove our main results.

Lemma 4.1 (See [5]). *For any $a_k, b_k \in [0, \infty)$, $k = 0, 1, \ldots, n$, we have*

$$\left| \bigvee_{k=0}^{n} a_k - \bigvee_{k=0}^{n} b_k \right| \leq \bigvee_{k=0}^{n} |a_k - b_k| \,.$$

Throughout the paper, we consider a non-negative regular summability method $\mathscr{A} = \{A^{\nu}\} = \{[a_{nk}^{\nu}]\}$ such that the transformed operator $\sum_{n=1}^{\infty} a_{jn}^{\nu} L_n$ is acting from $C(X, [0, \infty))$ into itself.

Now, for $y \in Y$, consider the test function $e_0(y) := 1$ and the moment function $\varphi_x(y) := d^2(y, x)$ for each fixed $x \in X$. Then, we get the next approximation result.

Theorem 4.2. *Let $\mathscr{A} = \{A^{\nu}\} = \{[a_{nk}^{\nu}]\}$ be a non-negative regular summability method. If the operators L_n given by* (4.1) *satisfy the following conditions*

$$\lim_{j\to\infty} \left\| \sum_{n=1}^{\infty} a_{jn}^{\nu} L_n(e_0) - e_0 \right\| = 0, \text{ uniformly in } \nu \tag{4.2}$$

and

$$\lim_{j\to\infty} \left\| \sum_{n=1}^{\infty} a_{jn}^{\nu} L_n(\varphi_x) \right\| = 0, \text{ uniformly in } \nu, \tag{4.3}$$

then, for all $f \in C(X, [0, \infty))$, we have

$$\lim_{j\to\infty} \left\| \sum_{n=1}^{\infty} a_{jn}^{\nu} L_n(f) - f \right\| = 0, \text{ uniformly in } \nu,$$

i.e., in other words, the sequence $\{L_n(f)\}$ is (uniformly) $\mathscr{A}$-summable to f on X.

Proof. Let $x \in X$ and $f \in C(X, [0,\infty))$ be given. By definition of the operators and using Lemma 4.1, we may write that, for all $j, \upsilon \in \mathbb{N}$,

$$\begin{aligned}\left|\sum_{n=1}^{\infty} a_{jn}^{\nu} L_n(f;x) - f(x)\right| &= \sum_{n=1}^{\infty} a_{jn}^{\nu} \left|\bigvee_{k=0}^{n} K_{n,k}(x) \cdot f(x_{n,k}) - \bigvee_{k=0}^{n} K_{n,k}(x) \cdot f(x)\right| \\ &\quad + |f(x)| \left|\sum_{n=1}^{\infty} a_{jn}^{\nu} \bigvee_{k=0}^{n} K_{n,k}(x) - 1\right| \\ &\leq \sum_{n=1}^{\infty} a_{jn}^{\nu} \bigvee_{k=0}^{n} K_{n,k}(x) \cdot |f(x_{n,k}) - f(x)| \\ &\quad + |f(x)| \left|\sum_{n=1}^{\infty} a_{jn}^{\nu} L_n(e_0;x) - e_0(x)\right|.\end{aligned}$$

By the uniform continuity of f on the compact set X, for a given $\varepsilon > 0$, one can find a $\delta > 0$ such that the following inequality

$$|f(x_{n,k}) - f(x)| \leq \varepsilon + \frac{2\|f\|}{\delta^2} \varphi_x(x_{n,k}).$$

holds for all $x, x_{n,k} \in X$. Then, we get

$$\begin{aligned}\left|\sum_{n=1}^{\infty} a_{jn}^{\nu} L_n(f;x) - f(x)\right| &\leq \varepsilon \sum_{n=1}^{\infty} a_{jn}^{\nu} L_n(e_0;x) + \frac{2\|f\|}{\delta^2} \sum_{n=1}^{\infty} a_{jn}^{\nu} L_n(\varphi_x;x) \\ &\quad + |f(x)| \left|\sum_{n=1}^{\infty} a_{jn}^{\nu} L_n(e_0;x) - e_0(x)\right| \\ &\leq \varepsilon + (\varepsilon + \|f\|) \left|\sum_{n=1}^{\infty} a_{jn}^{\nu} L_n(e_0;x) - e_0(x)\right| \\ &\quad + \frac{2\|f\|}{\delta^2} \left|\sum_{n=1}^{\infty} a_{jn}^{\nu} L_n(\varphi_x;x)\right|.\end{aligned}$$

Now, taking supremum over $x \in X$ on the both sides of the last inequality, we see that

$$\begin{aligned}\left\|\sum_{n=1}^{\infty} a_{jn}^{\nu} L_n(f) - f\right\| &\leq \varepsilon + (\varepsilon + \|f\|) \left\|\sum_{n=1}^{\infty} a_{jn}^{\nu} L_n(e_0) - e_0\right\| \\ &\quad + \frac{2\|f\|}{\delta^2} \left\|\sum_{n=1}^{\infty} a_{jn}^{\nu} L_n(\varphi_x)\right\|\end{aligned}$$

Finally, taking limit as $j \to \infty$ (uniformly in ν) and also using (4.2) and (4.3), the proof is completed.

4.3 Error Estimation in the Approximation

This section is devoted to obtain the error estimation of the summation process in Theorem 4.2. We first need the following lemma.

Lemma 4.3 (See [14]). *For every* $a_k, b_k \geq 0$ $(k = 0, 1, \ldots, n)$*, we have*

$$\bigvee_{k=0}^{n} a_k b_k \leq \sqrt{\bigvee_{k=0}^{n} a_k^2}\sqrt{\bigvee_{k=0}^{n} b_k^2}.$$

For the classical modulus of continuity it is well known that $\omega(f, \lambda\delta) \leq (\lambda + 1)\,\omega(f, \delta)$ for any $\lambda, \delta \in [0, \infty)$. Then we obtain the next result.

Theorem 4.4. *Let* $\mathscr{A} = \{A^{\nu}\} = \{(a_{nk}^{\nu})\}$ *be a non-negative regular summability method. Then, for all* $f \in C(X, [0, \infty))$*, we have*

$$\begin{aligned}\left\| \sum_{n=1}^{\infty} a_{jn}^{\nu} L_n(f) - f \right\| &\leq \omega(f, \delta_j^{\nu}) \left\| \sum_{n=1}^{\infty} a_{jn}^{\nu} L_n(e_0) \right\| \\ &\quad + \omega(f, \delta_j^{\nu}) \sqrt{\left\| \sum_{n=1}^{\infty} a_{jn}^{\nu} L_n(e_0) \right\|} \\ &\quad + \|f\| \left\| \sum_{n=1}^{\infty} a_{jn}^{\nu} L_n(e_0) - e_0 \right\|\end{aligned}$$

where

$$\delta_j^{\nu} := \sqrt{\left\| \sum_{n=1}^{\infty} a_{jn}^{\nu} L_n(\varphi_x) \right\|} \quad (j, \nu \in \mathbb{N}).$$

Proof. Let $x \in X$ and $f \in C(X, [0, \infty))$ be given. Then, as in proof of Theorem 4.2, we obtain that, for any $\delta > 0$,

$$\begin{aligned}\left| \sum_{n=1}^{\infty} a_{jn}^{\nu} L_n(f; x) - f(x) \right| &\leq \sum_{n=1}^{\infty} a_{jn}^{\nu} \bigvee_{k=0}^{n} K_{n,k}(x) \cdot |f(x_{n,k}) - f(x)| \\ &\quad + |f(x)| \left| \sum_{n=1}^{\infty} a_{jn}^{\nu} L_n(e_0; x) - e_0(x) \right|\end{aligned}$$

$$\leq \omega(f,\delta)\sum_{n=1}^{\infty} a_{jn}^{\nu} \bigvee_{k=0}^{n} K_{n,k}(x)\left(1+\frac{d(x_{n,k},x)}{\delta}\right)$$
$$+\,|f(x)|\left|\sum_{n=1}^{\infty} a_{jn}^{\nu} L_n(e_0;x)-e_0(x)\right|.$$

Then we get

$$\left|\sum_{n=1}^{\infty} a_{jn}^{\nu} L_n(f;x)-f(x)\right| \leq \omega(f,\delta)\sum_{n=1}^{\infty} a_{jn}^{\nu} L_n(e_0;x)$$
$$+\frac{\omega(f,\delta)}{\delta}\sum_{n=1}^{\infty} a_{jn}^{\nu} \bigvee_{k=0}^{n} K_{n,k}(x)\cdot d(x_{n,k},x)$$
$$+\,|f(x)|\left|\sum_{n=1}^{\infty} a_{jn}^{\nu} L_n(e_0;x)-e_0(x)\right|.$$

If we use Lemma 4.3, then we have

$$\left|\sum_{n=1}^{\infty} a_{jn}^{\nu} L_n(f;x)-f(x)\right| \leq \omega(f,\delta)\sum_{n=1}^{\infty} a_{jn}^{\nu} L_n(e_0;x)$$
$$+\frac{\omega(f,\delta)}{\delta}\sum_{n=1}^{\infty} a_{jn}^{\nu} \sqrt{L_n(e_0;x)}\sqrt{L_n(\varphi_x;x)}$$
$$+\,|f(x)|\left|\sum_{n=1}^{\infty} a_{jn}^{\nu} L_n(e_0;x)-e_0(x)\right|.$$

Now applying again the Cauchy–Schwarz inequality on the summation, we obtain that

$$\left|\sum_{n=1}^{\infty} a_{jn}^{\nu} L_n(f;x)-f(x)\right| \leq \omega(f,\delta)\sum_{n=1}^{\infty} a_{jn}^{\nu} L_n(e_0;x)$$
$$+\frac{\omega(f,\delta)}{\delta}\sqrt{\sum_{n=1}^{\infty} a_{jn}^{\nu} L_n(e_0;x)}\sqrt{\sum_{n=1}^{\infty} a_{jn}^{\nu} L_n(\varphi_x;x)}$$
$$+\,|f(x)|\left|\sum_{n=1}^{\infty} a_{jn}^{\nu} L_n(e_0;x)-e_0(x)\right|$$

Letting supremum over $x \in X$ and also taking $\delta := \delta_j^\nu = \sqrt{\left\| \sum_{n=1}^{\infty} a_{jn}^\nu L_n(\varphi_x) \right\|}$ we get

$$\left\| \sum_{n=1}^{\infty} a_{jn}^\nu L_n(f) - f \right\| \leq \omega(f, \delta_j^\nu) \left\| \sum_{n=1}^{\infty} a_{jn}^\nu L_n(e_0) \right\| + \omega(f, \delta_j^\nu) \sqrt{\left\| \sum_{n=1}^{\infty} a_{jn}^\nu L_n(e_0) \right\|} + \|f\| \left\| \sum_{n=1}^{\infty} a_{jn}^\nu L_n(e_0) - e_0 \right\|$$

which completes the proof.

4.4 An Application to Max-Product Bernstein Operators

Define the sequence (u_n) by

$$u_n := \begin{cases} 0, & \text{if } n \text{ is odd} \\ 2, & \text{if } n \text{ is even.} \end{cases} \tag{4.4}$$

Then observe that (u_n) is Cesàro mean convergent to 1 (denote this by $C_1 - \lim_n u_n = 1$); however, it is non-convergent in the usual sense.

Now take $X = [0, 1]$, $n \in \mathbb{N}$, $x_{n,k} = \frac{k}{n} \in [0, 1] (k = 0, 1, \ldots, n)$, and

$$K_{n,k}(x) := \frac{\binom{n}{k} x^k (1-x)^{n-k}}{\bigvee_{m=0}^{n} \binom{n}{m} x^m (1-x)^{n-m}}. \tag{4.5}$$

Using (4.5), Bede and Gal (see [4]) introduced the max-product Bernstein operators as follows:

$$B_n^{(M)}(f; x) := \bigvee_{k=0}^{n} K_{n,k}(x) . f(\frac{k}{n}) = \frac{\bigvee_{k=0}^{n} \binom{n}{k} x^k (1-x)^{n-k} f(\frac{k}{n})}{\bigvee_{m=0}^{n} \binom{n}{m} x^m (1-x)^{n-m}}. \tag{4.6}$$

We know from [4] that

$$\lim_{n\to\infty} \left\| B_n^{(M)}(f) - f \right\| = 0. \tag{4.7}$$

Now using (4.4) and (4.6) we consider the following operators:

$$L_n(f;x) := u_n B_n^{(M)}(f;x). \tag{4.8}$$

Then, observe that it is impossible to approximate f by means of $L_n(f)$ since (u_n) is a non-convergent sequence in the ordinary sense. However, in the summation process, if we take Cesàro matrix C_1 instead of $\mathscr{A}$, we claim that $L_n(f)$ given by (4.8) is Cesàro mean convergent to f on $[0, 1]$.

Indeed, for any $f \in C([0, 1], [0, \infty))$ we may write that

$$\begin{aligned}
\left\| \frac{1}{j}\sum_{n=1}^{j} L_n(f) - f \right\| &= \left\| \frac{1}{j}\sum_{n=1}^{j} u_n B_n^{(M)}(f) - f \right\| \\
&\le \frac{1}{j}\sum_{n=1}^{j} |u_n| \left\| B_n^{(M)}(f) - f \right\| + \|f\| \left| \frac{1}{j}\sum_{n=1}^{j} u_n - 1 \right| \\
&\le \frac{2}{j}\sum_{n=1}^{j} \left\| B_n^{(M)}(f) - f \right\| + \|f\| \left| \frac{1}{j}\sum_{n=1}^{j} u_n - 1 \right|
\end{aligned}$$

Finally, letting $j \to \infty$ and also using the regularity of C_1, we get

$$\left\| \frac{1}{j}\sum_{n=1}^{j} L_n(f) - f \right\| \to 0 \ \ (\text{as } j \to \infty)$$

which corrects our claim.

References

1. F. Altomare, M. Campiti, *Korovkin Type Approximation Theory and Its Application* (Walter de Gruyter, Berlin, 1994)
2. G.A. Anastassiou, O. Duman, *Towards Intelligent Modeling: Statistical Approximation Theory*. Intelligent Systems Reference Library, vol. 14 (Springer, Berlin, 2011)
3. G.A. Anastassiou, S.G. Gal, *Approximation Theory: Moduli of Continuity and Global Smoothness Preservation* (Birkhäuser, Boston, 2000)
4. B. Bede, S.G. Gal, Approximation by nonlinear Bernstein and Favard-Szász-Mirakjan operators of max-product kind. J. Concr. Appl. Math. **8**, 193–207 (2010)
5. B. Bede, H. Nobuhara, M. Daňková, A. Di Nola, Approximation by pseudo-linear operators. Fuzzy Sets Syst. **159**, 804–820 (2008)

6. B. Bede, E.D. Schwab, H. Nobuhara, I.J. Rudas, Approximation by Shepard type pseudo-linear operators and applications to image processing. Int. J. Approx. Reason. **50**, 21–36 (2009)
7. B. Bede, L. Coroianu, S.G. Gal, Approximation and shape preserving properties of the Bernstein operator of max-product kind. Int. J. Math. Math. Sci. **2009**, 26 pp. (2009). Art. ID 590589
8. B. Bede, L. Coroianu, S.G. Gal, Approximation and shape preserving properties of the nonlinear Baskakov operator of max-product kind. Stud. Univ. Babeş-Bolyai Math. **55**, 193–218 (2010)
9. B. Bede, L. Coroianu, S.G. Gal, Approximation and shape preserving properties of the nonlinear Favard-Szász-Mirakjan operator of max-product kind. Filomat **24**, 55–72 (2010)
10. B. Bede, L. Coroianu, S.G. Gal, Approximation and shape preserving properties of the nonlinear Bleimann-Butzer-Hahn operators of max-product kind. Comment. Math. Univ. Carol. **51**, 397–415 (2010)
11. B. Bede, L. Coroianu, S.G. Gal, Approximation and shape preserving properties of the nonlinear Meyer-König and Zeller operator of max-product kind. Numer. Funct. Anal. Optim. **31**, 232–253 (2010)
12. B. Bede, L. Coroianu, S.G. Gal, Approximation by truncated Favard-Szász-Mirakjan operator of max-product kind. Demonstratio Math. **44**, 105–122 (2011)
13. H.T. Bell, Order summability and almost convergence. Proc. Am. Math. Soc. **38**, 548–552 (1973)
14. O. Duman, Statistical convergence of max-product approximating operators. Turk. J. Math. **34**, 501–514 (2010)
15. H. Fast, Sur la convergence statistique. Colloq. Math. **2**, 241–244 (1951)
16. P.P. Korovkin, *Linear Operators and Theory of Approximation* (Hindustan Publ. Co, Delhi, 1960)
17. G.G. Lorentz, A contribution to the theory of divergent sequences. Acta Math. **80**, 167–190 (1948)

Chapter 5
Fixed Point in a Non-metrizable Space

Abdalla Tallafha and Suad A. Alhihi

Abstract In this paper, we shall define Lipschitz condition for functions and contraction functions on non-metrizable spaces. Finally, we ask the natural question: "Does every contraction have a unique fixed point?".

5.1 Introduction

A uniform space is a set with a uniform structure. Uniform spaces are topological spaces with additional structure that is used to define uniform properties such as completeness, uniform continuity and uniform convergence.

The notion of uniformity has been investigated by several mathematicians such as Weil [10–12], Cohen [3, 4] and Graves [6]. The theory of uniform spaces was given by Bourbaki in [2]. Also Weil's booklet [12] defines uniformly continuous mapping.

Contraction functions on complete metric spaces played an important role in the theory of fixed point (Banach fixed point theory). Lipschitz condition and contractions are usually discussed in metric and normed spaces, and have never been studied in a non-metrizable space. The object of this paper is to define Lipschitz condition, and contraction mapping on semi-linear uniform spaces, which enables us to study fixed point for such functions. We believe that the structure of semi-linear uniform spaces is very rich, and all the known results on fixed point theory can be generalized.

Let X be a non-empty set and D_X be a collection of all subsets of $X \times X$, such that each element V of D_X contains the diagonal $\Delta = \{(x,x) : x \in X\}$and $V = V^{-1} = \{(y,x) : (x,y) \in V\}$ for all $V \in D_X$. D_X is called the family of all entourages of the diagonal. Let Γ be a sub-collection of D_X. Then we have the following definition.

A. Tallafha (✉)
Department of Mathematics, The University of Jordan, Amman, Jordan
e-mail: a.tallafha@ju.edu.jo; atallafha@gmail.com

S.A. Alhihi
Department of Mathematics, Al-Balqa Applied University, Alsalt, Jordan
e-mail: suad.hihi@bau.edu.jo; suadalhihi@yahoo.com

G.A. Anastassiou, O. Duman (eds.), *Computational Analysis*, Springer Proceedings in Mathematics & Statistics 155, DOI 10.1007/978-3-319-28443-9_5

Definition 5.1 ([2]). The pair (X, Γ) is called a uniform space if

(i) $V_1 \cap V_2 \in \Gamma$ for all V_1, V_2 in Γ.
(ii) For every $V \in \Gamma$, there exists $U \in \Gamma$ such that $U \circ U \subset V$.
(iii) $\bigcap\{V : V \in \Gamma\} = \Delta$.
(iv) If $V \in \Gamma$ and $V \subseteq W \in D_X$, then $W \in \Gamma$.

If the condition $V = V^{-1}$ is omitted, then the space is quasiuniform.

5.2 Semi-linear Uniform Spaces

In [9] Tallafha and Khalil define a new type of uniform space, namely, semi-linear uniform space.

Definition 5.2 (See [9]). Let Γ be a sub-collection of D_X. Then, the pair (X, Γ) is called a semi-linear uniform space if

(i) Γ is a chain,
(ii) for every $V \in \Gamma$, there exists $U \in \Gamma$ such that $U \circ U \subseteq V$,
(iii) $\bigcap\limits_{V \in \Gamma} V = \Delta$,
(iv) $\bigcup\limits_{V \in \Gamma} V = X \times X$.

Definition 5.3 (See [9]). Let $(X\ , \Gamma)$ be a semi-linear uniform space, for $(x, y) \in X \times X$, and let $\Gamma_{(x,y)} = \{V \in \Gamma : (x, y) \in V\}$. Then, the set valued map ρ on $X \times X$ is defined by $\rho(x, y) = \bigcap \{V : V \in \Gamma_{(x,y)}\}$.

Semi-linear uniform spaces are stronger than topological spaces.
If $(X\ , \Gamma)$ is a semi-linear uniform space, then we have

Definition 5.4 (See [5]). For $x \in X$ and $V \in \Gamma$. The open ball of center x and radius V is defined by $B(x, V) = \{y : \rho(x, y) \subseteq V\}$.

Clearly, from the properties of Γ, if $y \in B(x, V)$, then there is a $W \in \Gamma$ such that $B(y, W) \subseteq B(x, V)$.
The family

$$\tau = \{G \subseteq X : \text{ for every } x \in G \text{ there is a } V \in \Gamma \text{ such that } B\,(x, V) \subseteq G\}$$

is a topology on X. That is, a set G is open if for every point x in G, there exist $V \in \Gamma$ such that $B(x, V) \subseteq G$. Also, metric spaces are stronger than semi-linear uniform spaces (see Theorem 5.24).

In [9], it was shown that open balls separate points, so if X is finite, then we have the discrete topology; therefore, interesting examples are given when X is infinite. Also, if X is infinite, then Γ should be infinite; otherwise, $\Delta \in \Gamma$, which implies that the topology is the discrete one. The elements of Γ may be assumed to be open in the topology on $X \times X$ (see [5, 7]).

Clearly, from Definition 5.3, for all $(x, y) \in X \times X$, we have $\rho(x, y) = \rho(y, x)$ and $\Delta \subseteq \rho(x, y)$. In [8], Tallafha defines a new set valued δ. Let

$$\Gamma \backslash \Gamma_{(x,y)} = \left(\Gamma_{(x,y)}\right)^c = \{V \in \Gamma : (x, y) \notin V\},$$

from now on, we shall denote $\Gamma \backslash \Gamma_{(x,y)}$ by $\Gamma^c_{(x,y)}$.

Definition 5.5 (See [8]). Let (X, Γ) be a semi-linear uniform space. Then, the set valued map δ on $X \times X$ is defined by

$$\delta(x, y) = \begin{cases} \bigcup\{V : V \in \Gamma^c_{(x,y)}\}, & \text{if } x \neq y \\ \phi, & \text{if } x = y. \end{cases}$$

In [8], Tallafha gave some important properties of semi-linear uniform spaces, using the set valued map ρ and δ. And he showed that if (X, Γ) is a semi-linear uniform space, then $(X, \boldsymbol{\delta})$, $(X, \boldsymbol{\rho})$ and $(X, \Gamma \cup \boldsymbol{\sigma} \cup \delta)$ are semi-linear uniform spaces, where

$$\boldsymbol{\delta} = \{\boldsymbol{\delta}\,(x, y) : (x, y) \in X \times X\}\,,$$
$$\boldsymbol{\rho} = \{\boldsymbol{\rho}\,(x, y) : (x, y) \in X \times X\}\,.$$

Also Alhihi in [1] gave more properties of semi-linear uniform spaces. Now, we shall give new properties of semi-linear uniform spaces used in the following definitions.

Proposition 5.6. *Let (X, Γ) be a semi-linear uniform space. If Λ is a sub-collection of Γ such that $\bigcap_{V \in \Lambda} V \neq \Delta$, then there exists $U \in \Gamma$ such that $U \subsetneqq \bigcap_{V \in \Lambda} V$.*

Proof. Since $\bigcap_{V \in \Lambda} V \neq \Delta$, there exists a point $(x, y) \in \bigcap_{V \in \Lambda} V$ such that $x \neq y$. Let $U \in \Gamma$ be such that $(x, y) \notin U$. Clearly since Γ is a chain, U is the required set.

The following is an immediate consequence of Proposition 5.6 and Proposition 3.2 in [9].

Corollary 5.7. *Let (X, Γ) be a semi-linear uniform space. If $\rho(x, y) \neq \Delta$, then,*

1. *there exist $U \in \Gamma$ such that $U \subsetneqq \rho(x, y)$,*
2. *$U \subseteq \delta(x, y)$.*

Let (X, Γ) be a semi-linear uniform space. If $V \in \Gamma$, then, for all $n \in \mathbb{N}$, by nV, we mean $V \circ V \circ \ldots \circ V$ (n-times).

Proposition 5.8. *Let (X, Γ) be a semi-linear uniform space. If $x \neq y$, then $n\delta(x, y) = \bigcup_{V \in \Gamma^c_{(x,y)}} nV$.*

Proof. Clearly $n\delta(x,y) \subseteq n \bigcup_{V\in\Gamma^c_{(x,y)}} V$. Let $(s_0, s_n) \in n\delta(x,y)$, then there exist $s_1, \ldots, s_{n-1}$ such that $(s_i, s_{i+1}) \in \delta(x,y)$, $i = 1, \ldots, n-1$. So there exist $U_1, U_2, \ldots, U_n \in \Gamma^c_{(x,y)}$ such that $(s_{i-1}, s_i) \in U_i$. Since Γ is a chain, there exist $U \in \Lambda$ such that $U_1 \circ U_2 \circ \ldots \circ U_n \subseteq U \circ \ldots \circ U = nU$. So

$$(s_0, s_n) \in U_1 \circ U_2 \circ \ldots \circ U_n \subseteq \bigcup_{U\in\Gamma^c_{(x,y)}} U \circ \ldots \circ U = \bigcup_{V\in\Gamma^c_{(x,y)}} nV.$$

By a similar idea one can prove the following proposition.

Proposition 5.9. *Let (X, Γ) be a semi-linear uniform space. If Λ is a sub-collection of Γ, then $\bigcup_{V\in\Lambda} nV = n \bigcup_{V\in\Lambda} V$.*

Proposition 5.8 is one of the important properties of δ. Now the question is whether ρ satisfies a similar property or not. It is clear that $n\rho(x,y) \subseteq \bigcap_{V\in\Gamma_{(x,y)}} nV$, so we ask the following question.

Question. $\bigcap_{V\in\Gamma_{(x,y)}} nV \subseteq n\rho(x,y)$?

Using the definition of δ, ρ, one can note that, for $(x,y) \in X \times X$, if $\rho(x,y) \neq \Delta$, then for all $r, n \in \mathbb{N}$, $r \leq n$, we have

1. $r\rho(x,y) \subseteq n\rho(x,y)$,
2. $r\delta(x,y) \subseteq n\delta(x,y)$.

5.3 Topological Properties of Uniform Spaces

Uniform spaces are stronger than topological spaces.

If (X, U) is a uniform space, then we have

Definition 5.10 (See [4]). For $x \in X$ and $u \in U$, the open ball of center x and radius u is defined by $B(x,u) = \{y : (x,y) \subseteq u\}$.

The family

$$\tau = \{G \subseteq X : \text{for every } x \in G,\ \exists\, u \in U \text{ such that } B(x,u) \subseteq G\},$$

is a topology on X. That is, a set G is open if for every point x in G, there exists $u \in U$ such that $B(x,u) \subseteq G$. Also, metric spaces are stronger than uniform spaces. If (X, τ) is a topological space induced by a metric d on X, then

$$U = \{u_\in : \in > 0\},\ u_\in = \{(x,y) : d(x,y) < \in\}$$

is a uniform structure on X and the topology induced by U on X is τ.

Definition 5.11 (See [5]). A topological space is called uniformizable if there is a uniform structure compatible with the topology.

Theorem 5.12 (See [5]).

1. *Every uniformizable space is a completely regular topological space,*
2. *For a uniformizable space X, the following are equivalent:*

   - *X is a Kolmogorov space,*
   - *X is a Hausdorff space,*
   - *X is a Tychonoff space.*

Definition 5.13 (See [5]). Let $(X, U_X), (Y, U_Y)$ be two uniform spaces and $f : (X, U_X) \to (Y, U_Y)$. Then f is uniformly continuous if $\forall u \in U_Y$, $\exists v \in U_X$ such that, for all $x, y \in X$, if $\rho_X(x, y) \subseteq v$, then $\rho_Y(f(x), f(y)) \subseteq u$.

Also it is known that, for any compatible uniform structure, the intersection of all entourages $\{(x, x) : x \in X\} = \Delta$. Some authors (see, for instance, [5]) add this last condition directly in the definition of a uniformizable space.

Conversely, each completely regular space is uniformizable. A uniformity compatible with the topology of a completely regular space X can be defined as the coarsest uniformity that makes all continuous real-valued functions on X uniformly continuous. A fundamental system of entourages for this uniformity is provided by all finite intersections of sets $(f \times f)^{-1}(V)$, where f is a continuous real-valued function on X and V is an entourage of the uniform space X. This uniformity defines a topology, which is coincides with the original topology (hence coincides with it) is a simple consequence of complete regularity: for any $x \in X$ and a neighborhood V of x, there is a continuous real-valued function $f : X \to \mathbb{R}$ with $f(x) = 0$ and $f(v^c) = 1$, where v^c is the complement of v.

In particular, a compact Hausdorff space is uniformizable. In fact, for a compact Hausdorff space X, the set of all neighborhoods of the diagonal in $X \times X$ form the unique uniformity compatible with the topology.

A Hausdorff uniform space is metrizable if its uniformity can be defined by a countable family of pseudometrics. Indeed, as discussed above, such a uniformity can be defined by a single pseudometric, which is necessarily a metric if the space is Hausdorff. In particular, if the topology of a vector space is Hausdorff and definable by a countable family of seminorms, it is metrizable.

Similar to continuous functions between topological spaces, "which preserve topological properties", are the uniform continuous functions between uniform spaces, which preserve uniform properties. Uniform spaces with uniform maps form a category. An isomorphism between uniform spaces is called a uniform isomorphism.

All uniformly continuous functions are continuous with respect to the induced topologies. Generalizing the notion of complete metric space, one can also define completeness for uniform spaces. Instead of working with Cauchy sequences, one works with Cauchy filters (or Cauchy nets).

A Cauchy filter F on a uniform space X is a filter F such that for every entourage u, there exists $o \in F$ with $o \times o \subseteq u$. In other words, a filter is Cauchy if it contains "arbitrarily small" sets. It follows from the definitions that each filter that converges (with respect to the topology defined by the uniform structure) is a Cauchy filter. A Cauchy filter is called minimal if it contains no smaller (i.e., coarser) Cauchy filter (other than itself). It can be shown that every Cauchy filter contains a unique minimal Cauchy filter. The neighborhood filter of each point (the filter consisting of all neighborhoods of the point) is a minimal Cauchy filter.

Conversely, a uniform space is called complete if every Cauchy filter converges. Any compact Hausdorff space is a complete uniform space with respect to the unique uniformity compatible with the topology.

Theorem 5.14 (See [5]). *Let X be a uniform space and let $f : A \to Y$ be a uniformly continuous function from a dense subset A of X into a complete uniform space Y. Then f can be extended (uniquely) into a uniformly continuous function on all of X.*

Definition 5.15 (See [5]). A topological space that can be made into a complete uniform space, whose uniformity induces the original topology, is called a completely uniformizable space.

Theorem 5.16 (See [5]). *Every uniform space X has a Hausdorff completion; that is, there exist a complete Hausdorff uniform space Y and a uniformly continuous map $i : X \to Y$ with the following property: for any uniformly continuous mapping f of X into a complete Hausdorff uniform space Z, there is a unique uniformly continuous map $g : Y \to Z$ such that $f = g \circ i$.*

The Hausdorff completion Y is unique up to isomorphism. As a set, Y can be taken to consist of the minimal Cauchy filters on X. As the neighborhood filter $B(x)$ of each point $x \in X$ is a minimal Cauchy filter, the map i can be defined by mapping x to $B(x)$.

In the definition of semi-linear uniform spaces condition (i) is stronger than condition (i) and condition (iv) is weaker than condition (iv) in the definition of uniform spaces, so we have the following question.

Question. Which of the above topological properties are satisfied in semi-linear uniform spaces?

5.4 Contractions

The following definitions are given in [5]. In [9] the authors modified the definitions of semi-linear uniform spaces.

Definition 5.17. Let (X, Γ) be a semi-linear uniform space. A sequence (x_n) in X converges to $x \in X$, if for all $V \in \Gamma$, there exists $N. \in \mathbb{N}$ such that $(x_n, x) \in V$ for all $n \geq N$.

Definition 5.18. Let (X, Γ) be a semi-linear uniform space. A sequence (x_n) in X is Cauchy, if for all $V \in \Gamma$, there exist $N \in \mathbb{N}$ such that $(x_n, x_m) \in V$ for all $n, m \geq N$.

The concept of uniform continuity is given by Weil [12]. Now, we shall rewrite the definition using our notation for semi-linear uniform spaces.

Definition 5.19. Let (X, Γ_X), (Y, Γ_Y) be two semi-linear uniform spaces and let $f : (X, \Gamma_X) \to (Y, \Gamma_Y)$ be any function. Then, f is called uniformly continuous, if $\forall U \in \Gamma_Y$, $\exists V \in \Gamma_X$ such that, for all $x, y \in X$, if $\rho_X(x, y) \subseteq V$, then $\rho_Y(f(x), f(y)) \subseteq U$.

The following Proposition shows that we may replace ρ by δ in Definition 5.19.

Proposition 5.20. *Let $f : (X, \Gamma_X) \to (Y, \Gamma_Y)$. Then, f is uniformly continuous if and only if $\forall U \in \Gamma_Y$, $\exists V \in \Gamma_X$ such that, for all $x, y \in X$, if $\delta_X(x, y) \subseteq V$, then $\delta_Y(f(x), f(y)) \subseteq U$.*

Proof. Let $f : (X, \Gamma_X) \to (Y, \Gamma_Y)$ be uniformly continuous. Let $U \in \Gamma_Y$, then $\exists W \in \Gamma_X$ such that, if $\rho_X(x, y) \subseteq W$, then $\rho_Y(f(x), f(y)) \subseteq U$. Let V be a proper subset of W. We want to show that V is the required set. Let x, y be such that $\delta_X(x, y) \subseteq V$. Then, $\delta_X(x, y) \subsetneqq W$. So, by Proposition 3.2 (*ii*) in [9], $\rho_X(x, y) \subseteq W$, and so $\delta(f(x), f(y)) \subseteq \rho_Y(f(x), f(y)) \subseteq W$.

Conversely, let $U \in \Gamma_Y$. Choose W a proper subset of U, by assumption $\exists V \in \Gamma_X$ such that for all $x, y \in X$, if $\delta_X(x, y) \subseteq V$, then $\delta_Y(f(x), f(y)) \subseteq W \subsetneqq U$. Also by Proposition 3.2 (*ii*) in [9], $\delta_Y(f(x), f(y)) \subsetneqq U$, therefore $\rho_Y(f(x), f(y)) \subseteq U$.

In [9], Tallafha gave the following example of a space which is semi-linear uniform spaces, but not metrizable.

Example 5.21. Let $X = \mathbb{R}$, $\Gamma = \{U_t : 0 < t < \infty\}$ where

$$U_t = \{(x, y) : x^2 + y^2 < t,\} \cup \{(x, x) : x \in \mathbb{R}\}.$$

Till now, to define a function f that satisfies Lipschitz condition, or to be a contraction, it should be defined on a metric space to another metric space. The main idea of this paper is to define such concepts without metric spaces, just we need a semi-linear uniform space, which is weaker as we mentioned before.

Definition 5.22. Let $f : (X, \Gamma) \to (X, \Gamma)$. Then, f satisfied Lipschitz condition if there exist $m, n \in \mathbb{N}$ such that $m\delta(f(x), f(y)) \subseteq n\delta(x, y)$. Moreover, if $m > n$, then we call f a contraction.

Remark 5.23. One may use the set valued function ρ, instead of δ in the above definition. But we use δ, since δ satisfies Proposition 5.8.

It is known that, every topological space (X, τ), whose topology induced by a metric or a norm on X, can be generated by a uniform space. In the following theorem we shall show that (X, τ) can be generated also by a semi-linear uniform space.

Theorem 5.24. *Every topological space whose topology induced by a metric or a norm on X can be generated by a semi-linear uniform space.*

Proof. Let (X, τ) be a topological space whose topology induced by a metric or a norm on X. Let

$$\Gamma = \left\{ V_{\in}, \in > 0 : V_{\in} = \bigcup_{x \in X} \{x\} \times B(x, \in) \right\}.$$

Clearly, Γ is a chain and, since $B(x, V_{\in}) = B(x, \in)$, the topology induced by Γ on X is τ. Moreover, we have

1. $\Delta \subseteq V_{\in}$, for all $\in > 0$,
2. if $(s, t) \in V_{\in}$, then $(s, t) \in \{s\} \times B(s, \in)$, hence $(t, s) \in \{t\} \times B(t, \in) \subseteq V_{\in}$,
3. $V_{\frac{\in}{2}} \circ V_{\frac{\in}{2}} \subseteq V_{\in}$,
4. $\bigcap_{V \in \Gamma} V = \bigcap_{\in > 0} V_{\in} = \Delta$,
5. for all $\in > 0$, $\bigcup_{n=1}^{\infty} nV_{\in} = \bigcup_{n=1}^{\infty} V_{n\in} = X \times X$.

Let $(X, d) \to (Y, \delta)$ metric spaces, and $f : (X, d) \to (Y, \delta)$. It is known that

1. if f is continuous at x and $x_n \to x$, then $f(x_n) \to f(x)$,
2. if f is uniformly continuous, f maps Cauchy sequences to Cauchy sequences.

Now we shall show that these statements are still valid if the metric spaces (X, d), (Y, δ) are replaced by a semi-linear uniform space (X, Γ_X), (Y, Γ_Y).

Theorem 5.25. *Let (X, Γ_X), (Y, Γ_Y) be two semi-linear uniform spaces, and $f : (X, \Gamma_X) \to (Y, \Gamma_Y)$. Then:*

1. *if f is continuous at x, then $x_n \to x$ implies $f(x_n) \to f(x)$;*
2. *if f is uniformly continuous, then f maps Cauchy sequences to Cauchy sequences.*

Proof. 1. Let $f : (X, \Gamma_X) \to (Y, \Gamma_Y)$ be continuous, and $x_n \to x$. Let $U \in \Gamma_Y$, so $\exists V \in \Gamma_X$ such that, for all $x, y \in X$, if $\rho_X(x, y) \subseteq V$, then $\rho_Y(f(x), f(y)) \subseteq U$. Now since $x_n \to x$, there exists k such that $(x_n, x) \in V$ for every $n \geq k$, which implies that $(f(x_n), f(x)) \in U$ for every $n \geq k$.

2. Let $f : (X, \Gamma_X) \to (Y, \Gamma_Y)$ be uniformly continuous, and x_n is Cauchy. Let $U \in \Gamma_Y$, by uniformly continuity $\exists V \in \Gamma_X$, such that for all $x, y \in X$, if $\rho_X(x, y) \subseteq V$, then $\rho_Y(f(x), f(y)) \subseteq U$. Now since x_n is Cauchy, there exists k such that $(x_n, x_m) \in V$ for every $m, n \geq k$, which implies that $(f(x_n), f(x_m)) \in U$, for every $n, m \geq k$.

In metric spaces the converse of part (1) is true, so is it still true in semi-linear uniform spaces?

Question. Let (X, Γ_X), (Y, Γ_Y) be two semi-linear uniform spaces, and $f : (X, \Gamma_X) \to (Y, \Gamma_Y)$. If $f(x_n) \to f(x)$ in (Y, Γ_Y), for all $x_n \to x$ in (X, Γ_X), is f continuous?

Let (X, U), (Y, V) be any two uniform spaces and $f : (X, U) \to (Y, V)$. We know that if f is uniformly continuous, then f is continuous [5].

Also, if the uniform spaces (X, U), (Y, V) are replaced by the metric spaces (X, d), (Y, δ) and if $f : (X, d) \to (Y, \delta)$, then we have

1. if f is a contraction, the f satisfied Lipschitz condition;
2. if f satisfied Lipschitz condition, then f is uniformly continuous;
3. if f is uniformly continuous, then f is continuous.

Now we shall show that these statements are still valid if the metric spaces (X, d), (Y, δ) are replaced by semi-linear uniform spaces (X, Γ_X), (Y, Γ_Y).

Theorem 5.26. *Let* (X, Γ_X) *be any semi-linear uniform space, and* $f : (X, \Gamma) \to (X, \Gamma)$. *Then:*

1. *if* f *is a contraction, then it satisfies Lipschitz condition;*
2. *if* f *satisfy Lipschitz condition, then it is uniformly continuous.*

Proof. Since (1) is trivial, we shall prove (2). Let $f : (X, \Gamma) \to (X, \Gamma)$ satisfies Lipschitz condition. Then there exist $m, n \in \mathbb{N}$ such that $m\delta(f(x), f(y)) \subseteq n\delta(x, y)$. Let $U \in \Gamma$. Since $\exists V \in \Gamma$ such that $nV \subseteq U$; for $x, y \in X$, if $\delta(x, y) \subseteq V$, then

$$\delta(f(x), f(y)) \subseteq m\delta(f(x), f(y)) \subseteq n\delta(x, y) \subseteq nV \subseteq U.$$

By Proposition 5.20, the result follows.

Definition 5.27 (See [9]). A semi-linear uniform space (X, Γ) is called complete, if every Cauchy sequence is convergent.

Fixed point theorems are well-known results in mathematics, and have useful applications in many applied fields such as game theory, mathematical economics, and the theory of quasi-variational inequalities. It states that every contraction from a complete metric space to itself has a unique fixed point. So the following question is natural.

Question. Let (X, Γ) be a complete semi-linear uniform space, and $f : (X, \Gamma) \to (X, \Gamma)$ be a contraction. Does f have a unique fixed point?

Remark 5.28. All the results which were obtained using contraction on metric spaces can be considered as open questions in semi-linear uniform spaces.

References

1. S.A. Alhihi, More properties of semi linear uniform spaces. Appl. Math. **6**, 1069–1075 (2015)
2. N. Bourbaki, *Topologie Générale* [General Topology] (Hermann, Paris, 1940)
3. L.W. Cohen, Uniformity properties in a topological space satisfying the first denumerability postulate. Duke Math. J. **3**, 610–615 (1937)
4. L.W. Cohen, On imbedding a space in a complete space. Duke Math. J. **5**, 174–183 (1939)

5. R. Engelking, *Outline of General Topology* (North-Holand, Amsterdam, 1968)
6. L.M. Graves, On the completing of a Housdroff space. Ann. Math. **38**, 61–64 (1937)
7. I.M. James, *Topological and Uniform Spaces*. Undergraduate Texts in Mathematics (Springer, New York, 1987)
8. A. Tallafha, Some properties of semi-linear uniform spaces. Bol. Soc. Paranaen. Mat. **29**(2), 9–14 (2011)
9. A. Tallafha, R. Khalil, Best approximation in uniformity type spaces. Eur. J. Pure Appl. Math. **2**(2), 231–238 (2009)
10. A. Weil, Les recouvrements des espaces topologiques: espaces complete, espaces bicompact. C. R. Acad. Paris **202**, 1002–1005 (1936)
11. A. Weil, *Sur les espaces a structure uniforme et sur la topologie generale*. Act. Sci. Ind., vol. 551 (Hermann, Paris, 1937)
12. A. Weil, *Sur les espaces à structure uniforme et sur la topologic générale* (Hermann, Paris, 1938)

Chapter 6
Second Hankel Determinant for New Subclass Defined by a Linear Operator

Aisha Ahmed Amer

Abstract In Amer and Darus (Missouri J Math Sci, to appear; Int J Math Anal 6(12):591–597, 2012), the author introduced and studied a linear operator defined on the class of normalized analytic function in the unit disk. This operator is motivated by many researchers. With this operator sharp bound for the nonlinear functional for the class of analytic functions in the open unit disk is obtained. In this paper we discuss sharp bound for the nonlinear functional for the class of analytic functions defined by a linear operator has been considered. Several other results are also considered. There are interesting properties of normalized function in the unit disk for sharp sconced hankel for linear operator. In addition, various other known results are also pointed out. We also find some interesting corollaries on the class of normalized analytic functions in the open unit disk. Our results certainly generalized several results obtained earlier. Therefore, many interesting results could be obtained and we also derive some interesting corollaries of this class. The operator defined can be extended and can solve many new results and properties.

6.1 Introduction

In 1976, Noonan and Thomas [16] defined the qth Hankel determinant of the function f for given by (6.1) and $q \in \mathbb{N} = \{1, 2, 3, \ldots\}$, by

$$H_q(k) = \begin{vmatrix} a_k & a_{k+1} \cdots & a_{k+q-1} \\ a_{k+1} & a_{k+2} \cdots & a_{k+q} \\ \vdots & \vdots \quad \cdots & \vdots \\ a_{k+q-1} & a_{k+q} \cdots & a_{k+2q-2} \end{vmatrix}.$$

A.A. Amer (✉)
Department of Mathematics, University of Al-Mergeb, Alkoms, Libya
e-mail: eamer_80@yahoo.com

G.A. Anastassiou, O. Duman (eds.), *Computational Analysis*, Springer Proceedings in Mathematics & Statistics 155, DOI 10.1007/978-3-319-28443-9_6

Further, Fekete–Szegö [7] considered the Hankel determinant of $f \in \mathscr{A}$ for $q = 2$ and $n = 1$, $H_2(1) = \begin{vmatrix} a_1 & a_2 \\ a_2 & a_3 \end{vmatrix}$. They made an early study for the estimates of $|a_2a_4 - a_3^2|$, when $a_1 = 1$ with μ real. The well-known result due to them states that if $f \in \mathscr{A}$, then

$$|a_3 - \mu a_2^2| \leq \begin{cases} 4\mu - 3, & \text{if } \mu \geq 1, \\ 1 + 2\exp\left(\frac{-2\mu}{1-\mu}\right), & \text{if } 0 \leq \mu \geq 1, \\ 3 - 4\mu, & \text{if } \mu \leq 0. \end{cases}$$

Furthermore, Hummel [8, 9] obtained sharp estimates for $|a_3 - \mu a_2^2|$ when f is convex function and also Keogh and Merkes [11] obtained sharp estimates for $|a_3 - \mu a_2^2|$ when f is close-to-convex, starlike, and convex in $\mathbb{U}$. Here we consider the Hankel determinant of $f \in \mathscr{A}$ for $q = 2$ and $n = 2$,

$$H_2(2) = \begin{vmatrix} a_2 & a_3 \\ a_3 & a_4 \end{vmatrix}.$$

In this article, by making use of the operator $D_l^{m,\lambda}(a,b)f(z)$ defined recently by the author, a class of analytic functions $\mathscr{R}^*(\lambda_1, \lambda_2, l, n, \alpha)$ of $\mathscr{A}$ is introduced. The sharp upper bound for the nonlinear functional $|a_2a_4 - a_3^2|$ is obtained. The rest of the article is organized as follows. In Sect. 6.2, we introduce some definitions and results that we use in our proofs. Section 6.3 contains the main results. The last section is devoted to several previous known results as special cases.

6.2 Background

As usual, in this section we introduce some notations, definitions, and lemmas which will be needed later.

Definition 6.1. The class of functions of the form

$$f(z) = z + \sum_{k=2}^{\infty} a_k z^k, \quad (z \in \mathbb{U}), \tag{6.1}$$

which are analytic in the open unit disc $\mathbb{U} = \{z : |z| < 1\}$ on the complex plane $\mathbb{C}$ will be denoted by $\mathscr{A}$.

Definition 6.2. In particular, let $\mathscr{P}$ be the family of all functions p analytic in $\mathbb{U}$ for which $\Re\{p(z)\} > 0$ and

$$p(z) = 1 + \sum_{k=1}^{\infty} c_k z^k, \quad (z \in \mathbb{U}). \tag{6.2}$$

In particular, for $f \in \mathscr{A}$ and $(z \in \mathbb{U}, b \neq 0, -1, -2, -3, \ldots)$, $\lambda \geq 0$, $m \in \mathbb{Z}$, $l \geq 0$, the authors (cf., [2, 3]) introduced the following linear operator:

Definition 6.3. For $f \in \mathscr{A}$ the operator $D_l^{m,\lambda}(a,b)f(z)$ is defined by $D_l^{m,\lambda}(a,b)f(z) : \mathscr{A} \to \mathscr{A}$ and let

$$\phi(z) := \frac{1+l-\lambda}{1+l}\frac{z}{1-z} + \frac{\lambda}{1+l}\frac{z}{(1-z)^2},$$

and

$$D_l^{m,\lambda}(a,b)f(z) = \underbrace{\phi(z) * \cdots * \phi(z)}_{(m)\text{-times}} * zF(a,1;b;z) * f(z),$$

if $m = 0, 1, 2, \ldots$; and

$$D_l^{m,\lambda}(a,b)f(z) = \underbrace{\phi(z) * \cdots * \phi(z)}_{(-m)\text{-times}} * zF(a,1;b;z) * f(z),$$

if $(m = -1, -2, \ldots)$. Thus we have

$$D_l^{m,\lambda}(a,b)f(z) := z + \sum_{k=2}^{\infty}\left(\frac{1+\lambda(k-1)+l}{1+l}\right)^m \frac{(a)_{k-1}}{(b)_{k-1}} a_k z^k,$$

where $f \in \mathscr{A}$ and $(z \in \mathbb{U}, b \neq 0, -1, -2, -3, \ldots)$, $\lambda \geq 0, m \in \mathbb{Z}, l \geq 0$.

This linear operator is the generalized form of the following operators:

- $D_0^{m,0}(a,b)f(z) = D_l^{0,\lambda}(a,b)f(z) = L(a,b)f(z)$ (see [4]).
- $D_0^{0,0}(\beta+1,1)f(z) = D^{\beta}f(z); \beta \geq -1$ (see [18]).
- $D_0^{m,1}(1,1)f(z)$, $m \in \mathbb{N}$ (see [19]).
- $D_0^{m,\lambda}(1,1)f(z)$, $m \in \mathbb{N}$ (see [1]).
- $D_l^{m,\lambda}(1,1)f(z) = D_l^{m,\lambda}$ (see [5]).
- $D_0^{0,0}(2,2-\gamma)f(z) = \Omega^{\gamma}f(z) = \Gamma(2-\gamma)z^{\gamma}D_z^{\gamma}f(z)$, where $D_z^{\gamma}f(z)$ is the fractional derivative of f of order γ; $\gamma \neq 2, 3, 4, \ldots$ (see [17]).

Definition 6.4. We consider the following subclass $\mathscr{R}^*(\lambda,,l,m,a,\alpha)$ of $\mathscr{A}$:

$$\mathscr{R}^*(\lambda,,l,m,a,\alpha) = \left\{f : \Re\left\{(1-\alpha)\frac{D_l^{m,\lambda}(a,b)f(z)}{z} + \alpha(D_l^{m,\lambda}(a,b)f(z))'\right\} > 0\right\},$$

where

$$(z \in \mathbb{U}, b \neq 0, -1, -2, -3, \ldots), \quad \lambda \geq 0,\ m \in \mathbb{Z},\ l \geq 0.$$

Remark 6.5. The subclass $\mathscr{R}^*(0,0,0,1,1) = \mathscr{R}$ was studied systematically by MacGregor [14] who indeed referred to numerous earlier investigations involving functions whose derivative has a positive real part.

Lemma 6.6 (See [6]). *If $p \in \mathscr{P}$, then $|c_k| \leq 2$ for all k.*

Lemma 6.7 ([13], See Also [12]). *Let the function $p \in \mathscr{P}$ be given by the power series (6.2). Then,*

$$2c_2 = c_1^2 + x(4 - c_2),$$

for some x, $|x| \leq 1$, and

$$4c_3 = c_1^3 + 2(4 - c_1^2)c_1x - c_1(4 - c_1^2)x^2 + 2(4 - c_1^2)2(1 - |x|^2)z,$$

for some z, $|z| \leq 1$.

Now, we will obtain sharp upper bound for the functional $|a_2a_4 - a_3^2|$, of $f \in \mathscr{R}^*(\lambda, , l, m, a, \alpha)$.

6.3 Main Results

In this section, we present and prove our main results.
Using the techniques of Libera and Zlotkiewicz [12, 13], we now prove the following theorem.

Theorem 6.8. *Let $\alpha > 0$. If $f \in \mathscr{R}^*(\lambda, , l, m, a, \alpha)$, then*

$$\left|a_2a_4 - a_3^2\right| \leq \frac{16(1+l)^{2(m-1)}}{(1+2\alpha)^2(1+2\lambda+l)^{2(m-1)}a^2(a+1)^2}.$$

The result is sharp.

Proof. Since $f \in \mathscr{R}^*(\lambda, , l, m, a, \alpha)$

$$(1-\alpha)\frac{D_l^{m,\lambda}(a,b)f(z)}{z} + \alpha(D_l^{m,\lambda}(a,b)f(z))' = p(z), \tag{6.3}$$

for some $p \in \mathscr{P}$. Equating coefficients in (6.3), we have

$$\begin{cases} a_2 = \dfrac{c_1(1+l)^{m-1}}{(1+\alpha)(1+\lambda+l)^{m-1}a}, \\ a_3 = \dfrac{2c_2(1+l)^{m-1}}{(1+2\alpha)(1+2\lambda+l)^{m-1}a(a+1)}, \\ a_4 = \dfrac{6c_3(1+l)^{m-1}}{(1+3\alpha)(1+3\lambda+l)^{m-1}a(a+1)(a+2)}. \end{cases} \tag{6.4}$$

From (6.4), it can be established that

$$\left|a_2a_4 - a_3^2\right| = \left|\frac{6c_1c_3(1+l)^{2(m-1)}}{(1+\alpha)(1+3\alpha)(1+\lambda+l)^{m-1}(1+3\lambda+l)^{m-1}a^2(a+1)(a+2)} - \frac{4c_2^2(1+l)^{2(m-1)}}{(1+2\alpha)^2(1+2\lambda+l)^{2(m-1)}a^2(a+1)^2}\right|.$$

Now assume that

$$X_1 = \frac{(1+\lambda+l)^{m-1}a}{(1+l)^{m-1}}, \quad X_2 = \frac{(1+2\lambda+l)^{m-1}a(a+1)(a+2)}{2(1+l)^{m-1}},$$

and

$$X_3 = \frac{(1+3\lambda+l)^{m-1}a(a+1)(a+2)}{(1+l)^{m-1}}.$$

Also, let $c_1 = c$ $(0 \leq c \leq 2)$, and making use of Lemma 6.7 we have

$$\left|a_2a_4 - a_3^2\right| = \begin{cases} \left| \dfrac{(c^4+4\alpha c^4)(X_2^2 - X_1X_3) + (4\alpha^2c^4X_2^2 - 3\alpha^2c^4X_1X_3)}{4(1+\alpha)(1+2\alpha)^2(1+3\alpha)X_1X_2^2X_3} \right. \\ + \dfrac{[(2c^2+8\alpha c^2)(X_2^2 - X_1X_3) + (8\alpha^2c^2X_2^2 - 6\alpha^2c^2X_1X_3)]x(4-c^2)}{4(1+\alpha)(1+2\alpha)^2(1+3\alpha)X_1X_2^2X_3} \\ - \dfrac{x^2(4-c^2)\left[(c^2+4\alpha c^2)(X_2^2-X_1X_3)+(4\alpha^2c^2X_2^2-3\alpha^2c^2X_1X_3)+4X_1X_3+16\alpha X_1X_3+12\alpha^2X_1X_3\right]}{4(1+\alpha)(1+2\alpha)^2(1+3\alpha)X_1X_2^2X_3} \\ \left. + \dfrac{c(4-c^2)(1-|x|^2)z}{2(1+\alpha)(1+3\alpha)X_1X_3} \right|. \end{cases}$$

An application of triangle inequality and replacement of $|x|$ by ρ is given by

$$\left|a_2a_4 - a_3^2\right| \leq \begin{cases} \dfrac{(c^4+4\alpha c^4)(X_2^2 - X_1X_3) + (4\alpha^2c^4X_2^2 - 3\alpha^2c^4X_1X_3)}{4(1+\alpha)(1+2\alpha)^2(1+3\alpha)X_1X_2^2X_3} \\ + \dfrac{[(c^2+8\alpha c^2)(X_2^2 - X_1X_3) + (4\alpha^2c^2X_2^2 - 3\alpha^2c^2X_1X_3)]\rho(4-c^2)}{2(1+\alpha)(1+2\alpha)^2(1+3\alpha)X_1X_2^2X_3} \\ + \dfrac{\rho^2(4-c^2)\left[(c^2+4\alpha c^2)(X_2^2 - X_1X_3) + (4\alpha^2c^2X_2^2 - 3\alpha^2c^2X_1X_3)\right]}{4(1+\alpha)(1+2\alpha)^2(1+3\alpha)X_1X_2^2X_3} \\ + \dfrac{\rho^2(4-c^2)\left[4X_1X_3 + 16\alpha X_1X_3 + 12\alpha^2X_1X_3 - 2cX_2^2(1+2\alpha)^2\right]}{4(1+\alpha)(1+2\alpha)^2(1+3\alpha)X_1X_2^2X_3} \\ + \dfrac{c(4-c^2)}{2(1+\alpha)(1+3\alpha)X_1X_3} \end{cases}$$

$$= F(\rho), \tag{6.5}$$

with $z = |x| \leq 1$ and $\alpha > 0$. We assume that the upper bound for (6.5) is attained at an interior point of the set $\{(\rho, c) : \rho \in [0,1], c \in [0,2]\}$, then

$$\begin{aligned}\frac{\partial F}{\partial \rho} &= \frac{[(c^2+4\alpha c^2)(X_2^2 - X_1X_3) + (4\alpha^2c^2X_2^2 - 3\alpha^2c^2X_1X_3)](4-c^2)}{2(1+\alpha)(1+2\alpha)^2(1+3\alpha)X_1X_2^2X_3} \\ &+ \frac{\rho(4-c^2)\left[(c^2+4\alpha c^2)(X_2^2 - X_1X_3) + (4\alpha^2c^2X_2^2 - 3\alpha^2c^2X_1X_3)\right]}{2(1+\alpha)(1+2\alpha)^2(1+3\alpha)X_1X_2^2X_3} \\ &+ \frac{\rho(4-c^2)\left[4X_1X_3 + 16\alpha X_1X_3 + 12\alpha^2X_1X_3 - 2cX_2^2(1+2\alpha)^2\right]}{2(1+\alpha)(1+2\alpha)^2(1+3\alpha)X_1X_2^2X_3}. \qquad (6.6)\end{aligned}$$

We note that $F'(\rho) > 0$ and consequently F is increasing and $\max_{(0\leq\mu\leq1)} F(\rho) = F(1)$, which contradicts with our assumption of having the maximum value at the interior of $\rho \in [0,1]$. Now let

$$G(c) = F(1) = \begin{cases} \dfrac{(c^4+4\alpha c^4)(X_2^2 - X_1X_3) + (4\alpha^2c^4X_2^2 - 3\alpha^2c^4X_1X_3)}{4(1+\alpha)(1+2\alpha)^2(1+3\alpha)X_1X_2^2X_3} \\ +\dfrac{[(c^2+8\alpha c^2)(X_2^2 - X_1X_3) + (4\alpha^2c^2X_2^2 - 3\alpha^2c^2X_1X_3)](4-c^2)}{2(1+\alpha)(1+2\alpha)^2(1+3\alpha)X_1X_2^2X_3} \\ +\dfrac{(4-c^2)\left[(c^2+4\alpha c^2)(X_2^2 - X_1X_3) + (4\alpha^2c^2X_2^2 - 3\alpha^2c^2X_1X_3)\right]}{4(1+\alpha)(1+2\alpha)^2(1+3\alpha)X_1X_2^2X_3} \\ +\dfrac{(4-c^2)\left[4X_1X_3 + 16\alpha X_1X_3 + 12\alpha^2X_1X_3 - 2cX_2^2(1+2\alpha)^2\right]}{4(1+\alpha)(1+2\alpha)^2(1+3\alpha)X_1X_2^2X_3} \\ +\dfrac{c(4-c^2)}{2(1+\alpha)(1+3\alpha)X_1X_3}, \end{cases}$$

then,

$$G'(c) = \begin{cases} \dfrac{(4c^3+16\alpha c^3)(X_2^2 - X_1X_3) + (16\alpha^2c^3X_2^2 - 12\alpha^2c^3X_1X_3)}{4(1+\alpha)(1+2\alpha)^2(1+3\alpha)X_1X_2^2X_3} \\ +\dfrac{[(2c+16\alpha c)(X_2^2 - X_1X_3) + (8\alpha^2cX_2^2 - 6\alpha^2cX_1X_3)](4-c^2)}{2(1+\alpha)(1+2\alpha)^2(1+3\alpha)X_1X_2^2X_3} \\ -\dfrac{2c[(c^2+8\alpha c^2)(X_2^2 - X_1X_3) + (4\alpha^2c^2X_2^2 - 3\alpha^2c^2X_1X_3)]}{2(1+\alpha)(1+2\alpha)^2(1+3\alpha)X_1X_2^2X_3} \\ +\frac{(4-c^2)\left[(2c+8\alpha c)(X_2^2 - X_1X_3) + (8\alpha^2cX_2^2 - 6\alpha^2cX_1X_3) - 2X_2^2(1+2\alpha)^2\right]}{4(1+\alpha)(1+2\alpha)^2(1+3\alpha)X_1X_2^2X_3} \\ -\dfrac{2c\left[(c^2+4\alpha c^2)(X_2^2 - X_1X_3) + (4\alpha^2c^2X_2^2 - 3\alpha^2c^2X_1X_3)\right]}{4(1+\alpha)(1+2\alpha)^2(1+3\alpha)X_1X_2^2X_3} \\ -\dfrac{2c\left[4X_1X_3 + 16\alpha X_1X_3 + 12\alpha^2X_1X_3 - 2cX_2^2(1+2\alpha)^2\right]}{4(1+\alpha)(1+2\alpha)^2(1+3\alpha)X_1X_2^2X_3} \\ +\dfrac{4-3c^2}{2(1+\alpha)(1+3\alpha)X_1X_3}. \end{cases}$$

Therefore the last equality implies $c = 0$, which is a contradiction. Thus any maximum points of G must be on the boundary of $c \in [0, 2]$. However, $G(c) \geq G(2)$ and thus G has maximum value at $c = 0$. The upper bound for (6.5) corresponds to $\rho = 1$ and $c = 0$, in which case

$$\left| \frac{6c_1c_3(1+l)^{2(m-1)}}{(1+\alpha)(1+3\alpha)(1+\lambda+l)^{m-1}(1+3\lambda+l)^{m-1}a^2(a+1)(a+2)} - \frac{4c_2^2(1+l)^{2(m-1)}}{(1+2\alpha)^2(1+2\lambda+l)^{2(m-1)}a^2(a+1)^2} \right| \leq \frac{16(1+l)^{2(m-1)}}{(1+2\alpha)^2(1+2\lambda+l)^{2(m-1)}a^2(a+1)^2}.$$

This completes the proof of Theorem 6.8.

6.4 Conclusions

Remark 6.9. If $\alpha > 0$, then we get the corresponding functional $|a_2a_4 - a_3^2|$, for the class $f \in \mathcal{R}^*(0, 0, 0, 0, \alpha) = R(\alpha)$, studied in [15] as in the following corollary.

Corollary 6.10. *If $f \in R(\alpha)$, then*

$$|a_2a_4 - a_3^2| \leq \frac{4}{(1+2\alpha)^2},$$

the result is sharp.

Remark 6.11. If $\alpha = 1$, then we get the corresponding functional $|a_2a_4 - a_3^2|$, for the class $f \in \mathcal{R}^*(0, 0, 0, 0, 1) = \mathcal{R}$, studied in [10] as in the following corollary.

Corollary 6.12. *If $f \in \mathcal{R}$, then*

$$|a_2a_4 - a_3^2| \leq \frac{4}{9},$$

the result is sharp.

References

1. F.M. Al-Oboudi, On univalent functions defined by a generalised Salagean operator. Int. J. Math. Math. Sci. **27**, 1429–1436 (2004)
2. A.A. Amer, M. Darus, On a property of a subclass of Bazilevic functions. Missouri J. Math. Sci. (to appear)

3. A.A. Amer, M. Darus, A distortion theorem for a certain class of Bazilevic function. Int. J. Math. Anal. **6**(12), 591–597 (2012)
4. B.C. Carlson, D.B. Shaffer, Starlike and prestarlike hypergeometric functions. SIAM J. Math. Anal. **15**(4), 737–745 (1984)
5. A. Catas, On a certain differential sandwich theorem associated with a new generalized derivative operator. Gen. Math. **4**, 83–95 (2009)
6. P.L. Duren, *Univalent Functions Grundlehren der Mathematischen Wissenschaften*, vol. 259 (Springer, New York, 1983)
7. M. Fekete, G. Szegö, Eine Bermerkung uber ungerade schlichte functionen. J. Lond. Math. Soc. **8**, 85–89 (1933)
8. J. Hummel, The coefficient regions of starlike functions. Pacific J. Math. **7**, 1381–1389 (1957)
9. J. Hummel, Extremal problems in the class of starlike functions. Proc. Am. Math. Soc. **11**, 741–749 (1960)
10. A. Janteng, S.A. Halim, M. Darus, Coefficient inequality for a function whose derivative has a positive real part. J. Inequal. Pure Appl. Math. **7**, 1–5 (2006)
11. F.R. Keogh, E.P. Merkes, A coefficient inequality for certain classes of analytic functions. Proc. Am. Math. Soc. **20**, 8–12 (1969)
12. R.J. Libera, E.J. Złotkiewicz, Early coefficients of the inverse of a regular convex function. Proc. Am. Math. Soc. **85**, 225–230 (1982)
13. R.J. Libera, E.J. Złotkiewicz, Coefficient bounds for the inverse of a function with derivative in P. Proc. Am. Math. Soc. **78**, 251–257 (1983)
14. T.H. MacGregor, Functions whose derivative has a positive real part. Trans. Am. Math. Soc. **104**, 532–537 (1962)
15. G. Murugusundaramoorthy, N. Magesh, Coefficient inequalities for certain classes of analytic functions associated with hankel determinant. Bull. Math. Anal. Appl. **1**, 85–89 (2009)
16. J.W. Noonan, D.K. Thomas, On the second Hankel determinant of areally mean p-valent functions. Trans. Am. Math. Soc. **223**, 337–346 (1976)
17. S. Owa, H.M. Srivastava, Univalent and starlike generalized hypergeometric functions. Can. J. Math. **39**, 1057–1077 (1987)
18. St. Ruscheweyh, New criteria for univalent functions. Proc. Am. Math. Soc. **49**, 109–115 (1975)
19. G.S. Salagean, *Subclasses of Univalent Functions*. Lecture Notes in Mathematics, vol. 1013 (Springer, Berlin, 1983), pp. 362–372

Chapter 7
Some Asymptotics for Extremal Polynomials

Gökalp Alpan, Alexander Goncharov, and Burak Hatinoğlu

Abstract We review some asymptotics for Chebyshev polynomials and orthogonal polynomials. Our main interest is in the behaviour of Widom factors for the Chebyshev and the Hilbert norms on small sets such as generalized Julia sets.

7.1 Introduction

Let $K \subset \mathbb{C}$ be a compact set containing an infinite number of points and $\mathrm{Cap}(K)$ stand for the logarithmic capacity of K. Given $n \in \mathbb{N}$, by $\mathcal{M}_n$ we denote the set of all monic polynomials of degree at most n.

Given probability measure μ with $\mathrm{supp}(\mu) = K$ and $1 \le p \le \infty$, we define the nth Widom factor associated with μ as $W_n^p(\mu) = \frac{\inf_{Q \in \mathcal{M}_n} ||Q||_p}{(\mathrm{Cap}(K))^n}$ where $|| \cdot ||_p$ is taken in the space $L^p(\mu)$. If K is polar, then let $W_n^p(\mu) := \infty$. Clearly, $W_n^p(\mu) \le W_n^r(\mu)$ for $1 \le p \le r \le \infty$; W_n^p is invariant under dilation and translation of μ.

We omit the upper index for the case $p = \infty$. Here the values $W_n(K) = \frac{||T_{n,K}||_\infty}{(\mathrm{Cap}(K))^n}$ provide us with information about behaviour of the Chebyshev polynomials $T_{n,K}$ on K. In Sect. 7.2 we review some results in this direction.

Another important case is $p = 2$, where $\inf_{\mathcal{M}_n} ||Q||_2$ is realized on the monic orthogonal polynomial with respect to μ. The sequence $(W_n^2(\mu))_{n=1}^\infty$ is rather convenient to describe measures that are regular in the Stahl–Totik sense and the Szegő class that provides the strong asymptotics of general orthogonal polynomials. In Sect. 7.3 we recall basic concepts of the theory, in Sect. 7.4 model examples of $W_n^2(\mu)$ are considered. The next sections are related to the results of the first two authors about orthogonal polynomials with respect to equilibrium measures on generalized Julia sets. All results of the authors mentioned in this review were recently published or submitted except Theorem 7.1, which is new.

G. Alpan • A. Goncharov (✉)
Department of Mathematics, Bilkent University, Ankara, Turkey
e-mail: gokalp@fen.bilkent.edu.tr; goncha@fen.bilkent.edu.tr

B. Hatinoğlu
Texas A&M University, College Station, TX, USA
e-mail: burakhatinoglu@mail.math.tamu.edu

G.A. Anastassiou, O. Duman (eds.), *Computational Analysis*, Springer Proceedings in Mathematics & Statistics 155, DOI 10.1007/978-3-319-28443-9_7

We suggest the name *Widom factor* for $W_n^p(\mu)$ because of the fundamental paper [42], where Widom systematically considered the corresponding ratios for finite unions of smooth Jordan curves and arcs.

For basic notions of logarithmic potential theory we refer the reader to [30], log denotes the natural logarithm, μ_K is the equilibrium measure of K. Introduction to the theory of general orthogonal polynomials can be found in [33, 34, 37, 40], see [27] for basic concepts of complex dynamics and [13] for a generalization of Julia sets. The symbol $\sim$ denotes the strong equivalence: $a_n \sim b_n$ means that $a_n = b_n(1 + o(1))$ for $n \to \infty$.

7.2 Widom Factors for the Sup-Norm

Given K as above, by $T_{n,K}$ we denote the nth Chebyshev polynomial and by $t_n(K)$ the corresponding Chebyshev number $t_n(K) := ||T_{n,K}||_\infty$. By M. Fekete and G. Szegő we have $t_n(K)^{\frac{1}{n}} \to \mathrm{Cap}(K)$ as $n \to \infty$. Bernstein–Walsh inequality (see, e.g., Theorem 5.5.7 in [30]) implies that $t_n(K) \geq (\mathrm{Cap}(K))^n$ for all n. Thus, $W_n(K) \geq 1$ and $(W_n(K))_{n=1}^\infty$ have subexponential growth (that is, $\log W_n/n \to 0$). We mention two important cases: $W_n(\partial\mathbb{D}) = 1$ and $W_n([-1, 1]) = 2$ for all $n \in \mathbb{N}$.

If K is a subarc of the unit circle with angle 2α, then $W_n(K) \sim 2\cos^2(\alpha/4)$ (see, e.g., p. 779 in [36]). The circle and the interval can be considered now as limit cases with $\alpha \to \pi$ and $\alpha \to 0$.

By Schiefermayr [31], $W_n(K) \geq 2$ if K lies on the real line.

The behaviour of $(W_n(K))_{n=1}^\infty$ may be rather irregular, even for simple compact sets. Achieser considered in [1, 2] the set $K = [a, b] \cup [c, d]$ and showed that $(W_n(K))_{n=1}^\infty$ has a finite number of accumulation points from which the smallest is 2 provided K is a polynomial preimage of an interval. Otherwise, the accumulation points of $(W_n(K))_{n=1}^\infty$ fill out an entire interval of which the left endpoint is 2.

In the generalization of this result the concept of Parreau–Widom sets is important. Let $K \subset \mathbb{R}$ be regular with respect to the Dirichlet problem. Then the Green function $g_{\mathbb{C}\setminus K}$ of $\mathbb{C} \setminus K$ with pole at infinity is continuous throughout $\mathbb{C}$. By $\mathscr{C}$ we denote the set of critical points of $g_{\mathbb{C}\setminus K}$, where its derivative vanishes. Clearly, $\mathscr{C}$ is at most countable. Then K is called *a Parreau–Widom set* if

$$\mathrm{PW}(K) := \sum_{z\in\mathscr{C}} g_{\mathbb{C}\setminus K}(z) < \infty.$$

It was shown recently in [18] that $W_n(K) \leq 2\exp(\mathrm{PW}(K))$ for a Parreau–Widom set K.

In extension of Widom's theory, Totik and Yuditskii considered in [39] the case when $K = \cup_{j=1}^p K_j$ is a union of p disjoint C^{2+} Jordan curves which are symmetric with respect to the real line. They showed that the accumulation points of $(W_n(K))_{n=1}^\infty$ lie in $[1, \exp(\mathrm{PW}(K))]$. Moreover, if the values $(\mu_K(K_j))_{j=1}^p$ are rationally independent, then the limit points of $W_n(K)$ fill out the whole interval above. We recall that $(x_j)_{j=1}^n \subset \mathbb{R}$ are rationally independent if $\sum_{j=1}^n \alpha_j x_j = 0$ with $a_j \in \mathbb{Z}$ implies that $a_j = 0$ for all j.

There are also new results [8, 38] for the case when $K = \cup_{j=1}^p K_j$ is a union of p disjoint Jordan curves or arcs (not necessarily smooth), where quasi-smoothness or Dini-smoothness is required instead of smoothness.

Parreau–Widom sets have positive Lebesgue measure (see, e.g., [14] for a proof). All finite gap sets (see, e.g., [15, 17]) and symmetric Cantor sets with positive length (see, e.g., [29]) are Parreau–Widom sets. Hence, in all cases considered above the sequence of Widom factors is bounded. The second and the third authors showed that any subexponential growth of $(W_n(K))_{n=1}^\infty$ can be achieved and presented a Cantor-type set with highly irregular behaviour of Widom factors, namely [21],

1. For each (M_n) of subexponential growth there is K with $W_n(K) \geq M_n$ for all n.
2. Given $\sigma_n \searrow 0$ and $M_n \to \infty$ (of subexponential growth), there is K such that $W_{n_j}(K) < 2(1+\sigma_{n_j})$ and $W_{m_j}(K) > M_{m_j}$ for some subsequences (n_j) and (m_j).

In the last section, we consider non-Parreau–Widom sets with slow growth of Widom factors.

7.3 General Orthogonal Polynomials

Given μ as above, the Gram–Schmidt process in $L^2(\mu)$ defines orthonormal polynomials $p_n(z,\mu) = \kappa_n z^n + \cdots$ with $\kappa_n > 0$. Let $q_n = \kappa_n^{-1} p_n$. Then $||q_n||_2 = \kappa_n^{-1} = \inf_{Q\in\mathscr{M}_n} ||Q||_2$. If $K \subset \mathbb{R}$, then a three-term recurrence relation

$$x\, q_n(x) = q_{n+1}(x) + b_n\, q_n(x) + a_{n-1}^2\, q_{n-1}(x)$$

is valid with the Jacobi parameters $a_n = \kappa_n/\kappa_{n+1}$ and $b_n = \int x p_n^2(x)\, \mathrm{d}\mu(x)$. Since $\mu(\mathbb{R}) = 1$, we have $p_0 = q_0 \equiv 1$, so $\kappa_0 = 1$ and $a_0 a_1 \cdots a_{n-1} = \kappa_n^{-1}$.

Thus, $W_n^2(\mu) = (\kappa_n \cdot \mathrm{Cap}^n(K))^{-1}$ and, in particular, for $K = [-1,1]$ we have $W_n^2(\mu) = a_0 a_1 \cdots a_{n-1} \cdot 2^n$.

For example, the equilibrium measure $\mathrm{d}\mu_{[-1,1]} = \frac{\mathrm{d}x}{\pi\sqrt{1-x^2}}$ generates the Chebyshev polynomials of the first kind with $W_n^2(\mu_{[-1,1]}) = \sqrt{2}$ for all n, whereas for the Chebyshev polynomials of the second kind $\mathrm{d}\nu = \frac{2}{\pi}\sqrt{1-x^2}\,\mathrm{d}x$ and $W_n^2(\nu) = 1$.

The Jacobi parameters generate the matrix

$$J = \begin{pmatrix} b_0 & a_0 & 0 & 0 & \dots \\ a_0 & b_1 & a_1 & 0 & \dots \\ 0 & a_1 & b_2 & a_2 & \dots \\ \vdots & \vdots & \vdots & \vdots & \ddots \end{pmatrix},$$

where μ is the spectral measure for the unit vector δ_1 and the self-adjoint operator J on $l_2(\mathbb{Z}_+)$, which is defined by this matrix.

Both (a_n) and (b_n) are bounded sequences. Conversely, if we are given bounded sequences (a_n) and (b_n) with $a_n > 0$ and $b_n \in \mathbb{R}$, then, as a result of the spectral theorem, there is a unique probability measure μ such that the associated recurrence coefficients are $(a_n, b_n)_{n=0}^{\infty}$.

For a wide class of measures the polynomials $p_n = p_n(\cdot, \mu)$ enjoy regular limit behaviour. Let $\Omega = \overline{\mathbb{C}} \setminus K$ and ν_{p_n} be the counting measure on the zeros of p_n. Suppose the set K is not polar. Let us consider the asymptotics:

1. $\kappa_n^{1/n} \to \mathrm{Cap}(K)^{-1}$
2. $|p_n|^{1/n} \rightrightarrows \exp g_\Omega$ (locally uniformly on $\overline{\mathbb{C}} \setminus \mathrm{Conv.hull}(K)$)
3. $\limsup |p_n(z)|^{1/n} \overset{\text{q.e.}}{=} 1$ on $\partial\Omega$
4. $\frac{1}{n}\nu_{p_n} \overset{w^*}{\to} \mu_K$.

By Theorem 3.1.1 in [34], the conditions (1)–(3) are pairwise equivalent. If, in addition, $K \subset \partial\Omega$ and the minimal carrier capacity of μ is positive, then (1) is equivalent to (4).

A measure μ with support K is called *regular in the Stahl–Totik* sense ($\mu \in$ **Reg**) if (1) is valid. This definition allows measures with polar support. In this case the equivalence of (1)–(3) is still valid if we take $g_\Omega \equiv \infty$ in (2).

Till now there is no complete description of regularity in terms of the size of μ. We will use the generalized version of the Erdös–Turán criterion for $K \subset \mathbb{R}$ ([34], Theorem 4.1.1): $\mu \in$ **Reg** provided $\mathrm{d}\mu/\mathrm{d}\mu_K > 0$, $\mu_K - a.e.$ Thus (see also [41] and [32]), equilibrium measures are regular in the Stahl–Totik sense.

We see that $\mu \in$ **Reg** if and only if $(W_n^2(\mu))_{n=1}^{\infty}$ has subexponential growth.

7.4 Strong Asymptotics

The conditions (1)–(4) from the previous section can be considered as weak asymptotics. For measures from the Szegő class stronger asymptotics are valid for the corresponding orthogonal polynomials.

Suppose $\mathrm{d}\mu = \omega(x)\mathrm{d}x$ on $K = [-1, 1]$. Then we say that μ is in *the Szegő class* ($\mu \in Sz[-1, 1]$) if

$$I(\omega) := \int_{-1}^{1} \frac{\log \omega(x)}{\pi\sqrt{1-x^2}}\,\mathrm{d}x = \int \log \omega(x)\,\mathrm{d}\mu_K(x) > -\infty,$$

which means that the integral converges for it cannot be $+\infty$. For such measures [35, p. 297]

$$p_n(z, \mu) = \kappa_n z^n + \cdots = (1 + o(1))\ (z + \sqrt{z^2 - 1})^n \frac{1}{\sqrt{2\pi}} D_\mu^{-1}(z),$$

where the Szegő function

$$D_\mu(z) = \exp\left(\frac{1}{2}\sqrt{z^2-1}\int \frac{\log[\omega(x)\sqrt{1-x^2}]}{z-x}\,d\mu_K(x)\right)$$

is a certain outer function in the Hardy space on $\mathbb{C}\setminus[-1,1]$. Here the square root $\sqrt{z^2-1}$ is taken such that $|z+\sqrt{z^2-1}|>1$ at $z\notin K$.

Now $z\to\infty$ implies not only that $\kappa_n^{1/n}\to 2$, so $\mu\in\mathbf{Reg}$, but also the existence of

$$\lim_n W_n^2(\mu) = \sqrt{\pi}\,\exp(I(\omega)/2)$$

((12.7.2) in [35]), which is essentially stronger than the fact of subexponential growth of the sequence.

The inverse implication is also valid: if $\lim_n W_n^2(\mu)$ exists in $(0,\infty)$, then we have $\mu\in Sz[-1,1]$ (see, e.g., T.2.4 in [16]).

The Szegő theory was extended first to the case of measures that generate a finite gap Jacobi matrix (see, e.g., [9, 16, 28, 42]) and then for measures on $\mathbb{R}$ such that the essential support of μ is a Parreau–Widom set.

Let $\{y_j\}_j$ be the set of all isolated points of the support of μ and $K=\text{ess supp}(\mu)$, so $\text{supp}(\mu)=K\cup\{y_j\}_j$. Suppose that K is a Parreau–Widom set, so it has positive Lebesgue measure. Let $\omega(x)\,dx$ be the absolutely continuous part of $d\mu$ in its Lebesgue decomposition. In addition, let $\sum g_{\mathbb{C}\setminus K}(y_j)<\infty$. Then, in our terms (see, e.g., Theorem 2 in [14]),

$$\int \log\omega(x)d\mu_K(x) > -\infty \iff \limsup_{n\to\infty} W_n^2(\mu) > 0. \tag{7.1}$$

Moreover, if one of the conditions above holds, then there is a positive number M such that

$$\frac{1}{M} < W_n^2(\mu) < M,$$

holds for all n. Thus, any of the conditions in (7.1) implies regularity of the corresponding measure.

We write $\mu\in Sz(K)$ if the Szegő condition on the left-hand side of (7.1) is valid. We see that this definition can be applied only to measures that have nontrivial absolutely continuous part. On the other hand, the Widom condition (on the right side) is applicable to any measure.

For each Parreau–Widom set K, its equilibrium measure μ_K belongs to $Sz(K)$ [14] and the sequence $(W_n^2(\mu_K))$ is bounded above [18]. In [5, 7] the first two authors presented non-polar sets with unbounded above sequence $(W_n^2(\mu_K))$.

The Widom condition is the main candidate to characterize the Szegő class in the general case. In [5] it was conjectured that the equilibrium measure always is in the Szegő class and the following form of the Szegő condition was suggested

$$\int \log(\mathrm{d}\mu/\mathrm{d}\mu_K)\mathrm{d}\mu_K(t) > -\infty$$

that can be used for all non-polar sets.

7.5 Widom Factors for the Hilbert Norm

Here we consider some model examples of Widom–Hilbert factors (see [7] for more details).

1. Jacobi weight. For $-1 < \alpha, \beta < \infty$ let

$$\mathrm{d}\mu_{\alpha,\beta} = C_{\alpha,\beta}^{-1}(1-x)^{\alpha}(1+x)^{\beta}\mathrm{d}x$$

with

$$C_{\alpha,\beta} = \int_{-1}^{1}(1-x)^{\alpha}(1+x)^{\beta}\,\mathrm{d}x.$$

Set $W_{\alpha,\beta} := \sqrt{\dfrac{\pi}{2^{\alpha+\beta}\,C_{\alpha,\beta}}}$. Then $W_n^2(\mu_{\alpha,\beta}) \to W_{\alpha,\beta}$. Here, $W_{\alpha,\beta} \to 0$ as (α,β) approaches the boundary of the domain $(-1,\infty)^2$ and

$$\sup_{-1<\alpha,\beta<\infty} W_{\alpha,\beta} = W_{-1/2,-1/2} = \sqrt{2}.$$

We see that, in the class of Jacobi polynomials, the maximal value of $I(\omega)$ is attained on the equilibrium measure. By Jensen's inequality, $\mu_{[-1,1]}$ gives the maximum of the Szegő integral in the whole class $Sz[-1,1]$. Indeed,

$$\int \log(\omega/\omega_e)\,\mathrm{d}\mu_{[-1,1]} \le \log \int \omega/\omega_e\,\mathrm{d}\mu_{[-1,1]} = \log\int_{-1}^{1}\omega(x)\,\mathrm{d}x = 0,$$

where $\mu \in Sz[-1,1]$ with $\mathrm{d}\mu = \omega(x)\mathrm{d}x$ and $\omega_e(x) = \dfrac{1}{\pi\sqrt{1-x^2}}$.

2. Regular measure beyond the Szegő class. A typical example of such measure is given by the density

$$\omega(x) = \frac{1+a}{2\pi}\exp(-2\,t\cdot\arcsin x)\cdot|\,\Gamma(1/2+i\,t)\,|^2$$

with $t = \dfrac{ax+b}{2\sqrt{1-x^2}}$, where $a, b \in \mathbb{R}$, $a \geq |b|$, $a+|b| > 0$. The measure generates the Pollaczek polynomials. Here, μ is regular, as $\omega > 0$ for $|x| < 1$, but since $\omega \to 0$ exponentially fast near ± 1, the integral $I(\omega)$ diverges, so $\mu \notin Sz[-1, 1]$. In this case,

$$\lim_n W_n^2(\mu) \cdot n^{a/2} = \Gamma\left(\frac{a+1}{2}\right),$$

so the Widom factors go to zero but not very fast.

3. $\mu \notin$ **Reg**. Using techniques from [34], one can show that any rate of decrease, as fast as we wish, can be achieved for the sequence $(W_n^2(\mu))$. Namely, ([7], Example 5) for each sequence $\sigma_n \searrow 0$ there exists a measure μ such that $W_n^2(\mu) < \sigma_n$ for all n. Here, $\mathrm{Cap}(\mathrm{supp}(\mu))$ does not coincide with the minimal carrier capacity of μ.
4. Jacobi matrix with periodic coefficients (a_n) and zero (or slowly oscillating) main diagonal. The periodic coefficients give a Jacobi matrix in the Szegő class. We follow [26] here.
   Let $a_{2n-1} = a, a_{2n} = b$ for $n \in \mathbb{N}$ with $b > 0$ and $a = b + 2$. These values with $b_n = 0$ define a Jacobi matrix B_0 with spectrum

$$\sigma(B_0) = [-b-a, b-a] \cup [a-b, a+b].$$

   The same values $(a_n)_{n=1}^{\infty}$ with $b_n = \sin n^{\gamma}$ for $0 < \gamma < 1$ give a matrix B with

$$\sigma(B) = [-b-a-1, b-a+1] \cup [a-b-1, a+b+1].$$

   Then $\mathrm{Cap}(\sigma(B_0)) = \sqrt{ab}$, $\mathrm{Cap}(\sigma(B)) = \sqrt{a(b+1)}$. Let μ_0 and μ be spectral measures for B_0 and B correspondingly. Then $W_{2n}^2(\mu_0) = 1$ and $W_{2n-1}^2(\mu_0) = \sqrt{a/b}$. Hence, $\mu_0 \in Sz(\sigma(B_0))$, as we expected. On the other hand,

$$W_{2n}^2(\mu) = \left(\frac{b}{b+1}\right)^n$$

   and

$$W_{2n+1}^2(\mu) = \left(\frac{b}{b+1}\right)^n \sqrt{\frac{a}{b+1}}.$$

   Thus, $W_n^2(\mu) \to 0$ as $n \to \infty$, $\mu \notin Sz(\sigma(B))$ and, moreover, $\mu \notin$ **Reg**.
5. Julia sets generated by $T(z) = z^3 - \lambda z$ with $\lambda > 3$ [11].
   Iterations $T_0 = z$, $T_n = T_{n-1}(T)$ define a Cantor-type Julia set $J = \mathrm{supp}(\mu_J)$. Let $W_k := W_k^2(\mu_J)$. Then $W_{3^n} = 1$, whereas $W_{3^n-1} \to \infty$. Also,

$$W_{3^n+1} \to \sqrt{2\lambda/3},\ W_{3^n+2} \to \sqrt{2}\,\lambda/3, \text{ etc.}$$

7.6 Weakly Equilibrium Cantor Sets

The theory of orthogonal polynomials is well developed for measures that are absolutely continuous with respect to the Lebesgue measure ($\mu = \mu_a$), at least for the finite gap case. There are also numerous results for measures ($\mu = \mu_a + \mu_p$) that allow nontrivial point spectrum. Here in the description of the Szegő class a condition of Blaschke-type is added. But there are only a few results for concrete singular continuous measures, mainly they are concerned with orthogonal polynomials for equilibrium measures on Julia sets. As we mentioned above, Parreau–Widom sets (in particular homogeneous sets in the sense of Carleson) may have Cantor structure, but their Lebesgue measure is positive.

There are only particular results for a prescribed measure μ supported on a Cantor set with zero Lebesgue measure. For example, if μ is the Cantor–Lebesgue measure or the equilibrium measure on the Cantor ternary set K_0, then a little is known except some conjectures depending on numerical results. For this case and other attractors of iterated function systems, we refer the reader to [22, 23, 25].

The first two authors found in [5] a new family of orthogonal polynomials with respect to the equilibrium measure on the so-called weakly equilibrium Cantor sets, that were suggested in [20]. Here we recall the construction. Given $\gamma = (\gamma_s)_{s=1}^{\infty}$ with $0 < \gamma_s < \frac{1}{4}$, let $r_0 = 1$ and $r_s = \gamma_s r_{s-1}^2$. We define recursively polynomials

$$P_2(x) = x(x-1)$$

and

$$P_{2^{s+1}} = P_{2^s} \cdot (P_{2^s} + r_s).$$

We consider the complex level domains

$$D_s = \{z \in \mathbb{C} : |P_{2^s}(z) + r_s/2| < r_s/2\}$$

with $D_s \searrow$, which allows, by the Harnack Principle, to get a good representation of the Green function for the intersection of domains, and

$$E_s := \{x \in \mathbb{R} : |P_{2^s}(x) + r_s/2| \le r_s/2\} = \cup_{j=1}^{2^s} I_{j,s}.$$

Then the set

$$K(\gamma) := \bigcap_{s=1}^{\infty} \overline{D}_s = \bigcap_{s=1}^{\infty} E_s = \bigcap_{s=1}^{\infty} \left(\frac{2}{r_s} P_{2^s} + 1\right)^{-1} ([-1,1])$$

is an intersection of polynomial preimages that provides some additional useful features. In particular, $P_{2^s} + r_s/2$ is the 2^sth Chebyshev polynomial on $K(\gamma)$.

At least for small γ, the set $K(\gamma)$ is weakly equilibrium in the following sense. Let us distribute uniformly the mass 2^{-s} on each $I_{j,s}$ for $1 \leq j \leq 2^s$. This defines a measure λ_s supported on E_s with $d\lambda_s = (2^s l_{j,s})^{-1}dt$ on $I_{j,s}$. Then $\lambda_s \overset{*}{\to} \mu_{K(\gamma)}$ provided $\gamma_n \leq 1/32$ and $K(\gamma)$ is not polar.

In [21] the Widom–Chebyshev factors for $K(\gamma)$ were calculated and the result mentioned in Sect. 7.2 was obtained.

In [4] it was shown that, provided some restriction on the sequence γ, the equilibrium measure on $K(\gamma)$ and the corresponding Hausdorff measure are mutually absolutely continuous. This is not valid for geometrically symmetric Cantor-type sets, where these measures are essentially different. Makarov and Volberg proved in [24] a surprising result: the equilibrium measure for the classical Cantor set is supported by a set whose Hausdorff dimension is strictly smaller than $\log 2/\log 3$. Therefore, μ_{K_0} is mutually singular with the Hausdorff measure of the set. Later this was generalized to Cantor-type sets of higher dimension and to Cantor repellers that appear in complex dynamics.

The set $K(\gamma)$ has positive Lebesgue measure if γ_s are rather closed to $\frac{1}{4}$. Moreover, in the limit case $\gamma_s = \frac{1}{4}$ for all s we have $K(\gamma) = [0, 1]$.

7.7 Orthogonal Polynomials on $K(\gamma)$

The set $K(\gamma)$ is non-polar if and only if

$$\sum_{n=1}^{\infty} 2^{-n} \log \frac{1}{\gamma_n} < \infty,$$

where the series represents the Robin constant of the set. Orthogonal polynomials with respect to the equilibrium measure on non-polar $K(\gamma)$ were considered in [5]. It is proven that the monic orthogonal polynomials Q_{2^s} coincide with the Chebyshev polynomials of the set. Procedures were suggested to find orthogonal polynomials Q_n of all degrees and to calculate the corresponding Jacobi parameters. In addition, it was shown that the sequence of Widom factors is bounded below by a positive number (in confirmation of our hypothesis that equilibrium measures always belong to the Szegő class in its Widom characterization).

First the authors used a technique of decomposition of zeros of $P_{2^s} + r_s/2$ into certain groups and the approximation of the equilibrium measure $\mu_{K(\gamma)}$ by the normalized counting measure at zeros of the Chebyshev polynomials of the set. Namely, let $\nu_s = 2^{-s} \sum_{k=1}^{2^s} \delta_{x_k}$, where $(x_k)_{k=1}^{2^s}$ are the zeros of $P_{2^s} + r_s/2$ (they are simple and real). Then for $s > m$ it is possible to decompose all zeros $(x_k)_{k=1}^{2^s}$ into 2^{s-m-1} groups, on which we can control the value of $P_{2^m} + r_m/2$. This allows to show that

$$\int \left(P_{2^m} + \frac{r_m}{2}\right) d\nu_s = 0.$$

Since $\nu_s \to \mu_{K(\gamma)}$ in the weak-star topology, we have that the integral

$$\int \left(P_{2^m} + \frac{r_m}{2}\right) \mathrm{d}\mu_{K(\gamma)}$$

also is zero.

Similarly it was shown that

$$\int \left(P_{2^{i_1}} + \frac{r_{i_1}}{2}\right)\left(P_{2^{i_2}} + \frac{r_{i_2}}{2}\right)\dots\left(P_{2^{i_n}} + \frac{r_{i_n}}{2}\right) \mathrm{d}\nu_s = 0$$

for $0 \le i_1 < i_2 < \cdots < i_n < s$. Each polynomial P of degree less than 2^s is a linear combination of polynomials of the type

$$\left(P_{2^{s-1}} + \frac{r_{s-1}}{2}\right)^{n_{s-1}} \dots \left(P_2 + \frac{r_1}{2}\right)^{n_1} \left(x - \frac{1}{2}\right)^{n_0},$$

with $n_i \in \{0, 1\}$. Therefore, Q_{2^s} coincides with $P_{2^s} + r_s/2$. In addition, the norm $||Q_{2^s}||_2$ has a simple representation in terms of $(\gamma_k)_{k=1}^{s+1}$ ((3.1) in [5]).

In the next step, A-type and B-type polynomials were introduced. In particular, for $2^m \le n < 2^{m+1}$ with the binary representation $n = i_m 2^m + \cdots + i_0$, the second polynomial is

$$B_n = (Q_{2^m})^{i_m} (Q_{2^{m-1}})^{i_{m-1}} \dots (Q_1)^{i_1}.$$

The polynomials $B_{(2k+1)\cdot 2^s}$ and $B_{(2j+1)\cdot 2^m}$ are orthogonal for all $j, k, m, s \in \mathbb{Z}_+$ with $s \neq m$. They can be considered as a basis in the set of polynomials: for each $n \in \mathbb{N}$ with $n = 2^s(2k + 1)$, the polynomial Q_n has a unique representation as a linear combination of

$$B_{2^s}, B_{3\cdot 2^s}, B_{5\cdot 2^s} \dots, B_{(2k-1)\cdot 2^s}, B_{(2k+1)\cdot 2^s}.$$

This allows to present formulas to express coefficients of each Q_n and the corresponding Jacobi parameters in terms of $(\gamma_k)_{k=1}^{\infty}$. Some asymptotics of Jacobi parameters were presented in Theorem 4.7 in [5]: *Let* $\gamma_s \le 1/6$ *for all* s. *Then* $\lim_{s\to\infty} a_{j\cdot 2^s+n} = a_n$ *for* $j \in N$ *and* $n \in Z_+$. *Here,* $a_0 := 0$. *In particular,* $\liminf a_n = 0$.

In the last section the Widom factors for $\mu_{K(\gamma)}$ were evaluated. If $\gamma_k \le \frac{1}{6}$ for all k, then

$$\liminf_{n\to\infty} W_n = \liminf_{s\to\infty} W_{2^s} \ge \sqrt{2}$$

and

$$\limsup_{n\to\infty} W_n = \infty.$$

The following examples illustrate the behaviour of Widom factors:

1. If $\gamma_n \to 0$, then $W_{2^s} \to \infty$. Therefore $W_n \to \infty$.
2. There exists $\gamma_n \nrightarrow 0$ with $W_n \to \infty$. One can take $\gamma_{2k} = 1/6,\ \gamma_{2k-1} = 1/k$.
3. If $\gamma_n \geq c > 0$ for all n, then $\liminf_{n\to\infty} W_n \leq 1/2c$.
4. There exists γ with $\inf \gamma_n = 0$ and $\liminf_{n\to\infty} W_n < \infty$. Here we can take $\gamma_n = 1/6$ for $n \neq n_k$ and $\gamma_{n_k} = 1/k$ for a sparse sequence $(n_k)_{k=1}^{\infty}$. Then $(W_{2^{n_k}})_{k=1}^{\infty}$ is bounded.

Later, in [6], it was shown that $K(\gamma)$ is a Parreau–Widom set if and only if

$$\sum_{n=1}^{\infty} \sqrt{\frac{1}{4} - \gamma_n} < \infty.$$

7.8 Generalized Julia Sets

In [6] the first two authors generalized some of the results [10–12] by Barnsley et al. obtained for autonomous Julia sets to more general class of sets. Also, [6] is a generalization of Alpan and Goncharov [5] as $K(\gamma)$ can be considered as a generalized Julia set.

We recall some basic definitions.

Let $(f_n(z))_{n=1}^{\infty}$ be a sequence of rational functions with $\deg f_n \geq 2$. in $\overline{\mathbb{C}}$. Let us define $F_n(z) := f_n \circ F_{n-1}(z)$ recursively for $n \geq 1$ and $F_0(z) = z$. Then domain of normality for $(F_n)_{n=1}^{\infty}$ in the sense of Montel is called the *Fatou set* for (f_n). The complement of the Fatou set in $\overline{\mathbb{C}}$ is called the *Julia set* for (f_n). We denote them by $F_{(f_n)}$ and $J_{(f_n)}$, respectively. In particular, if $f_n = f$ for some fixed rational f for all n, then we use the notations $F(f)$ and $J(f)$. To distinguish this last case, the word *autonomous* is used.

We consider only polynomial Julia sets. In order to have an appropriate Julia set in terms of orthogonal polynomials and potential theory, we need to put some restrictions on the given polynomials. Let $f_n(z) = \sum_{j=0}^{d_n} a_{n,j} \cdot z^j$ where $d_n \geq 2$ and $a_{n,d_n} \neq 0$ for all $n \in \mathbb{N}$. Following [13], we say that (f_n) is a *regular polynomial sequence* if the following properties are satisfied:

- There exists a real number $A_1 > 0$ such that $|a_{n,d_n}| \geq A_1$, for all $n \in \mathbb{N}$.
- There exists a real number $A_2 \geq 0$ such that $|a_{n,j}| \leq A_2 |a_{n,d_n}|$ for $j = 0, 1, \ldots, d_n - 1$ and $n \in \mathbb{N}$.
- There exists a real number A_3 such that

$$\log |a_{n,d_n}| \leq A_3 \cdot d_n,$$

for all $n \in \mathbb{N}$.

If (f_n) is a regular polynomial sequence, then we write $(f_n) \in \mathscr{R}$. If this is the case then, by Brück and Büger [13], $J_{(f_n)}$ is a compact subset of $\mathbb{C}$ that is regular with respect to the Dirichlet problem. Thus, $\mathrm{Cap}(J_{(f_n)}) > 0$. Moreover, $J_{(f_n)}$ is just the boundary of

$$\mathscr{A}_{(f_n)}(\infty) := \{z \in \overline{\mathbb{C}} : (F_n(z))_{n=1}^{\infty} \text{ goes locally uniformly to } \infty\}.$$

Let $K = J_{(f_n)}$ with $(f_n) \in \mathscr{R}$. In [6], it was shown that, for each integer n, the monic orthogonal polynomial associated with μ_K of degree $d_1 \cdots d_n$ can be written explicitly in terms of F_n. In [3], it was proven that the Chebyshev polynomials of degree $d_1 \cdots d_n$ on K are same up to constant terms with the orthogonal polynomials for μ_K.

In some cases the set $J_{(f_n)}$ is real. For example, this is valid for admissible (in the sense of Geronimo and Van Assche [19]) polynomials. Then a three-term recurrence relation is valid for orthogonal polynomials and the corresponding Jacobi coefficients can be found by a recursive procedure that is depicted.

Let a sequence γ be the same as in Sect. 7.6. If we take

$$f_n(z) = \frac{1}{2\gamma_n}(z^2 - 1) + 1$$

for all n, then $K_1(\gamma) := J_{(f_n)}$ is a stretched version of the set $K(\gamma)$. Let $\varepsilon_k = \frac{1}{4} - \gamma_k$.

By Theorem 8 in [6], the Green function $g_{\mathbb{C}\setminus K_1(\gamma)}$ has optimal smoothness (is Hölder continuous with the exponent $1/2$) if and only if $\sum_{k=1}^{\infty} \varepsilon_k < \infty$. This completes the analysis of smoothness of $g_{\mathbb{C}\setminus K(\gamma)}$ for the case of small γ in [20].

By Theorem 9 in [6], $K_1(\gamma)$ is a Parreau–Widom set if and only if $\sum_{k=1}^{\infty} \sqrt{\varepsilon_k} < \infty$.

It is interesting to analyse the character of growth of Widom's factor for non-Parreau–Widom sets.

7.9 Widom's Factor for Non-Parreau–Widom Sets

Here we return to Widom factors for the Chebyshev norm on $K(\gamma)$. As above, let $\varepsilon_k = \frac{1}{4} - \gamma_k$. Clearly, $0 < 1 - 4\varepsilon_k < 1$. Suppose

$$\sum_{k=1}^{\infty} \varepsilon_k < \infty \quad \text{but} \quad \sum_{k=1}^{\infty} \sqrt{\varepsilon_k} = \infty. \tag{7.2}$$

By C we denote the product $2 \prod_{k=1}^{\infty} (1 - 4\varepsilon_k)^{-1}$, which is finite by (7.2). Also this condition implies that the set $K(\gamma)$ is not polar and is not Parreau–Widom.

Theorem 7.1. *Let* $\gamma = (\gamma_k)_{k=1}^{\infty}$ *be a monotone sequence satisfying* (7.2). *Then the bound* $W_n(K(\gamma)) \le Cn$ *holds for all* $n \in \mathbb{N}$.

Proof. By [21], for all $s \in \mathbb{Z}_+$ we have

$$W_{2^s}(K(\gamma)) = \frac{1}{2}\exp\left(2^s \sum_{k=s+1}^{\infty} 2^{-k}\log\frac{1}{\gamma_k}\right).$$

Since $(\gamma_k)_{k=1}^{\infty}$ monotonically increases, we get the inequality

$$W_{2^s}(K(\gamma)) \le \frac{1}{2\gamma_{s+1}} = \frac{2}{1-4\varepsilon_{s+1}}. \tag{7.3}$$

Given $n \in \mathbb{N}$, take $s \in \mathbb{Z}_+$ with $2^s \le n < 2^{s+1}$. If $n = 2^s$ then, by (7.3),

$$W_n(K(\gamma)) \le \frac{2}{1-4\varepsilon_{s+1}} < C.$$

If $n \ne 2^s$, then there are integer numbers $0 \le p_1 < p_2 < \cdots < p_m \le s-1$ with $m \le s$ such that $n = 2^s + 2^{p_m} + \cdots + 2^{p_1}$. Widom factors are logarithmic subadditive, that is $W_{n+r}(K) \le W_n(K) \cdot W_r(K)$. Therefore,

$$W_n(K(\gamma)) \le W_{2^s}(K(\gamma)) \cdot W_{2^{p_m}}(K(\gamma)) \cdots W_{2^{p_1}}(K(\gamma)).$$

By (7.3) we see that

$$\begin{aligned} W_n(K(\gamma)) &\le \frac{2}{1-4\varepsilon_{s+1}}\,\frac{2}{1-4\varepsilon_{p_m+1}}\cdots\frac{2}{1-4\varepsilon_{p_1+1}} \\ &\le 2^{s+1}C/2 < nC. \end{aligned}$$

This completes the proof.

Acknowledgements The first two authors are partially supported by TÜBİTAK (Scientific and Technological Research Council of Turkey), Project 115F199.

References

1. N. Achyeser, Über einige Funktionen, welche in zwei gegebenen Interwallen am wenigsten von Null abweichen. I Teil, Bulletin de l'Académie des Sciences de l'URSS. Classe des sciences mathématiques et na (9), 1163–1202 (1932)
2. N. Achyeser, Über einige Funktionen, welche in zwei gegebenen Intervallen am wenigsten von Null abweichen. II Teil, Bulletin de l'Académie des Sciences de l'URSS. Classe des sciences mathématiques et na (3), 309–344 (1933)

3. G. Alpan, Chebyshev polynomials on generalized Julia sets. Comput. Methods Funct. Theory (2015). doi:10.1007/s40315-015-0145-8
4. G. Alpan, A. Goncharov, Two measures on Cantor sets. J. Approx. Theory **186**, 28–32 (2014)
5. G. Alpan, A. Goncharov, Orthogonal polynomials for the weakly equilibrium Cantor sets. Proc. Am. Math. Soc. (accepted for publication)
6. G. Alpan, A. Goncharov, Orthogonal polynomials on generalized Julia sets (preprint). arXiv: 1503.07098v3
7. G. Alpan, A. Goncharov, Widom Factors for the Hilbert Norm, vol. 107 (Banach Center Publications), pp. 9–16
8. V.V. Andrievskii, Chebyshev Polynomials on a system of continua. Constr. Approx. (2015). doi:10.1007/s00365-015-9280-8
9. A.I. Aptekarev, Asymptotic properties of polynomials orthogonal on a system of contours, and periodic motions of Toda lattices. Mat. Sb. **125**, 231–258 (1984). English translations in Math. USSR Sb. 53 (1986), 233–260
10. M.F. Barnsley, J.S. Geronimo, A.N. Harrington, Orthogonal polynomials associated with invariant measures on Julia sets. Bull. Am. Math. Soc. **7**, 381–384 (1982)
11. M.F. Barnsley, J.S. Geronimo, A.N. Harrington, Infinite-dimensional Jacobi matrices associated with Julia sets. Proc. Am. Math. Soc. **88**, 625–630 (1983)
12. M.F. Barnsley, J.S. Geronimo, A.N. Harrington, Almost periodic Jacobi matrices associated with Julia sets for polynomials. Commun. Math. Phys. **99**, 303–317 (1985)
13. R. Brück, M. Büger, Generalized iteration. Comput. Methods Funct. Theory **3**, 201–252 (2003)
14. J.S. Christiansen, Szegő's theorem on Parreau-Widom sets. Adv. Math. **229**, 1180–1204 (2012)
15. J.S. Christiansen, B. Simon, M. Zinchenko, Finite gap Jacobi matrices. I. The isospectral torus. Constr. Approx. **32**, 1–65 (2010)
16. J.S. Christiansen, B. Simon, M. Zinchenko, Finite gap Jacobi matrices, II. The Szegö class. Constr. Approx. **33**, 365–403 (2011)
17. J.S. Christiansen, B. Simon, M. Zinchenko, Finite gap Jacobi matrices: a review, in *Proceedings of Symposia in Pure Mathematics* (American Mathematical Society, Providence, RI, 2013), pp. 87–103
18. J.S. Christiansen, B. Simon, M. Zinchenko, Asymptotics of Chebyshev polynomials, I. Subsets of *R* (2015). http://arxiv.org/pdf/1505.02604v1.pdf
19. J.S. Geronimo, W. Van Assche, Orthogonal polynomials on several intervals via a polynomial mapping. Trans. Am. Math. Soc. **308**, 559–581 (1988)
20. A. Goncharov, Weakly equilibrium Cantor type sets. Potential Anal. **40**, 143–161 (2014)
21. A. Goncharov, B. Hatinoğlu, Widom factors. Potential Anal. **42**, 671–680 (2015)
22. S.M. Heilman, P. Owrutsky, R. Strichartz, Orthogonal polynomials with respect to self-similar measures. Exp. Math. **20**, 238–259 (2011)
23. H. Krüger, B. Simon, Cantor polynomials and some related classes of OPRL. J. Approx. Theory **191**, 71–93 (2015)
24. N.G. Makarov, A.L. Volberg, On the harmonic measure of discontinuous fractals. LOMI Preprint, E-6-86, Steklov Mathematical Institute, Leningrad (1986)
25. G. Mantica, A stable Stieltjes technique for computing orthogonal polynomials and Jacobi matrices associated with a class of singular measures. Constr. Approx. **12**, 509–530 (1996)
26. C. Martínez, The spectrum of periodic Jacobi matrices with slowly oscillating diagonal terms. Proc. Edinb. Math. Soc. **51**, 751–763 (2008)
27. J. Milnor, *Dynamics in One Complex Variables*. Annals of Mathematics Studies, vol. 160 (Princeton University Press, Princeton, NJ, 2006)
28. F. Peherstorfer, Orthogonal and extremal polynomials on several intervals. J. Comput. Appl. Math. **48**, 187–205 (1993)
29. F. Peherstorfer, P. Yuditskii, Asymptotic behavior of polynomials orthonormal on a homogeneous set. J. Anal. Math. **89**, 113–154 (2003)
30. T. Ransford, *Potential Theory in the Complex Plane* (Cambridge University Press, Cambridge, 1995)

31. K. Schiefermayr, A lower bound for the minimum deviation of the Chebyshev polynomial on a compact real set. East J. Approx. **14**, 223–233 (2008)
32. B. Simon, Equilibrium measures and capacities in spectral theory. Inverse Probl. Imaging **1**, 713–772 (2007)
33. B. Simon, *Szegő's Theorem and Its Descendants: Spectral Theory for L^2 Perturbations of Orthogonal Polynomials* (Princeton University Press, Princeton, NJ, 2011)
34. H. Stahl, V. Totik, *General Orthogonal Polynomials*. Encyclopedia of Mathematics, vol. 43 (Cambridge University Press, New York, 1992)
35. G. Szegő, *Orthogonal Polynomials*, 4th edn. (American Mathematical Society, Providence, RI, 1975). American Mathematical Society, Colloquium Publications, vol. XXIII
36. J.P. Thiran, C. Detaille, Chebyshev polynomials on circular arcs and in the complex plane, in *Progress in Approximation Theory* (Academic, Boston, MA, 1991), pp. 771–786
37. V. Totik, Orthogonal polynomials. Surv. Approx. Theory **1**, 70–125 (2005)
38. V. Totik, T. Varga, Chebyshev and fast decreasing polynomials. Proc. Lond. Math. Soc. (2015). doi:10.1112/plms/pdv014
39. V. Totik, P. Yuditskii, On a conjecture of Widom. J. Approx. Theory **190**, 50–61 (2015)
40. W. Van Assche, *Asymptotics for Orthogonal Polynomials*. Lecture Notes in Mathematics, vol. 1265 (Springer, Berlin, 1987)
41. H. Widom, Polynomials associated with measures in the complex plane. J. Math. Mech. **16**, 997–1013 (1967)
42. H. Widom, Extremal polynomials associated with a system of curves in the complex plane. Adv. Math. **3**, 127–232 (1969)

Chapter 8
Some Differential Subordinations Using Ruscheweyh Derivative and a Multiplier Transformation

Alina Alb Lupaş

Abstract In this paper the author derives several interesting differential subordination results. These subordinations are established by means of a differential operator obtained using Ruscheweyh derivative $R^m f(z)$ and the multiplier transformations $I(m, \lambda, l) f(z)$, namely $RI^{\alpha}_{m,\lambda,l}$ the operator given by

$$RI^{\alpha}_{m,\lambda,l} : \mathscr{A} \to \mathscr{A},$$
$$RI^{\alpha}_{m,\lambda,l} f(z) = (1-\alpha) R^m f(z) + \alpha I(m, \lambda, l) f(z),$$

for $z \in U$, $m \in \mathbb{N}$, $\lambda, l \geq 0$ and

$$\mathscr{A}_n = \{f \in \mathscr{H}(U) : f(z) = z + a_{n+1} z^{n+1} + \cdots, \ z \in U\},$$

with $\mathscr{A}_1 = \mathscr{A}$. A number of interesting consequences of some of these subordination results are discussed. Relevant connections of some of the new results obtained in this paper with those in earlier works are also provided.

8.1 Introduction

Denote by U the unit disc of the complex plane,

$$U = \{z \in \mathbb{C} : |z| < 1\}$$

and $\mathscr{H}(U)$ the space of holomorphic functions in U.

Let

$$\mathscr{A}_n = \{f \in \mathscr{H}(U) : f(z) = z + a_{n+1} z^{n+1} + \cdots, \ z \in U\}$$

A. Alb Lupaş (✉)
Department of Mathematics and Computer Science, University of Oradea,
1 Universitatii Street, Romania
e-mail: dalb@uoradea.ro; alblupas@gmail.com

G.A. Anastassiou, O. Duman (eds.), *Computational Analysis*, Springer Proceedings in Mathematics & Statistics 155, DOI 10.1007/978-3-319-28443-9_8

with $\mathscr{A}_1 = \mathscr{A}$; and let

$$\mathscr{H}[a,n] = \{f \in \mathscr{H}(U) : f(z) = a + a_n z^n + a_{n+1} z^{n+1} + \cdots,\ z \in U\}$$

for $a \in \mathbb{C}$ and $n \in \mathbb{N}$.

Denote by

$$K = \left\{ f \in \mathscr{A} : \operatorname{Re} \frac{zf''(z)}{f'(z)} + 1 > 0,\ z \in U \right\},$$

the class of normalized convex functions in U.

If f and g are analytic functions in U, we say that f is subordinate to g, written $f \prec g$, if there is a function w analytic in U, with $w(0) = 0$, $|w(z)| < 1$, for all $z \in U$, such that $f(z) = g(w(z))$ for all $z \in U$. If g is univalent, then $f \prec g$ if and only if $f(0) = g(0)$ and $f(U) \subseteq g(U)$.

Let $\psi : \mathbb{C}^3 \times U \to \mathbb{C}$ and h an univalent function in U. If p is analytic in U and satisfies the (second-order) differential subordination

$$\psi(p(z), zp'(z), z^2 p''(z); z) \prec h(z), \quad z \in U, \tag{8.1}$$

then p is called a solution of the differential subordination. The univalent function q is called a dominant of the solutions of the differential subordination, or more simply a dominant, if $p \prec q$ for all p satisfying (8.1).

A dominant $\tilde{q}$ that satisfies $\tilde{q} \prec q$ for all dominants q of (8.1) is said to be the best dominant of (8.1). The best dominant is unique up to a rotation of U.

Definition 8.1 (See Ruscheweyh [13]). For $f \in \mathscr{A}$, $m \in \mathbb{N}$, the operator R^m is defined by $R^m : \mathscr{A} \to \mathscr{A}$,

$$R^0 f(z) = f(z),\ R^1 f(z) = zf'(z),\ \ldots$$
$$(m+1) R^{m+1} f(z) = z\left(R^m f(z)\right)' + mR^m f(z),\ z \in U.$$

Remark 8.2. If $f \in \mathscr{A}$, $f(z) = z + \sum_{j=2}^{\infty} a_j z^j$, then

$$R^m f(z) = z + \sum_{j=2}^{\infty} \frac{(n+j-1)!}{n!\,(j-1)!} a_j z^j, \quad z \in U.$$

Definition 8.3 (See [4]). For $f \in \mathscr{A}$, $m \in \mathbb{N}$, $\lambda, l \geq 0$, the operator $I(m, \lambda, l) f(z)$ is defined by the following infinite series

$$I(m, \lambda, l) f(z) = z + \sum_{j=2}^{\infty} \left(\frac{\lambda(j-1) + l + 1}{l+1} \right)^m a_j z^j.$$

Remark 8.4. It follows from the above definition that

$$I(0,\lambda,l)f(z) = f(z),$$
$$(l+1)I(m+1,\lambda,l)f(z) = (l+1-\lambda)I(m,\lambda,l)f(z) + \lambda z(I(m,\lambda,l)f(z))'$$
$$(z \in U).$$

Remark 8.5. For $l = 0$, $\lambda \geq 0$, the operator $D_\lambda^m = I(m,\lambda,0)$ was introduced and studied by Al-Oboudi [10], which is reduced to the Sălăgean differential operator [14] for $\lambda = 1$.

Definition 8.6 (See [3]). Let $\alpha, \lambda, l \geq 0$, $m \in \mathbb{N}$. Denote by $RI_{m,\lambda,l}^{\alpha}$ the operator given by $RI_{m,\lambda,l}^{\alpha} : \mathscr{A} \to \mathscr{A}$,

$$RI_{m,\lambda,l}^{\alpha} f(z) = (1-\alpha)R^m f(z) + \alpha I(m,\lambda,l)f(z), \quad z \in U.$$

Remark 8.7. If $f \in \mathscr{A}, f(z) = z + \sum_{j=2}^{\infty} a_j z^j$, then

$$RI_{m,\lambda,l}^{\alpha} f(z) = z + \sum_{j=2}^{\infty} \left\{ \begin{array}{c} \alpha \left(\frac{1+\lambda(j-1)+l}{l+1}\right)^m \\ +(1-\alpha)\frac{(n+j-1)!}{n!(j-1)!} \end{array} \right\} a_j z^j$$

for $z \in U$.

Remark 8.8. For $\alpha = 0$, $RI_{m,\lambda,l}^{0} f(z) = R^m f(z)$, where $z \in U$ and for $\alpha = 1$, $RI_{m,\lambda,l}^{1} f(z) = I(m,\lambda,l)f(z)$, where $z \in U$, which was studied in [2, 9].

For $l = 0$, we obtain $RI_{m,\lambda,0}^{\alpha} f(z) = RD_{\lambda,\alpha}^{m} f(z)$ which was studied in [5–7, 11] and for $l = 0$ and $\lambda = 1$, we obtain $RI_{m,1,0}^{\alpha} f(z) = L_\alpha^m f(z)$ which was studied in [1, 8].

For $n = 0$,

$$\begin{aligned} RI_{0,\lambda,l}^{\alpha} f(z) &= (1-\alpha)R^0 f(z) + \alpha I(0,\lambda,l)f(z) \\ &= f(z) = R^0 f(z) = I(0,\lambda,l)f(z), \end{aligned}$$

where $z \in U$.

Lemma 8.9 (See Miller and Mocanu [12]). *Let g be a convex function in U and let $h(z) = g(z) + n\alpha z g'(z)$, for $z \in U$, where $\alpha > 0$ and n is a positive integer. If*

$$p(z) = g(0) + p_n z^n + p_{n+1} z^{n+1} + \cdots, \quad z \in U,$$

is holomorphic in U and

$$p(z) + \alpha z p'(z) \prec h(z), \qquad z \in U,$$

then $p(z) \prec g(z)$, $z \in U$, and this result is sharp.

Lemma 8.10 (See Hallenbeck and Ruscheweyh [12, Theorem 3.1.6, p. 71]). *Let h be a convex function with $h(0) = a$, and let $\gamma \in \mathbb{C}\backslash\{0\}$ be a complex number with $Re\,\gamma \geq 0$. If $p \in \mathscr{H}[a, n]$ and*

$$p(z) + \frac{1}{\gamma} zp'(z) \prec h(z), \qquad z \in U,$$

then

$$p(z) \prec g(z) \prec h(z), \; z \in U,$$

where

$$g(z) = \frac{\gamma}{nz^{\gamma/n}} \int_0^z h(t) t^{\gamma/n-1} \mathrm{d}t, \;\; z \in U.$$

8.2 Main Results

Theorem 8.11. *Let g be a convex function, $g(0) = 1$ and let h be the function*

$$h(z) = g(z) + \frac{z}{\delta} g'(z), \;\; z \in U.$$

If $\alpha, \lambda, l, \delta \geq 0$, $m \in \mathbb{N}$, $f \in \mathscr{A}$ and satisfies the differential subordination

$$\left(\frac{RI^{\alpha}_{m,\lambda,l} f(z)}{z} \right)^{\delta-1} \left(RI^{\alpha}_{m,\lambda,l} f(z) \right)' \prec h(z), \;\; z \in U, \tag{8.2}$$

then

$$\left(\frac{RI^{\alpha}_{m,\lambda,l} f(z)}{z} \right)^{\delta} \prec g(z), \;\; z \in U,$$

and this result is sharp.

Proof. Consider

$$\begin{aligned} p(z) &= \left(\frac{RI^{\alpha}_{m,\lambda,l} f(z)}{z} \right)^{\delta} \\ &= \left(\frac{z + \sum_{j=2}^{\infty} \left\{ \alpha \left(\frac{1+\lambda(j-1)+l}{l+1} \right)^m + (1-\alpha) \frac{(n+j-1)!}{n!(j-1)!} \right\} a_j z^j}{z} \right)^{\delta} \\ &= 1 + p_1 z + p_2 z^2 + \cdots, \;\; z \in U. \end{aligned}$$

We deduce that $p \in \mathscr{H}[1,1]$. Differentiating we obtain

$$\left(\frac{RI^{\alpha}_{m,\lambda,l}f(z)}{z}\right)^{\delta-1}\left(RI^{\alpha}_{m,\lambda,l}f(z)\right)' = p(z) + \frac{1}{\delta}zp'(z), \quad z \in U.$$

Then (8.2) becomes

$$p(z) + \frac{1}{\delta}zp'(z) \prec h(z) = g(z) + \frac{z}{\delta}g'(z), \; z \in U.$$

By using Lemma 8.9, we have

$$p(z) \prec g(z), \; z \in U, \text{ i.e., } \left(\frac{RI^{\alpha}_{m,\lambda,l}f(z)}{z}\right)^{\delta} \prec g(z), \; z \in U.$$

Theorem 8.12. *Let h be a holomorphic function which satisfies the inequality*

$$Re\left(1 + \frac{zh''(z)}{h'(z)}\right) > -\frac{1}{2}, \quad z \in U,$$

and $h(0) = 1$. If $\alpha, \lambda, l, \delta \geq 0$, $m \in \mathbb{N}$, $f \in \mathscr{A}$ and satisfies the differential subordination

$$\left(\frac{RI^{\alpha}_{m,\lambda,l}f(z)}{z}\right)^{\delta-1}\left(RI^{\alpha}_{m,\lambda,l}f(z)\right)' \prec h(z), \quad z \in U, \tag{8.3}$$

then

$$\left(\frac{RI^{\alpha}_{m,\lambda,l}f(z)}{z}\right)^{\delta} \prec q(z), \quad z \in U,$$

where $q(z) = \frac{\delta}{z^{\delta}}\int_0^z h(t)t^{\delta-1}\mathrm{d}t.$

Proof. Let

$$\begin{aligned}
p(z) &= \left(\frac{RI^{\alpha}_{m,\lambda,l}f(z)}{z}\right)^{\delta}\\
&= \left(\frac{z + \sum_{j=2}^{\infty}\left\{\alpha\left(\frac{1+\lambda(j-1)+l}{l+1}\right)^{m} + (1-\alpha)\frac{(n+j-1)!}{n!(j-1)!}\right\}a_j z^j}{z}\right)^{\delta}\\
&= \left(1 + \sum_{j=2}^{\infty}\left\{\alpha\left(\frac{1+\lambda(j-1)+l}{l+1}\right)^{m} + (1-\alpha)\frac{(n+j-1)!}{n!\,(j-1)!}\right\}a_j z^{j-1}\right)^{\delta}\\
&= 1 + \sum_{j=2}^{\infty} p_j z^{j-1}
\end{aligned}$$

for $z \in U$, $p \in \mathscr{H}[1,1]$. Differentiating, we obtain

$$\left(\frac{RI^{\alpha}_{m,\lambda,l}f(z)}{z}\right)^{\delta-1}\left(RI^{\alpha}_{m,\lambda,l}f(z)\right)' = p(z) + \frac{1}{\delta}zp'(z), \ \ z \in U,$$

and (8.3) becomes

$$p(z) + \frac{1}{\delta}zp'(z) \prec h(z), \ \ z \in U.$$

Using Lemma 8.10, we have

$$p(z) \prec q(z), \quad z \in U, \text{ i.e.}$$

i.e.,

$$\left(\frac{RI^{\alpha}_{m,\lambda,l}f(z)}{z}\right)^{\delta} \prec q(z) = \frac{\delta}{z^{\delta}}\int_0^z h(t)t^{\delta-1}\mathrm{d}t, \ z \in U,$$

and q is the best dominant.

Corollary 8.13. *Let*

$$h(z) = \frac{1+(2\beta-1)z}{1+z}$$

be a convex function in U, where $0 \leq \beta < 1$. If $\alpha, \delta, l, \lambda \geq 0$, $m \in \mathbb{N}$, $f \in \mathscr{A}$ and satisfies the differential subordination

$$\left(\frac{RI^{\alpha}_{m,\lambda,l}f(z)}{z}\right)^{\delta-1}\left(RI^{\alpha}_{m,\lambda,l}f(z)\right)' \prec h(z), \ \ z \in U, \tag{8.4}$$

then

$$\left(\frac{RI^{\alpha}_{m,\lambda,l}f(z)}{z}\right)^{\delta} \prec q(z), \ \ z \in U,$$

where q is given by

$$q(z) = (2\beta-1) + \frac{2(1-\beta)\delta}{z^{\delta}}\int_0^z \frac{t^{\delta-1}}{1+t}\mathrm{d}t, \ z \in U.$$

The function q is convex and it is the best dominant.

Proof. Following the same steps as in the proof of Theorem 8.12 and considering

$$p(z) = \left(\frac{RI^{\alpha}_{m,\lambda,l}f(z)}{z}\right)^{\delta},$$

the differential subordination (8.4) becomes

$$p(z) + \frac{z}{\delta}p'(z) \prec h(z) = \frac{1 + (2\beta - 1)z}{1 + z}, \quad z \in U.$$

By using Lemma 8.10 for $\gamma = \delta$, we have $p(z) \prec q(z)$, i.e.

$$\left(\frac{RI^{\alpha}_{m,\lambda,l}f(z)}{z}\right)^{\delta} \prec q(z) = \frac{\delta}{z^{\delta}}\int_0^z h(t)t^{\delta-1}\mathrm{d}t$$

$$= \frac{\delta}{z^{\delta}}\int_0^z t^{\delta-1}\frac{1 + (2\beta - 1)t}{1 + t}\mathrm{d}t$$

$$= \frac{\delta}{z^{\delta}}\int_0^z \left[(2\beta - 1)\,t^{\delta-1} + 2\,(1 - \beta)\,\frac{t^{\delta-1}}{1 + t}\right]\mathrm{d}t$$

$$= (2\beta - 1) + \frac{2\,(1 - \beta)\,\delta}{z^{\delta}}\int_0^z \frac{t^{\delta-1}}{1 + t}\mathrm{d}t,$$

for $z \in U$.

Theorem 8.14. *Let g be a convex function such that $g(0) = 1$ and let h be the function*

$$h(z) = g(z) + \frac{z}{\delta}g'(z), \quad z \in U.$$

If $\alpha, \lambda, l, \delta \geq 0$, $m \in \mathbb{N}$, $f \in \mathcal{A}$ and the differential subordination

$$z\frac{\delta + 1}{\delta}\frac{RI^{\alpha}_{m,\lambda,l}f(z)}{\left(RI^{\alpha}_{m+1,\lambda,l}f(z)\right)^2} + \frac{z^2}{\delta}\frac{RI^{\alpha}_{m,\lambda,l}f(z)}{\left(RI^{\alpha}_{m+1,\lambda,l}f(z)\right)^2}\left[\frac{\left(RI^{\alpha}_{m,\lambda,l}f(z)\right)'}{RI^{\alpha}_{m,\lambda,l}f(z)} - 2\frac{\left(RI^{\alpha}_{m+1,\lambda,l}f(z)\right)'}{RI^{\alpha}_{m+1,\lambda,l}f(z)}\right] \prec h(z), \tag{8.5}$$

$z \in U$, holds, then

$$z\frac{RI^{\alpha}_{m,\lambda,l}f(z)}{\left(RI^{\alpha}_{m+1,\lambda,l}f(z)\right)^2} \prec g(z), \; z \in U,$$

and this result is sharp.

Proof. Consider

$$p(z) = z\frac{RI^{\alpha}_{m,\lambda,l}f(z)}{\left(RI^{\alpha}_{m+1,\lambda,l}f(z)\right)^2}$$

and we obtain

$$p(z) + \frac{z}{\delta}p'(z) = z\frac{\delta+1}{\delta}\frac{RI^{\alpha}_{m,\lambda,l}f(z)}{\left(RI^{\alpha}_{m+1,\lambda,l}f(z)\right)^2} + \frac{z^2}{\delta}\frac{RI^{\alpha}_{m,\lambda,l}f(z)}{\left(RI^{\alpha}_{m+1,\lambda,l}f(z)\right)^2}$$
$$\times\left[\frac{\left(RI^{\alpha}_{m,\lambda,l}f(z)\right)'}{RI^{\alpha}_{m,\lambda,l}f(z)} - 2\frac{\left(RI^{\alpha}_{m+1,\lambda,l}f(z)\right)'}{RI^{\alpha}_{m+1,\lambda,l}f(z)}\right].$$

Relation (8.5) becomes

$$p(z) + \frac{z}{\delta}p'(z) \prec h(z) = g(z) + \frac{z}{\delta}g'(z), \; z \in U.$$

By using Lemma 8.9, we have

$$p(z) \prec g(z), \quad z \in U,$$

i.e.,

$$z\frac{RI^{\alpha}_{m,\lambda,l}f(z)}{\left(RI^{\alpha}_{m+1,\lambda,l}f(z)\right)^2} \prec g(z), \quad z \in U.$$

Theorem 8.15. *Let h be a holomorphic function which satisfies the inequality*

$$Re\left(1 + \frac{zh''(z)}{h'(z)}\right) > -\frac{1}{2}, \quad z \in U,$$

and $h(0) = 1$. If $\alpha, \lambda, l, \delta \geq 0$, $m \in \mathbb{N}$, $f \in \mathcal{A}$ and satisfies the differential subordination

$$z\frac{\delta+1}{\delta}\frac{RI^{\alpha}_{m,\lambda,l}f(z)}{\left(RI^{\alpha}_{m+1,\lambda,l}f(z)\right)^{2}}$$
$$+\frac{z^{2}}{\delta}\frac{RI^{\alpha}_{m,\lambda,l}f(z)}{\left(RI^{\alpha}_{m+1,\lambda,l}f(z)\right)^{2}}\left[\frac{\left(RI^{\alpha}_{m,\lambda,l}f(z)\right)'}{RI^{\alpha}_{m,\lambda,l}f(z)}-2\frac{\left(RI^{\alpha}_{m+1,\lambda,l}f(z)\right)'}{RI^{\alpha}_{m+1,\lambda,l}f(z)}\right]$$
$$\prec h(z), \tag{8.6}$$

$z\in U$, then

$$z\frac{RI^{\alpha}_{m,\lambda,l}f(z)}{\left(RI^{\alpha}_{m+1,\lambda,l}f(z)\right)^{2}}\prec q(z),\quad z\in U,$$

where $q(z)=\frac{\delta}{z^{\delta}}\int\limits_{0}^{z}h(t)t^{\delta-1}\mathrm{d}t$.

Proof. Let

$$p(z)=z\frac{RI^{\alpha}_{m,\lambda,l}f(z)}{\left(RI^{\alpha}_{m+1,\lambda,l}f(z)\right)^{2}},\quad z\in U,\ p\in\mathscr{H}[1,1].$$

Differentiating, we obtain

$$p(z)+\frac{z}{\delta}p'(z)=z\frac{\delta+1}{\delta}\frac{RI^{\alpha}_{m,\lambda,l}f(z)}{\left(RI^{\alpha}_{m+1,\lambda,l}f(z)\right)^{2}}+\frac{z^{2}}{\delta}\frac{RI^{\alpha}_{m,\lambda,l}f(z)}{\left(RI^{\alpha}_{m+1,\lambda,l}f(z)\right)^{2}}$$
$$\times\left[\frac{\left(RI^{\alpha}_{m,\lambda,l}f(z)\right)'}{RI^{\alpha}_{m,\lambda,l}f(z)}-2\frac{\left(RI^{\alpha}_{m+1,\lambda,l}f(z)\right)'}{RI^{\alpha}_{m+1,\lambda,l}f(z)}\right],$$

$z\in U$, and (8.6) becomes

$$p(z)+\frac{z}{\delta}p'(z)\prec h(z),\quad z\in U.$$

Using Lemma 8.10, we have

$$p(z)\prec q(z),\quad z\in U,$$

i.e.,

$$z\frac{RI^{\alpha}_{m,\lambda,l}f(z)}{\left(RI^{\alpha}_{m+1,\lambda,l}f(z)\right)^{2}}\prec q(z)=\frac{\delta}{z^{\delta}}\int\limits_{0}^{z}h(t)t^{\delta-1}\mathrm{d}t,\ z\in U,$$

and q is the best dominant.

Theorem 8.16. *Let g be a convex function such that $g(0) = 0$ and let h be the function*

$$h(z) = g(z) + \frac{z}{\delta} g'(z), \quad z \in U.$$

If $\alpha, \lambda, l, \delta \geq 0$, $m \in \mathbb{N}$, $f \in \mathscr{A}$ and the differential subordination

$$z^2 \frac{\delta+2}{\delta} \frac{\left(RI_{m,\lambda,l}^{\alpha} f(z)\right)'}{RI_{m,\lambda,l}^{\alpha} f(z)} + \frac{z^3}{\delta} \left[\frac{\left(RI_{m,\lambda,l}^{\alpha} f(z)\right)''}{RI_{m,\lambda,l}^{\alpha} f(z)} - \left(\frac{\left(RI_{m,\lambda,l}^{\alpha} f(z)\right)'}{RI_{m,\lambda,l}^{\alpha} f(z)} \right)^2 \right] \prec h(z), \; z \in U, \tag{8.7}$$

holds, then

$$z^2 \frac{\left(RI_{m,\lambda,l}^{\alpha} f(z)\right)'}{RI_{m,\lambda,l}^{\alpha} f(z)} \prec g(z), \quad z \in U.$$

This result is sharp.

Proof. Let

$$p(z) = z^2 \frac{\left(RI_{m,\lambda,l}^{\alpha} f(z)\right)'}{RI_{m,\lambda,l}^{\alpha} f(z)}.$$

We deduce that $p \in \mathscr{H}[0, 1]$. Differentiating, we obtain

$$p(z) + \frac{z}{\delta} p'(z) = z^2 \frac{\delta+2}{\delta} \frac{\left(RI_{m,\lambda,l}^{\alpha} f(z)\right)'}{RI_{m,\lambda,l}^{\alpha} f(z)} + \frac{z^3}{\delta} \left[\frac{\left(RI_{m,\lambda,l}^{\alpha} f(z)\right)''}{RI_{m,\lambda,l}^{\alpha} f(z)} - \left(\frac{\left(RI_{m,\lambda,l}^{\alpha} f(z)\right)'}{RI_{m,\lambda,l}^{\alpha} f(z)} \right)^2 \right],$$

$z \in U$. Using the notation in (8.7), the differential subordination becomes

$$p(z) + \frac{1}{\delta} z p'(z) \prec h(z) = g(z) + \frac{z}{\delta} g'(z).$$

By using Lemma 8.9, we have

$$p(z) \prec g(z), \quad z \in U,$$

i.e.,

$$z^2\frac{\left(RI^{\alpha}_{m,\lambda,l}f(z)\right)'}{RI^{\alpha}_{m,\lambda,l}f(z)} \prec g(z),\quad z\in U,$$

and this result is sharp.

Theorem 8.17. *Let h be a holomorphic function which satisfies the inequality*

$$Re\left(1+\frac{zh''(z)}{h'(z)}\right) > -\frac{1}{2},\quad z\in U,$$

and $h(0)=0$. *If* $\alpha,\lambda,l,\delta\geq 0$, $m\in\mathbb{N}$, $f\in\mathscr{A}$ *and satisfies the differential subordination*

$$\begin{aligned} & z^2\frac{\delta+2}{\delta}\frac{\left(RI^{\alpha}_{m,\lambda,l}f(z)\right)'}{RI^{\alpha}_{m,\lambda,l}f(z)} \\ & +\frac{z^3}{\delta}\left[\frac{\left(RI^{\alpha}_{m,\lambda,l}f(z)\right)''}{RI^{\alpha}_{m,\lambda,l}f(z)}-\left(\frac{\left(RI^{\alpha}_{m,\lambda,l}f(z)\right)'}{RI^{\alpha}_{m,\lambda,l}f(z)}\right)^2\right] \\ & \prec h(z),\quad z\in U, \end{aligned} \tag{8.8}$$

then

$$z^2\frac{\left(RI^{\alpha}_{m,\lambda,l}f(z)\right)'}{RI^{\alpha}_{m,\lambda,l}f(z)} \prec q(z),\quad z\in U,$$

where $q(z)=\frac{\delta}{z^{\delta}}\int_0^z h(t)t^{\delta-1}\mathrm{d}t$.

Proof. Let

$$p(z)=z^2\frac{\left(RI^{\alpha}_{m,\lambda,l}f(z)\right)'}{RI^{\alpha}_{m,\lambda,l}f(z)},\quad z\in U,\ p\in\mathscr{H}[0,1].$$

Differentiating, we obtain

$$\begin{aligned} p(z)+\frac{z}{\delta}p'(z) &= z^2\frac{\delta+2}{\delta}\frac{\left(RI^{\alpha}_{m,\lambda,l}f(z)\right)'}{RI^{\alpha}_{m,\lambda,l}f(z)} \\ &+\frac{z^3}{\delta}\left[\frac{\left(RI^{\alpha}_{m,\lambda,l}f(z)\right)''}{RI^{\alpha}_{m,\lambda,l}f(z)}-\left(\frac{\left(RI^{\alpha}_{m,\lambda,l}f(z)\right)'}{RI^{\alpha}_{m,\lambda,l}f(z)}\right)^2\right], \end{aligned}$$

$z \in U$, and (8.8) becomes

$$p(z) + \frac{1}{\delta} z p'(z) \prec h(z), \ z \in U.$$

Using Lemma 8.10, we have

$$p(z) \prec q(z), \ \ z \in U,$$

i.e.,

$$z^2 \frac{\left(RI^{\alpha}_{m,\lambda,l} f(z) \right)'}{RI^{\alpha}_{m,\lambda,l} f(z)} \prec q(z) = \frac{\delta}{z^{\delta}} \int_0^z h(t) t^{\delta - 1} \mathrm{d}t, \ z \in U,$$

and q is the best dominant.

Theorem 8.18. *Let g be a convex function such that $g(0) = 1$ and let h be the function*

$$h(z) = g(z) + z g'(z), \quad z \in U.$$

If $\alpha, \lambda, l \geq 0$, $m \in \mathbb{N}$, $f \in \mathscr{A}$ and the differential subordination

$$1 - \frac{RI^{\alpha}_{m,\lambda,l} f(z) \cdot \left(RI^{\alpha}_{m,\lambda,l} f(z) \right)''}{\left[\left(RI^{\alpha}_{m,\lambda,l} f(z) \right)' \right]^2} \prec h(z), \ \ z \in U \tag{8.9}$$

holds, then

$$\frac{RI^{\alpha}_{m,\lambda,l} f(z)}{z \left(RI^{\alpha}_{m,\lambda,l} f(z) \right)'} \prec g(z), \ \ z \in U.$$

This result is sharp.

Proof. Let

$$p(z) = \frac{RI^{\alpha}_{m,\lambda,l} f(z)}{z \left(RI^{\alpha}_{m,\lambda,l} f(z) \right)'}.$$

We deduce that $p \in \mathscr{H}[1, 1]$. Differentiating, we obtain

$$1 - \frac{RI^{\alpha}_{m,\lambda,l} f(z) \cdot \left(RI^{\alpha}_{m,\lambda,l} f(z) \right)''}{\left[\left(RI^{\alpha}_{m,\lambda,l} f(z) \right)' \right]^2} = p(z) + z p'(z), \ \ z \in U.$$

Using the notation in (8.9), the differential subordination becomes

$$p(z) + zp'(z) \prec h(z) = g(z) + zg'(z).$$

By using Lemma 8.9, we have

$$p(z) \prec g(z), \quad z \in U,$$

i.e.,

$$\frac{RI^{\alpha}_{m,\lambda,l}f(z)}{z\left(RI^{\alpha}_{m,\lambda,l}f(z)\right)'} \prec g(z), \quad z \in U,$$

and this result is sharp.

Theorem 8.19. *Let h be a holomorphic function which satisfies the inequality*

$$Re\left(1 + \frac{zh''(z)}{h'(z)}\right) > -\frac{1}{2}, \quad z \in U,$$

and $h(0) = 1$. *If* $\alpha, \lambda, l \geq 0$, $m \in \mathbb{N}$, $f \in \mathcal{A}$ *and satisfies the differential subordination*

$$1 - \frac{RI^{\alpha}_{m,\lambda,l}f(z) \cdot \left(RI^{\alpha}_{m,\lambda,l}f(z)\right)''}{\left[\left(RI^{\alpha}_{m,\lambda,l}f(z)\right)'\right]^2} \prec h(z), \quad z \in U, \tag{8.10}$$

then

$$\frac{RI^{\alpha}_{m,\lambda,l}f(z)}{z\left(RI^{\alpha}_{m,\lambda,l}f(z)\right)'} \prec q(z), \quad z \in U,$$

where $q(z) = \frac{1}{z}\int_0^z h(t)\mathrm{d}t$.

Proof. Let

$$p(z) = \frac{RI^{\alpha}_{m,\lambda,l}f(z)}{z\left(RI^{\alpha}_{m,\lambda,l}f(z)\right)'}, \quad z \in U, \ p \in \mathcal{H}[1,1].$$

Differentiating, we obtain

$$1 - \frac{RI^{\alpha}_{m,\lambda,l}f(z) \cdot \left(RI^{\alpha}_{m,\lambda,l}f(z)\right)''}{\left[\left(RI^{\alpha}_{m,\lambda,l}f(z)\right)'\right]^2} = p(z) + zp'(z), \quad z \in U,$$

and (8.10) becomes

$$p(z) + zp'(z) \prec h(z),\ z \in U.$$

Using Lemma 8.10, we have

$$p(z) \prec q(z), \quad z \in U,$$

i.e.,

$$\frac{RI^{\alpha}_{m,\lambda,l}f(z)}{z\left(RI^{\alpha}_{m,\lambda,l}f(z)\right)'} \prec q(z) = \frac{1}{z}\int_0^z h(t)\mathrm{d}t,\ z \in U,$$

and q is the best dominant.

Corollary 8.20. *Let*

$$h(z) = \frac{1+(2\beta-1)z}{1+z}$$

be a convex function in U, where $0 \le \beta < 1$. If $\alpha, \lambda, l \ge 0$, $m \in \mathbb{N}$, $f \in \mathscr{A}$ and satisfies the differential subordination

$$1 - \frac{RI^{\alpha}_{m,\lambda,l}f(z) \cdot \left(RI^{\alpha}_{m,\lambda,l}f(z)\right)''}{\left[\left(RI^{\alpha}_{m,\lambda,l}f(z)\right)'\right]^2} \prec h(z), \quad z \in U, \tag{8.11}$$

then

$$\frac{RI^{\alpha}_{m,\lambda,l}f(z)}{z\left(RI^{\alpha}_{m,\lambda,l}f(z)\right)'} \prec q(z),\ \ z \in U,$$

where q is given by

$$q(z) = (2\beta-1) + 2(1-\beta)\frac{\ln(1+z)}{z},\ \ z \in U.$$

The function q is convex and it is the best dominant.

Proof. Following the same steps as in the proof of Theorem 8.19 and considering

$$p(z) = \frac{RI^{\alpha}_{m,\lambda,l}f(z)}{z\left(RI^{\alpha}_{m,\lambda,l}f(z)\right)'},$$

the differential subordination (8.11) becomes

$$p(z) + zp'(z) \prec h(z) = \frac{1+(2\beta-1)z}{1+z},\ \ z \in U.$$

By using Lemma 8.10 for $\gamma = 1$, we have $p(z) \prec q(z)$, i.e.

$$\frac{RI^{\alpha}_{m,\lambda,l}f(z)}{z\left(RI^{\alpha}_{m,\lambda,l}f(z)\right)'} \prec q(z) = \frac{1}{z}\int_0^z h(t)\mathrm{d}t$$

$$= \frac{1}{z}\int_0^z \frac{1+(2\beta-1)t}{1+t}\mathrm{d}t$$

$$= \frac{1}{z}\int_0^z \left[(2\beta-1) + \frac{2(1-\beta)}{1+t}\right]\mathrm{d}t$$

$$= (2\beta-1) + 2(1-\beta)\frac{\ln(1+z)}{z}$$

for $z \in U$.

Example 8.21. Let $h(z) = \frac{1-z}{1+z}$ a convex function in U with $h(0) = 1$ and

$$Re\left(\frac{zh''(z)}{h'(z)} + 1\right) > -\frac{1}{2}.$$

Let $f(z) = z + z^2$, $z \in U$. For $n = 1$, $m = 1$, $l = 2$, $\lambda = 1$, $\alpha = \frac{1}{2}$, we obtain

$$RI^{\frac{1}{2}}_{1,1,2}f(z) = \frac{1}{2}R^1 f(z) + \frac{1}{2}I(1,1,2)f(z)$$

$$= \frac{1}{3}f(z) + \frac{2}{3}zf'(z)$$

$$= z + \frac{5}{3}z^2, \quad z \in U.$$

Then

$$\left(RI^{\frac{1}{2}}_{1,1,2}f(z)\right)' = 1 + \frac{10}{3}z$$

and

$$\left(RI^{\frac{1}{2}}_{1,1,2}f(z)\right)'' = \frac{10}{3},$$

$$\frac{RI^{\frac{1}{2}}_{1,1,2}f(z)}{z\left(RI^{\frac{1}{2}}_{1,1,2}f(z)\right)'} = \frac{z + \frac{5}{3}z^2}{z\left(1 + \frac{10}{3}z\right)} = \frac{3+5z}{3+10z},$$

$$1 - \frac{RI^{\frac{1}{2}}_{1,1,2}f(z) \cdot \left(RI^{\frac{1}{2}}_{1,1,2}f(z)\right)''}{\left[\left(RI^{\frac{1}{2}}_{1,1,2}f(z)\right)'\right]^2} = 1 - \frac{\left(z + \frac{5}{3}z^2\right) \cdot \frac{10}{3}}{\left(1 + \frac{10}{3}z\right)^2}$$

$$= \frac{50z^2 + 30z + 9}{(3 + 10z)^2}.$$

We have

$$q(z) = \frac{1}{z}\int_0^z \frac{1-t}{1+t}\mathrm{d}t = -1 + \frac{2\ln(1+z)}{z}.$$

By using Theorem 8.19 we obtain

$$\frac{50z^2 + 30z + 9}{(3 + 10z)^2} \prec \frac{1-z}{1+z}, \quad z \in U,$$

induce

$$\frac{3 + 5z}{3 + 10z} \prec -1 + \frac{2\ln(1+z)}{z}, \; z \in U.$$

Theorem 8.22. *Let g be a convex function such that $g(0) = 1$ and let h be the function*

$$h(z) = g(z) + zg'(z), \quad z \in U.$$

If $\alpha, \lambda, l \geq 0$, $m \in \mathbb{N}$, $f \in \mathscr{A}$ and the differential subordination

$$\left[\left(RI^{\alpha}_{m,\lambda,l}f(z)\right)'\right]^2 + RI^{\alpha}_{m,\lambda,l}f(z) \cdot \left(RI^{\alpha}_{m,\lambda,l}f(z)\right)'' \prec h(z), \quad z \in U \tag{8.12}$$

holds, then

$$\frac{RI^{\alpha}_{m,\lambda,l}f(z) \cdot \left(RI^{\alpha}_{m,\lambda,l}f(z)\right)'}{z} \prec g(z), \quad z \in U.$$

This result is sharp.

Proof. Let

$$p(z) = \frac{RI^{\alpha}_{m,\lambda,l}f(z) \cdot \left(RI^{\alpha}_{m,\lambda,l}f(z)\right)'}{z}.$$

We deduce that $p \in \mathscr{H}[1,1]$. Differentiating, we obtain

$$\left[\left(RI^{\alpha}_{m,\lambda,l}f(z)\right)'\right]^2 + RI^{\alpha}_{m,\lambda,l}f(z) \cdot \left(RI^{\alpha}_{m,\lambda,l}f(z)\right)'' = p(z) + zp'(z), \quad z \in U.$$

Using the notation in (8.12), the differential subordination becomes

$$p(z) + zp'(z) \prec h(z) = g(z) + zg'(z).$$

By using Lemma 8.9, we have

$$p(z) \prec g(z), \quad z \in U,$$

i.e.,

$$\frac{RI^{\alpha}_{m,\lambda,l}f(z) \cdot \left(RI^{\alpha}_{m,\lambda,l}f(z)\right)'}{z} \prec g(z), \;\; z \in U,$$

and this result is sharp.

Theorem 8.23. *Let h be a holomorphic function which satisfies the inequality*

$$Re\left(1 + \frac{zh''(z)}{h'(z)}\right) > -\frac{1}{2}, \quad z \in U,$$

and $h(0) = 1$. *If* $\alpha, \lambda, l \geq 0$, $m \in \mathbb{N}$, $f \in \mathscr{A}$ *and satisfies the differential subordination*

$$\left[\left(RI^{\alpha}_{m,\lambda,l}f(z)\right)'\right]^2 + RI^{\alpha}_{m,\lambda,l}f(z) \cdot \left(RI^{\alpha}_{m,\lambda,l}f(z)\right)'' \prec h(z), \;\; z \in U, \tag{8.13}$$

then

$$\frac{RI^{\alpha}_{m,\lambda,l}f(z) \cdot \left(RI^{\alpha}_{m,\lambda,l}f(z)\right)'}{z} \prec q(z), \;\; z \in U,$$

where $q(z) = \dfrac{1}{z}\int\limits_0^z h(t)\mathrm{d}t$.

Proof. Let

$$p(z) = \frac{RI^{\alpha}_{m,\lambda,l}f(z) \cdot \left(RI^{\alpha}_{m,\lambda,l}f(z)\right)'}{z}, \;\; z \in U, \;\; p \in \mathscr{H}[1,1].$$

Differentiating, we obtain

$$\left[\left(RI^{\alpha}_{m,\lambda,l}f(z)\right)'\right]^2 + RI^{\alpha}_{m,\lambda,l}f(z) \cdot \left(RI^{\alpha}_{m,\lambda,l}f(z)\right)'' = p(z) + zp'(z),$$

$z \in U$, and (8.13) becomes

$$p(z) + zp'(z) \prec h(z), \; z \in U.$$

Using Lemma 8.10, we have

$$p(z) \prec q(z), \quad z \in U,$$

i.e.,

$$\frac{RI^{\alpha}_{m,\lambda,l}f(z) \cdot \left(RI^{\alpha}_{m,\lambda,l}f(z)\right)'}{z} \prec q(z) = \frac{1}{z}\int_0^z h(t)\mathrm{d}t, \ z \in U,$$

and q is the best dominant.

Corollary 8.24. *Let*

$$h(z) = \frac{1 + (2\beta - 1)z}{1 + z}$$

be a convex function in U*, where* $0 \leq \beta < 1$*. If* $\alpha, \lambda, l \geq 0$*,* $m \in \mathbb{N}$*,* $f \in \mathcal{A}$ *and satisfies the differential subordination*

$$\left[\left(RI^{\alpha}_{m,\lambda,l}f(z)\right)'\right]^2 + RI^{\alpha}_{m,\lambda,l}f(z) \cdot \left(RI^{\alpha}_{m,\lambda,l}f(z)\right)'' \prec h(z), \ z \in U, \tag{8.14}$$

then

$$\frac{RI^{\alpha}_{m,\lambda,l}f(z) \cdot \left(RI^{\alpha}_{m,\lambda,l}f(z)\right)'}{z} \prec q(z), \ z \in U,$$

where q *is given by*

$$q(z) = (2\beta - 1) + 2(1 - \beta)\frac{\ln(1 + z)}{z}, \ z \in U.$$

The function q *is convex and it is the best dominant.*

Proof. Following the same steps as in the proof of Theorem 8.23 and considering

$$p(z) = \frac{RI^{\alpha}_{m,\lambda,l}f(z) \cdot \left(RI^{\alpha}_{m,\lambda,l}f(z)\right)'}{z},$$

the differential subordination (8.14) becomes

$$p(z) + zp'(z) \prec h(z) = \frac{1 + (2\beta - 1)z}{1 + z}, \quad z \in U.$$

By using Lemma 8.10 for $\gamma = 1$, we have $p(z) \prec q(z)$, i.e.

$$\frac{RI^{\alpha}_{m,\lambda,l}f(z)\cdot\left(RI^{\alpha}_{m,\lambda,l}f(z)\right)'}{z}\prec q(z)=\frac{1}{z}\int_0^z h(t)\mathrm{d}t$$

$$=\frac{1}{z}\int_0^z\frac{1+(2\beta-1)t}{1+t}\mathrm{d}t$$

$$=\frac{1}{z}\int_0^z\left[(2\beta-1)+\frac{2(1-\beta)}{1+t}\right]\mathrm{d}t$$

$$=(2\beta-1)+2(1-\beta)\frac{\ln(1+z)}{z},\quad z\in U.$$

Example 8.25. Let $h(z)=\dfrac{1-z}{1+z}$ a convex function in U with $h(0)=1$ and

$$Re\left(\frac{zh''(z)}{h'(z)}+1\right)>-\frac{1}{2}.$$

Let $f(z)=z+z^2$, $z\in U$. For $n=1$, $m=1$, $l=2$, $\lambda=1$, $\alpha=\frac{1}{2}$, we obtain

$$RI^{\frac{1}{2}}_{1,1,2}f(z)=\frac{1}{2}R^1f(z)+\frac{1}{2}I(1,1,2)f(z)$$

$$=\frac{1}{3}f(z)+\frac{2}{3}zf'(z)=z+\frac{5}{3}z^2,\quad z\in U.$$

Then

$$\left(RI^{\frac{1}{2}}_{1,1,2}f(z)\right)'=1+\frac{10}{3}z,\ \left(RI^{\frac{1}{2}}_{1,1,2}f(z)\right)''=\frac{10}{3},$$

$$\frac{RI^{\frac{1}{2}}_{1,1,2}f(z)\cdot\left(RI^{\frac{1}{2}}_{1,1,2}f(z)\right)'}{z}$$

$$=\frac{\left(z+\frac{5}{3}z^2\right)\left(1+\frac{10}{3}z\right)}{z}$$

$$=\frac{50}{9}z^2+5z+1,$$

$$\left[\left(RI^{\frac{1}{2}}_{1,1,2}f(z)\right)'\right]^2+RI^{\frac{1}{2}}_{1,1,2}f(z)\cdot\left(RI^{\frac{1}{2}}_{1,1,2}f(z)\right)''$$

$$=\left(1+\frac{10}{3}z\right)^2+\left(z+\frac{5}{3}z^2\right)\cdot\frac{10}{3}$$

$$=\frac{50}{3}z^2+10z+1.$$

We have

$$q(z) = \frac{1}{z}\int_0^z \frac{1-t}{1+t}\mathrm{d}t = -1 + \frac{2\ln(1+z)}{z}.$$

By using Theorem 8.23 we obtain

$$\frac{50}{3}z^2 + 10z + 1 \prec \frac{1-z}{1+z}, \quad z \in U,$$

induce

$$\frac{50}{9}z^2 + 5z + 1 \prec -1 + \frac{2\ln(1+z)}{z}, \quad z \in U.$$

Theorem 8.26. *Let g be a convex function such that $g(0) = 1$ and let h be the function*

$$h(z) = g(z) + \frac{z}{1-\delta}g'(z), \quad z \in U.$$

If $\alpha, \lambda, l \geq 0$, $\delta \in (0,1)$, $m \in \mathbb{N}$, $f \in \mathscr{A}$ and the differential subordination

$$\left(\frac{z}{RI^{\alpha}_{m,\lambda,l}f(z)}\right)^{\delta}\frac{RI^{\alpha}_{m+1,\lambda,l}f(z)}{1-\delta}\left(\frac{\left(RI^{\alpha}_{m+1,\lambda,l}f(z)\right)'}{RI^{\alpha}_{m+1,\lambda,l}f(z)} - \delta\frac{\left(RI^{\alpha}_{m,\lambda,l}f(z)\right)'}{RI^{\alpha}_{m,\lambda,l}f(z)}\right)$$
$$\prec h(z), \quad z \in U \tag{8.15}$$

holds, then

$$\frac{RI^{\alpha}_{m+1,\lambda,l}f(z)}{z}\cdot\left(\frac{z}{RI^{\alpha}_{m,\lambda,l}f(z)}\right)^{\delta} \prec g(z), \quad z \in U.$$

This result is sharp.

Proof. Let

$$p(z) = \frac{RI^{\alpha}_{m+1,\lambda,l}f(z)}{z}\cdot\left(\frac{z}{RI^{\alpha}_{m,\lambda,l}f(z)}\right)^{\delta}.$$

We deduce that $p \in \mathscr{H}[1,1]$. Differentiating, we obtain

$$\left(\frac{z}{RI^{\alpha}_{m,\lambda,l}f(z)}\right)^{\delta}\frac{RI^{\alpha}_{m+1,\lambda,l}f(z)}{1-\delta}\left(\frac{\left(RI^{\alpha}_{m+1,\lambda,l}f(z)\right)'}{RI^{\alpha}_{m+1,\lambda,l}f(z)} - \delta\frac{\left(RI^{\alpha}_{m,\lambda,l}f(z)\right)'}{RI^{\alpha}_{m,\lambda,l}f(z)}\right)$$
$$= p(z) + \frac{1}{1-\delta}zp'(z), \quad z \in U.$$

Using the notation in (8.15), the differential subordination becomes

$$p(z) + \frac{1}{1-\delta} zp'(z) \prec h(z) = g(z) + \frac{z}{1-\delta} g'(z).$$

By using Lemma 8.9, we have

$$p(z) \prec g(z), \quad z \in U,$$

i.e.,

$$\frac{RI^{\alpha}_{m+1,\lambda,l} f(z)}{z} \cdot \left(\frac{z}{RI^{\alpha}_{m,\lambda,l} f(z)} \right)^{\delta} \prec g(z), \ \ z \in U,$$

and this result is sharp.

Theorem 8.27. *Let h be a holomorphic function which satisfies the inequality $Re\left(1 + \frac{zh''(z)}{h'(z)}\right) > -\frac{1}{2}$, $z \in U$, and $h(0) = 1$. If $\alpha, \lambda, l \geq 0$, $\delta \in (0,1)$, $m \in \mathbb{N}$, $f \in \mathscr{A}$ and satisfies the differential subordination*

$$\left(\frac{z}{RI^{\alpha}_{m,\lambda,l} f(z)} \right)^{\delta} \frac{RI^{\alpha}_{m+1,\lambda,l} f(z)}{1-\delta} \left(\frac{\left(RI^{\alpha}_{m+1,\lambda,l} f(z)\right)'}{RI^{\alpha}_{m+1,\lambda,l} f(z)} - \delta \frac{\left(RI^{\alpha}_{m,\lambda,l} f(z)\right)'}{RI^{\alpha}_{m,\lambda,l} f(z)} \right)$$
$$\prec h(z), \ \ z \in U, \tag{8.16}$$

then

$$\frac{RI^{\alpha}_{m+1,\lambda,l} f(z)}{z} \cdot \left(\frac{z}{RI^{\alpha}_{m,\lambda,l} f(z)} \right)^{\delta} \prec q(z), \ \ z \in U,$$

where $q(z) = \frac{1-\delta}{z^{1-\delta}} \int\limits_0^z h(t) t^{-\delta} \mathrm{d}t$.

Proof. Let

$$p(z) = \frac{RI^{\alpha}_{m+1,\lambda,l} f(z)}{z} \cdot \left(\frac{z}{RI^{\alpha}_{m,\lambda,l} f(z)} \right)^{\delta}, \ \ z \in U, \ p \in \mathscr{H}[1,1].$$

Differentiating, we obtain

$$\left(\frac{z}{RI^{\alpha}_{m,\lambda,l} f(z)} \right)^{\delta} \frac{RI^{\alpha}_{m+1,\lambda,l} f(z)}{1-\delta} \left(\frac{\left(RI^{\alpha}_{m+1,\lambda,l} f(z)\right)'}{RI^{\alpha}_{m+1,\lambda,l} f(z)} - \delta \frac{\left(RI^{\alpha}_{m,\lambda,l} f(z)\right)'}{RI^{\alpha}_{m,\lambda,l} f(z)} \right)$$
$$= p(z) + \frac{1}{1-\delta} zp'(z), \ \ z \in U,$$

and (8.16) becomes

$$p(z) + \frac{1}{1-\delta} z p'(z) \prec h(z), \quad z \in U.$$

Using Lemma 8.10, we have

$$p(z) \prec q(z), \quad z \in U,$$

i.e.,

$$\frac{RI^{\alpha}_{m+1,\lambda,l} f(z)}{z} \cdot \left(\frac{z}{RI^{\alpha}_{m,\lambda,l} f(z)} \right)^{\delta} \prec q(z) = \frac{1-\delta}{z^{1-\delta}} \int_0^z h(t) t^{-\delta} \mathrm{d}t, \ z \in U,$$

and q is the best dominant.

References

1. A. Alb Lupaş, On special differential subordinations using Sălăgean and Ruscheweyh operators. Math. Inequal. Appl. **12**(4), 781–790 (2009)
2. A. Alb Lupas, A special comprehensive class of analytic functions defined by multiplier transformation. J. Comput. Anal. Appl. **12**(2), 387–395 (2010)
3. A. Alb Lupas, On special differential subordinations using multiplier transformation and Ruscheweyh derivative. Rom. J. **6**(2), 1–14 (2010)
4. A. Alb Lupas, A new comprehensive class of analytic functions defined by multiplier transformation. Math. Comput. Model. **54**, 2355–2362 (2011)
5. A. Alb Lupaş, On special differential subordinations using a generalized Sălăgean operator and Ruscheweyh derivative. J. Comput. Anal. Appl. **13**(1), 98–107 (2011)
6. D.A. Alb Lupaş, *Subordinations and Superordinations* (Lap Lambert Academic Publishing, Saarbrücken, 2011)
7. A. Alb Lupaş, On a certain subclass of analytic functions defined by a generalized Sălăgean operator and Ruscheweyh derivative. Carpathian J. Math. **28**(2), 183–190 (2012)
8. A. Alb Lupas, Some differential subordinations using Ruscheweyh derivative and Sălăgean operator. Adv. Differ. Equ. **2013**, 150 (2013)
9. A. Alb Lupas, Some differential subordinations using multiplier transformation. Analele Universităţii din Oradea, Fascicola Matematica **XXI**(2), 133–142 (2014)
10. F.M. Al-Oboudi, On univalent functions defined by a generalized Sălăgean operator. Ind. J. Math. Math. Sci. **27**, 1429–1436 (2004)
11. L. Andrei, Differential subordinations using Ruscheweyh derivative and generalized Sălăgean operator. Adv. Differ. Equ. **2013**, 252 (2013)
12. S.S. Miller, P.T. Mocanu, *Differential Subordinations*. Theory and Applications (Dekker, New York, Basel, 2000)
13. St. Ruscheweyh, New criteria for univalent functions. Proc. Am. Math. Soc. **49**, 109–115 (1975)
14. G.St. Sălăgean, *Subclasses of Univalent Functions*. Lecture Notes in Mathematics, vol. 1013 (Springer, Berlin, 1983), pp. 362–372

Chapter 9
Some Results on the Bivariate Laguerre Polynomials

Mehmet Ali Özarslan and Cemaliye Kürt

Abstract In this paper, we consider the general class of bivariate Laguerre polynomials introduced in Ozarslan and Kürt (On a double integral equation including a set of two variables polynomials suggested by Laguerre polynomials. In: Proceedings of international conference on recent advances in pure and applied mathematics (ICRAPAM 2014), 2014). We first obtain linear and mixed multilateral generating functions for the above-mentioned classes. We further derive a finite summation formula for our polynomials. Finally, by using the fractional derivative operator, we give a series relation between the bivariate Laguerre polynomials and a product of confluent hypergeometric functions.

9.1 Introduction

Recently, a class of polynomials $Z^{(\alpha)}_{n_1,\ldots,n_j}\left(x_1,\ldots,x_j;\rho_1,\ldots,\rho_j\right)$ (see [5]) suggested by the multivariate Laguerre polynomials were introduced as

$$\begin{aligned} &Z^{(\alpha)}_{n_1,\ldots,n_j}\left(x_1,\ldots,x_j;\rho_1,\ldots,\rho_j\right) \\ &= \frac{\Gamma\left(\rho_1 n_1+\cdots+\rho_j n_j+\alpha+1\right)}{n_1!\ldots n_j!} \\ &\times \sum_{k_1,\ldots,k_j=0}^{n_1,\ldots,n_j} \frac{(-n_1)_{k_1}\ldots\left(-n_j\right)_{k_j} x_1^{\rho_1 k_1}\ldots x_j^{\rho_j k_j}}{\Gamma\left(\rho_1 k_1+\cdots+\rho_j k_j+\alpha+1\right)k_1!\ldots k_j!}. \qquad (9.1) \\ &\left(\alpha,\rho_1,\ldots,\rho_j\in\mathbb{C},\ Re(\rho_i)>0,\ i=1,\ldots,j\right) \end{aligned}$$

and investigated by the first author.

M.A. Özarslan • C. Kürt (✉)
Department of Mathematics, Eastern Mediterranean University, Famagusta, Mersin 10, Turkey
e-mail: mehmetali.ozarslan@emu.edu.tr; cemaliye.kurt@emu.edu.tr

G.A. Anastassiou, O. Duman (eds.), *Computational Analysis*, Springer Proceedings in Mathematics & Statistics 155, DOI 10.1007/978-3-319-28443-9_9

Obviously $Z_{n_1,\ldots,n_j}^{(\alpha)}\left(x_1,\ldots,x_j;\rho_1,\ldots,\rho_j\right)$ gives $L_{n_1,\ldots,n_j}^{(\alpha)}\left(x_1,\ldots,x_j\right)$ when $\rho_1 = \cdots = \rho_j = 1$, where

$$L_{n_1,\ldots,n_j}^{(\alpha)}\left(x_1,\ldots,x_j\right) = \frac{\Gamma\left(n_1+\cdots+n_j+\alpha+1\right)}{n_1!\ldots n_j!} \times \sum_{k_1,\ldots,k_j=0}^{n_1,\ldots,n_j} \frac{(-n_1)_{k_1}\ldots(-n_j)_{k_j} x_1^{k_1}\ldots x_j^{k_j}}{\Gamma\left(k_1+\cdots+k_j+\alpha+1\right) k_1!\ldots k_j!} \tag{9.2}$$

is the multivariable Laguerre polynomial defined by Carlitz [2] (see also [1]).

In [7], the authors introduce a new general class of bivariate Laguerre polynomials as follows:

$$L_{n,m}^{(\alpha,\beta,\gamma,\eta,\xi)}(x,y) = \frac{\Gamma(\alpha n+\beta m+\gamma+1)}{\Gamma(\xi+\eta m)} \times \sum_{k_1=0}^{n}\sum_{k_2=0}^{m} \frac{(-n)_{k_1}(-m)_{k_2} x^{\alpha k_1} y^{\beta k_2}}{\Gamma(\alpha k_1+\beta k_2+\gamma+1)\Gamma(\eta k_2+\xi)k_1!k_2!} \tag{9.3}$$

where $\alpha,\beta,\gamma,\eta,\xi\in\mathbb{C}$, $Re(\alpha), Re(\beta), Re(\eta), Re(\xi)>0$, $Re(\gamma)>-1$.

They investigated double fractional integral and derivative properties of the bivariate Laguerre polynomials $L_{n,m}^{(\alpha,\beta,\gamma,\eta,\xi)}(x,y)$. Furthermore, they obtained linear generating functions for $L_{n,m}^{(\alpha,\beta,\gamma,\eta,\xi)}(x,y)$ in terms of $E_{(\gamma_1,\gamma_2)}^{(\alpha,\beta,\gamma,\eta,\xi)}(x,y)$, which is defined by Ozarslan and Kürt [7]

$$E_{\gamma_1,\gamma_2}^{(\alpha,\beta,\eta,\xi,\lambda)}(x,y) = \sum_{k_1=0}^{\infty}\sum_{k_2=0}^{\infty} \frac{(\gamma_1)_{k_1}(\gamma_2)_{k_2} x^{k_1} y^{k_2}}{\Gamma(\alpha k_1+\beta k_2+\lambda)\Gamma(\eta k_2+\xi)k_1!k_2!}$$

where $\gamma_1,\gamma_2,\alpha,\beta,\lambda,\eta,\xi\in\mathbb{C}$, $Re(\alpha+\eta)>0$, $Re(\beta)>0$.

Finally, they found double Laplace transforms of $L_{n,m}^{(\alpha,\beta,\gamma,\eta,\xi)}(x,y)$ and gave the solution of a double integral equation involving the bivariate Laguerre polynomials in the kernel.

Now, we consider a general double hypergeometric series which were defined in [10]:

$$S\begin{matrix} A:B;B' \\ \\ C:D;D' \end{matrix}\begin{pmatrix} x \\ \\ y \end{pmatrix} \equiv S\begin{matrix} A:B;B' \\ \\ C:D;D' \end{matrix}\begin{pmatrix} [(a):\vartheta,\varphi]:[(b):\psi];[(b'):\psi']; & \\ & x,y \\ [(c):\delta,\varepsilon]:\ [(d):\eta];\ [(d'):\eta']; & \end{pmatrix}$$

$$= \sum_{m=0}^{\infty}\sum_{n=0}^{\infty}\left[\frac{\prod_{j=1}^{A}\Gamma[a_j+m\vartheta_j+n\varphi_j]\prod_{j=1}^{B}\Gamma[b_j+m\psi_j]}{\prod_{j=1}^{C}\Gamma[c_j+m\delta_j+n\varepsilon_j]\prod_{j=1}^{D}\Gamma[d_j+m\eta_j]} \frac{\prod_{j=1}^{B'}\Gamma[b_j'+n\psi_j']}{\prod_{j=1}^{D'}\Gamma[d_j'+n\eta_j']}\frac{x^m}{m!}\frac{y^n}{n!}\right],$$

where the coefficients

$$\begin{cases} \vartheta_1, \ldots, \vartheta_A; \varphi_1, \ldots, \varphi_A; \psi_1, \ldots, \psi_B; \psi_1^{'}, \ldots, \psi_B^{'}; \delta_1, \ldots, \delta_C; \\ \varepsilon_1, \ldots, \varepsilon_C; \eta_1, \ldots, \eta_D; \eta_1^{'}, \ldots, \eta_{D'}^{'}; \end{cases}$$

are real and positive. Here, (a) denotes the sequence of parameters $a_1, a_2, \ldots, a_A$ with a similar manner as for (b), $(b^{'})$, etc.

In the present paper, we consider the following special cases of the above functions, which we give their name by ${}_0\Psi_2^*$ and ${}_2\Psi_4^*$ as follows:

$${}_0\Psi_2^*\left(\begin{matrix} - \\ (\alpha, \beta, \gamma+1), (\eta, \xi) \end{matrix}; -x^\alpha t_1, -y^\beta t_2\right) := S\begin{matrix} 0:0;0 \\ \\ 1:1;0 \end{matrix}\left(\begin{matrix} -: & -; & -; \\ [(\gamma+1):\alpha,\beta] : & [(\xi):\eta]; & -; \end{matrix} \; -x^\alpha t_1, -y^\beta t_2\right).$$

and

$${}_2\Psi_4^*\left(\begin{matrix} (1,\lambda_1), (1,\lambda_2) \\ (\alpha, \beta, \gamma+1), (\eta, \xi), (1, \mu_1+1), (1, \mu_2+1) \end{matrix}; -x^\alpha t_1, -y^\beta t_2\right)$$

$$:= S\begin{matrix} 0:1;1 \\ \\ 1:2;0 \end{matrix}\left(\begin{matrix} -: & [(\lambda_1):1]; & [(\lambda_2):1]; \\ [(\gamma+1):\alpha,\beta] : & [(\xi,\mu_2+1):\eta,1]; & [(\mu_1+1):1]; \end{matrix} \; -x^\alpha t_1, -y^\beta t_2\right)$$

Note that for the absolute convergence of the functions

$${}_0\Psi_2^*\left(\begin{matrix} - \\ (\alpha, \beta, \gamma+1), (\eta, \xi) \end{matrix}; -x^\alpha t_1, -y^\beta t_2\right)$$

we need $Re(\alpha + \eta) > -1$ and $Re(\beta) > -1$ (see [10] and also see [11, 12]). In the same way, for the absolute convergence of

$${}_2\Psi_4^*\left(\begin{matrix} (1,\lambda_1), (1,\lambda_2) \\ (\alpha, \beta, \gamma+1), (\eta, \xi), (1, \mu_1+1), (1, \mu_2+1) \end{matrix}; -x^\alpha t_1, -y^\beta t_2\right),$$

we need $Re(\beta + \eta) > -2$ and $Re(\alpha) > -2$ (see [10] and also see [11, 12]).

9.2 Linear Generating Function and a Summation Formula

In this section, we give a linear, mixed multilinear generating functions and a summation formula for the polynomials $L_{n,m}^{(\alpha,\beta,\gamma,\eta,\xi)}(x,y)$.

Theorem 9.1. *The polynomials* $L_{n,m}^{(\alpha,\beta,\gamma,\eta,\xi)}(x,y)$ *have the generating function as follows:*

$$\sum_{n=0}^{\infty}\sum_{m=0}^{\infty}\frac{L_{n,m}^{(\alpha,\beta,\gamma,\eta,\xi)}(x,y)\Gamma(\xi+\eta m)}{\Gamma(\alpha n+\beta m+\gamma+1)n!m!}t_1^n t_2^m$$
$$= e^{t_1+t_2}\Psi^*\left(\begin{matrix} - \\ (\alpha,\beta,\gamma+1),(\eta,\xi)\end{matrix};-x^{\alpha}t_1,-y^{\beta}t_2\right). \tag{9.4}$$

Proof. Direct calculations yield that

$$\sum_{n=0}^{\infty}\sum_{m=0}^{\infty}\frac{L_{n,m}^{(\alpha,\beta,\gamma,\eta,\xi)}(x,y)\Gamma(\xi+\eta m)}{\Gamma(\alpha n+\beta m+\gamma+1)n!m!}t_1^n t_2^m$$
$$=\sum_{n,m=0}^{\infty}\sum_{k_1,k_2=0}^{n,m}\frac{(-n)_{k_1}(-m)_{k_2}x^{\alpha k_1}y^{\beta k_2}}{\Gamma(\alpha k_1+\beta k_2+\gamma+1)\Gamma(\xi+\eta k_2)n!m!k_1!k_2!}t_1^n t_2^m$$
$$=\sum_{n,m=0}^{\infty}\sum_{k_1,k_2=0}^{n,m}\frac{(-1)^{k_1+k_2}x^{\alpha k_1}y^{\beta k_2}}{\Gamma(\alpha k_1+\beta k_2+\gamma+1)\Gamma(\xi+\eta k_2)(n-k_1)!(m-k_2)!k_1!k_2!}t_1^n t_2^m.$$

Letting $n=n+k_1$ and $m=m+k_2$, we have

$$\sum_{n=0}^{\infty}\sum_{m=0}^{\infty}\frac{L_{n,m}^{(\alpha,\beta,\gamma,\eta,\xi)}(x,y)\Gamma(\xi+\eta m)}{\Gamma(\alpha n+\beta m+\gamma+1)n!m!}t_1^n t_2^m$$
$$=\sum_{n,m=0}^{\infty}\frac{t_1^n t_2^m}{n!m!}\sum_{k_1,k_2=0}^{\infty}\frac{(-1)^{k_1+k_2}x^{\alpha k_1}y^{\beta k_2}}{\Gamma(\alpha k_1+\beta k_2+\gamma+1)\Gamma(\xi+\eta k_2)k_1!k_2!}t_1^{k_1}t_2^{k_2}$$
$$=e^{t_1+t_2}{}_0\Psi_2^*\left(\begin{matrix} - \\ (\alpha,\beta,\gamma+1),(\eta,\xi)\end{matrix};-x^{\alpha}t_1,-y^{\beta}t_2\right).$$

Whence the result.

Now, we aim to obtain mixed multilateral generating functions for the polynomials $L_{n,m}^{(\alpha,\beta,\gamma,\eta,\xi)}(x,y)$ by using the same method considered in [13] (see also [4–6]). Let $(\gamma):=(\gamma_1,\gamma_2)$, $(\lambda):=(\lambda_1,\lambda_2)$, $(\eta):=(\eta_1,\eta_2)$, $(\psi):=(\psi_1,\psi_2)$, $(\rho):=(\rho_1,\rho_2)$ be complex 2-tuples. By making use of the above theorem, we have the following result.

Theorem 9.2. *Corresponding to an identically non-vanishing function*

$$\Omega_{(\eta)}(\xi_1, \xi_2, \ldots, \xi_s)$$

of complex variables $\xi_1, \xi_2, \ldots, \xi_s$ $(s \in \mathbb{N})$, *let*

$$\begin{aligned}
&\Lambda_{(\eta),(\psi)}(\xi_1, \xi_2, \ldots, \xi_s; \varsigma_1, \varsigma_2)\\
&:= \sum_{\substack{k_1,k_2=0\\(a_{k_1,k_{2j}} \neq 0)}}^{\infty} a_{k_1,k_2} \Omega_{\eta_1+\psi_1 k_1, \eta_2+\psi_2 k_2}(\xi_1, \xi_2, \ldots, \xi_s) \varsigma_1^{k_1} \varsigma_2^{k_2}.
\end{aligned} \tag{9.5}$$

Suppose also that

$$\begin{aligned}
&\Theta_{n,m;q_1,q_2}^{(\gamma),(\lambda),(\eta),(\psi),\alpha}(\xi_1, \xi_2, \ldots, \xi_s; x_1, x_2; (\alpha, \beta); \varsigma_1, \varsigma_2)\\
&= \sum_{k_1,k_2=0}^{\left[\frac{n}{q_1}\right],\left[\frac{m}{q_2}\right]} a_{k_1,k_2} \Omega_{\eta_1+\psi_1 k_1, \eta_2+\psi_2 k_2}(\xi_1, \xi_2, \ldots, \xi_s)\\
&\times \frac{L_{n-q_1k_1, m-q_2k_2}^{(\alpha,\beta,\gamma,\xi,\eta)}(x,y)\Gamma(\eta(m-q_2k_2)+\xi)}{\Gamma(\alpha(n-q_1k_1)+\beta(m-q_2k_2)+\gamma+1)(n-q_1k_1)!(m-q_2k_2)!} \varsigma_1^{k_1} \varsigma_2^{k_2}.\\
&(q_1, q_2 \in \mathbb{N})
\end{aligned} \tag{9.6}$$

Then,

$$\begin{aligned}
&\sum_{n,m=0}^{\infty} \Theta_{n,m;q_1,q_2}^{(\gamma),(\lambda),(\eta),(\psi),\alpha}\left(\xi_1, \xi_2, \ldots, \xi_s; x_1, x_2; (\alpha, \beta); \frac{\varsigma_1}{t_1^{q_1}}, \frac{\varsigma_2}{t_2^{q_2}}\right) t^n t^m\\
&= e^{t_1+t_2} \Lambda_{(\eta),(\psi)}(\xi_1, \xi_2, \ldots, \xi_s; \varsigma_1, \varsigma_2)\\
&\times {}_0\Psi_2^*\left(\begin{matrix} - \\ (\alpha, \beta, \gamma+1), (\eta, \xi) \end{matrix}; -x^{\alpha} t_1, -y^{\beta} t_2 \right)
\end{aligned} \tag{9.7}$$

provided that each member of Eq. (9.7) exists, where $|t_1| < 1$ *and* $|t_2| < 1$.

Proof. Let $\mathscr{F}$ denote the left member of (9.7). Substituting the polynomials

$$\Theta_{n,m;q_1,q_2}^{(\gamma),(\lambda),(\eta),(\psi),\alpha}(\xi_1, \xi_2, \ldots, \xi_s; x_1, x_2; (\alpha, \beta); \varsigma_1, \varsigma_2)$$

from the definition (9.6) into the left-hand side of (9.7), we have

$$\begin{aligned}
\mathscr{F} = &\sum_{n,m=0}^{\infty} \sum_{k_1,k_2=0}^{\left[\frac{n}{q_1}\right],\left[\frac{m}{q_2}\right]} a_{k_1,k_2} \Omega_{n+\psi_1 k_1, m+\psi_2 k_2}(\xi_1, \xi_2, \ldots, \xi_s) \varsigma_1^{k_1} \varsigma_2^{k_2}\\
&\times \frac{L_{n-q_1k_1, m-q_2k_2}^{(\alpha,\beta,\gamma,\xi,\eta)}(x,y)\Gamma(\eta(m-q_2k_2)+\xi)}{\Gamma(\alpha(n-q_1k_1)+\beta(m-q_2k_2)+\gamma+1)(n-q_1k_1)!(m-q_2k_2)!}
\end{aligned}$$

$$\times t_1^{n-q_1k_1} t_2^{m-q_2k_2}$$

$$= \sum_{k_1,k_2=0}^{\infty} a_{k_1,k_2} \Omega_{n+\psi_1k_1,m+\psi_2k_2}(\xi_1,\xi_2,\ldots,\xi_s)\varsigma_1^{k_1}\varsigma_2^{k_2}$$

$$\times \sum_{n,m=0}^{\infty} \frac{L_{n,m}^{(\alpha,\beta,\gamma,\xi,\eta)}(x,y)\Gamma(\eta m+\xi)}{\Gamma(\alpha n+\beta m+\gamma+1)} t_1^n t_2^m.$$

Using Theorem 9.1 with $\gamma_1 \to \gamma_1 + \lambda_1 k_1$ and $\gamma_2 \to \gamma_2 + \lambda_2 k_2$, we get

$$\mathscr{F} = a_{k_1,k_2} \Omega_{n+\psi_1k_1,m+\psi_2k_2}(\xi_1,\xi_2,\ldots,\xi_s)\varsigma_1^{k_1}\varsigma_2^{k_2} e^{t_1+t_2}$$

$$\times {}_0\Psi_2^* \left(\begin{matrix} - \\ (\alpha,\beta,\gamma+1),(\eta,\xi) \end{matrix} ; -x^\alpha t_1, -y^\beta t_2 \right).$$

Using (9.5), the result follows.

Using (9.4) and a technique used by Srivastava [8, 9], we give the following interesting summation formula for the bivariate Laguerre polynomials $L_{n,m}^{(\alpha,\beta,\gamma,\eta,\xi)}(x,y)$ by the following theorem:

Theorem 9.3. *The bivariate Laguerre polynomials defined in (9.2) satisfy the following summation formula:*

$$L_{n,m}^{(\alpha,\beta,\gamma,\eta,\xi)}(x,y) = \frac{\Gamma(\alpha n+\beta m+\gamma+1)}{\Gamma(\xi+\eta m)}$$

$$\times \sum_{j,l=0}^{n,m} \binom{n}{j}\binom{m}{l} \frac{L_{n-j,m-l}^{(\alpha,\beta,\gamma,\eta,\xi)}(t,k)}{\Gamma(\alpha(n-j)+\beta(m-l)+\gamma+1)},$$

$$\times \left(\frac{x^\alpha}{t^\alpha}\right)^n \left(\frac{y^\beta}{k^\beta}\right)^m \left(\frac{t^\alpha}{x^\alpha}-1\right)^j \left(\frac{k^\beta}{y^\beta}-1\right)^l. \tag{9.8}$$

Proof. Setting $t_1 = [-t^\alpha]z_1$ and $t_2 = [-k^\beta]z_2$ in (9.4), we have

$$\sum_{n=0}^{\infty}\sum_{m=0}^{\infty} \frac{L_{n,m}^{(\alpha,\beta,\gamma,\eta,\xi)}(x,y)\Gamma(\xi+\eta m)}{\Gamma(\alpha n+\beta m+\gamma+1)n!m!}([-t^\alpha]z_1)^n([-k^\beta]z_2)^m$$

$$= e^{[-t^\alpha]z_1+[-k^\beta]z_2} {}_0\Psi_2^* \left(\begin{matrix} - \\ (\alpha,\beta,\gamma+1),(\eta,\xi) \end{matrix} ; -x^\alpha[-t^\alpha]z_1. - y^\beta[-k^\beta]z_2 \right). \tag{9.9}$$

Interchange x by t and y by k to get

$$\sum_{n=0}^{\infty}\sum_{m=0}^{\infty}\frac{L_{n,m}^{(\alpha,\beta,\gamma,\eta,\xi)}(t,k)\Gamma(\xi+\eta m)}{\Gamma(\alpha n+\beta m+\gamma+1)n!m!}([-x^{\alpha}]z_1)^n([-y^{\beta}]z_2)^m$$
$$= e^{[-x^{\alpha}]z_1+[-y^{\beta}]z_2}{}_0\Psi_2^*\left(\begin{matrix}-\\(\alpha,\beta,\gamma+1),(\eta,\xi)\end{matrix};-x^{\alpha}[-t^{\alpha}]z_1,-y^{\beta}[-k^{\beta}]z_2\right). \tag{9.10}$$

Comparing (9.9) and (9.10), we have

$$\sum_{n=0}^{\infty}\sum_{m=0}^{\infty}\frac{L_{n,m}^{(\alpha,\beta,\gamma,\eta,\xi)}(x,y)\Gamma(\xi+\eta m)}{\Gamma(\alpha n+\beta m+\gamma+1)n!m!}(-t^{\alpha}z_1)^n(-k^{\beta}z_2)^m$$
$$= e^{-t^{\alpha}z_1-k^{\beta}z_2+x^{\alpha}z_1+y^{\beta}z_2}$$
$$\times\sum_{n=0}^{\infty}\sum_{m=0}^{\infty}\frac{L_{n,m}^{(\alpha,\beta,\gamma,\eta,\xi)}(t,k)\Gamma(\xi+\eta m)}{\Gamma(\alpha n+\beta m+\gamma+1)n!m!}(-x^{\alpha}z_1)^n(-y^{\beta}z_2)^m$$
$$=\sum_{n,m=0}^{\infty}\sum_{j,l=0}^{\infty}\frac{L_{n,m}^{(\alpha,\beta,\gamma,\eta,\xi)}(t,k)\Gamma(\xi+\eta m)}{\Gamma(\alpha n+\beta m+\gamma+1)n!m!j!l!}$$
$$\times(-x^{\alpha}z_1)^n(-y^{\beta}z_2)^m(-t^{\alpha}z_1+x^{\alpha}z_1)^j(-k^{\beta}z_2+y^{\beta}z_2)^l,$$

and hence

$$=\sum_{n,m=0}^{\infty}\sum_{j,l=0}^{n,m}\frac{L_{n-j,m-l}^{(\alpha,\beta,\gamma,\eta,\xi)}(t,k)\Gamma(\xi+\eta m)}{\Gamma(\alpha(n-j)+\beta(m-l)+\gamma+1)(n-j)!(m-l)!j!l!}$$
$$\times(-x^{\alpha}z_1)^{n-j}(-y^{\beta}z_2)^{m-l}(-t^{\alpha}z_1+x^{\alpha}z_1)^j(-k^{\beta}z_2+y^{\beta}z_2)^l$$
$$=\sum_{n,m=0}^{\infty}\sum_{j,l=0}^{n,m}\binom{n}{j}\binom{m}{l}\frac{L_{n-j,m-l}^{(\alpha,\beta,\gamma,\eta,\xi)}(t,k)\Gamma(\xi+\eta m)}{\Gamma(\alpha(n-j)+\beta(m-l)+\gamma+1)}$$
$$\times(-x^{\alpha}z_1)^{n-j}(-y^{\beta}z_2)^{m-l}(-t^{\alpha}z_1+x^{\alpha}z_1)^j(-k^{\beta}z_2+y^{\beta}z_2)^l$$

from which, on comparing the coefficients $z_1^n z_2^m$ on both sides, we get (9.8).

9.3 A Series Relation for $L_{n,m}^{(\alpha,\beta,\gamma,\eta,\xi)}(x,y)$

In this section, we recall the definition of Riemann–Liouville fractional derivative.

Definition 9.4 (See [3]). The Riemann–Liouville fractional derivative of order $\mu\in\mathbb{C}$ $(Re(\mu)\geq 0)$ is defined by

$$
{}_xD^{\mu}_{a+}[f] = \left(\frac{d}{dx}\right)^n \frac{1}{\Gamma(n-\alpha)} \int_a^x (x-\xi)^{\alpha-n-1} f(\xi)d\xi
$$
$$
(n = [Re(\mu)] + 1,\ x > a)
$$

where, as usual, $[Re(\mu)]$ means the integral part of $Re(\mu)$.

Proposition 9.5. *The following property holds true:*

$$
D^{\lambda}_w(w^{\mu-1}) = \frac{d^{\lambda}}{dw^{\lambda}}(w^{\mu-1}) = \frac{\Gamma(\mu)}{\Gamma(\mu-\lambda)} w^{\mu-\lambda-1} \ for\ \lambda \neq \mu, \tag{9.11}
$$

where μ is an arbitrary complex number.

Theorem 9.6. *The following series relation holds true between the bivariate Laguerre polynomials and the confluent hypergeometric functions:*

$$
\begin{aligned}
&\sum_{n=0}^{\infty}\sum_{m=0}^{\infty} \frac{L_{n,m}^{(\alpha,\beta,\gamma,\eta,\xi)}(x,y)\Gamma(\xi+\eta m)(\lambda_1)_m(\lambda_2)_m}{\Gamma(\alpha n+\beta m+\gamma+1)(\mu_1+1)_n(\mu_2+1)_m n!m!} \\
&\quad \times {}_1F_1(\mu_1-\lambda_1+1, n+\mu_1+1; t_1)\,{}_1F_1(\mu_2-\lambda_2+1, m+\mu_2+1; t_2) t_1^n t_2^m \\
&= e^{t_1+t_2} \frac{\Gamma(\mu_1+1)\Gamma(\mu_2+1)}{\Gamma(\lambda_1)\Gamma(\lambda_2)} \\
&\quad \times {}_2\Psi_4^{*}\left(\begin{matrix} (1,\lambda_1),(1,\lambda_2) \\ (\alpha,\beta,\gamma+1),(\eta,\xi),(1,\mu_1+1),(1,\mu_2+1) \end{matrix} ; -x^{\alpha}t_1, -y^{\beta}t_2 \right).
\end{aligned}
$$

Proof. If we rewrite (9.4) as

$$
\begin{aligned}
&\sum_{n=0}^{\infty}\sum_{m=0}^{\infty} \frac{L_{n,m}^{(\alpha,\beta,\gamma,\eta,\xi)}(x,y)\Gamma(\xi+\eta m)}{\Gamma(\alpha n+\beta m+\gamma+1)n!m!} t_1^n t_2^m e^{-t_1-t_2} \\
&\quad = {}_0\Psi_2^{*}\left(\begin{matrix} - \\ (\alpha,\beta,\gamma+1),(\eta,\xi) \end{matrix} ; -x^{\alpha}t_1, -y^{\beta}t_2 \right)
\end{aligned}
$$

and expand the exponential function to a series, we get

$$
\begin{aligned}
&\sum_{n,m=0}^{\infty}\sum_{r,k=0}^{\infty} \frac{L_{n,m}^{(\alpha,\beta,\gamma,\eta,\xi)}(x,y)\Gamma(\xi+\eta m)}{\Gamma(\alpha n+\beta m+\gamma+1)n!m!} \frac{(-t_1)^r}{r!} \frac{(-t_2)^k}{k!} t_1^n t_2^m \\
&\quad = \sum_{n=0}^{\infty}\sum_{m=0}^{\infty} \frac{(-x)^{\alpha n}(-y)^{\beta n}}{\Gamma(\alpha n+\beta m+\gamma+1)\Gamma(\eta m+\xi)n!m!} t_1^n t_2^m.
\end{aligned}
$$

Multiplying both sides by $t_1^{\lambda_1-1}$ and $t_2^{\lambda_2-1}$, we obtain

$$\sum_{n,m=0}^{\infty}\sum_{r,k=0}^{\infty}\frac{L_{n,m}^{(\alpha,\beta,\gamma,\eta,\xi)}(x,y)\Gamma(\xi+\eta m)}{\Gamma(\alpha n+\beta m+\gamma+1)n!m!}\frac{(-1)^r}{r!}\frac{(-1)^k}{k!}t_1^{n+\lambda_1+r-1}t_2^{m+\lambda_2+k-1}$$
$$=\sum_{n=0}^{\infty}\sum_{m=0}^{\infty}\frac{(-x^\alpha t_1)^n(-y^\beta t_2)^m t_1^{\lambda_1-1}t_2^{\lambda_2-1}}{\Gamma(\alpha n+\beta m+\gamma+1)\Gamma(\eta m+\xi)n!m!}.$$

Now apply the operator $D_{t_1}^{\lambda_1-\mu_1-1}$ and $D_{t_2}^{\lambda_2-\mu_2-1}$, we get

$$\sum_{n,m=0}^{\infty}\sum_{r,k=0}^{\infty}\frac{L_{n,m}^{(\alpha,\beta,\gamma,\eta,\xi)}(x,y)\Gamma(\xi+\eta m)}{\Gamma(\alpha n+\beta m+\gamma+1)n!m!}\frac{(-1)^r}{r!}\frac{(-1)^k}{k!}$$
$$\times D_{t_1}^{\lambda_1-\mu_1-1}[t_1^{n+\lambda_1+r-1}]D_{t_2}^{\lambda_2-\mu_2-1}[t_2^{m+\lambda_2+k-1}]$$
$$=\sum_{n=0}^{\infty}\sum_{m=0}^{\infty}\frac{(-x^\alpha t_1)^m(-y^\beta t_2)^m t_2^{\lambda_2-1}}{\Gamma(\alpha n+\beta m+\gamma+1)\Gamma(\eta m+\xi)n!m!}$$
$$\times D_{t_1}^{\lambda_1-\mu_1-1}[t_1^{\lambda_1-1}]D_{t_2}^{\lambda_2-\mu_2-1}[t_2^{\lambda_2-1}].$$

Using (9.11), we have

$$\sum_{n,m=0}^{\infty}\sum_{r,k=0}^{\infty}\frac{L_{n,m}^{(\alpha,\beta,\gamma,\eta,\xi)}(x,y)\Gamma(\xi+\eta m)\Gamma(n+\lambda_1+r)\Gamma(m+\lambda_2+k)}{\Gamma(\alpha n+\beta m+\gamma+1)\Gamma(n+\mu_1+r+1)\Gamma(m+\mu_2+k+1)n!m!}$$
$$\times\frac{(-1)^r}{r!}\frac{(-1)^k}{k!}t_1^{n+r}t_2^{m+k}$$
$$=\sum_{n=0}^{\infty}\sum_{m=0}^{\infty}\frac{(-x^\alpha)^n(-y^\beta)^m\Gamma(n+\lambda_1)\Gamma(m+\lambda_2)t_1^n t_2^m}{\Gamma(\alpha n+\beta m+\gamma+1)\Gamma(\eta m+\xi)\Gamma(n+\mu_1+1)\Gamma(m+\mu_2+1)n!m!}$$
$$={}_2\Psi_4^*\left(\begin{matrix}(1,\lambda_1),(1,\lambda_2)\\(\alpha,\beta,\gamma+1),(\eta,\xi),(1,\mu_1+1),(1,\mu_2+1)\end{matrix};-x^\alpha t_1,-y^\beta t_2\right).$$

Therefore, we obtain

$$\sum_{n,m=0}^{\infty}\sum_{r,k=0}^{\infty}\left[\frac{L_{n,m}^{(\alpha,\beta,\gamma,\eta,\xi)}(x,y)\Gamma(\xi+\eta m)\Gamma(\lambda_1)\Gamma(\lambda_2)}{\Gamma(\alpha n+\beta m+\gamma+1)\Gamma(\mu_1+1)\Gamma(\mu_2+1)}\right.$$
$$\left.\times\frac{(n+\lambda_1)_r(m+\lambda_2)_k(\lambda_1)_n(\lambda_2)_m}{(n+\mu_1+1)_r(\mu_1+1)_n(m+\mu_2+1)_k(\mu_2+1)_m}\times\frac{(-t_1)^r t_1^n(-t_2)^k t_2^m}{n!m!r!k!}\right]$$
$$={}_2\Psi_4^*\left(\begin{matrix}(1,\lambda_1),(1,\lambda_2)\\(\alpha,\beta,\gamma+1),(\eta,\xi),(1,\mu_1+1),(1,\mu_2+1)\end{matrix};-x^\alpha t_1,-y^\beta t_2\right). \tag{9.12}$$

For the sake of brevity, let $\mathscr{S}$ denote the left member of (9.12). It follows that

$$\mathscr{S} = \frac{\Gamma(\lambda_1)\Gamma(\lambda_2)}{\Gamma(\mu_1+1)\Gamma(\mu_2+1)}$$
$$\times \sum_{n=0}^{\infty}\sum_{m=0}^{\infty} \frac{L_{n,m}^{(\alpha,\beta,\gamma,\eta,\xi)}(x,y)\Gamma(\xi+\eta m)(\lambda_1)_n(\lambda_2)_m t_1^n t_2^m}{\Gamma(\alpha n+\beta m+\gamma+1)(\mu_1+1)_n(\mu_2+1)_m n!m!}$$
$$\times \sum_{r=0}^{\infty} \frac{(n+\lambda_1)_r}{(n+\mu_1+1)_r r!}(-t_1)^r \sum_{k=0}^{\infty} \frac{(m+\lambda_2)_k}{(m+\mu_2+1)_k k!}(-t_2)^k,$$

which gives

$$\mathscr{S} = \frac{\Gamma(\lambda_1)\Gamma(\lambda_2)}{\Gamma(\mu_1+1)\Gamma(\mu_2+1)}$$
$$\times \sum_{n=0}^{\infty}\sum_{m=0}^{\infty} \frac{L_{n,m}^{(\alpha,\beta,\gamma,\eta,\xi)}(x,y)\Gamma(\xi+\eta m)(\lambda_1)_n(\lambda_2)_m}{\Gamma(\alpha n+\beta m+\gamma+1)(\mu_1+1)_n(\mu_2+1)_m n!m!}$$
$$\times {}_1F_1(n+\lambda_1, n+\mu_1+1; -t_1)\,{}_1F_1(m+\lambda_2, m+\mu_2+1; -t_2).$$

Finally, since ${}_1F_1(a;b;z) = e^z {}_1F_1(b-a;b;-z)$, we get

$$\mathscr{S} = \frac{\Gamma(\lambda_1)\Gamma(\lambda_2)}{\Gamma(\mu_1+1)\Gamma(\mu_2+1)} e^{-t_1-t_2}$$
$$\times \sum_{n=0}^{\infty}\sum_{m=0}^{\infty} \frac{L_{n,m}^{(\alpha,\beta,\gamma,\eta,\xi)}(x,y)\Gamma(\xi+\eta m)(\lambda_1)_n(\lambda_2)_m t_1^n t_2^m}{\Gamma(\alpha n+\beta m+\gamma+1)(\mu_1+1)_n(\mu_2+1)_m n!m!}$$
$$\times {}_1F_1(\mu_1-\lambda_1+1, n+\mu_1+1; t_1)\,{}_1F_1(\mu_2-\lambda_2+1, m+\mu_2+1; t_2). \tag{9.13}$$

Comparing (9.12) and (9.13), we get the desired result.

References

1. P. Appell, J. Kampé de Fériet, *Fonctions Hyperg éométriques et Hypersphériques* (Gauthier-Villars, Paris, 1926)
2. L. Carlitz, Bilinear generating functions for Laguerre and Lauricella polynomials in several variables. Rend. Sem. Mat. Univ. Padova **43**, 269–276 (1970)
3. A.A. Kılbas, H.M. Srivastava, J.J. Trujillo, *Theory and Applications of Fractional Differential Equations*. North-Holland Mathematics Studies, vol. 204 (Elsevier, Amsterdam, 2006)
4. S.D. Lin, H.M. Srivastava, P.Y. Wang, Some families of hypergeometric transformations and generating relations. Math. Comput. Model. **36**, 445–459 (2002)
5. M.A. Ozarslan, On a singular integral equation including a set of multivariate polynomials suggested by Laguerre polynomials. Appl. Math. Comput. **229**, 350–358 (2014)

6. M.A. Ozarslan, A. Altın, Some families of generating functions for the multiple orthogonal polynomials associated with modified Bessel K-functions. J. Math. Anal. Appl. **297**, 186–193 (2004)
7. M.A. Ozarslan, C. Kürt, On a double integral equation including a set of two variables polynomials suggested by Laguerre polynomials, in *Proceedings of International Conference on Recent Advances in Pure and Applied Mathematics (ICRAPAM 2014)* (2014)
8. H.M. Srivastava, Finite summation formulas associated with a class of generalized hypergeometric polynomial. J. Math. Anal. Appl. **23**, 453–458 (1968)
9. H.M. Srivastava, On the Konhauser sets of biorthogonal polynomials suggested by the Laguerre polynomials. Pac. J. Math. **49**, 489–492 (1973)
10. H.M. Srivastava, M.C. Daoust, Certain generalized Neumann expansion associated with Kampe' de Feriet function. Nederl. Akad. Wetensch. Proc. Ser. A **72**(31), 449–457 (1969) (Indag. Math.)
11. H.M. Srivastava, M.C. Daoust, A note on the convergence of Kampé de Fériet's double hypergeometric series. Math. Nachr. **53**, 151–159 (1972)
12. H.M. Srivastava, P.W. Karlsson, Multiple Gaussian Hypergeometric Series (Ellis Horwood Limited, New York, 1985)
13. H.M. Srivastava, H.L. Manocha, *A Treatise on Generating Functions* (Halsted Press/Wiley/Ellis Horwood Limited, New York/Chicester/Brisbane/Toronto, 1984)

Chapter 10
Inner Product Spaces and Quadratic Functional Equations

Choonkil Park, Won-Gil Park, and Themistocles M. Rassias

Abstract In this paper, we prove that the norm defined over a real vector space V is induced by an inner product if and only if for a fixed integer $n \geq 2$

$$n \left\| \sum_{i=1}^{n} x_i \right\|^2 + \sum_{i=1}^{n} \left\| n x_i - \sum_{j=1}^{n} x_j \right\|^2 = n^2 \sum_{i=1}^{n} \|x_i\|^2$$

holds for all $x_1, \ldots, x_n \in V$. Let V, W be real vector spaces. It is shown that if a mapping $f : V \to W$ satisfies

$$nf\left(\sum_{i=1}^{n} x_i\right) + \sum_{i=1}^{n} f\left(n x_i - \sum_{j=1}^{n} x_j\right) = n^2 \sum_{i=1}^{n} f(x_i), \quad (n > 2)$$

or

$$nf\left(\sum_{i=1}^{n} x_i\right) + \sum_{i=1}^{n} f\left(n x_i - \sum_{j=1}^{n} x_j\right)$$
$$= \frac{n^2 + n}{2} \sum_{i=1}^{n} f(x_i) + \frac{n^2 - n}{2} \sum_{i=1}^{n} f(-x_i), \quad (n \geq 2)$$

C. Park (✉)
Department of Mathematics, Hanyang University, Seoul, Republic of Korea
e-mail: baak@hanyang.ac.kr

W.-G. Park
Mokwon University, Daejeon, Republic of Korea
e-mail: wgpark@mokwon.ac.kr

T.M. Rassias
National Technical University of Athens, Athens, Greece
e-mail: trassias@math.ntua.gr

G.A. Anastassiou, O. Duman (eds.), *Computational Analysis*, Springer Proceedings in Mathematics & Statistics 155, DOI 10.1007/978-3-319-28443-9_10

for all $x_1, \ldots, x_n \in V$, then the mapping $f : V \to W$ is Cauchy additive-quadratic. Furthermore, we prove the Hyers–Ulam stability of the above quadratic functional equations in Banach spaces.

Keywords Inner product space • Quadratic mapping • Quadratic Functional equation • Hyers–Ulam stability.

2010 AMS Math. Subject Classification: Primary 39B72, 46C05; Secondary 46C05

10.1 Introduction and Preliminaries

The stability problem of functional equations originated from a question of Ulam [21] concerning the stability of group homomorphisms. Hyers [6] gave a first affirmative partial answer to the question of Ulam for Banach spaces. Hyers' theorem was generalized by Aoki [1] for additive mappings and by Th. M. Rassias [12] for linear mappings by considering an unbounded Cauchy difference. A generalization of the Th. M. Rassias theorem was obtained by Găvruta [5] by replacing the unbounded Cauchy difference by a general control function in the spirit of Th. M. Rassias' approach.

A square norm on an inner product space satisfies the parallelogram equality

$$\|x + y\|^2 + \|x - y\|^2 = 2\|x\|^2 + 2\|y\|^2.$$

The functional equation

$$f(x + y) + f(x - y) = 2f(x) + 2f(y)$$

is called a *quadratic functional equation*. In particular, every solution of the quadratic functional equation is said to be a *quadratic mapping*. A Hyers–Ulam stability problem for the quadratic functional equation was proved by Skof [20] for mappings $f : X \to Y$, where X is a normed space and Y is a Banach space. Cholewa [3] noticed that the theorem of Skof is still true if the relevant domain X is replaced by an Abelian group. In [4], Czerwik proved the Hyers–Ulam stability of the quadratic functional equation. Several functional equations have been investigated in [2, 7–11, 14–19].

In [13], Th. M. Rassias proved that the norm defined over a real vector space V is induced by an inner product if and only if for a fixed integer $n \geq 2$

$$n\left\|\frac{1}{n}\sum_{i=1}^{n}x_i\right\|^2+\sum_{i=1}^{n}\left\|x_i-\frac{1}{n}\sum_{j=1}^{n}x_j\right\|^2=\sum_{i=1}^{n}\|x_i\|^2 \tag{10.1}$$

holds for all $x_1,\ldots,x_n \in V$.

Throughout this paper, let X be a real normed vector space with norm $||\cdot||$, and Y a real Banach space with norm $\|\cdot\|$.

In this paper, we investigate the quadratic functional equations

$$nf\left(\sum_{i=1}^{n}x_i\right)+\sum_{i=1}^{n}f\left(nx_i-\sum_{j=1}^{n}x_j\right)=n^2\sum_{i=1}^{n}f(x_i),\quad (n>2) \tag{10.2}$$

and

$$nf\left(\sum_{i=1}^{n}x_i\right)+\sum_{i=1}^{n}f\left(nx_i-\sum_{j=1}^{n}x_j\right)$$
$$=\frac{n^2+n}{2}\sum_{i=1}^{n}f(x_i)+\frac{n^2-n}{2}\sum_{i=1}^{n}f(-x_i),\quad (n\geq 2), \tag{10.3}$$

and prove the Hyers–Ulam stability of the quadratic functional equations (10.2) and (10.3) in Banach spaces.

10.2 On the Stability of a Cauchy Quadratic Functional Equation Associated with Inner Product Spaces

Throughout this section, assume that V and W are real vector spaces and that n is a fixed integer greater than 1.

Theorem 10.1. *A norm $\|\cdot\|: V\to \mathbf{R}$ is induced by an inner product if and only if*

$$n\left\|\sum_{i=1}^{n}x_i\right\|^2+\sum_{i=1}^{n}\left\|nx_i-\sum_{j=1}^{n}x_j\right\|^2=n^2\sum_{i=1}^{n}\|x_i\|^2 \tag{10.4}$$

holds for all $x_1,\ldots,x_n \in V$.

Proof. Assume that $\|\cdot\|$ satisfies (10.4).

Letting $x_1=\cdots=x_n=x$ in (10.4), we get

$$n\|nx\|^2=n^3\|x\|^2 \tag{10.5}$$

for all $x\in V$.

By (10.5), replacing x_k by $\frac{x_k}{n}$ $(k = 1, \ldots, n)$ in (10.4), we get the equality (10.1). By [13, Theorem 2], the norm $\| \cdot \|$ is induced by an inner product. The converse follows from an easy computation.

We investigate the quadratic functional equation (10.2).

Lemma 10.2. *If a mapping $f : V \to W$ satisfies*

$$nf\left(\sum_{i=1}^{n} x_i\right) + \sum_{i=1}^{n} f\left(nx_i - \sum_{j=1}^{n} x_j\right) = n^2 \sum_{i=1}^{n} f(x_i) \tag{10.6}$$

for all $x_1, \ldots, x_n \in V$ and a fixed integer n greater than 2, then the mapping $f : V \to W$ satisfies the Cauchy additive-quadratic functional equation

$$\begin{aligned} &2f(x_1 + x_2) + f(x_1 - x_2) + f(x_2 - x_1) \\ &\quad = 3f(x_1) + 3f(x_2) + f(-x_1) + f(-x_2) \end{aligned} \tag{10.7}$$

for all $x_1, x_2 \in V$.

Proof. Assume that $f : V \to W$ satisfies (10.6).

Letting $x_1 = \cdots = x_n = 0$ in (10.6), $nf(0) + nf(0) = n^3 f(0)$. So $f(0) = 0$. Letting $x_1 = \cdots = x_n = x$ in (10.6),

$$nf(nx) = nf(nx) + nf(0) = n^3 f(x) \tag{10.8}$$

for all $x \in V$.

By (10.8), replacing x_k by $\frac{(n-1)x_k}{n}$ $(k = 1, \ldots, n-1)$ and x_n by $\frac{\sum_{j=1}^{n-1} x_j}{n}$ in (10.6), we get

$$(n-1)f\left(\sum_{i=1}^{n-1} x_i\right) + \sum_{i=1}^{n-1} f\left((n-1)x_i - \sum_{j=1}^{n-1} x_j\right) = \sum_{i=1}^{n-1} f((n-1)x_i) \tag{10.9}$$

for all $x_1, \ldots, x_{n-1} \in V$.

Replacing x_k by $\frac{(n-2)x_k}{n-1}$ $(k = 1, \ldots, n-2)$ and x_{n-1} by $\frac{\sum_{j=1}^{n-2} x_j}{n-1}$ in (10.9), we get

$$(n-2)f\left(\sum_{i=1}^{n-2} x_i\right) + \sum_{i=1}^{n-2} f\left((n-2)x_i - \sum_{j=1}^{n-2} x_j\right) = \sum_{i=1}^{n-2} f((n-2)x_i)$$

for all $x_1, \ldots, x_{n-2} \in V$. Applying continuously this method $n-4$ times, we get

$$2f(x_1 + x_2) + f(x_1 - x_2) + f(x_2 - x_1) = f(2x_1) + f(2x_2) \tag{10.10}$$

for all $x_1, x_2 \in V$. Letting $x_2 = 0$ in (10.10), we get

$$2f(x_1) + f(x_1) + f(-x_1) = f(2x_1)$$

for all $x_1 \in V$. Similarly, we obtain

$$3f(x_2) + f(-x_2) = f(2x_2)$$

for all $x_2 \in V$. It follows from (10.10) that

$$2f(x_1 + x_2) + f(x_1 - x_2) + f(x_2 - x_1) = 3f(x_1) + 3f(x_2) + f(-x_1) + f(-x_2)$$

for all $x_1, x_2 \in V$, as desired.

One can easily show that an even mapping $f : V \to W$ satisfies (10.7) if and only if the even mapping $f : V \to W$ is a quadratic mapping, i.e.,

$$f(x + y) + f(x - y) = 2f(x) + 2f(y),$$

and that an odd mapping $f : V \to W$ satisfies (10.7) if and only if the odd mapping $f : V \to W$ is a Cauchy additive mapping, i.e.,

$$f(x + y) = f(x) + f(y).$$

From now on, assume that n is a fixed integer greater than 2. For a given mapping $f : X \to Y$, we define

$$Df(x_1, \dots, x_n) := nf\left(\sum_{i=1}^{n} x_i\right) + \sum_{i=1}^{n} f\left(nx_i - \sum_{j=1}^{n} x_j\right) - n^2 \sum_{i=1}^{n} f(x_i)$$

for all $x_1, \dots, x_n \in X$.

Now we prove the Hyers–Ulam stability of the quadratic functional equation $Df(x_1, \dots, x_n) = 0$ in real Banach spaces.

Theorem 10.3. *Let $f : X \to Y$ be a mapping satisfying $f(0) = 0$ for which there exists a function $\varphi : X^n \to [0, \infty)$ such that*

$$\tilde{\varphi}(x_1, \dots, x_n) := \sum_{j=1}^{\infty} n^{2j} \varphi\left(\frac{x_1}{n^j}, \dots, \frac{x_n}{n^j}\right) < \infty, \tag{10.11}$$

$$\|Df(x_1, \dots, x_n)\| \le \varphi(x_1, \dots, x_n) \tag{10.12}$$

for all $x_1, \dots, x_n \in X$. Then there exists a unique quadratic mapping $Q : X \to Y$ satisfying (10.6) *and*

$$\|f(x) + f(-x) - Q(x)\| \le \frac{1}{n^3}\tilde{\varphi}(\underbrace{x, \dots, x}_{n \text{ times}}) + \frac{1}{n^3}\tilde{\varphi}(\underbrace{-x, \dots, -x}_{n \text{ times}}) \tag{10.13}$$

for all $x \in X$.

Proof. Letting $x_1 = \cdots = x_n = x$ in (10.12), we get

$$\left\| nf(nx) - n^3 f(x) \right\| \le \varphi(\underbrace{x, \ldots, x}_{n \text{ times}}) \tag{10.14}$$

for all $x \in X$. Replacing x by $-x$ in (10.14), we get

$$\left\| nf(-nx) - n^3 f(-x) \right\| \le \varphi(\underbrace{-x, \ldots, -x}_{n \text{ times}}) \tag{10.15}$$

for all $x \in X$. Let $g(x) := f(x) + f(-x)$ for all $x \in X$. It follows from (10.14) and (10.15) that

$$\left\| ng\,(nx) - n^3 g(x) \right\| \le \varphi(\underbrace{x, \ldots, x}_{n \text{ times}}) + \varphi(\underbrace{-x, \ldots, -x}_{n \text{ times}}) \tag{10.16}$$

for all $x \in X$. So

$$\left\| g(x) - n^2 g\left(\frac{x}{n}\right) \right\| \le \frac{1}{n}\varphi\left(\underbrace{\frac{x}{n}, \ldots, \frac{x}{n}}_{n \text{ times}}\right) + \frac{1}{n}\varphi\left(\underbrace{-\frac{x}{n}, \ldots, -\frac{x}{n}}_{n \text{ times}}\right)$$

for all $x \in X$. Hence

$$\begin{aligned} \| n^{2l} g\left(\frac{x}{n^l}\right) - n^{2m} g\left(\frac{x}{n^m}\right) \| \le & \sum_{j=l+1}^{m} \frac{n^{2j}}{n^3} \varphi\left(\underbrace{\frac{x}{n^j}, \ldots, \frac{x}{n^j}}_{n \text{ times}}\right) \\ & + \sum_{j=l+1}^{m} \frac{n^{2j}}{n^3} \varphi\left(\underbrace{-\frac{x}{n^j}, \ldots, -\frac{x}{n^j}}_{n \text{ times}}\right) \end{aligned} \tag{10.17}$$

for all nonnegative integers m and l with $m > l$ and all $x \in X$. It follows from (10.11) and (10.17) that the sequence $\{n^{2k} g(\frac{x}{n^k})\}$ is Cauchy for all $x \in X$. Since Y is complete, the sequence $\{n^{2k} g(\frac{x}{n^k})\}$ converges. So one can define the mapping $Q : X \to Y$ by

$$Q(x) := \lim_{k\to\infty} n^{2k} g\left(\frac{x}{n^k}\right)$$

for all $x \in X$. By (10.11) and (10.12),

$$\begin{aligned}
&\|DQ(x_1,\ldots,x_n)\| \\
&= \lim_{k\to\infty} n^{2k} \left\| Dg\left(\frac{x_1}{n^k},\ldots,\frac{x_n}{n^k}\right)\right\| \\
&\le \lim_{k\to\infty} n^{2k} \left(\varphi\left(\frac{x_1}{n^k},\ldots,\frac{x_n}{n^k}\right) + \varphi\left(-\frac{x_1}{n^k},\ldots,-\frac{x_n}{n^k}\right)\right) = 0
\end{aligned}$$

for all $x_1,\ldots,x_n \in X$. So $DQ(x_1,\ldots,x_n) = 0$. Since g is even, the mapping Q is even. By Lemma 10.2, the mapping $Q : X \to Y$ is quadratic. Moreover, letting $l = 0$ and passing the limit $m \to \infty$ in (10.17), we get (10.13). So there exists a quadratic mapping $Q : X \to Y$ satisfying (10.6) and (10.13).

Now, let $Q' : X \to Y$ be another quadratic mapping satisfying (10.6) and (10.13). Then we have

$$\begin{aligned}
&\|Q(x) - Q'^{2q} \left\| Q\left(\frac{x}{n^q}\right) - Q'\left(\frac{x}{n^q}\right)\right\| \\
&\le n^{2q} \begin{pmatrix} \left\|Q\left(\frac{x}{n^q}\right) - f\left(\frac{x}{n^q}\right) - f\left(\frac{-x}{n^q}\right)\right\| \\ + \left\|Q'\left(\frac{x}{n^q}\right) - f\left(\frac{x}{n^q}\right) - f\left(\frac{-x}{n^q}\right)\right\| \end{pmatrix} \\
&\le \frac{2\cdot n^{2q}}{n^3}\tilde{\varphi}\left(\underbrace{\frac{x}{n^q},\ldots,\frac{x}{n^q}}_{n \text{ times}}\right) + \frac{2\cdot n^{2q}}{n^3}\tilde{\varphi}\left(\underbrace{\frac{-x}{n^q},\ldots,\frac{-x}{n^q}}_{n \text{ times}}\right),
\end{aligned}$$

which tends to zero as $q \to \infty$ for all $x \in X$. So we can conclude that $Q(x) = Q'(x)$ for all $x \in X$. This proves the uniqueness of Q.

Corollary 10.4. *Let $p > 2$ and θ be positive real numbers, and let $f : X \to Y$ be a mapping such that*

$$\|Df(x_1,\ldots,x_n)\| \le \theta \sum_{j=1}^{n} ||x_j||^p \tag{10.18}$$

for all $x_1,\ldots,x_n \in X$. Then there exists a unique quadratic mapping $Q : X \to Y$ satisfying (10.6) *and*

$$\|f(x) + f(-x) - Q(x)\| \le \frac{2\theta}{n^p - n^2}||x||^p$$

for all $x \in X$.

Proof. Define $\varphi(x_1,\ldots,x_n) = \theta \sum_{j=1}^{n} ||x_j||^p$, and apply Theorem 10.3 to get the desired result.

Theorem 10.5. *Let $f : X \to Y$ be a mapping satisfying $f(0) = 0$ for which there exists a function $\varphi : X^n \to [0,\infty)$ satisfying* (10.12) *such that*

$$\tilde{\varphi}(x_1,\ldots,x_n) := \sum_{j=0}^{\infty} n^{-2j}\varphi(n^j x_1,\ldots,n^j x_n) < \infty \tag{10.19}$$

for all $x_1, \ldots, x_n \in X$. *Then there exists a unique quadratic mapping* $Q : X \to Y$ *satisfying* (10.6) *and*

$$\|f(x) + f(-x) - Q(x)\| \leq \frac{1}{n^3}\tilde{\varphi}(\underbrace{x, \ldots, x}_{n \text{ times}}) + \frac{1}{n^3}\tilde{\varphi}(\underbrace{-x, \ldots, -x}_{n \text{ times}}) \tag{10.20}$$

for all $x \in X$.

Proof. It follows from (10.16) that

$$\left\|g(x) - \frac{1}{n^2}g(nx)\right\| \leq \frac{1}{n^3}\varphi(\underbrace{x, \ldots, x}_{n \text{ times}}) + \frac{1}{n^3}\varphi(\underbrace{-x, \ldots, -x}_{n \text{ times}})$$

for all $x \in X$. So

$$\left\|\frac{1}{n^{2l}}g(n^l x) - \frac{1}{n^{2m}}g(n^m x)\right\| \leq \sum_{j=l}^{m-1} \frac{1}{n^{2j+3}}\varphi(\underbrace{n^j x, \ldots, n^j x}_{n \text{ times}})$$
$$+ \sum_{j=l}^{m-1} \frac{1}{n^{2j+3}}\varphi(\underbrace{-n^j x, \ldots, -n^j x}_{n \text{ times}}) \tag{10.21}$$

for all nonnegative integers m and l with $m > l$ and all $x \in X$. It follows from (10.19) and (10.21) that the sequence $\{\frac{1}{n^{2k}}g(n^k x)\}$ is Cauchy for all $x \in X$. Since Y is complete, the sequence $\{\frac{1}{n^{2k}}g(n^k x)\}$ converges. So one can define the mapping $Q : X \to Y$ by

$$Q(x) := \lim_{k\to\infty} \frac{1}{n^{2k}}g(n^k x)$$

for all $x \in X$.

By (10.12) and (10.19),

$$\begin{aligned}
&\|DQ(x_1, \ldots, x_n)\| \\
&= \lim_{k\to\infty} \frac{1}{n^{2k}}\|Dg(n^k x_1, \ldots, n^k x_n)\| \\
&\leq \lim_{k\to\infty} \frac{1}{n^{2k}}(\varphi(n^k x_1, \ldots, n^k x_n) + \varphi(-n^k x_1, \ldots, -n^k x_n)) = 0
\end{aligned}$$

for all $x_1, \ldots, x_n \in X$. So $DQ(x_1, \ldots, x_n) = 0$. Since g is even, the mapping Q is even. By Lemma 10.2, the mapping $Q : X \to Y$ is quadratic. Moreover, letting $l = 0$ and passing the limit $m \to \infty$ in (10.21), we get (10.20). So there exists a quadratic mapping $Q : X \to Y$ satisfying (10.6) and (10.20). The rest of the proof is similar to the proof of Theorem 10.3.

Corollary 10.6. *Let $p < 2$ and θ be positive real numbers, and let $f : X \to Y$ be a mapping satisfying* (10.18). *Then there exists a unique quadratic mapping $Q : X \to Y$ satisfying* (10.6) *and*

$$\|f(x) + f(-x) - Q(x)\| \le \frac{2\theta}{n^2 - n^p}||x||^p$$

for all $x \in X$.

Proof. Define $\varphi(x_1, \ldots, x_n) = \theta \sum_{j=1}^{n} ||x_j||^p$, and apply Theorem 10.5 to get the desired result.

10.3 On the Stability of a Cauchy Additive-Quadratic Functional Equation Associated with Inner Product Spaces

Throughout this section, assume that V and W are real vector spaces and that n is a fixed integer greater than 1.

One can generalize the functional equation (10.7) to the functional equation (10.3).

We investigate the quadratic functional equation (10.3).

Lemma 10.7. *If a mapping $f : V \to W$ satisfies*

$$nf\left(\sum_{i=1}^{n} x_i\right) + \sum_{i=1}^{n} f\left(nx_i - \sum_{j=1}^{n} x_j\right) = \frac{n^2+n}{2}\sum_{i=1}^{n} f(x_i) + \frac{n^2-n}{2}\sum_{i=1}^{n} f(-x_i) \tag{10.22}$$

for all $x_1, \ldots, x_n \in V$, then the mapping $f : V \to W$ satisfies the Cauchy additive-quadratic functional equation (10.7).

Proof. The proof is similar to the proof of Lemma 10.2.

For a given mapping $f : X \to Y$, we define

$$Cf(x_1, \ldots, x_n) := nf\left(\sum_{i=1}^{n} x_i\right) + \sum_{i=1}^{n} f\left(nx_i - \sum_{j=1}^{n} x_j\right) - \frac{n^2+n}{2}\sum_{i=1}^{n} f(x_i) - \frac{n^2-n}{2}\sum_{i=1}^{n} f(-x_i)$$

for all $x_1, \ldots, x_n \in X$.

Now we prove the Hyers–Ulam stability of the quadratic functional equation $Cf(x_1, \ldots, x_n) = 0$ in real Banach spaces.

Theorem 10.8. *Let $f : X \to Y$ be a mapping satisfying $f(0) = 0$ for which there exists a function $\varphi : X^n \to [0, \infty)$ satisfying* (10.11) *and*

$$\|Cf(x_1, \ldots, x_n)\| \le \varphi(x_1, \ldots, x_n) \tag{10.23}$$

for all $x_1, \ldots, x_n \in X$. Then there exists a unique quadratic mapping $Q : X \to Y$ satisfying (10.22) *and*

$$\|f(x) + f(-x) - Q(x)\| \le \frac{1}{n^3}\tilde{\varphi}(\underbrace{x, \ldots, x}_{n \text{ times}}) + \frac{1}{n^3}\tilde{\varphi}(\underbrace{-x, \ldots, -x}_{n \text{ times}})$$

for all $x \in X$.

Proof. Letting $x_1 = \cdots = x_n = x$ in (10.23), we get

$$\left\| nf(nx) - \frac{n^3 + n^2}{2} f(x) - \frac{n^3 - n^2}{2} f(-x) \right\| \le \varphi(\underbrace{x, \ldots, x}_{n \text{ times}}) \tag{10.24}$$

for all $x \in X$. Replacing x by $-x$ in (10.24), we get

$$\left\| nf(-nx) - \frac{n^3 + n^2}{2} f(-x) - \frac{n^3 - n^2}{2} f(x) \right\| \le \varphi(\underbrace{-x, \ldots, -x}_{n \text{ times}}) \tag{10.25}$$

for all $x \in X$. Let $g(x) := f(x) + f(-x)$ for all $x \in X$. It follows from (10.24) and (10.25) that

$$\left\| ng\,(nx) - n^3 g(x) \right\| \le \varphi(\underbrace{x, \ldots, x}_{n \text{ times}}) + \varphi(\underbrace{-x, \ldots, -x}_{n \text{ times}}) \tag{10.26}$$

for all $x \in X$. The rest of the proof is the same as in the proof of Theorem 10.3.

Corollary 10.9. *Let $p > 2$ and θ be positive real numbers, and let $f : X \to Y$ be a mapping such that*

$$\|Cf(x_1, \ldots, x_n)\| \le \theta \sum_{j=1}^{n} ||x_j||^p \tag{10.27}$$

for all $x_1, \ldots, x_n \in X$. Then there exists a unique quadratic mapping $Q : X \to Y$ satisfying (10.22) *and*

$$\|f(x) + f(-x) - Q(x)\| \le \frac{2\theta}{n^p - n^2} ||x||^p$$

for all $x \in X$.

Proof. Define $\varphi(x_1, \ldots, x_n) = \theta \sum_{j=1}^{n} ||x_j||^p$, and apply Theorem 10.8 to get the desired result.

Theorem 10.10. *Let $f : X \to Y$ be a mapping satisfying $f(0) = 0$ for which there exists a function $\varphi : X^n \to [0, \infty)$ satisfying* (10.19) *and* (10.23)*. Then there exists a unique quadratic mapping $Q : X \to Y$ satisfying* (10.22) *and*

$$\|f(x) + f(-x) - Q(x)\| \leq \frac{1}{n^3}\tilde{\varphi}(\underbrace{x, \ldots, x}_{n \text{ times}}) + \frac{1}{n^3}\tilde{\varphi}(\underbrace{-x, \ldots, -x}_{n \text{ times}})$$

for all $x \in X$.

Proof. It follows from (10.26) that

$$\left\| g(x) - \frac{1}{n^2} g(nx) \right\| \leq \frac{1}{n^3}\varphi(\underbrace{x, \ldots, x}_{n \text{ times}}) + \frac{1}{n^3}\varphi(\underbrace{-x, \ldots, -x}_{n \text{ times}})$$

for all $x \in X$. The rest of the proof is similar to the proofs of Theorems 10.3 and 10.8.

Corollary 10.11. *Let $p < 2$ and θ be positive real numbers, and let $f : X \to Y$ be a mapping satisfying* (10.27)*. Then there exists a unique quadratic mapping $Q : X \to Y$ satisfying* (10.22) *and*

$$\|f(x) + f(-x) - Q(x)\| \leq \frac{2\theta}{n^2 - n^p}||x||^p$$

for all $x \in X$.

Proof. Define $\varphi(x_1, \ldots, x_n) = \theta \sum_{j=1}^{n} ||x_j||^p$, and apply Theorem 10.10 to get the desired result.

Theorem 10.12. *Let $f : X \to Y$ be a mapping satisfying $f(0) = 0$ for which there exists a function $\varphi : X^n \to [0, \infty)$ satisfying* (10.23) *and*

$$\Psi(x_1, \ldots, x_n) := \sum_{j=1}^{\infty} n^j \varphi\left(\frac{x_1}{n^j}, \ldots, \frac{x_n}{n^j}\right) < \infty \tag{10.28}$$

for all $x_1, \ldots, x_n \in X$. Then there exists a unique Cauchy additive mapping $A : X \to Y$ satisfying (10.22) *and*

$$\|f(x) - f(-x) - A(x)\| \leq \frac{1}{n^2}\Psi(\underbrace{x, \ldots, x}_{n \text{ times}}) + \frac{1}{n^2}\Psi(\underbrace{-x, \ldots, -x}_{n \text{ times}})$$

for all $x \in X$.

Proof. Let $h(x) := f(x) - f(-x)$ for all $x \in X$. Then h is odd. It follows from (10.24) and (10.25) that

$$\left\| nh\,(nx) - n^2 h(x) \right\| \le \varphi(\underbrace{x, \ldots, x}_{n\ \text{times}}) + \varphi(\underbrace{-x, \ldots, -x}_{n\ \text{times}}) \tag{10.29}$$

for all $x \in X$. The rest of the proof is similar to the proof of Theorem 10.3.

Corollary 10.13. *Let $p > 1$ and θ be positive real numbers, and let $f : X \to Y$ be a mapping satisfying* (10.27)*. Then there exists a unique Cauchy additive mapping $A : X \to Y$ satisfying* (10.22) *and*

$$\|f(x) - f(-x) - A(x)\| \le \frac{2\theta}{n^p - n} ||x||^p$$

for all $x \in X$.

Proof. Define $\varphi(x_1, \ldots, x_n) = \theta \sum_{j=1}^{n} ||x_j||^p$, and apply Theorem 10.12 to get the desired result.

Theorem 10.14. *Let $f : X \to Y$ be a mapping satisfying $f(0) = 0$ for which there exists a function $\varphi : X^n \to [0, \infty)$ satisfying* (10.23) *such that*

$$\Psi(x_1, \ldots, x_n) := \sum_{j=0}^{\infty} n^{-j} \varphi(n^j x_1, \ldots, n^j x_n) < \infty \tag{10.30}$$

for all $x_1, \ldots, x_n \in X$. Then there exists a unique Cauchy additive mapping $A : X \to Y$ satisfying (10.22) *and*

$$\|f(x) - f(-x) - A(x)\| \le \frac{1}{n^2} \Psi(\underbrace{x, \ldots, x}_{n\ \text{times}}) + \frac{1}{n^2} \Psi(\underbrace{-x, \ldots, -x}_{n\ \text{times}})$$

for all $x \in X$.

Proof. It follows from (10.29) that

$$\left\| h(x) - \frac{1}{n} h(nx) \right\| \le \frac{1}{n^2} \varphi(\underbrace{x, \ldots, x}_{n\ \text{times}}) + \frac{1}{n^2} \varphi(\underbrace{-x, \ldots, -x}_{n\ \text{times}})$$

for all $x \in X$. The rest of the proof is similar to the proof of Theorem 10.3.

Corollary 10.15. *Let $p < 1$ and θ be positive real numbers, and let $f : X \to Y$ be a mapping satisfying* (10.27)*. Then there exists a unique Cauchy additive mapping $A : X \to Y$ satisfying* (10.22) *and*

$$\|f(x) - f(-x) - A(x)\| \le \frac{2\theta}{n - n^p} ||x||^p$$

for all $x \in X$.

Proof. Define $\varphi(x_1,\ldots,x_n) = \theta\sum_{j=1}^{n}||x_j||^p$, and apply Theorem 10.14 to get the desired result.

Note that

$$\sum_{j=1}^{\infty} n^j\varphi\left(\frac{x_1}{n^j},\ldots,\frac{x_n}{n^j}\right) \le \sum_{j=1}^{\infty} n^{2j}\varphi\left(\frac{x_1}{n^j},\ldots,\frac{x_n}{n^j}\right).$$

Combining Theorems 10.8 and 10.12, we obtain the following result.

Theorem 10.16. *Let $f : X \to Y$ be a mapping satisfying $f(0) = 0$ for which there exists a function $\varphi : X^n \to [0,\infty)$ satisfying* (10.11) *and* (10.23). *Then there exists a unique Cauchy additive mapping $A : X \to Y$ satisfying* (10.22) *and a unique quadratic mapping $Q : X \to Y$ satisfying* (10.22) *such that*

$$\begin{aligned}\|2f(x) - A(x) - Q(x)\| \le{}& \frac{1}{n^3}\tilde{\varphi}(\underbrace{x,\ldots,x}_{n\ \text{times}}) + \frac{1}{n^3}\tilde{\varphi}(\underbrace{-x,\ldots,-x}_{n\ \text{times}}) \\ &+\frac{1}{n^2}\Psi(\underbrace{x,\ldots,x}_{n\ \text{times}}) + \frac{1}{n^2}\Psi(\underbrace{-x,\ldots,-x}_{n\ \text{times}})\end{aligned}$$

for all $x \in X$, where $\tilde{\varphi}$ and Ψ are defined in (10.11) *and* (10.28), *respectively.*

Corollary 10.17. *Let $p > 2$ and θ be positive real numbers, and let $f : X \to Y$ be a mapping satisfying* (10.27). *Then there exists a unique Cauchy additive mapping $A : X \to Y$ satisfying* (10.22) *and a unique quadratic mapping $Q : X \to Y$ satisfying* (10.22) *such that*

$$\|2f(x) - A(x) - Q(x)\| \le \left(\frac{2}{n^p - n} + \frac{2}{n^p - n^2}\right)\theta||x||^p$$

for all $x \in X$.

Proof. Define $\varphi(x_1,\ldots,x_n) = \theta\sum_{j=1}^{n}||x_j||^p$, and apply Theorem 10.16 to get the desired result.

Note that

$$\sum_{j=0}^{\infty} n^{-2j}\varphi(n^jx_1,\ldots,n^jx_n) \le \sum_{j=0}^{\infty} n^{-j}\varphi(n^jx_1,\ldots,n^jx_n).$$

Combining Theorems 10.10 and 10.14, we obtain the following result.

Theorem 10.18. *Let $f : X \to Y$ be a mapping satisfying $f(0) = 0$ for which there exists a function $\varphi : X^n \to [0,\infty)$ satisfying* (10.23) *and* (10.30). *Then there exist a unique Cauchy additive mapping $A : X \to Y$ satisfying* (10.22) *and a unique quadratic mapping $Q : X \to Y$ satisfying* (10.22) *such that*

$$\|2f(x) - A(x) - Q(x)\| \le \frac{1}{n^3}\tilde{\varphi}(\underbrace{x, \dots, x}_{n \text{ times}}) + \frac{1}{n^3}\tilde{\varphi}(\underbrace{-x, \dots, -x}_{n \text{ times}})$$
$$+\frac{1}{n^2}\Psi(\underbrace{x, \dots, x}_{n \text{ times}}) + \frac{1}{n^2}\Psi(\underbrace{-x, \dots, -x}_{n \text{ times}})$$

for all $x \in X$, *where* $\tilde{\varphi}$ *and* Ψ *are defined in* (10.19) *and* (10.30), *respectively.*

Corollary 10.19. *Let* $p < 1$ *and* θ *be positive real numbers, and let* $f : X \to Y$ *be a mapping satisfying* (10.27)*. Then there exist a unique Cauchy additive mapping* $A : X \to Y$ *satisfying* (10.22) *and a unique quadratic mapping* $Q : X \to Y$ *satisfying* (10.22) *such that*

$$\|2f(x) - A(x) - Q(x)\| \le \left(\frac{2}{n - n^p} + \frac{2}{n^2 - n^p}\right)\theta||x||^p$$

for all $x \in X$.

Proof. Define $\varphi(x_1, \dots, x_n) = \theta \sum_{j=1}^{n} ||x_j||^p$, and apply Theorem 10.18 to get the desired result.

References

1. T. Aoki, On the stability of the linear transformation in Banach spaces. J. Math. Soc. Jpn. **2**, 64–66 (1950)
2. C. Baak, M.S. Moslehian, On the stability of J^*-homomorphisms. Nonlinear Anal. Theory Methods Appl. **63**, 42–48 (2005)
3. P.W. Cholewa, Remarks on the stability of functional equations. Aequationes Math. **27**, 76–86 (1984)
4. S. Czerwik, On the stability of the quadratic mapping in normed spaces. Abh. Math. Sem. Univ. Hamburg **62**, 59–64 (1992)
5. G. Găvruta, A generalization of the Hyers-Ulam-Rassias stability of approximately additive mappings. J. Math. Anal. Appl. **184**, 431–436 (1994)
6. D.H. Hyers, On the stability of the linear functional equation. Proc. Natl. Acad. Sci. U.S.A. **27**, 222–224 (1941)
7. K. Jun, H. Kim, Ulam stability problem for quadratic mappings of Euler-Lagrange. Nonlinear Anal. Theory Methods Appl. **61**, 1093–1104 (2005)
8. K. Jun, H. Kim, On the generalized A-quadratic mappings associate with the variance of a discrete-type distribution. Nonlinear Anal. Theory Methods Appl. **62**, 975–987 (2005)
9. S. Jung, *Hyers-Ulam-Rassias Stability of Functional Equations in Mathematical Analysis* (Hadronic Press, Palm Harbor, 2001)
10. C. Park, Generalized quadratic mappings in several variables. Nonlinear Anal. Theory Methods Appl. **57**, 713–722 (2004)
11. C. Park, Cauchy-Rassias stability of a generalized Trif's mapping in Banach modules and its applications. Nonlinear Anal. Theory Methods Appl. **62**, 595–613 (2005)
12. T.M. Rassias, On the stability of the linear mapping in Banach spaces. Proc. Am. Math. Soc. **72**, 297–300 (1978)

13. T.M. Rassias, New characterizations of inner product spaces. Bull. Sci. Math. **108**, 95–99 (1984)
14. T.M. Rassias, On the stability of the quadratic functional equation and its applications. Stud. Univ. Babes-Bolyai **XLIII**, 89–124 (1998)
15. T.M. Rassias, The problem of S.M. Ulam for approximately multiplicative mappings. J. Math. Anal. Appl. **246**, 352–378 (2000)
16. T.M. Rassias, On the stability of functional equations in Banach spaces. J. Math. Anal. Appl. **251**, 264–284 (2000)
17. T.M. Rassias, On the stability of functional equations and a problem of Ulam. Acta Appl. Math. **62**, 23–130 (2000)
18. T.M. Rassias, P. Šemrl, On the Hyers-Ulam stability of linear mappings. J. Math. Anal. Appl. **173**, 325–338 (1993)
19. T.M. Rassias, K. Shibata, Variational problem of some quadratic functionals in complex analysis. J. Math. Anal. Appl. **228**, 234–253 (1998)
20. S. Skof, Proprietà locali e approssimazione di operatori. Rend. Sem. Mat. Fis. Milano **53**, 113–129 (1983)
21. S.M. Ulam, *A Collection of the Mathematical Problems* (Interscience Publishers, New York, 1960)

Chapter 11
Fuzzy Partial Metric Spaces

Fariha Jumaa Amer

Abstract We define fuzzy partial metric space by following the development of fuzzy metric space. The concept of partial metric is investigated to generalize metric space. In particular, the self-distance for any point need not be equal to zero. The main part of this research concentrates on how the idea of fuzzy partial metric space can be defined by following the same idea in the development of fuzzy metric space. We will bring together the necessary basic concepts to generalize the fuzzy metric spaces and their topological properties into fuzzy partial metric spaces, under the bewildering axiom that the self-distance of any point need not to be zero.

11.1 Introduction

The theory of fuzzy sets was introduced by Zadeh in 1965, and since then there has been tremendous interest in the subject due to its diverse applications ranging from engineering and computer science to social behavior studies. Partial metric spaces were originally developed by Mattews [6] to provide mechanism generalizing metric space theories. Many authors have introduced the concepts of fuzzy metric in different ways (see [1, 2, 5]). In particular, George and Veeramani [3] generalized the concept of probabilistic metric space given by Menger [7]. George and Veeramani [3] modified the concept of fuzzy metric space introduced by Kramosiland and Michalek [4] and obtained a Housdorff and first countable topology on this modified fuzzy metric space, (X, τ_M) is a Hausdorff first countable topological space.

In this paper we will define fuzzy partial metric space by the following the development of fuzzy metric space.

Definition 11.1 (See [9]). Fuzzy sets are considered with respect to a nonempty set X of elements of interest. The essential idea is that each element $x \in X$ is assigned a membership grade $u(x)$ taking values in $[0, 1]$, with $u(x) = 0$ corresponding to non-membership $0 < u(x) < 1$ to partial membership, and $u(x) = 1$ to full membership. And he gave a binary operation $* : [0, 1] \times [0, 1] \to [0, 1]$ is a continuous t-norm, if

F.J. Amer (✉)
Department of Mathematics, University of Al-Mergeb, Al-Khoms, Libya
e-mail: fa.fa82@yahoo.com

G.A. Anastassiou, O. Duman (eds.), *Computational Analysis*, Springer Proceedings in Mathematics & Statistics 155, DOI 10.1007/978-3-319-28443-9_11

$([0, 1], *)$ is an Abelian (topological) monoid with the unit 1 such that $a * b \leq c * d$ whenever $a \leq c$ and $b \leq d$ for all $a, b, c, d \in [0, 1]$. Two typical examples of continuous t-norm are $a * b = ab$ and $a * b = \mathrm{Min}(a, b)$ for all $a, b \in [0, 1]$

Definition 11.2. A metric on a set space X is a function $d : X \times X \to R$ with the following properties:

(i) $d(x, y) \geq 0$ for all $x, y \in X$; equality holds if and only if $x = y$,
(ii) $d(x, y) = d(y, x)$ for all $x, y \in X$,
(iii) $d(x, y) + d(y, z) \geq d(x, z)$ for all $x, y, z \in X$ (the triangle inequality).

We call $d(x, y)$ the distance between x and y , and we call pair (X, d) consisting of the set X and the metric d, a metric space.

Definition 11.3. Let be (X, d) a metric space. For $x \in X$ and $\epsilon > 0$ define the open ball of radius ϵ centered at x to be the set

$$B_d(x, \epsilon) = y \in X | d(x, y) < \epsilon,$$

and define the closed ball of radius ϵ centered at x to be the set

$$\overline{B}_d(x, \epsilon) = y \in X | d(x, y) \leq \epsilon.$$

Theorem 11.4. *Every metric space is Hausdorff.*

Definition 11.5 (See [6]). A partial metric space is a pair $(X, p : X \times X \to R)$ such that

(i) $p(x, x) \leq p(x, y)$,
(ii) if $p(x, x) = p(y, y) = p(x, y)$, then $x = y$,
(iii) $p(x, y) = p(y, x)$,
(iv) $p(x, z) \leq p(x, y) + p(y, z) - p(y, y)$ for all $x, y, z \in X$.

Note that the self-distance of any point need not be zero, hence the idea of generalizing metrics so that a metric on a non-empty set X is precisely a partial metric p on X such that for any $x \in X$, $p(x, x) = 0$.

Similar to the case of metric space, a partial metric space p on X.

Definition 11.6. Let (X, p) be a partial metric space. For any $x \in X$ and $\epsilon > 0$, we define, respectively, the open and closed ball for the partial metric p by setting

$$B_\epsilon(x) = y \in X : p(x, y) < \epsilon,$$

$$\overline{B}_\epsilon(x) = y \in X : p(x, y) \leq \epsilon.$$

Example 11.7 (See [8]). Consider the function $p : R^- \times R^- \to R^+$ defined by $p(x,y) = -\min\{x,y\}$ for any $x, y \in X$. The pair (R^-, p) is a partial metric space for which p is called the usual partial metric on R^-, and where the self-distance for any point $x \in R^-$ is its absolute value. Indeed, for any $x, y, z \in R^-$,

1. $\min\{x,y\} \leq x$, so $p(x,y) \geq p(x,x) = -x$.
2. Suppose that $p(x,x) = p(x,y) = p(y,y)$, it then follows that $-x = -y$, hence $x = y$.
3. It is obvious that $p(x,y) = p(y,x)$.
4. One verifies that

$$\min\{x,z\}\min\{x,y\} + \min\{y,z\} - \min\{y,y\}$$

by considering the cases $y \leq x \leq z, x \leq y \leq z$ and $x \leq z \leq y$; hence

$$p(x,z) \leq p(x,y) + p(y,z) - p(y,y).$$

The open balls are of the form $B_\epsilon(x) = y \in R^- : -\min\{x,y\} < \epsilon = (-\epsilon, 0)$ with $x \geq -\epsilon$ otherwise, if $x < -\epsilon$, then $p(x,x) = -x > \epsilon$ and $B_\epsilon(x) = \varphi$. Suppose that $y \in B_\epsilon(x)$, then $-\min\{x,y\} < \epsilon$ which implies that $\min\{x,y\} > \epsilon$, hence $y > -\epsilon$.

Definition 11.8. A sequence (x_n) in a partial metric space (X,p) converges to $x \in X$, and one writes $\lim\limits_{n\to\infty} x_n = x$ if for any $\epsilon > 0$ such that $x \in B_\epsilon(x)$, there exists $N \geq 1$ so that for any $n \geq N$, $x_n \in B_\epsilon(x)$.

Proposition 11.9. *Suppose that (x_n) is a sequence in a partial metric space (X,p) and $x \in X$. Then $x_n \to x$ if and only if $\lim_{n\to\infty} p(x_n, x) = p(x,x)$.*

Definition 11.10 (See [8]). A sequence (x_n) in a partial metric space (X,p) is a Cauchy sequence if it is a Cauchy sequence in the induced metric space (X, p_m). A partial metric space is said to be complete if its induced metric space is complete.

Lemma 11.11 (See [8]). *Suppose that (x_n) is a sequence in a partial metric space (X,p). Then (x_n) is a Cauchy sequence if and only if $\lim\limits_{k,n\to\infty} p(x_n, x_k)$ exists.*

The concept of partial metric spaces is investigated to generalize metric spaces. In particular, the self-distance for any point need not be equal to zero. This idea of nonzero self-distance is motivated by experience from computer science, and this is a relatively new field and has vast application potentials in the study of computer domains and semantics. Moreover, there have been different approaches in this area when it comes to applying the developing mathematical concepts to computer science.

11.2 Fuzzy Metric Space

Many authors have introduced the concepts of fuzzy metric in different ways (see [1, 2, 4]). In particular, George and Veeramani [3] generalized the concept of probabilistic metric space given by Menger [7].

We shall recall some different definitions of fuzzy metric space which were given by different authors and the topology in such space as well as the open ball and closed ball that are related to this topology. In addition, we will mention some properties of fuzzy metric space.

Definition 11.12 (See [3]). A 3-tuple $(X, M, *)$ is said to be a fuzzy metric space if X is a nonempty set, $*$ is a continuous t-norm, and M is a fuzzy set, $M : X \times X \times [0, \infty) \to [0, 1]$ is a mapping (called fuzzy metric) which satisfies the following properties: for every $x, y, z \in X$ and $t, s > 0$,

(i) $M(x, y, t) > 0$,
(ii) $M(x, y, t) = 1$ if and only if $x = y$,
(iii) $M(x, y, t) = M(y, x, t)$,
(iv) $M(x, y, t) * M(y, z, s) = M(x, z, t + s)$,
(v) $M(x, y, .) : [0, \infty) \to [0, 1]$ is continuous. If $(X, M, *)$ is a fuzzy metric space, we will say that M is a fuzzy metric on X.

In metric space (X, d) if we define $a * b = ab$ and $M(x, y, t) = \dfrac{t}{t + d(x, y)}$, then $(X, M, *)$ is a fuzzy metric space. We call this M as the standard fuzzy metric space induced by d. Even if we take $a * b = \min(a, b)$, $(X, M, *)$ will be a fuzzy metric space.

Definition 11.13 (See [3]). Let $(X, M, *)$ be a fuzzy metric space. We define open ball $B(x, r, t)$ for $t > 0$ with center $x \in X$ and radius r, $0 < r < 1$, as $B(x, r, t) = y \in X : M(x, y, t) > 1 - r$.

Definition 11.14 (See [3]). Let $(X, M, *)$ be a fuzzy metric space defined by

$$\tau = \{A \subset X : x \in A \;\Leftrightarrow\; \exists r, t > 0, 0 < r < 1, \text{ such that } B(x, r, t) \subset A\}.$$

Then τ is a topology on X.

And every fuzzy metric M on X generates a topology τ_M on X which has a base the family of sets of the form $B_x(r, t) : x \in X, r \in (0, 1), t > 0$, where

$$B_x(x, r, t) = y \in X : M(x, y, t) > 1 - r$$

is a neighborhood of $x \in X$ for all $r \in (0, 1)$ and $t > 0$, $(X, M, *)$ is a Hausdorff first countable topological space. Moreover, if (X, d) is a metric space, then the topology generated by d coincides with the topology M_d generated by the induced fuzzy metric M_d. If $(X, M, *)$ is a fuzzy metric space and τ is the topology induced by the fuzzy metric then for a sequence (x_n) in X, (x_n) converges to x in X if and only if $M(x_n, x, t)$ tends to 1 as n tends to ∞ for $t > 0$.

Definition 11.15 (See [3]). A sequence (x_n) in a fuzzy metric space $(X, M, *)$ is said to be a Cauchy sequence if for each ϵ, $0 < \epsilon < 1$ and $t > 0$, there exists $n_0 \in N$ such that $M(x_n, x_m, t) > 1 - \epsilon$ for all $n, m \geq n_0$.

Definition 11.16 (See [3]). A fuzzy metric space is said to be complete if every Cauchy sequence is convergent.

The induced metric space $(X, M, *)$ is complete if and only if the metric (X, d) is complete where $M(x, y, t) = \dfrac{t}{t + d(x, y)}$ for all $x, y \in X$ and $t \in (0, \infty)$.

Definition 11.17. Let $(X_1, M_1, *)$ and $(X_2, M_2, *)$ be the given fuzzy metric spaces. For (x_1, x_2), $(y_1, y_2) \in X_1 \times X_2$, $t > 0$, if we define

$$M((x_1, x_2), (y_1, y_2), t) = M_1(x_1, y_1, t) * M_1(x_2, y_2, t),$$

then M is a fuzzy metric on $X_1 \times X_2$. Further if X_1 and X_2 are complete fuzzy metric spaces, then the product space $X_1 \times X_2$ is also a complete fuzzy metric space.

11.3 Main Result

We explained the definition of metric space and partial metric space, and we have noted that a partial metric space is generalizing of metric space. In this chapter we will give the concept of fuzzy partial metric space which is generalizing of fuzzy metric space.

Definition 11.18. Let p be partial metric space, the triple $(X, M_p, *)$ is said to be a fuzzy partial metric space if X is a nonempty set, $*$ is a continuous t-norm and M_p is a fuzzy set on $X \times X \times (0, \infty)$ satisfying the following conditions, for all $x, y, z \in X$, $s, t \geq 0$:

(i) $M_p(x, y, 0) = 0$,
(ii) $M_p(x, y, t) = M_p(y, x, t)$,
(iii) $M_p(x, y, t) * M_p(y, z, s) \leq M_p(x, z, t + s)$,
(iv) $M_p(x, y, t) \leq 1$ for all $t > 0$ and $M_p(x, y, t) = 1$ if and only if $p(x, y) = 0$,
(v) $M_p(x, y, .) : [0, \infty) \rightarrow [0, 1]$ is continuous,

where $M_p(x, y, t) = \dfrac{t}{t + p(x, y)}$. If $(X, M_p, *)$ is a fuzzy partial metric space, we will say that M_p is a fuzzy partial metric on X.

Definition 11.19. In a partial metric space (X,p) the 3-tuple $(X, M_p, *)$ where $M_p(x,y,t) = \dfrac{t}{t+p(x,y)}$ and $a * b = ab$ is a fuzzy partial metric space. This M_p is called the standard fuzzy metric induced by p.

Example 11.20. Let $X = R^-$ and (X,p) be a partial metric space. Let $p(x,y) = -\min\{x,y\}$, $x,y \in R^-$. Denote $a * b = ab$ for all a, $b \in [0,1]$ and let M_p be fuzzy sets on $X \times X \times (0,\infty)$ defined as follows: $M_p(x,y,t) = \dfrac{t}{t+p(x,y)}$. Then the triple $(X, M_p, *)$ is a fuzzy partial metric space where the self-distance for any point is its absolute value.

Proof. It is clear that $0 \leq M_p(x,y,t) \leq 1$ because $p(x,y) \geq 0$.

(i) $M_p(x,y,0) = \dfrac{0}{0+p(x,y)} = 0.$

(ii)

$$\begin{aligned} M_p(x,y,t) &= \frac{t}{t+p(x,y)} = \frac{t}{t+(-\min\{x,y\})} \\ &= \frac{t}{t+(-\min\{y,x\})} = \frac{t}{t+p(y,x)} \\ &= M_p(y,x,t). \end{aligned}$$

(iii)

$$\begin{aligned} M_p(x,y,t) * M_p(y,z,s) &= \frac{t}{t+p(x,y)} * \frac{s}{s+p(y,z)} \\ &= \frac{t}{t+p(x,y)} . \frac{s}{s+p(y,z)} \\ &\leq \frac{t+s}{t+p(x,y)+s+p(y,z)} \\ &= \frac{t+s}{t+s+(-\min\{x,y\})+(-\min\{y,z\})}. \end{aligned}$$

Since $-\min\{x,z\} \leq -\min\{x,y\} + -\min\{y,z\} - -\min\{y,y\}$,

$$\frac{1}{-\min\{x,z\}} \geq \frac{1}{-\min\{x,y\} + -\min\{y,z\} - \min\{y,y\}},$$

$$\begin{aligned} M_p(x,y,t) * M_p(y,z,s) &\leq \frac{t+s}{(t+s+(-\min\{x,z\})} \\ &= \frac{t+s}{t+s+p(x,z)} \\ &= M_p(x,z,t+s). \end{aligned}$$

Thus

$$M_p(x, y, t) * M_p(y, z, s) \leq M_p(x, z, t + s).$$

(iv) $M_p(x, y, t) = 1$ if and only if $p(x, y) = 0$.

Example 11.21. Let $X = R^+$ and (X, p) be a partial metric space. Let $p(x, y) = \max\{x, y\}$, $x, y \in R^-$. Denote $a * b = ab$ for all $a, b \in [0, 1]$ and let M_p be fuzzy sets on $X \times X \times (0, \infty)$ defined as follows: $M_p(x, y, t) = \dfrac{t}{t + p(x, y)}$. Then the triple $(X, M_p, *)$ is a fuzzy partial metric space where the self-distance for any point is its value itself.

Example 11.22. Let P_ω denote the power set of the natural numbers $\omega = N^* = N\ \{0\}$ different from zero, with the subset ordering. The function $p : P_\omega \times P_\omega \to [0, 1]$ such that $p(x, y) = 1 - \sum_{(n \in x \cap y)} 2^{-n}$ for any $x, y \in P_\omega$ is a partial metric on P_ω. Denoted $a * b = ab$ for all $a, b \in [0, 1]$ and let M_p be fuzzy sets on $P_\omega \times P_\omega \times (0, \infty)$ defined as follows: $M_p(x, y, t) = \dfrac{t}{t + p(x, y)}$. Then the triple $(X, M_p, *)$ is a fuzzy partial metric space.

(i)

$$M_p(x, y, 0) = \frac{0}{0 + p(x, y)} = \frac{0}{0 + 1 - \sum_{(n \in x \cap y)} 2^{-n}} = 0.$$

(ii)

$$\begin{aligned} M_p(x, y, t) &= \frac{t}{t + p(x, y)} = \frac{t}{t + (1 - \sum_{(n \in x \cap y)} 2^{-n})} \\ &= \frac{t}{t + p(y, x)} = M_p(y, x, t) \end{aligned}$$

(iii)

$$\begin{aligned} M_p(x, y, t) * M_p(y, z, s) &= \frac{t}{t + p(x, y)} * \frac{s}{s + p(y, z)} \\ &= \frac{t}{t + p(x, y)} . \frac{s}{s + p(y, z)} \\ &\leq \frac{t + s}{t + p(x, y) + s + p(y, z)} \\ &= \frac{t + s}{t + s + \left(1 - \sum_{(n \in x \cap y)} 2^{-n} + 1 - \sum_{(n \in x \cap y)} 2^{-n}\right)} \end{aligned}$$

Let $x, y, z \in P_\omega$. Using the property of sets, one has that

$$(x \cap y) \cup (y \cap z) = (x \cap y \cap z) \cup (y \cap (x \cup z)).$$

Thus, we get

$$\sum_{(n \in x \cap y)} 2^{-n} + \sum_{(n \in y \cap z)} 2^{-n} = \sum_{(n \in x \cap z \cap y)} 2^{-n} + \sum_{n \in y \cap (x \cup z)} 2^{-n}.$$

Consequently,

$$\begin{aligned} p(x, y) + p(y, z) &= p(x \cap z, y) + p(y, x \cup z) \\ &\geq p(x, z) + p(y, y) \end{aligned}$$

Hence,

$$p(x, z) \leq p(x, y) + p(y, z) - p(y, y)$$

for any $x, y, z \in P_\omega$. Then

$$\begin{aligned} M_p(x, y, t) * M_p(y, z, s) &\leq \frac{t + s}{t + s + p(x, y) + p(y, z) - p(y, y)} \\ &\leq \frac{t + s}{t + s + p(x, z)} = M_p(x, z, t + s) \end{aligned}$$

(iv) $M_p(x, y, t) = 1$ if and only if $p(x, y) = 0$. It is clear that $1 \leq M_p(x, y, t) \leq 0$, that is

$$p \to [0, 1] \Rightarrow 1 \leq \frac{t}{t + p(x, y)} \leq 0.$$

Definition 11.23. Let $(X, M_p, *)$ be a fuzzy partial metric space. We define open ball $B(x, r, t)$ for $t > 0$ with center $x \in X$ and radius r, $0 < r < 1$,

$$B(x, r, t) = \{y \in X : M_p(x, y, t) > 1 - r\}.$$

Definition 11.24. Let $(X, M_p, *)$ be a fuzzy partial metric space. Define

$$\tau = \{A \subset X : x \in A \Leftrightarrow \exists\, r, t > 0,\ 0 < r < 1,\ \text{such that } B(x, r, t) \subset A\}.$$

Then τ is a topology on X.

Definition 11.25. A sequence (x_n) in a fuzzy partial metric space $(X, M_p, *)$ is said to be a Cauchy sequence if for each ϵ, $0 < \epsilon < 1$ and $t > 0$, there exists $n_0 \in N$ such that $M_p(x_n, x_m, t) > 1 - \epsilon$ for all $n, m \geq n_0$.

Definition 11.26. A fuzzy partial metric space is said to be complete if every Cauchy sequence is convergent.

Definition 11.27. Let $(X_1, M_{p_1}, *)$ and $(X_2, M_{p_2}, *)$ be the given fuzzy partial metric spaces. For $(x_1, x_2), (y_1, y_2) \in X_1 \times X_2, t > 0$, if we define

$$M_p((x_1, x_2), (y_1, y_2), t) = M_{p_1}(x_1, y_1, t) * M_{p_2}(x_2, y_2, t),$$

then M_p is a fuzzy partial metric space on $X_1 \times X_2$.

11.4 Conclusion

In this research we introduced a new concept for fuzzy partial metric space in order to understand the intricacies involved in incorporating fuzziness in terms of partial metric space, whose classical mathematics requires that all mathematical notions must be exact, otherwise precise reasoning would be impossible. That is, mathematical tools may not be successfully used. But there is a need for a mathematical apparatus to describe the vague notions to overcome the obstacles in modeling and such an apparatus is provided by fuzzy set theory.

References

1. Z. Deng, Fuzzy pseudometric spaces. J. Math. Anal. Appl. **86**, 74–95 (1982)
2. M.A. Erceg, Metric spaces in fuzzy set theory. J. Math. Anal. Appl. **69**, 205–230 (1979)
3. A. George, P. Veeramani, On some results in fuzzy metric spaces. Fuzzy Sets Syst. **64**, 395–399 (1994)
4. O. Kaleva, S. Seikkala, On fuzzy metric spaces. Fuzzy Sets Syst. **12**, 215–229 (1984)
5. O. Kramosil, J. Michalek, Fuzzy metric and statistical metric spaces. Kybernetica **11**, 326–334 (1975)
6. S.G. Mattews, Partial metric topology, in *Proceedings of the 8th Summer Conference on Topology and Its Applications*. Annals of the New York Academy of Sciences, vol. 728 (New York Academy of Sciences, New York, 1994), pp. 183–197
7. K. Menger, Statistical metrics. Proc. Natl. Acad. Sci. U.S.A. **28**, 535–537 (1942)
8. S.J. O'Neill, Two topologies are better than one. Report technical, University of Warwick, Coventry (1995)
9. L.A. Zadeh, Fuzzy sets. Inf. Control **8**, 338–353 (1965)

Chapter 12
Bounded and Unbounded Fundamental Solutions in MAC Models

Igor Neygebauer

Abstract Many linear partial differential equations in mathematical physics have the fundamental solutions with singularities. This does not correspond to the real physical situation. The additional terms were introduced into the classical equations using the constitutive laws for internal body interactions and so the MAC models were created. This paper analyzes the boundedness of the fundamental solutions of some MAC models with local internal body forces. The 1D, 2D, and 3D steady state problems are considered. The mechanical models are an elastic string, heat conduction, membrane, plate, linear isotropic elasticity. The Fourier transform is used. The new strength criteria is given. It is shown that the displacements under applied force are finite for membrane, plate and in 2D and 3D elasticity. The bending stresses are finite in plate. The stresses are zero in elasticity problem at the point of applied force but the new strength criteria is working in this case too. The temperatures are finite in case of 2D and 3D point source of heat flux.

12.1 Introduction

The steady state models for the linear elastic string, beam, membrane, plate, elasticity, and heat conduction are taken into consideration [7]. These models are widely used to consider many applied and theoretical problems. We are interested in a test problem when the continuum model occupies an infinite domain and only one force or a heat source is applied at one point only. The solutions of such kind of problems are called in mathematics the fundamental solutions of the correspondent differential equation. Unfortunately all classical fundamental solutions of the above problems do not satisfy the natural physical behavior of that solutions. That natural behavior can be described first of all as the boundedness of the solutions in the whole domain of consideration and the tendency to zero at the infinity.

I. Neygebauer (✉)
Department of Mathematics, University of Dodoma, Tanzania
e-mail: newigor52r@yahoo.com

G.A. Anastassiou, O. Duman (eds.), *Computational Analysis*, Springer Proceedings in Mathematics & Statistics 155, DOI 10.1007/978-3-319-28443-9_12

We will generalize the classical models in order to obtain the physically correct fundamental solutions. The method to get the generalized model is just inserting the constitutive laws for internal body forces or fluxes into the classical model. These new equations will be called the MAC (method of additional conditions) models [6, 8]. The analysis of fundamental solutions of that equations will be of interest and the point singularities especially. If the constitutive law for internal body interactions is chosen, then the solution of the generalized problem can be found. That solutions are not presented in this paper but they can be the purpose of the next paper.

This paper describes how the models can be tested. The main tool of testing is the Fourier transform [4]. The Fourier transform can be applied because the awaiting solution should have the physical behavior. If the stated problem is physically well, then the obtained solution will be well too. If the stated problem is not physically correct, then the fundamental solution will create singularities, or it will destroy the tendency to zero at infinity, or the solution is not physically awaited.

The problem is stated as following: Find the constitutive law for the internal body forces or heat fluxes which is presented using operator B in the linear steady state problems for the elastic string, beam, membrane, plate, heat conduction, and isotropic elasticity. The equation is

$$Lu + Bu = \delta, \tag{12.1}$$

where L is the classical differential operator, δ is the Dirac δ function, u is the transversal displacement in the string, beam, membrane, plate problems, or it is the temperature in the heat conduction problem, or it is a displacement vector in the elasticity problem. The δ function is a vector in the last case and this vector has one component δ function and other component or components are zero. The solution of Eq. (12.1) should satisfy the conditions:

$$|u(0)| < \infty, \tag{12.2}$$

$$u(x) \neq 0, \text{ for any } x, \tag{12.3}$$

$$u(x) \to 0 \text{ as } |x| \to \infty. \tag{12.4}$$

It is well known that there are no solutions of the stated problem in case $B = 0$ which satisfy the conditions (12.2)–(12.4). The operator B could be suggested in many different forms. It can be local and nonlocal, linear and nonlinear, deterministic and stochastic [1]. There can be used classical or fractional derivatives and integrals. This paper considers local deterministic linear ordinary differential operator.

12.2 String

12.2.1 Differential Equation of the Problem

The equation of one-dimensional steady state problem of the string is taken in the form [2, 9–11]

$$T\frac{d^2u}{dx^2} + q^i(x) + q^e(x) = 0, \tag{12.5}$$

where T is the tension applied to the string, x is a Cartesian coordinate of a cross-section, $-\infty \leq x \leq +\infty$, u is the transversal displacement of a cross-section, $q^i(x), q^e(x)$ are the density of the transversal internal and external body forces per unit length correspondingly. Equation (12.5) can be written in the form

$$\frac{d^2u}{dx^2} = p^i(x) + p^e(x), \tag{12.6}$$

where

$$p^i(x) = -\frac{q^i(x)}{T}, \quad p^e(x) = -\frac{q^e(x)}{T}. \tag{12.7}$$

The following constitutive laws for the internal body forces will be considered

$$\begin{aligned} q^i &= 0; \\ q^i &= -cu. \\ c &> 0. \end{aligned} \tag{12.8}$$

It will be shown that the law (12.8) only is physically acceptable.

12.2.2 String: Case 1

The fundamental solution for classical string problem satisfies the following equations:

$$\frac{d^2u}{dx^2} = \delta(0), \tag{12.9}$$

and

$$u(x) \to 0$$

as $x \to -\infty$ and $x \to +\infty$. $\delta(x)$ is the δ function. Because it is supposed that the Fourier transform of the function $u(x)$ should exist in case of physical solution, then the Fourier transform can be applied to this function

$$U(s) = \frac{1}{\sqrt{2\pi}} \int_{-\infty}^{+\infty} u(x) e^{-ixs} dx. \tag{12.10}$$

If the Fourier transform is applied to Eq. (12.9), then the following algebraic equation will be obtained

$$-s^2 U = 1 \tag{12.11}$$

and then the solution of Eq. (12.11) gives

$$U(s) = -\frac{1}{s^2}. \tag{12.12}$$

The inverse Fourier transform of the function (12.12) is

$$u(x) = \frac{1}{\sqrt{2\pi}} \int_{-\infty}^{+\infty} U(s) e^{ixs} ds. \tag{12.13}$$

Then

$$u(0) = \frac{1}{\sqrt{2\pi}} \int_{-\infty}^{+\infty} \left(-\frac{1}{s^2}\right) ds = -\infty. \tag{12.14}$$

Equation (12.14) shows that the obtained solution is not bounded and it means that the model of a string should be improved. The generalized or the MAC model can be developed introducing the constitutive law for internal body forces.

12.2.3 String: Case 2

Let us consider the following law for internal body forces

$$q^i(x) = -cu(x),$$

where $c > 0$ is a material constant which should be determined experimentally. Let

$$k = \frac{c}{T}.$$

Then Eq. (12.5) will take the form

$$\frac{d^2u}{dx^2} - ku = p^e(x),$$

The fundamental solution for the stated string problem satisfies the following equation:

$$\frac{d^2u}{dx^2} - ku = \delta(0), \tag{12.15}$$

and

$$u(x) \to 0$$

as $x \to -\infty$ and $x \to +\infty$. If the Fourier transform (12.10) is applied to Eq. (12.15), then the following algebraic equation will be obtained

$$-s^2U - kU = 1 \tag{12.16}$$

and then the solution of Eq. (12.16) gives

$$U(s) = -\frac{1}{s^2 + k}. \tag{12.17}$$

The inverse Fourier transform (12.13) of Eq. (12.17) at $x = 0$ creates

$$u(0) = \frac{1}{\sqrt{2\pi}} \int_{-\infty}^{+\infty} \left(-\frac{1}{s^2 + k} \right) ds = -\frac{1}{\sqrt{2\pi k}} \arctan\left(\frac{s}{\sqrt{k}} \right) |_{-\infty}^{+\infty} = -\sqrt{\frac{\pi}{2k}}. \tag{12.18}$$

Equation (12.18) shows that the obtained solution is bounded and it means that this MAC model of a string could be used to get the physically accepted solution.

12.3 Beam

Consider an elastic beam [5]. The equation of steady state problem of the bending of the beam with internal body forces could be written in the form

$$\mathrm{EI}\frac{d^4u}{dx^4} = q^i(x) + q^e(x), \tag{12.19}$$

where EI is the bending stiffness of the beam, x is the Cartesian coordinate of a cross-section, $-\infty \leq x \leq +\infty$, u is the transversal displacement of a cross-section, $q^i(x)$ is the density of the transversal internal body forces per unit length, $q^e(x)$ is the density of the transversal external body forces per unit length. Equation (12.19) can be written in the form

$$\frac{d^4u}{dx^4} = p^i(x) + p^e(x),$$

where

$$p^i(x) = \frac{q^i(x)}{\mathrm{EI}}, p^e(x) = \frac{q^e(x)}{\mathrm{EI}}.$$

The following constitutive laws for the internal body forces were considered

$$\begin{aligned} q^i &= 0; \\ q^i &= -cu; \\ q^i &= b\frac{d^2u}{dx^2}; \\ q^i &= b\frac{d^2u}{dx^2} - cu. \\ b &> 0, c > 0. \end{aligned} \tag{12.20}$$

It could be shown that the law (12.20) only is physically acceptable.

12.4 Membrane

Consider an elastic membrane [6]. The equation of steady state problem for the membrane with internal body forces could be written in the form

$$T\left(\frac{\partial^2 w}{\partial x^2} + \frac{\partial^2 w}{\partial y^2}\right) + q^i(x,y) + q^e(x,y) = 0, \tag{12.21}$$

where the membrane lies in the plane (x, y) in its natural state, $-\infty < x < \infty, -\infty < y < \infty$, T is its tension per unit of length, $w(x, y)$ is the transversal displacement of the point (x, y) of the initially plane membrane, $q^i(x, y)$ is the density of the transversal internal body forces per unit area, $q^e(x, y)$ is the density of the transversal external body forces per unit area. The tension T is supposed to be a constant in this statement of the problem. Equation (12.21) can be written in the form

$$\frac{\partial^2 w}{\partial x^2} + \frac{\partial^2 w}{\partial y^2} = p^i(x,y) + p^e(x,y),$$

where

$$p^i(x,y) = -\frac{q^i(x,y)}{T}, \quad p^e(x,y) = -\frac{q^e(x,y)}{T}.$$

The following constitutive laws for the internal body forces were considered

$$\begin{aligned} q^i &= 0; \\ q^i &= -cw; \\ q^i &= -a\nabla^4 w; \\ q^i &= -a\nabla^4 w - cw. \end{aligned} \tag{12.22}$$

where

$$a > 0, \; c > 0.$$

It could be shown that the law (12.22) only is physically acceptable.

12.5 Plate

Consider the bending of an elastic plate. The equation of steady state problem for the plate with internal body forces could be written in the form [3, 12]

$$D\nabla^4 w = q^i(x,y) + q^e(x,y), \tag{12.23}$$

where the middle reference plane of the plate lies in the plane (x, y) in its natural state, $-\infty < x < \infty, -\infty < y < \infty$, E is the Young modulus, ν is the Poisson ratio, h is the plate thickness, $w(x, y)$ is the transversal displacement of the point (x, y) of the initially plane reference plane, $q^i(x, y)$ is the density of the transversal internal body forces per unit area, $q^e(x, y)$ is the density of the transversal external body forces per unit area. The flexural rigidity

$$D = \frac{Eh^3}{12(1-\nu^2)}$$

is supposed to be a constant in this statement of the problem. Equation (12.23) can be written in the form

$$D\nabla^4 w = p^i(x) + p^e(x),$$

where

$$p^i(x,y) = \frac{q^i(x,y)}{D}, \quad p^e(x,y) = \frac{q^e(x,y)}{D}.$$

The following constitutive laws for the internal body forces were considered

$$\begin{aligned}
q^i &= 0; \\
q^i &= -cw; \\
q^i &= b\nabla^2 w; \\
q^i &= a\nabla^6 w; \\
q^i &= a\nabla^6 w + b\nabla^2 w; \\
q^i &= a\nabla^6 w - cw; \\
q^i &= b\nabla^2 w - cw; \\
q^i &= a\nabla^6 w + b\nabla^2 w - cw, \qquad (12.24)
\end{aligned}$$

where

$$a > 0,\ b > 0,\ c > 0.$$

It could be shown that the law (12.24) only is physically acceptable. That case gives the finite displacement and the finite bending moments under applied force. Then the strength criteria for the finite stresses can be used.

12.6 Heat Conduction

Consider the 2D heat conduction problem. The equation of the steady state problem is

$$k\left(\frac{\partial^2 u}{\partial x^2} + \frac{\partial^2 u}{\partial y^2}\right) + q^i(x,y) + q^e(x,y) = 0, \qquad (12.25)$$

where $u(x,y)$ is the temperature of the point (x,y) of the plane, k is the coefficient of thermal conduction, $q^i(x,y)$ is the rate of internal heat generation per unit volume produced in the body, $q^e(x,y)$ is the rate of external heat generation per unit volume. Equation (12.25) will coincide with the membrane equation (12.21) if the parameter k and variable u will be replaced by the parameter T and the variable w in the membrane equation. Then it could be concluded that the physically accepted equation for the 2D steady state heat conduction problem is

$$a\nabla^4 u - k\nabla^2 u + cu = q^e,$$

where $a > 0, c > 0$ are physical constants of the heat conduction problem which should be determined experimentally and the constitutive equation for the rate of internal heat generation per unit volume produced in the body is taken in the form

$$q^i(x,y) = -a\nabla^4 u(x,y) - cu(x,y).$$

Consider now the 3D heat conduction problem. The equation of the steady state problem is

$$k\left(\frac{\partial^2 u}{\partial x^2} + \frac{\partial^2 u}{\partial y^2} + \frac{\partial^2 u}{\partial z^2}\right) + q^i(x,y,z) + q^e(x,y,z) = 0, \tag{12.26}$$

where $u(x,y,z)$ is the temperature of the point (x,y,z) of the plane, k is the coefficient of thermal conduction, $q^i(x,y,z)$ is the rate of internal heat generation per unit volume produced in the body, $q^e(x,y,z)$ is the rate of external heat generation per unit volume. Equation (12.26) can be written in the form

$$u_{xx} + u_{yy} + u_{zz} = p^i(x,y,z) + p^e(x,y,z),$$

where

$$p^i(x,y,z) = -\frac{q^i(x,y,z)}{k}, \quad p^e(x,y,z) = -\frac{q^e(x,y,z)}{k}.$$

The following constitutive laws for the rate of internal heat generation per unit volume produced in the body were considered

$$\begin{aligned} q^i &= 0; \\ q^i &= -cu; \\ q^i &= -a\nabla^4 u; \\ q^i &= -a\nabla^4 u - cu, \end{aligned} \tag{12.27}$$

where

$$a > 0, \ c > 0.$$

It could be shown that the law (12.27) only is physically acceptable in all 1D, 2D, and 3D cases.

12.7 Elasticity and 1D Elasticity

12.7.1 Differential Equation of the Problem

Consider a linear isotropic elastic body. The equation of the steady state problem for the elastic body with internal body forces could be written in the form [6]

$$(\lambda + \mu)\nabla e + \mu\nabla^2\mathbf{u} + \mathbf{q}^{\mathrm{i}} + \mathbf{q}^{\mathrm{e}} = 0, \tag{12.28}$$

where dilatation e equals

$$e = \mathrm{div}\,\mathbf{u}$$

and $\mathbf{u}$ is the displacement vector with the Cartesian components u, v, w, x, y, z are Cartesian components of the position vector of some point in the body, $\mathbf{q}^{\mathrm{i}}$ is the internal body force per unit volume, $\mathbf{q}^{\mathrm{e}}$ is the external body force per unit volume, λ and μ are Lame's coefficients or Lame's constants, ∇ is the gradient, ∇^2 is the Laplacian. Equation (12.28) can be written in the form

$$(\lambda + \mu)\nabla \mathtt{div}\,\mathbf{u} + \mu\nabla^2\mathbf{u} + \mathbf{q}^{\mathrm{i}} + \mathbf{q}^{\mathrm{e}} = 0, \tag{12.29}$$

The following constitutive laws for the internal body forces were considered

$$\begin{aligned}
\mathbf{q}^{\mathrm{i}} &= \mathbf{0}; \\
\mathbf{q}^{\mathrm{i}} &= -c\mathbf{u}; \\
\mathbf{q}^{\mathrm{i}} &= -a\nabla^4\mathbf{u}; \\
\mathbf{q}^{\mathrm{i}} &= -a\nabla^4\mathbf{u} - c\mathbf{u}; \\
\mathbf{q}^{\mathrm{i}} &= +b\nabla^6\mathbf{u} - a\nabla^4\mathbf{u} - c\mathbf{u}.
\end{aligned} \tag{12.30}$$

It is supposed that a, b, c are positive constants. It could be shown that the law (12.30) only is physically acceptable in all cases of 1D, 2D, 3D problems.

12.7.2 1D Elasticity: Case 4

Let us consider the following law for internal body forces

$$q^i(x) = -a\frac{d^4u}{dx^4}(x) - cu(x),$$

where $a > 0, c > 0$ are the material constants which should be determined experimentally. Then Eq. (12.29) will take the form

$$-m\frac{d^4u}{dx^4} + \frac{d^2u}{dx^2} - nu = p^e(x),$$

where

$$m = \frac{a}{\lambda + 2\mu}, \quad n = \frac{c}{\lambda + 2\mu}.$$

The fundamental solution for the stated beam problem satisfies the following equation:

$$m\frac{d^4u}{dx^4} - \frac{d^2u}{dx^2} + nu = \delta(0), \tag{12.31}$$

and the conditions

$$|u(0)| < \infty,$$

$$u(x) \neq 0 \text{ for } -\infty < x < +\infty, \tag{12.32}$$

$$u(x) \to 0 \text{ as } x \to -\infty \text{ and } x \to +\infty. \tag{12.33}$$

If the Fourier transform is applied to Eq. (12.31), then the following algebraic equation will be obtained

$$ms^4U + s^2U + nU = 1 \tag{12.34}$$

and then the solution of Eq. (12.34) gives

$$U(s) = \frac{1}{ms^4 + s^2 + n}. \tag{12.35}$$

The inverse Fourier transform of Eq. (12.35) at the point $x = 0$ creates

$$\begin{aligned} u(0) &= \frac{1}{\sqrt{2\pi}} \int_{-\infty}^{+\infty} \frac{1}{ms^4 + s^2 + n} ds \\ &\leq \frac{1}{\sqrt{2\pi}} \int_{-\infty}^{+\infty} \frac{1}{s^2 + n} ds \\ &= \frac{1}{\sqrt{2\pi n}} \arctan\left(\frac{s}{\sqrt{n}}\right)\Big|_{-\infty}^{+\infty} = \frac{1}{2}\sqrt{\frac{\pi}{n}}, \end{aligned} \tag{12.36}$$

The estimation (12.36) shows that the obtained solution is bounded. The exact solution of Eq. (12.31) can show that the conditions (12.32), (12.33) are fulfilled too. And it means that this MAC model of the 1D elasticity can be used to get the physically accepted solution. There are two more parameters with respect to the classical model of elasticity and that could allow to accept the bigger set of experimental data.

12.7.3 Strength Criteria in 1D Elasticity

Consider the density of the elastic energy in 1D elastic model considered in the previous section "1D Elasticity: Case 4". Then

$$E = \frac{1}{2}\left[a\left(\frac{d^2u}{dx^2}\right)^2 + (\lambda + 2\mu)\left(\frac{du}{dx}\right)^2 + cu^2\right] \tag{12.37}$$

is the density of the elastic energy per unit length. The strength criteria could be suggested in the following form

$$E \leq E_{\text{lim}}, \tag{12.38}$$

where E_{lim} is the maximum possible density of elastic energy in the body. It means that if the value of E reaches the limit value E_{lim} then the material will change its behavior from the elastic state to the plastic one for plastic materials or there could be initiated a crack for the brittle materials. The energy in Eq. (12.37) consists of two parts

$$E = E_1 + E_2,$$

where

$$E_1 = \frac{1}{2}(\lambda + 2\mu)\left(\frac{du}{dx}\right)^2 \tag{12.39}$$

is the classical elastic energy corresponding to the surface interactions, and

$$E_2 = \frac{1}{2}\left[a\left(\frac{d^2u}{dx^2}\right)^2 + cu^2\right] \tag{12.40}$$

is the elastic energy corresponding to the internal body interactions. Consider the density of the elastic energy (12.37) at the point $x = 0$. The finite value of $u(0)$ can be evaluated using (12.36). Then

$$\frac{du}{dx}(0) = -\frac{i}{\sqrt{2\pi}} \int_{-\infty}^{+\infty} \frac{sds}{ms^4 + s^2 + n} = 0. \quad (12.41)$$

Equation (12.41) shows that the classical elastic energy according to Eq. (12.39) is zero at the point $x = 0$. Consider now

$$\frac{d^2u}{dx^2}(0) = -\frac{1}{\sqrt{2\pi}} \int_{-\infty}^{+\infty} \frac{s^2ds}{ms^4 + s^2 + n} = -\sqrt{\frac{2}{\pi}} \int_0^{+\infty} \frac{s^2ds}{ms^4 + s^2 + n}.$$

Then

$$\left|\frac{d^2u}{dx^2}(0)\right| \leq \sqrt{\frac{2}{\pi}} \int_0^{+\infty} \frac{ds}{ms^2 + 1} = \sqrt{\frac{\pi}{2m}}. \quad (12.42)$$

Equations (12.37), (12.42) show that the density of elastic energy E_2 corresponding to internal body interactions is bounded and it could be used to determine the strength of material under an applied force. The considered strength criteria (12.38) can be applied in other cases for 1D elasticity. But that cases are not physically accepted in cases of 2D and 3D elasticity which will be analyzed in the following sections.

12.7.4 3D Elasticity: Case 5

Let us consider the following law for the internal body forces

$$\mathbf{q}^{\mathbf{i}}(x, y, z) = b\nabla^6\mathbf{u} - a\nabla^4\mathbf{u}(x, y, z) - c\mathbf{u},$$

where $a > 0$, $b > 0$, $c > 0$ is a material constant which should be determined experimentally. Consider the fundamental solution of the 3D elasticity problem consisting of the following equations and additional conditions:

$$u = u(x, y, z),\ v = v(x, y, z),\ w = w(x, y, z),$$

$$b\Delta^3 u - a\Delta^2 u + (\lambda + \mu)\left(u_{xx} + v_{xy} + w_{xz}\right)$$

$$+\mu\Delta u - cu = P\delta(0, 0, 0), \quad (12.43)$$

$$b\Delta^3 v - a\Delta^2 v + (\lambda + \mu)\left(u_{xy} + v_{yy} + w_{yz}\right) + \mu\Delta v - cv = 0, \quad (12.44)$$

$$b\Delta^3 w - a\Delta^2 w + (\lambda + \mu)\left(u_{xz} + v_{yz} + w_{zz}\right) + \mu\Delta w - cw = 0, \quad (12.45)$$

$$|u(0, 0, 0)| < \infty,\ |v(0, 0, 0)| < \infty,\ |w(0, 0, 0)| < \infty,$$

$$u(x,y,z) \to 0,\ \ v(x,y,z) \to 0,$$
$$w(x,y,z) \to 0 \text{ as } x^2+y^2+z^2 \to +\infty,$$

$\delta(x,y,z)$ is the δ function. Because it is supposed that the Fourier transform of the functions $u(x,y,z)$, $v(x,y,z)$, $w(x,y,z)$ should exist in case of physical solution, then the Fourier transform can be applied to these functions. If the Fourier transform is applied to Eqs. (12.43)–(12.45), then the obtained algebraic equations will be solved and we get

$$\begin{aligned}
&u(0,0,0)\\
&= \frac{p_1}{(2\pi)^{\frac{3}{2}}} \int\int\int_{-\infty}^{+\infty} \frac{\left(t^2+p^2+\alpha R^2+\gamma R^4+fR^6+\beta\right) dsdtdp}{[\alpha R^2+\gamma R^4+fR^6+\beta][(1+\alpha)R^2+\gamma R^4+fR^6+\beta]}\\
&= \frac{p_1}{(2\pi)^{\frac{3}{2}}} \int_0^{\pi} d\theta \int_0^{2\pi} d\varphi \int_0^{+\infty} R^2 dR \sin\theta\\
&\quad \times \frac{R^2(\sin^2\theta\sin^2\varphi+\cos^2\theta)+\alpha R^2+\gamma R^4+fR^6+\beta}{(\alpha R^2+\gamma R^4+fR^6+\beta)[(1+\alpha)R^2+\gamma R^4+fR^6+\beta]}\\
&= \frac{p_1}{(2\pi)^{\frac{3}{2}}} \int_0^{\pi} d\theta \int_0^{2\pi} d\varphi \int_0^{+\infty} \frac{R^2}{(\alpha R^2+\gamma R^4+fR^6+\beta)} dR\\
&\le \frac{p_1}{(2\pi)^{\frac{3}{2}}} \int_0^{\pi} d\theta \int_0^{2\pi} d\varphi \int_0^{+\infty} \frac{1}{\alpha+\gamma R^2} dR < +\infty
\end{aligned} \tag{12.46}$$

where the following change of variables was used

$$s = R\sin\theta\cos\varphi,\ t = R\sin\theta\sin\varphi,\ p = R\cos\theta,$$

$$\begin{aligned}
&v(0,0,0)\\
&= -\frac{p_1}{(2\pi)^{\frac{3}{2}}} \int\int\int_{-\infty}^{+\infty} \frac{stdsdtdp}{[\alpha R^2+\gamma R^4+fR^6+\beta][(1+\alpha)R^2+\gamma R^4+fR^6+\beta]}\\
&= 0.
\end{aligned} \tag{12.47}$$

Similarly, we get

$$w(0,0,0) = 0. \tag{12.48}$$

Equations (12.46)–(12.48) show that the obtained solution is bounded and it means that this model of the 3D elasticity satisfies the first strength criteria of the bounded displacements. Let us check now the second strength criteria, which requires the boundedness of the density of the elastic energy.

12.7.5 Strength Criteria in 3D Elasticity: Case 5

Consider the density of the elastic energy in 3D elastic model considered in Sect. 12.7.4. Then

$$\begin{aligned} E = {} & 0.5\sigma_{ij}e_{ij} \\ & +0.5b\left[u_{xxx}^2 + u_{yyy}^2 + u_{zzz}^2\right. \\ & \left. + 3\left(u_{xxy}^2 + u_{xyy}^2 + u_{xxz}^2 + u_{zzx}^2 + u_{yyz}^2 + u_{yzz}^2\right) + 6u_{xyz}^2\right] \\ & +0.5b\left[v_{xxx}^2 + v_{yyy}^2 + v_{zzz}^2\right. \\ & \left. + 3\left(v_{xxy}^2 + v_{xyy}^2 + v_{xxz}^2 + v_{zzx}^2 + v_{yyz}^2 + v_{yzz}^2\right) + 6v_{xyz}^2\right] \\ & +0.5b\left[w_{xxx}^2 + w_{yyy}^2 + w_{zzz}^2\right. \\ & \left. +3\left(w_{xxy}^2 + w_{xyy}^2 + w_{xxz}^2 + w_{zzx}^2 + w_{yyz}^2 + w_{yzz}^2\right) + 6w_{xyz}^2\right] \\ & +0.5a\left(u_{xx}^2 + 2u_{xy}^2 + 2u_{xz}^2 + 2u_{yz}^2 + u_{yy}^2 + u_{zz}^2\right) \\ & +0.5a\left(v_{xx}^2 + 2v_{xy}^2 + 2v_{xz}^2 + 2v_{yz}^2 + v_{yy}^2 + v_{zz}^2\right) \\ & +0.5a\left(w_{xx}^2 + 2w_{xy}^2 + 2w_{xz}^2 + 2w_{yz}^2 + w_{yy}^2 + w_{zz}^2\right) \\ & +0.5c(u^2 + v^2 + w^2). \end{aligned} \tag{12.49}$$

is the density of the elastic energy per unit volume, stress-components are σ_{ij}, strain-components are e_{ij} and $i, j = 1, 2, 3$. The strain-displacement relations are taken in the form

$$e_{ij} = \frac{1}{2}\left(\frac{\partial u_i}{\partial x_j} + \frac{\partial u_j}{\partial x_i}\right).$$

The stress–strain equation is

$$\sigma_{ij} = \lambda e_{\mathrm{mm}}\delta_{ij} + 2\mu e_{ij}.$$

The strength criteria could be suggested in the following form

$$E \leq E_{\mathrm{lim}}, \tag{12.50}$$

where $E_{\lim}$ is the maximum possible density of elastic energy in the body. It means that if the value of E reaches the limit value $E_{\lim}$ then the material will change its behavior from the elastic state to the plastic one for plastic materials or there could be initiated a crack for the brittle materials. The energy in (12.49) consists of two parts

$$E = E_1 + E_2,$$

where E_1 is the classical elastic energy corresponding to the surface interactions, and E_2 is the elastic energy corresponding to the internal body interactions. Consider the density of the elastic energy (12.49) at the point $x = 0, y = 0, z = 0$. The finite values of $u(0,0,0)$, $v(0,0,0)$, $w(0,0,0)$ are evaluated by using (12.46)–(12.48). We can easily obtain also that

$$E_1(0,0,0) = 0, \ |E_2(0,0,0)| < +\infty.$$

Then the strength criteria (12.50) can be used.

References

1. G.A. Anastassiou, O. Duman, *Towards Intelligent Modeling: Statistical Approximation Theory* (Springer, Berlin, 2011)
2. S. Antman, *Nonlinear Problems of Elasticity* (Springer, Berlin, 2005)
3. S. Chakraverty, *Vibration of Plates* (CRC Press, New York, 2009)
4. D.G. Duffy, *Transform Methods for Solving Partial Differential Equations* (Chapman and Hall/CRC, Boca Raton, 2004)
5. P. Howell, G. Kozyreff, J. Ockendon, *Applied Solid Mechanics* (Cambridge University Press, Cambridge, 2008)
6. I. Neygebauer, Differential MAC models in continuum mechanics and physics. J. Appl. Funct. Anal. **8**(1), 100–124 (2013)
7. I.N. Neygebauer, Mechanical models with internal body forces. J. Concr. Appl. Math. **12**(3–4), 181–200 (2014)
8. IN. Neygebauer, MAC models in thermoelasticity. Trans. Nenjing Univ. Aeronaut. Astronaut. **31**(2), 127–131 (2014)
9. I. Neygebauer, Elastic string with decreasing influence of internal body forces, J. Basic Appl. Res. Int. **6**(4), 197–209 (2015)
10. I. Neygebauer, Elastic string with internal body forces. J. Concr. Appl. Math. **13**(3–4), 312–321 (2015)
11. I.G. Petrovsky, *Lectures on Partial Differential Equations* (Dover, New York, 1991)
12. A.P.S. Selvadurai, *Partial Differential Equations in Mechanics* (Springer, New York, 2010)

Chapter 13
Schurer Generalization of q-Hybrid Summation Integral Type Operators

İlker Vural, Birol Altın, and İsmet Yüksel

Abstract In this study, using the q-generalization of the well-known hybrid summation integral type operators, we generalize these operators to Schurer type operators. We give weighted approximation and obtain rate of convergence of these operators.

13.1 Introduction

The well-known hybrid operators are defined as follows:

$$M_n(f,x) = (n-1)\sum_{k=1}^{\infty} s_{n,k}(x)\int_0^{\infty} p_{n,k-1}(t)f(t)\,dt + e^{-nx}f(0), \tag{13.1}$$

where

$$s_{n,k}(x) = e^{-nx}\frac{(nx)^k}{k!}$$

and

$$p_{n,k}(x) = \binom{n+k-1}{k}\frac{x^k}{(1+x)^{n+k}}$$

are, respectively, well-known Szász and Baskakov basis functions. These operators were studied in [8, 16]. Several interesting q-generalizations of certain summation integral type operators were studied in [1, 7, 9, 10, 14, 15, 17]. In this work, our goal

İ. Vural (✉)
Domaniç Vocational School, Dumlupınar University, Kütahya, Turkey
e-mail: ilker.vural@dpu.edu.tr

B. Altın • İ. Yüksel
Department of Mathematics, Faculty of Science, Gazi University, Ankara, Turkey
e-mail: birola@gazi.edu.tr; iyuksel@gazi.edu.tr

G.A. Anastassiou, O. Duman (eds.), *Computational Analysis*, Springer Proceedings in Mathematics & Statistics 155, DOI 10.1007/978-3-319-28443-9_13

is to give generalization to q-calculus of hybrid summation integral type operators. Throughout the paper we will need the following notations and definitions which can be founded in [2, 3, 11–13]: For $n \in \mathbb{N}$ and $a, b \in \mathbb{R}$, the q-integer and the q-factorial are defined by

$$[n]_q = 1 + q + \cdots + q^{n-1};\ [0]_q = 0 \tag{13.2}$$

and

$$[n]_q! = [1]_q[2]_q \ldots [n]_q,\ n \in \mathbb{N}\backslash\{0\};\ [0]_q! = 1. \tag{13.3}$$

The q-binomial coefficients are given by

$$\begin{bmatrix} n \\ v \end{bmatrix}_q = \frac{[n]_q!}{[v]_q![n-v]_q!}, 0 \leq v \leq n. \tag{13.4}$$

The q-derivative D_qf of a function is given by

$$(D_qf)(x) = \frac{f(x) - f(qx)}{(1-q)x},\ \text{for } x \neq 0 \tag{13.5}$$

and $(D_qf)(0) = f'(0)$ provided that $f'(0)$ exists.

The two q-analogues of the exponential function are defined by

$$e_q^x = \sum_{n=0}^{\infty} \frac{x^n}{[n]_q!} = \frac{1}{(1-(1-q)x)_q^{\infty}} \tag{13.6}$$

and

$$E_q^x = \sum_{n=0}^{\infty} q^{n(n-1)/2} \frac{x^n}{[n]_q!} = (1 + (1-q)x)_q^{\infty} \tag{13.7}$$

where

$$(1+a)_q^{\infty} = \prod_{j=1}^{\infty} (1 + q^{j-1}a).$$

The improper q-Jackson integral is defined as

$$\int_0^{\infty/A} f(x) d_qx = (1-q) \sum_{n \in \mathbb{Z}} f\left(\frac{q^n}{A}\right) \frac{q^n}{A}, A > 0. \tag{13.8}$$

The q-Gamma function and the q-Beta function are defined as

$$\Gamma_q(u) = K(A,u) \int_0^{\infty/A(1-q)} x^{u-1} e_q^{-x} d_q x \tag{13.9}$$

and

$$B_q(u,v) = K(A,u) \int_0^{\infty/A} \frac{x^{u-1}}{(1+x)_q^{u+v}} d_q x = \frac{\Gamma_q(u)\Gamma_q(v)}{\Gamma_q(u+v)} \tag{13.10}$$

where

$$K(A,u) = \frac{A^u}{1+A} \left(1+\frac{1}{A}\right)_q^u (1+A)_q^{1-u}$$

and

$$(a+b)_q^n = \prod_{j=1}^{n} (a+q^{j-1}b).$$

In particular, for $u \in \mathbb{Z}$, $K(A,u) = q^{u(u-1)/2}$ and $K(A,0) = 1$.

13.2 Schurer Generalization of q-Hybrid Summation Integral Type Operators

In [5], Dinlemez et al. give a generalization to q-calculus of hybrid summation integral type operators and they obtain approximation properties of these operators. In this study we give a Schurer type generalization of these operators. Let $k, p \in \mathbb{N}$, $n \in \mathbb{N}\backslash\{0\}$, $A > 0$ and f be a real valued continuous function defined on the interval $[0,\infty)$. Using the formulas and the notations between (13.2) and (13.10), we introduce q-hybrid summation integral Schurer type linear positive operators for $0 < q \leq 1$ as

$$M_{n,p,q}^{(\alpha,\beta)}(f,x) = [n+p-1]_q \sum_{k=1}^{\infty} s_{n,p,k}^q(x) \times \int_0^{\infty/A} p_{n,p,k-1}^q(t) f\left(\frac{[n+p]_q t + \alpha}{[n+p]_q + \beta}\right) d_q t \tag{13.11}$$

$$+e^{-[n+p]_q x} f\left(\frac{\alpha}{[n+p]_q+\beta}\right)$$

where

$$s^q_{n,p,k}(x) = \left([n+p]_q x\right)^k \frac{e^{-[n+p]_q x}}{[k]_q!},$$

and

$$p^q_{n,p,k}(x) = \begin{bmatrix} n+p+k-1 \\ k \end{bmatrix}_q q^{k(k+1)} \frac{x^k}{(1+x)_q^{n+p+k}}.$$

If we write $p = 0$ in (13.11), then the operators $M_{n,q}^{(\alpha,\beta)}$ are reduced to hybrid summation integral type operators given in [5].

In this context, let us start to give the following lemma for the Korovkin test functions.

Lemma 13.1. *Let* $e_m(t) = t^m$, $m = 0, 1, 2$. *We get*

(i) $M_{n,p,q}^{(\alpha,\beta)}(e_0, x) = 1,$

(ii)

$$M_{n,p,q}^{(\alpha,\beta)}(e_1, x) = \frac{[n+p]_q^2}{q\left([n+p]_q+\beta\right)[n+p-2]_q} x + \frac{\alpha}{[n+p]_q+\beta},$$

(iii)

$$\begin{aligned} M_{n,p,q}^{(\alpha,\beta)}(e_2, x) &= \frac{[n+p]_q^4}{q^4\left([n+p]_q+\beta\right)^2[n+p-3]_q[n+p-2]_q} x^2 \\ &+ \left\{ \frac{[2]_q[n+p]_q^3}{q^3\left([n+p]_q+\beta\right)^2[n+p-3]_q[n+p-2]_q} \right. \\ &+ \left. \frac{2[n+p]_q^2\alpha}{q[n+p-2]_q\left([n+p]_q+\beta\right)^2} \right\} x \\ &+ \frac{\alpha^2}{\left([n+p]_q+\beta\right)^2}. \end{aligned}$$

Proof. Using (13.9) and (13.10), we can obtain the estimate

$$\int_0^{\infty/A} p_{n,p,k-1}^q(t)t^m d_q t$$

$$= \begin{bmatrix} n+p+k-2 \\ k-1 \end{bmatrix}_q \int_0^{\infty/A} q^{k(k-1)} \frac{t^{k-1+m}}{(1+t)_q^{n+p+k-1}} d_q t$$

$$= \frac{[n+p+k-2]_q! B_q(k+m, n+p-m-1) q^{k(k-1)}}{[k-1]_q! [n+p-1]_q! K(A, k+m)}$$

$$= \frac{[m+k-1]_q! [n+p-m-2]_q! q^{\{-(k+m)(k+m-1)+2k(k-1)\}/2}}{[n+p-1]_q! [k-1]_q!}. \quad (13.12)$$

then, for $m = 0$, we get

$$M_{n,p,q}^{(\alpha,\beta)}(e_0, x)$$

$$= [n+p-1]_q \sum_{k=1}^{\infty} s_{n,p,k}^q(x) \int_0^{\infty/A} p_{n,p,k-1}^q(t) d_q t + e^{-[n+p]_q x}$$

$$= e^{-[n+p]_q x} \sum_{k=0}^{\infty} \frac{\left([n+p]_q x\right)^k}{[k]_q!} q^{k(k-1)/2}$$

$$= e^{-[n+p]_q x} E_q^{[n+p]_q x}$$

$$= 1,$$

which completes the proof of (i). By a direct computation

$$M_{n,p,q}^{(\alpha,\beta)}(e_1, x)$$

$$= [n+p-1]_q \sum_{k=1}^{\infty} s_{n,p,k}^q(x) \int_0^{\infty/A} p_{n,p,k-1}^q(t) \left(\frac{[n+p]_q t + \alpha}{[n+p]_q + \beta} \right) d_q t$$

$$+ e^{-[n+p]_q x} \frac{\alpha}{[n+p]_q + \beta}$$

$$= \frac{[n+p]_q}{\left([n+p]_q + \beta\right) [n+p-2]_q}$$

$$\times \sum_{k=1}^{\infty} \frac{\left([n+p]_q x\right)^k}{[k-1]_q!} q^{k(k-3)/2} e^{-[n+p]_q x}$$

$$+\frac{\alpha}{[n+p]_q+\beta}\sum_{k=1}^{\infty}\frac{\left([n+p]_q x\right)^k}{[k]_q!}q^{k(k-1)/2}e^{-[n+p]_q x}$$
$$+e^{-[n+p]_q x}\frac{\alpha}{[n+p]_q+\beta}.$$

Then we get

$$M_{n,p,q}^{(\alpha,\beta)}(e_1,x)$$
$$=\frac{[n+p]_q^2 x}{q\left([n+p]_q+\beta\right)[n+p-2]_q}e^{-[n+p]_q x}E_q^{[n+p]_q x}$$
$$+\frac{\alpha}{[n+p]_q+\beta}e^{-[n+p]_q x}E_q^{[n+p]_q x}$$
$$=\frac{[n+p]_q^2 x}{q\left([n+p]_q+\beta\right)[n+p-2]_q}+\frac{\alpha}{[n+p]_q+\beta},$$

which gives proof of (ii). Using the equality

$$[n]_q=[s]_q+q^s[n-s]_q, 0\le s\le n, \tag{13.13}$$

we yield

$$M_{n,p,q}^{(\alpha,\beta)}(e_2,x)$$
$$=[n+p-1]_q\sum_{k=1}^{\infty}s_{n,p,k}^q(x)\int_0^{\infty/A}p_{n,p,k-1}^q(t)\left(\frac{[n+p]_q t+\alpha}{[n+p]_q+\beta}\right)^2 d_qt$$
$$+e^{-[n+p]x}\left(\frac{\alpha}{[n+p]_q+\beta}\right)^2$$
$$=\frac{[n+p-1]\,[n+p]_q^2}{\left([n+p]_q+\beta\right)^2}\,{}_q\sum_{k=1}^{\infty}s_{n,p,k}^q(x)\int_0^{\infty/A}p_{n,p,k-1}^q(t)t^2d_qt$$
$$+\frac{2\alpha\,[n+p-1]_q\,[n+p]_q}{\left([n+p]_q+\beta\right)^2}\sum_{k=1}^{\infty}s_{n,p,k}^q(x)\int_0^{\infty/A}p_{n,p,k-1}^q(t)td_{q_n}t$$
$$+\frac{\alpha^2\,[n+p-1]_q}{\left([n+p]_q+\beta\right)^2}\sum_{k=1}^{\infty}s_{n,p,k}^q(x)$$
$$\times\int_0^{\infty/A}p_{n,p,k-1}^q(t)d_qt+e^{-[n+p]x}\left(\frac{\alpha}{[n+p]_q+\beta}\right)^2.$$

Then we get

$$M_{n,p,q}^{(\alpha,\beta)}(e_2,x) = \frac{[n+p]_q^4}{q^4\left([n+p]_q+\beta\right)^2[n+p-3]_q[n+p-2]_q}x^2$$
$$+\left\{\frac{[2]_q[n+p]_q^3}{q^3\left([n+p]_q+\beta\right)^2[n+p-3]_q[n+p-2]_q}\right.$$
$$\left.+\frac{2[n+p]_q^2\alpha}{q\left([n+p]_q+\beta\right)^2[n+p-2]_q}\right\}x$$
$$+\frac{\alpha^2}{\left([n+p]_q+\beta\right)^2},$$

which gives proof of *(iii)*.

To obtain our main result we need computing second moment

Lemma 13.2. *Let $q \in (0,1)$, $p \in \mathbb{N}$ and $n > 3$. Then we have the following inequality*

$$M_{n,p,q}^{(\alpha,\beta)}((t-x)^2,x) \le \left(\frac{2(1-q^3)}{q^4}+\frac{512(\alpha+\beta+1)^2[n+p]_q}{q^4[n+p-3]_q[n+p-2]_q}\right)x(x+1)$$
$$+\frac{\alpha^2}{\left([n+p]_q+\beta\right)^2}.$$

Proof. From linearity of $M_{n,p,q}^{(\alpha,\beta)}$ operators and Lemma 13.1, we can write the second moment as

$$M_{n,p,q}^{(\alpha,\beta)}((t-x)^2,x) = M_{n,p,q}^{(\alpha,\beta)}(t^2;x) - 2xM_{n,p,q}^{(\alpha,\beta)}(t;x) + x^2M_{n,p,q}^{(\alpha,\beta)}(1;x)$$
$$=\left\{\frac{[n+p]_q^4}{q^4\left([n+p]_q+\beta\right)^2[n+p-3]_q[n+p-2]_q}\right.$$
$$\left.-\frac{2[n+p]_q^2}{q\left([n+p]_q+\beta\right)[n+p-2]_q}+1\right\}x^2$$
$$+\left\{\frac{[2]_q[n+p]_q^3}{q^3\left([n+p]_q+\beta\right)^2[n+p-3]_q[n+p-2]_q}\right.$$
$$\left.+\frac{2[n+p]_q^2\alpha}{q\left([n+p]_q+\beta\right)^2[n+p-2]_q}-\frac{2\alpha}{[n+p]_q+\beta}\right\}x$$
$$+\frac{\alpha^2}{\left([n+p]_q+\beta\right)^2},$$

and hence

$$
\begin{aligned}
&M_{n,p,q}^{(\alpha,\beta)}((t-x)^2,x)\\
&\le \Bigg\{\Bigg|\frac{[n+p]_q^4}{q^4\left([n+p]_q+\beta\right)^2[n+p-3]_q[n+p-2]_q}\\
&\quad -\frac{2[n+p]_q^2}{q\left([n+p]_q+\beta\right)[n+p-2]_q}+1\Bigg|\\
&\quad +\Bigg|\frac{[2]_q[n+p]_q^3}{q^3\left([n+p]_q+\beta\right)^2[n+p-3]_q[n+p-2]_q}\\
&\quad +\frac{2[n+p]_q^2\alpha}{q\left([n+p]_q+\beta\right)^2[n+p-2]_q}\\
&\quad -\frac{2\alpha}{[n+p]_q+\beta}\Bigg|\Bigg\}x(x+1)+\frac{\alpha^2}{\left([n+p]_q+\beta\right)^2}.
\end{aligned}
$$

Then we get

$$
\begin{aligned}
&M_{n,p,q}^{(\alpha,\beta)}((t-x)^2,x)\\
&\le \Bigg\{\frac{\left|[n+p]_q^4(1+q^4)-2q^3[n+p-3]_q^4\right|+2q^4\beta[n+p]_q[n+p-3]_q[n+p-2]_q}{q^4\left([n+p]_q+\beta\right)^2[n+p-3]_q[n+p-2]_q}\\
&\quad +\frac{q^4\beta^2[n+p-3]_q[n+p-2]_q+q[2]_q[n+p]_q^3+2q^3\alpha[n+p]_q^2[n+p-3]_q}{q^4\left([n+p]_q+\beta\right)^2[n+p-3]_q[n+p-2]_q}\Bigg\}x(x+1)\\
&\quad +\frac{\alpha^2}{\left([n+p]_q+\beta\right)^2}.
\end{aligned}
$$

From (13.13) we have

$$
\begin{aligned}
&M_{n,p,q}^{(\alpha,\beta)}((t-x)^2,x)\\
&\le \Bigg\{\frac{[n+p-3]_q^4\left|q^{12}+q^{16}-2q^3\right|}{q^4\left([n+p]_q+\beta\right)^2[n+p-3]_q[n+p-2]_q}\\
&\quad +\frac{(1+q^4)\left\{4q^9[n+p-3]_q^3[3]_q+6q^6[n+p-3]_q^2[3]_q^2+4q^3[n+p-3]_q[3]_q^3+[3]_q^4\right\}}{q^4\left([n+p]_q+\beta\right)^2[n+p-3]_q[n+p-2]_q}\\
&\quad +\frac{2[n+p]_q^3(\beta+\alpha+1)^2}{q^4\left([n+p]_q+\beta\right)^2[n+p-3]_q[n+p-2]_q}\Bigg\}x(x+1)\\
&\quad +\frac{\alpha^2}{\left([n+p]_q+\beta\right)^2},
\end{aligned}
$$

which implies

$$\begin{aligned} M_{n,p,q}^{(\alpha,\beta)}((t-x)^2,x) \\ \leq \left(\frac{2(1-q^3)}{q^4}+\frac{512(\alpha+\beta+1)^2[n+p]_q}{q^4[n+p-3]_q[n+p-2]_q}\right)x(x+1) \\ +\frac{\alpha^2}{\left([n+p]_q+\beta\right)^2}. \end{aligned}$$

And the proof of the Lemma is now finished.

Corollary 13.3. *Let* (q_n) *be a sequence in* $(0,1)$ *and let* $q_n\to 1$ *as* $n\to\infty$*, Then*

$$\lim_{n\to\infty} M_{n,p,q_n}^{(\alpha,\beta)}((t-x)^2,x)=0 \text{ for each } p\in\mathbb{N}.$$

13.3 Direct Results

Now, we need some definitions of elementary function spaces $B[0,\infty)$ and $C_B[0,\infty)$. $B[0,\infty)$ is the space of all bounded functions from $[0,\infty)$ to $\mathbb{R}$. $B[0,\infty)$ is a normed space with the norm

$$\|f\|_B=\sup\{|f(x)|:x\in[0,\infty)\}.$$

$C_B[0,\infty)$ consists of all continuous functions in $B[0,\infty)$. We denote first modulus of continuity on finite interval $[0,b]$, $b>0$

$$\omega_{[0,b]}(f,\delta)=\sup_{0<h\leq\delta,x\in[0,b]}|f(x+h)-f(x)|\,. \tag{13.14}$$

The well-known Peetre's K-functional is defined by

$$K_2(f,\delta)=\inf\left\{\|f-g\|_B+\delta\left\|g''\right\|_B:g\in W_\infty^2\right\},\ \delta>0$$

where

$$W_\infty^2=\left\{g\in C_B[0,\infty):g',g''\in C_B[0,\infty)\right\}.$$

In [4, p. 177, Theorem 2.4], there exists a positive constant C such that

$$K_2(f,\delta)\leq C\omega_2(f,\sqrt{\delta}) \tag{13.15}$$

where

$$\omega_2(f,\sqrt{\delta}) = \sup_{0<h\leq\sqrt{\delta}} \sup_{x\in[0,\infty)} |f(x+2h) - 2f(x+h) - f(x)|\,.$$

The weighted Korovkin-type theorems were firstly proved by Gadzhiev in [6]. We give some results similar to Gadzhiev's theorems in weighted spaces. Let $\rho(x) = 1 + x^2$. $B_\rho[0,\infty)$ denotes the set of all functions f, from $[0,\infty)$ to $\mathbb{R}$, satisfying growth condition $|f(x)| \leq N_f\rho(x)$, where N_f is a constant depending only on f. $B_\rho[0,\infty)$ is a normed space with the norm

$$\|f\|_\rho = \sup\left\{\frac{|f(x)|}{\rho(x)} : x \in \mathbb{R}\right\}.$$

$C_\rho[0,\infty)$ denotes the subspace of all continuous functions in $B_\rho[0,\infty)$ and $C^*_\rho[0,\infty)$ denotes the subspace of all functions $f \in C_\rho[0,\infty)$ for which $\lim\limits_{|x|\to\infty} \dfrac{|f(x)|}{\rho(x)}$ exists finitely.

The following lemma is routine and its proof is omitted.

Lemma 13.4. *Let*

$$\begin{aligned}\overline{M}^{(\alpha,\beta)}_{n,p,q}(f,x) = M^{(\alpha,\beta)}_{n,p,q}(f,x)\\ -f\left(\frac{[n+p]_q^2\,x}{q\left([n+p]_q+\beta\right)[n+p-2]_q} + \frac{\alpha}{[n+p]_q+\beta}\right)\\ +f\,(x)\,.\end{aligned} \tag{13.16}$$

Then the following assertions are hold:

(*i*) $\overline{M}^{(\alpha,\beta)}_{n,p,q}(1,x) = 1,$
(*ii*) $\overline{M}^{(\alpha,\beta)}_{n,p,q}(t,x) = x,$
(*iii*) $\overline{M}^{(\alpha,\beta)}_{n,p,q}(t-x,x) = 0.$

Lemma 13.5. *Let $q \in (0,1)$, $p \in \mathbb{N}$ and $n > 3$. Then, for every $x \in [0,\infty)$ and $f'' \in C_B[0,\infty)$ we have the inequality*

$$\left|\overline{M}^{(\alpha,\beta)}_{n,p,q}(f,x) - f(x)\right| \leq \delta^{(\alpha,\beta)}_{n,p,q}(x)\,\|f''\|_B$$

where

$$\delta^{(\alpha,\beta)}_{n,p,q}(x) = \left(\frac{2(1-q^3)}{q^4} + \frac{514(\alpha+\beta+1)^2[n+p]_q}{q^4[n+p-3]_q[n+p-2]_q}\right)x(x+1) + \frac{\alpha^2}{\left([n+p]_q+\beta\right)^2}.$$

Proof. Using Taylor's expansion

$$f(t) = f(x) + (t-x)f'(x) + \int_x^t (t-u)f''(u)du$$

and from Lemma 13.4, we obtain

$$\overline{M}_{n,p,q}^{(\alpha,\beta)}(f,x) - f(x) = \overline{M}_{n,p,q}^{(\alpha,\beta)}\left(\int_x^t (t-u)f''(u)du; x\right).$$

Then, using Lemma 13.1 and the inequality

$$\left|\int_x^t (t-u)f''(u)du\right| \leq \|f''\|_B \frac{(t-x)^2}{2}$$

we get

$$\begin{aligned}
&\left|\overline{M}_{n,p,q}^{(\alpha,\beta)}(f,x) - f(x)\right| \\
&\leq \left| M_{n,p,q}^{(\alpha,\beta)}\left(\int_x^t (t-u)f''(u)du, x\right) \right. \\
&\quad \left. - \int_x^{\frac{[n+p]_q^2 x}{q([n+p]_q+\beta)[n+p-2]_q}+\frac{\alpha}{[n+p]_q+\beta}} \left(\frac{[n+p]_q^2 x}{q([n+p]_q+\beta)[n+p-2]_q} + \frac{\alpha}{[n+p]_q+\beta} - u\right) f''(u)du \right| \\
&\leq \frac{\|f''\|_B}{2} M_{n,p,q}^{(\alpha,\beta)}((t-x)^2, x) \\
&\quad + \frac{\|f''\|_B}{2}\left(\left(\frac{[n+p]_q^2 x}{q\left([n+p]_q+\beta\right)[n+p-2]_q} - 1\right)x \right. \\
&\quad \left. + \frac{\alpha}{[n+p]_q+\beta}\right)^2,
\end{aligned}$$

which implies

$$\begin{aligned}
&\left|\overline{M}_{n,p,q}^{(\alpha,\beta)}(f,x) - f(x)\right| \\
&\leq \frac{\|f''\|_B}{2}\left\{\left(\frac{2\left(1-q^3\right)}{q^4} + \frac{512(\alpha+\beta+1)^2[n+p]_q}{q^4[n+p-3]_q[n+p-2]_q}\right)x(x+1) + \frac{\alpha^2}{\left([n+p]_q+\beta\right)^2}\right\}
\end{aligned}$$

$$+\frac{\|f''\|_B}{2}\left\{\frac{[n+p]_q^4x^2}{q^2\left([n+p]_q+\beta\right)^2[n+p-2]_q^2}-2\frac{[n+p]_q^2x^2}{q\left([n+p]_q+\beta\right)[n+p-2]_q}+x^2\right.$$

$$\left.+2\frac{\alpha}{[n+p]_q+\beta}\left(\frac{[n+p]_q^2x}{q\left([n+p]_q+\beta\right)[n+p-2]_q}-1\right)x+\left(\frac{\alpha}{[n+p]_q+\beta}\right)^2\right\}$$

$$\leq\left\{\left(\frac{2(1-q^3)}{q^4}+\frac{514(\alpha+\beta+1)^2[n+p]_q}{q^4[n+p-3]_q[n+p-2]_q}\right)x(x+1)+\frac{\alpha^2}{\left([n+p]_q+\beta\right)^2}\right\}\|f''\|_B.$$

And the proof of the lemma is now completed.

Theorem 13.6. *Let $(q_n)\subset(0,1)$ be a sequence such that $q_n\to1$ as $n\to\infty$. Then for every $n>3$, $x\in[0,\infty)$ and $f\in C_B[0,\infty)$ we have the inequality*

$$\left|M_{n,p,q_n}^{(\alpha,\beta)}(f,x)-f(x)\right|\leq 3C\omega_2\left(f,\sqrt{\delta_{n,p,q_n}^{(\alpha,\beta)}(x)}\right)+w\left(f,\eta_{n,p,q_n}^{(\alpha,\beta)}(x)\right)$$

where

$$\eta_{n,p,q_n}^{(\alpha,\beta)}(x)=\left(\frac{[n+p]_{q_n}^2}{q_n\left([n+p]_{q_n}+\beta\right)[n+p-2]_{q_n}}-1\right)x$$
$$+\frac{\alpha}{[n+p]_{q_n}+\beta}.$$

Proof. Using (13.16) for any $g\in W_\infty^2$, we obtain the equality

$$\begin{aligned}&\left|M_{n,p,q_n}^{(\alpha,\beta)}(f,x)-f(x)\right|\\&\leq\left|\overline{M}_{n,p,q_n}^{(\alpha,\beta)}(f-g,x)-(f-g)(x)+\overline{M}_{n,p,q_n}^{(\alpha,\beta)}(g,x)-g(x)\right|\\&\quad+\left|f\left(\frac{[n+p]_{q_n}^2x}{q\left([n+p]_{q_n}+\beta\right)[n+p-2]_{q_n}}+\frac{\alpha}{[n+p]_{q_n}+\beta}\right)-f(x)\right|.\end{aligned}$$

From Lemma 13.5, we get

$$\begin{aligned}&\left|M_{n,p,q_n}^{(\alpha,\beta)}(f,x)-f(x)\right|\\&\leq 3\,||f-g||_B+\delta_{n,p,q_n}^{(\alpha,\beta)}(x)\,\left\|g''\right\|\\&\quad+\left|f\left(\frac{[n+p]_{q_n}^2x}{q_n\left([n+p]_{q_n}+\beta\right)[n+p-2]_{q_n}}+\frac{\alpha}{[n+p]_{q_n}+\beta}\right)-f(x)\right|.\end{aligned}$$

By equality (13.14) we have

$$\left|M_{n,p,q_n}^{(\alpha,\beta)}(f,x)-f(x)\right| \leq 3\,||f-g||_B+\delta_{n,p,q_n}^{(\alpha,\beta)}(x)\left\|g''\right\|_B \\ + w\left(f,\eta_{n,p,q_n}^{(\alpha,\beta)}(x)\right).$$

Now taking infimum over $g \in W_\infty^2$ on the right side of the above inequality and using the inequality (13.16), we get the desired result.

Theorem 13.7. *Let $(q_n) \subset (0,1)$ be a sequence such that $q_n \to 1$ as $n \to \infty$. Then $f \in C_\rho^*[0,\infty)$, we have*

$$\lim_{n\to\infty} \|M_{n,p,q_n}^{(\alpha,\beta)}(f)-f\|_\rho = 0.$$

Proof. From Lemma 13.1, it is obvious that $\|M_{n,p,q_n}^{(\alpha,\beta)}(e_0)-e_0\|_\rho = 0$. Since

$$\left|\frac{[n+p]_{q_n}^2}{q_n\left([n+p]_{q_n}+\beta\right)[n+p-2]_{q_n}}x+\frac{\alpha}{[n+p]_{q_n}+\beta}-x\right| \leq (x+1)o(1)$$

and $\frac{x+1}{1+x^2}$ is positive and bounded from above for each $x \geq 0$, we obtain

$$\|M_{n,p,q_n}^{(\alpha,\beta)}(e_1)-e_1\|_\rho \leq \frac{x+1}{1+x^2}o(1).$$

And then

$$\lim_{n\to\infty} \|M_{n,p,q_n}^{(\alpha,\beta)}(e_1)-e_1\|_\rho = 0.$$

Similarly, for every $n > 3$, we write

$$\|M_{n,p,q_n}^{(\alpha,\beta)}(e_2)-e_2\|_\rho$$

$$= \sup_{x\in[0,\infty)}\left(\frac{\frac{[n+p]_{q_n}^4}{q_n^4\left([n+p]_{q_n}+\beta\right)^2[n+p-3]_{q_n}[n+p-2]_{q_n}}x^2}{1+x^2}\right.$$

$$+\frac{\left\{\frac{[2]_{q_n}[n+p]_{q_n}^3}{q_n^3\left([n+p]_{q_n}+\beta\right)^2[n+p-3]_{q_n}[n+p-2]_{q_n}}+\frac{2[n+p]_{q_n}^2\alpha}{q_n[n+p-2]_{q_n}\left([n+p]_{q_n}+\beta\right)^2}\right\}x}{1+x^2}$$

$$\left. + \frac{+\frac{\alpha^2}{\left([n+p]_{q_n}+\beta\right)^2} - x^2}{1+x^2} \right)$$

$$\leq \sup_{x\in[0,\infty)} \frac{1+x+x^2}{1+x^2} o(1),$$

we get

$$\lim_{n\to\infty} \|M_{n,p,q_n}^{(\alpha,\beta)}(e_2) - e_2\|_\rho = 0.$$

Thus, from Gadzhiev's theorem in [6], we obtain the desired result.

Lemma 13.8. *Let* $f \in C_\rho[0,\infty)$, $(q_n) \subset (0,1)$ *be a sequence such that* $q_n \to 1$ *as* $n \to \infty$ *and* $\omega_{[0,b+1]}(f,\delta)$ *be its modulus of continuity on the finite interval* $[0,b+1]$, $b > 0$. *Then for every* $n > 3$, *there exists a constant* $C > 0$ *such that the inequality holds*

$$\|M_{n,p,q_n}^{(\alpha,\beta)}(f,x) - f(x)\|_{C[0,b]}$$
$$\leq C\left\{(b+1)^2 \xi_{n,p,q_n}^{(\alpha,\beta)}(b) + \omega_{[0,b+1]}\left(f; \sqrt{\xi_{n,p,q_n}^{(\alpha,\beta)}(b)}\right)\right\},$$

where

$$\xi_{n,p,q_n}^{(\alpha,\beta)}(b) = \left(\frac{2(1-q_n^3)}{q_n^4} + \frac{512(\alpha+\beta+1)^2[n+p]_{q_n}}{q_n^4[n+p-3]_{q_n}[n+p-2]_{q_n}}\right) b(b+1)$$
$$+ \frac{\alpha^2}{\left([n+p]_{q_n}+\beta\right)^2}$$

Proof. Let $x \in [0,b]$ and $t > b+1$. Since $t - x > 1$, we have

$$\begin{aligned} |f(t) - f(x)| &\leq N_f(2 + (t-x+x)^2 + x^2) \\ &\leq 3N_f(1+b)^2(t-x)^2. \end{aligned} \tag{13.17}$$

Let $x \in [0,b]$, $t < b+1$ and $\delta > 0$. Then, we have

$$|f(t) - f(x)| \leq \left(1 + \frac{|t-x|}{\delta}\right)\omega_{[0,b+1]}(f,\delta). \tag{13.18}$$

Due to (13.17) and (13.18), we can write

$$|f(t) - f(x)| \leq 3N_f(1+b)^2(t-x)^2 + \left(1 + \frac{|t-x|}{\delta}\right)\omega_{[0,b+1]}(f,\delta).$$

Then, using Cauchy–Schwarz's inequality and Lemma 13.2, we get

$$\begin{aligned}
&\left|M_{n,p,q_n}^{(\alpha,\beta)}(f,x)-f(x)\right|\\
&\leq 3N_f(1+b)^2 M_{n,p,q_n}^{(\alpha,\beta)}\left((t-x)^2,x\right)\\
&\quad +\omega_{[0,b+1]}(f;\delta)\left[1+\frac{1}{\delta}\left(M_{n,p,q_n}^{(\alpha,\beta)}\left((t-x)^2,x\right)\right)^{1/2}\right]\\
&\leq 3N_f(1+b)^2\xi_{n,p,q_n}^{(\alpha,\beta)}(x)\\
&\quad +\omega_{[0,b+1]}(f;\delta)\left[1+\frac{1}{\delta}\left(\xi_{n,p,q_n}^{(\alpha,\beta)}(x)\right)^{1/2}\right],
\end{aligned}$$

where

$$\begin{aligned}
\xi_{n,p,q_n}^{(\alpha,\beta)}(x) &= \left(\frac{2(1-q_n^3)}{q_n^4}+\frac{512(\alpha+\beta+1)^2[n+p]_{q_n}}{q_n^4[n+p-3]_{q_n}[n+p-2]_{q_n}}\right)x(x+1)\\
&\quad +\frac{\alpha^2}{\left([n+p]_{q_n}+\beta\right)^2}.
\end{aligned}$$

Choosing,

$$\begin{aligned}
\delta^2 &:= \xi_{n,p,q_n}^{(\alpha,\beta)}(b)\\
&= \left(\frac{2\left(1-q_n^3\right)}{q_n^4}+\frac{512(\alpha+\beta+1)^2[n+p]_{q_n}}{q_n^4[n+p-3]_{q_n}[n+p-2]_{q_n}}\right)b(b+1)\\
&\quad +\frac{\alpha^2}{\left([n+p]_{q_n}+\beta\right)^2}
\end{aligned}$$

and $C=\max\{3N_f,2\}$. We reach the proof of Lemma.

Theorem 13.9. *Let $\gamma>0$, $(q_n)\subset(0,1)$ be a sequence such that $q_n\to 1$ as $n\to\infty$ and $f\in C_\rho^*[0,\infty)$. Then, we have*

$$\lim_{n\to\infty}\sup_{x\geq 0}\frac{\left|M_{n,p,q_n}^{(\alpha,\beta)}(f,x)-f(x)\right|}{1+x^{2+\gamma}}=0.$$

Proof. For $\gamma>0, f\in C_\rho^*[0,\infty)$ and $b\geq 1$ the following inequality is satisfied

$$\sup_{x\geq 0}\frac{\left|M_{n,p,q_n}^{(\alpha,\beta)}(f,x)-f(x)\right|}{1+x^{2+\gamma}}\leq \sup_{0\leq x<b}\frac{\left|M_{n,p,q_n}^{(\alpha,\beta)}(f,x)-f(x)\right|}{1+x^{2+\gamma}}$$

$$+\sup_{b\leq x}\frac{\left|M_{n,p,q_n}^{(\alpha,\beta)}(f,x)-f(x)\right|}{1+x^{2+\gamma}}$$
$$\leq \left\|M_{n,p,q_n}^{(\alpha,\beta)}(f)-f\right\|_{C[0,b]}$$
$$+\sup_{b\leq x}\frac{\left|M_{n,p,q_n}^{(\alpha,\beta)}(f,x)-f(x)\right|}{1+x^{2+\gamma}}$$
$$\leq \left\|M_{n,p,q_n}^{(\alpha,\beta)}(f)-f\right\|_{C[0,b]}$$
$$+\left\|M_{n,p,q_n}^{(\alpha,\beta)}(f)-f\right\|_{\rho}$$

Using Lemma 13.8 and Theorem 13.7, we complete the proof of Theorem 13.9.

References

1. A. Aral, V. Gupta, On the Durrmeyer type modification of the q-Baskakov type operators. Nonlinear Anal. **72**(3–4), 1171–1180 (2010)
2. A. Aral, V. Gupta, R.P. Agarwal, *Applications of q-Calculus in Operator Theory* (Springer, New York, 2013)
3. A. De Sole, V.G. Kac, On integral representations of q-gamma and q-beta functions. Atti Accad. Naz. Lincei Cl. Sci. Fis. Mat. Natur. Rend. Lincei (9) Mat. Appl. **16**(1), 11–29 (2005)
4. R.A. De Vore, G.G. Lorentz, *Constructive Approximation* (Springer, Berlin, 1993)
5. Ü. Dinlemez, İ. Yüksel, B. Altın, A note on the approximation by the q-hybrid summation integral type operators. Taiwanese. J. Math. **18**(3), 781–792 (2014)
6. A.D. Gadzhiev, Theorems of the type of P. P. Korovkin's type theorems. Math. Zametki **20**(5), 781–786 (1976); English Translation, Math. Notes **20**(5/6), 996–998 (1976)
7. V. Gupta, A. Aral, Convergence of the q-analogue of Szász-beta operators. Appl. Math. Comput. **216**(2), 374–380 (2010)
8. V. Gupta, E. Erkuş, On hybrid family of summation integral type operators. JAIPAM. J. Inequal. Pure Appl. Math. **7**(1) (2006). Article 23, 11 pp
9. V. Gupta, W. Heping, The rate of convergence of q-Durrmeyer operators for $0 < q < 1$. Math. Methods Appl. Sci. **31**(16), 1946–1955 (2008)
10. V. Gupta, H. Karsli, Some approximation properties by q-Szász-Mirakyan-Baskakov-Stancu operators. Lobachevskii J. Math. **33**(2), 175–182 (2012)
11. F.H. Jackson, On q-definite integrals. Q. J. Pure Appl. Math. **41**, 193–203 (1910)
12. V.G. Kac, P. Cheung, *Quantum Calculus*. Universitext (Springer, New York, 2002)
13. H.T. Koelink, T.H. Koorwinder, q-Special functions, a tutorial, in *Deformation Theory and Quantum Groups with Applications to Mathematical Physics* , Amherst, MA, 1990, vol. 134. Contemporary Mathematics (American Mathematical Society, Providence, RI, 1992), pp. 141–142
14. A. Lupaş, A q-analogue of the Bernstein operator, in *Seminar on Numerical and Statistical Calculus*, University of Cluj-Napoca 9 (1987), pp. 85–92
15. G.M. Phillips, Bernstein polynomials based on the q-integers. Ann. Numer. Math. **4**, 511–518 (1997)
16. J. Sinha, V.K. Singh, Rate of convergence on the mixed summation integral type operators. Gen. Math. **14**(4), 29–36 (2006)
17. İ. Yüksel, Direct results on the q-mixed summation integral type operators. J. Appl. Funct. Anal. **8**(2), 235–245 (2013)

Chapter 14
Approximation by q-Baskakov–Durrmeyer Type Operators of Two Variables

İsmet Yüksel, Ülkü Dinlemez, and Birol Altın

Abstract In this study, we investigate approximation properties of q-Baskakov–Durrmeyer type operators with two variables. We give a Voronovskaja type theorem for these operators. In addition, we obtain the rate of convergence for these operators.

14.1 Introduction

In their paper [5], Aral and Gupta introduced a new class of operators named q-Baskakov–Durrmeyer type operators of one variable which are defined as

$$D_n^q(f;x) = [n-1]_q \sum_{k=0}^{\infty} p_{n,k}^q(x) \int_0^{\infty/A} p_{n,k}^q(t) f(t)\, d_q t, \tag{14.1}$$

where

$$p_{n,k}^q(x) = \begin{bmatrix} n+k-1 \\ k \end{bmatrix}_q q^{\frac{k^2}{2}} \frac{x^k}{(1+x)_q^{n+k}}.$$

These type of operators were studied in [4, 6, 14, 19]. For $q = 1$, these operators are reduced to Baskakov–Durrmeyer type operators introduced in [18] and the approximation of these operators was studied in [13, 21]. We generalize these operators to two variables operators. Now we give necessary notations and definitions. Let $D = \{(x,y) \in \mathbb{R}^2 : 0 \le x,\ 0 \le y\}$, $C(D)$ be the set of real valued continuous functions on D and

$$C_\rho(D) = \{f \in C(D) : |f((x,y))| \le N_f\left(1 + x^2 + y^2\right) \text{ for some } N_f \in \mathbb{R}^+\}$$

İ. Yüksel (✉) • Ü. Dinlemez • B. Altın
Department of Mathematics, Faculty of Science, Gazi University, Ankara, Turkey
e-mail: iyuksel@gazi.edu.tr; ulku@gazi.edu.tr; birola@gazi.edu.tr

G.A. Anastassiou, O. Duman (eds.), *Computational Analysis*, Springer Proceedings in Mathematics & Statistics 155, DOI 10.1007/978-3-319-28443-9_14

and

$$C_{\rho}^{*}(D)=\left\{f\in C_{\rho}(D):\lim_{(x,y)\to(\infty,\infty)}\frac{f(x,y)}{1+x^2+y^2}\text{ exists}\right\}$$

with the norm $\|f\|_{\rho}=\sup_{0\leq x,\,0\leq y}\frac{|f(x,y)|}{1+x^2+y^2}$. Let us introduce two variables of Baskakov–Durrmeyer operators $M_n^q: C(D)\to C(D)$ as follows: for $f\in C(D)$ and $(x,y),(s,t)\in D$,

$$M_n^q(f;(x,y))=[n-1]_q^2\sum_{k,l\in\mathbb{N}}b_{n,k,l}^q(x,y)\int_0^{\infty/A}\int_0^{\infty/A}b_{n,k,l}^q(s,t)f(s,t)d_qsd_qt,\tag{14.2}$$

where

$$b_{n,k,l}^q(x,y)=p_{n,k}^q(x)p_{n,l}^q(y).$$

Approximation functions of one or two variables by certain positive linear operators in weighted spaces can be found in [2, 8–10, 12, 15–17, 22–26]. In all undefined terminology concerning approximation theory we will adhere to [6].

14.2 Generalization of q-Baskakov–Durrmeyer Type Operators

In this section, we give some classical approximation properties of these operators. We use the following notations, for brevity,

$$e_{ij}(s,t)=s^it^j,\ e_{i+j}(s,t)=s^i+t^j\text{ and }F_{m,q}(k)=[k]_q[k+1]_q\dots[k+m]_q$$

for $i,j,k,m\in\mathbb{N}$.

We give the following lemma to obtain a relation between the operators M_n^q in (14.1) and D_n^q in (14.2).

Lemma 14.1. *For $n>1+\max\{i,j\}$, we have the equation*

$$M_n^q(e_{ij};(x,y))=D_n^q(s^i;x)D_n^q(t^j;y).$$

Proof. By Fubini's Theorem, we may write that

$$M_n^q(e_{ij};(x,y))$$
$$=[n-1]_q^2\sum_{k,l\in\mathbb{N}}b_{n,k,l}^q(x,y)\int_0^{\infty/A}\int_0^{\infty/A}b_{n,k,l}^q(s,t)e_{ij}(s,t)d_qsd_qt,$$

$$= [n-1]_q^2 \sum_{k,l\in\mathbb{N}} p_{n,k}^q(x)p_{n,l}^q(y) \left\{ \int_0^{\infty/A} \begin{bmatrix} n+k-1 \\ k \end{bmatrix}_q \frac{s^{k+i}q^{\frac{k^2}{2}}}{(1+s)_q^{n+k}} d_qs \right.$$

$$\left. \times \int_0^{\infty/A} \begin{bmatrix} n+l-1 \\ l \end{bmatrix}_q \frac{t^{l+j}q^{\frac{l^2}{2}}}{(1+t)_q^{n+l}} d_qt \right\},$$

which gives

$$M_n^q(e_{ij};(x,y))$$

$$= D_n^q(s^i;x)\,[n-1]_q\, p_{n,0}^q(x) \int_0^{\infty/A} \begin{bmatrix} n-1 \\ 0 \end{bmatrix}_q \frac{t^jq^{\frac{l^2}{2}}}{(1+t)_q^n} d_qt$$

$$+ D_n^q(s^i;x)\,[n-1]_q\, p_{n,1}(x) \int_0^{\infty/A} \begin{bmatrix} n \\ 1 \end{bmatrix}_q \frac{t^{1+j}q^{\frac{l^2}{2}}}{(1+t)_q^{n+1}} d_qt + \cdots$$

$$= D_n^q(s^i;x)D_n^q(t^j;y).$$

The equations (*i*)–(*iii*) in the following lemma follow from the paper in [5].

Lemma 14.2. *The following statements hold for $n > 5$:*

(*i*) $M_n^q(e_{00};(x,y)) = 1$,

(*ii*)

$$M_n^q(e_{10};(x_1,x_2)) = \frac{F_{0,q}(n)\,x}{q^2F_{0,q}(n-2)} + \frac{1}{qF_{0,q}(n-2)},$$

(*iii*)

$$M_n^q(e_{20};(x_1,x_2)) = \frac{F_{1,q}(n)}{q^6F_{1,q}(n-3)}x^2 + \frac{[2]_q^2\,F_{0,q}(n)}{q^5F_{1,q}(n-3)}x + \frac{[2]_q}{q^3F_{1,q}(n-3)},$$

(*iv*)

$$M_n^q(e_{30};(x,y))$$

$$= \frac{F_{2,q}(n)}{q^{12}F_{2,q}(n-4)}x^3 + \left\{[5]_q + q[2]_q^2\right\} \frac{F_{1,q}(n)}{q^{11}F_{2,q}(n-4)}x^2$$

$$+\left\{[4]_q[2]_q^2+q^2[2]_q\right\}\frac{F_{0,q}(n)}{q^9F_{2,q}(n-4)}x+\frac{[2]_q[3]_q}{q^6F_{2,q}(n-4)},$$

(v)

$$\begin{aligned}
&M_n^q(e_{40};(x,y))\\
&=\frac{F_{3,q}(n)}{q^{20}F_{3,q}(n-5)}x^4+\left\{[7]_q+q[5]_q+q^2[2]_q^2\right\}\frac{F_{2,q}(n)}{q^{19}F_{3,q}(n-5)}x^3\\
&+\left\{[6]_q\left([5]_q+q[2]_q^2\right)+q^2[2]_q^2[4]_q+q^4[2]_q\right\}\frac{F_{1,q}(n)}{q^{17}F_{3,q}(n-5)}x^2\\
&+\left\{[5]_q\left([2]_q^2[4]_q+q^2[2]_q\right)+q^3[2]_q[3]_q\right\}\frac{[n]_q}{q^{14}F_{3,q}(n-5)}x\\
&+\frac{[2]_q[3]_q[4]_q}{q^{10}F_{3,q}(n-5)}.
\end{aligned}$$

Proof. We obtain the estimate

$$\begin{aligned}
&\int_0^{\infty/A}\int_0^{\infty/A}p_{n,k}^q(s)p_{n,l}^q(t)e_{m0}(s,t)e_{0v}(s,t)d_qsd_qt\\
&=\frac{[k+m]_q!\,[n-m-2]_q!\,[l+v]_q!\,[n-v-2]_q!}{\left([n-1]_q!\right)^2[k]_q!\,[l]_q!}\\
&\times q^{\frac{-(k+m)(k+m+1)-(l+v)(l+v+1)+k^2+l^2}{2}}. \qquad (14.3)
\end{aligned}$$

Then taking into account Lemma 14.1 and (14.3), we have

$$\begin{aligned}
M_n^q(e_{30};(x,y)) &= D_n^q(s^3;x)D_n^q(1;y)\\
&=[n-1]_q\sum_{k=0}^{\infty}p_{n,k}^q(x)\begin{bmatrix}n+k-1\\k\end{bmatrix}_q\frac{q^{\frac{k^2}{2}}B_q(k+4,n-4)}{q^{(k+3)(k+4)/2}}\\
&=\sum_{k=0}^{\infty}\begin{bmatrix}n+k-1\\k\end{bmatrix}_q q^{\frac{k^2-7k-12}{2}}\frac{F_{2,q}(k+1)}{F_{2,q}(n)}. \qquad (14.4)
\end{aligned}$$

Replacing the following equation in (14.4),

$$\begin{aligned}
F_{2,q}(k+1) &= F_{2,q}(k-2)+\left\{[5]_q+q[2]_q^2\right\}F_{1,q}(k-1)\\
&\quad+\left\{[4]_q[2]_q^2+q^2[2]_q\right\}[k]_q+[2]_q[3]_q,
\end{aligned}$$

we get

$$\begin{aligned}
&M_n^q(e_{30};(x,y)) \\
&= \frac{F_{2,q}(n)}{F_{2,q}(n-4)} \sum_{k=3}^{\infty} \begin{bmatrix} n+k-1 \\ k-3 \end{bmatrix}_q q^{\frac{k^2-7k-12}{2}} \frac{x^k}{(1+x)_q^{n+k}} \\
&+ \left\{[5]_q + q[2]_q^2\right\} \frac{F_{1,q}(n)}{F_{2,q}(n-4)} \\
&\times \sum_{k=2}^{\infty} \begin{bmatrix} n+k-1 \\ k-2 \end{bmatrix}_q q^{\frac{k^2-5k-16}{2}} \frac{x^k}{(1+x)_q^{n+k}} \\
&+ \left\{[4]_q[2]_q^2 + q^2[2]_q\right\} \frac{[n]_q}{F_{2,q}(n-4)} \\
&\times \sum_{k=1}^{\infty} \begin{bmatrix} n+k-1 \\ k-1 \end{bmatrix}_q q^{\frac{k^2-3k-16}{2}} \frac{x^k}{(1+x)_q^{n+k}} \\
&+ \frac{[2]_q[3]_q}{F_{2,q}(n-4)} \sum_{k=0}^{\infty} \begin{bmatrix} n+k-1 \\ k \end{bmatrix}_q q^{\frac{k^2-k-12}{2}} \frac{x^k}{(1+x)_q^{n+k}},
\end{aligned}$$

as desired (iv). Using similar steps and the following equation

$$\begin{aligned}
&F_{3,q}(k+1) \\
&= F_{3,q}(k-3) + \left\{[7]_q + q[5]_q + q^2[2]_q^2\right\} F_{2,q}(k-2) \\
&+ \left\{[6]_q\left([5]_q + q[2]_q^2\right) + q^2[2]_q^2[4]_q + q^4[2]_q\right\} q^3 F_{1,q}(k-1) \\
&+ \left\{[5]_q\left([2]_q^2[4]_q + q^2[2]_q\right) + q^3[2]_q[3]_q\right\} q^3[k]_q \\
&+ [2]_q[3]_q[4]_q
\end{aligned}$$

one can prove (v) easily.

The following theorem gives us the Baskakov type theorem (see [7]) to get the uniform approximation to the functions in $C_\rho^*(D)$ by the sequence of the positive linear operators M_n^q.

Theorem 14.3. *Let* $0 < q_n \leq 1$, $(q_n) \to 1$ *as* $n \to \infty$ *and* f *be belong to* $C_\rho^*(D)$. *Then*

$$\lim_n \left\| M_n^{q_n}(f) - f \right\|_\rho = 0$$

if and only if the following statements are satisfied:

(i) $\lim_{n\to\infty} \left\| M_n^{q_n}(e_{00}) - e_{00} \right\|_\rho = 0,$
(ii) $\lim_{n\to\infty} \left\| M_n^{q_n}(e_{10}) - e_{10} \right\|_\rho = 0,$
(iii) $\lim_{n\to\infty} \left\| M_n^{q_n}(e_{01}) - e_{01} \right\|_\rho = 0,$
(iv) $\lim_{n\to\infty} \left\| M_n^{q_n}(e_{20} + e_{02}) - (e_{20} + e_{02}) \right\|_\rho = 0.$

Proof. The necessity part is trivial. The sufficient part needs proof. Let

$$D_a = \left\{(x,y) \in \mathbb{R}^2 : 0 \le x \le a, 0 \le y \le a\right\},$$

$(x,y) \in D_a$, $r > a$, and $f \in C_\rho^*(D)$. Choose $\varepsilon > 0$. Since f is uniformly continuous on D_r, $r > a$, there exists some $\delta > 0$ such that $\sqrt{(s-x)^2 + (t-y)^2} < \delta$ implies

$$|f(x,y) - f(s,t)| < \varepsilon. \tag{14.5}$$

On the other hand, if $\sqrt{(s-x)^2 + (t-y)^2} \ge \delta$ for $(s,t) \in D$, we have

$$|f(s,t) - f(x,y)| \le N_1 \frac{(s-x)^2 + (t-y)^2}{\delta^2} \text{ for some } N_1 > 0. \tag{14.6}$$

Combining (14.5) and (14.6) we get

$$|f(s,t) - f(x,y)| < \varepsilon + N_1 \frac{(s-x)^2 + (t-y)^2}{\delta^2}. \tag{14.7}$$

If we apply the positive operators $M_n^{q_n}$ to the equation

$$f(s,t) = (f - f(x,y)e_{00})(s,t) + f(x,y),$$

then we obtain the following inequality:

$$\begin{aligned}
&\left|M_n^{q_n}(f;(x,y)) - f(x,y)\right| \\
&= \left|M_n^{q_n}(f - f(x,y)e_{00} + f(x,y)e_{00};(x,y)) - f(x,y)\right| \\
&\le M_n^{q_n}(|f - f(x,y)e_{00}|;(x,y)) + \|f\|_\rho \left|M_n^{q_n}(e_{00};(x,y)) - 1)\right|.
\end{aligned} \tag{14.8}$$

Again applying the operators $M_n^{q_n}$ to (14.7), (14.8) and from Lemma 14.2, we get

$$\begin{aligned}
&\left|M_n^{q_n}(f;(x,y)) - f(x,y)\right| \\
&\le \varepsilon + \frac{2K_f}{\delta^2}\left\{M_n^{q_n}\left((x-s)^2;(x,y)\right)\right.
\end{aligned}$$

$$+ M_n^{q_n}\left((y-t)^2;(x,y)\right)\Big\} + \|f\|_\rho \left|M_n^{q_n}(e_{00};(x,y)) - e_{00}(x,y)\right|$$

$$= \varepsilon + \frac{2K_f}{\delta^2}\left\{\left(\frac{F_{1,q_n}(n)}{q_n^6 F_{1,q_n}(n-3)} - \frac{2F_{0,q_n}(n)}{q_n^2 F_{0,q_n}(n-2)} + 1\right)(x^2+y^2)\right.$$

$$\left.+\left(\frac{[2]_{q_n}^2 F_{0,q_n}(n)}{q_n^5 F_{1,q_n}(n-3)} - \frac{2}{q_n F_{0,q_n}(n-2)}\right)(x+y) + \frac{[2]_{q_n}}{q_n^3 F_{1,q_n}(n-3)}\right\}.$$

Hence the proof of theorem is completed.

We need Bohman–Korovkin's Theorem in [1, 3, 20] for giving our following Remark 14.5.

Theorem 14.4 (Bohman–Korovkin). *Let K be a compact Hausdorff topological space which contains at least two distinct points and let $2m$ functions $f_1, f_2, \ldots, f_m$, $a_1, a_2, \ldots, a_m$ in $C(K,\mathbb{R})$ be such that*

$$P(x,y) := \sum_{j=1}^{m} f_j(x)a_j(y) \geq 0, \ \forall (x,y) \in K,$$

and $P(x,y) = 0$ if and only if $x = y$.

If (H_n), $H_n : C(K,\mathbb{R}) \to C(K,\mathbb{R})$, is a sequence of linear positive operators such that $H_n(f_j) \to f_j$, $n \to \infty$, $j = 1,\ldots,m$, then we have $H_n(f) \to f$, for each $f \in C(K,\mathbb{R})$.

Remark 14.5. If we take $C(D_a)$ instead of $C_\rho^*(D)$ in the hypothesis of Theorem 14.3, then we obtain alternative proof of Theorem 14.3 by means of Bohman–Korovkin's result. Indeed, take $K = D_a, f_1 = e_{00}, f_2 = e_{10}, f_3 = e_{01}, f_4 = e_{22}$, $a_1 = e_{22}, a_2 = -2e_{10}, a_3 = -2e_{01}$ and $a_4 = e_{00}, P((x,y),(s,t)) = (x-s)^2+(y-t)^2$.

14.3 Voronovskaja-Type Theorem

Let us define the moment functions $V_{i,j}$ by

$$V_{ij}(s,t) = (s-x)^i\,(t-y)^j\,,\ i,j \in \mathbb{N}.$$

By simple calculations, the following lemma can be obtained easily.

Lemma 14.6. *For each $(x_1,x_2) \in D$ and $n > 5$, the following equations are hold:*

(i)

$$M_n^q(V_{10};(x,y)) = \frac{[2]_q x}{q^2 F_{0,q}(n-2)} + \frac{1}{qF_{0,q}(n-2)},$$

(ii)

$$M_n^q(V_{11};(x,y)) = \left(\frac{1}{F_{0,q}(n-2)}\right)^2 \left(\frac{[2]_q^2 xy}{q^4} + \frac{[2]_q (x+y)}{q^3} + \frac{1}{q^2}\right),$$

(iii)

$$\begin{aligned} M_n^q(V_{20};(x,y)) &= \left(\frac{F_{1,q}(n)}{q^6 F_{1,q}(n-3)} - \frac{2F_{0,q}(n)}{q^2 F_{0,q}(n-2)} + 1\right) x^2 \\ &+ \left(\frac{[2]_q^2 F_{0,q}(n)}{q^5 F_{1,q}(n-3)} - \frac{2}{qF_{0,q}(n-2)}\right) x + \frac{[2]_q}{q^3 F_{1,q}(n-3)}, \end{aligned}$$

(iv)

$$\begin{aligned} &M_n^q(V_{40};(x,y)) \\ &= \left(\frac{F_{3,q}(n) - 4q^8 F_{2,q}(n)F_{0,q}(n-5) + 6q^{14} F_{1,q}(n)F_{1,q}(n-5)}{q^{20} F_{3,q}(n-5)}\right. \\ &\quad \left. \frac{-4q^{18} F_{0,q}(n)F_{2,q}(n-5) + q^{20} F_{3,q}(n-5)}{q^{20} F_{3,q}(n-5)}\right) x^4 \\ &\quad + \left(\frac{\left\{[7]_q + q[5]_q + q^2 [2]_q^2\right\} F_{2,q}(n) - 4q^8 \left\{[5]_q + q[2]_q^2\right\} F_{1,q}(n)F_{0,q}(n-5)}{q^{19} F_{3,q}(n-5)}\right. \\ &\quad \left. + \frac{6q^{14} [2]_q^2 F_{0,q}(n)F_{1,q}(n-5) - 4q^{18} F_{2,q}(n-5)}{q^{19} F_{3,q}(n-5)}\right) x^3 \\ &\quad + \left(\frac{\left\{[6]_q \left([5]_q + q[2]_q^2\right) + q^2 [2]_q^2 [4]_q + q^4 [2]_q\right\} F_{1,q}(n)}{q^{17} F_{3,q}(n-5)}\right. \\ &\quad \left. + \frac{-4q^8 \left([2]_q^2 [4]_q + q^2 [2]_q\right) F_{0,q}(n)F_{0,q}(n-5) + 6q^{14} [2]_q F_{1,q}(n-5)}{q^{17} F_{3,q}(n-5)}\right) x^2 \\ &\quad + \left(\frac{\left\{[5]_q \left([2]_q^2 [4]_q + q^2 [2]_q\right) + q^3 [2]_q [3]_q\right\} F_{0,q}(n) - 4q^8 [2]_q [3]_q F_{0,q}(n-5)}{q^{14} F_{3,q}(n-5)}\right) x \\ &\quad + \frac{[2]_q [3]_q [4]_q}{q^{10} F_{5,q}(n)}, \end{aligned}$$

On the other hand, by means of the following equality

$$[n+m]_q = [m]_q + q^m [n]_q, \tag{14.9}$$

one can rearrange some terms, in $M_n^q(V_{20};(x,y))$ and $M_n^q(V_{40};(x,y))$ in Lemma 14.6, as

$$\begin{aligned} M_n^q(V_{20};(x,y)) &= \frac{\left(q^3+q^6\right)F_{0,q}(n-3)+[2]_q[3]_q}{q^6F_{1,q}(n-3)}x^2 \\ &+\frac{\left(q^3+q^5\right)F_{0,q}(n-3)+[2]_q^2[3]_q}{q^5F_{1,q}(n-3)}x \\ &+\frac{[2]_q}{q^3F_{1,q}(n-3)} \end{aligned} \tag{14.10}$$

and

$$\begin{aligned} &M_n^q(V_{40};(x,y)) \\ &= \left(\frac{\left(F_{0,q}(n-5)\right)^3(1-q)^2q^{18}(1+q)\left(1-q+q^2-q^3+q^4\right)}{q^{20}F_{3,q}(n-5)}\right. \\ &\quad+\left(F_{0,q}(n-5)\right)^2q^{11}(1+q) \\ &\quad\quad\times\frac{\left(1+2q+q^2-q^3+q^4-q^5+q^6-4q^7+5q^8-5q^9+6q^{10}-3q^{11}+3q^{12}\right)}{q^{20}F_{3,q}(n-5)} \\ &\quad+F_{0,q}(n-5)q^5\left(q+1\right) \\ &\quad\quad\times\left\{\frac{3q+7q^2+9q^3+13q^4+15q^5+17q^6+15q^7+13q^8+15q^9+13q^{10}}{q^{20}F_{3,q}(n-5)}\right. \\ &\quad\left.+\frac{13q^{12}+14q^{13}+8q^{14}+9q^{15}+3q^{16}+3q^{17}+1}{q^{20}F_{3,q}(n-5)}\right\}+\left.\frac{[5]_q[7]_q[8]_q[6]_q}{q^{20}F_{3,q}(n-5)}\right)x^4 \\ &\quad+\left(\frac{\left(F_{0,q}(n-5)\right)^3(1-q)^2q^{18}(1+q^4)}{q^{19}F_{3,q}(n-5)}\right. \\ &\quad+\left(F_{0,q}(n-5)\right)^2q^{12}\left\{\frac{\left(2q-5q^3-7q^4-6q^5-6q^6-5q^7-q^8+6q^9+2q^{10}+q^{11}+1\right)}{q^{19}F_{3,q}(n-5)}\right. \\ &\quad+F_{0,q}(n-5)q^5\left\{\frac{\left(5q+14q^2+27q^3+41q^4+51q^5+56q^6+55q^7+48q^8+43q^9\right)}{q^{19}F_{3,q}(n-5)}\right. \\ &\quad\left.+\frac{46q^{10}+47q^{11}+43q^{12}+37q^{13}+28q^{14}+17q^{15}+7q^{16}+3q^{17}+1}{q^{19}F_{3,q}(n-5)}\right\} \\ &\quad\left.+\frac{[5]_q[6]_q[7]_q\left([7]_q+q^2[2]_q^2+q[5]_q\right)}{q^{19}F_{3,q}(n-5)}\right)x^3 \end{aligned}$$

$$+\left(\frac{\left\{[6]_q\left([5]_q+q\,[2]_q^2\right)+q^2\,[2]_q^2\,[4]_q+q^4\,[2]_q\right\}F_{1,q}(n)}{q^{17}F_{3,q}(n-5)}\right.$$
$$\left.+\frac{-4q^8\left([2]_q^2\,[4]_q+q^2\,[2]_q\right)F_{0,q}(n)F_{0,q}(n-5)+6q^{14}\,[2]_q\,F_{1,q}(n-5)}{q^{17}F_{3,q}(n-5)}\right)nx^2$$
$$+\left(\frac{\left\{[5]_q\left([2]_q^2\,[4]_q+q^2\,[2]_q\right)+q^3\,[2]_q\,[3]_q\right\}F_{0,q}(n)-4q^8\,[2]_q\,[3]_q\,F_{0,q}(n-5)}{q^{14}F_{3,q}(n-5)}\right)x$$
$$+\frac{[2]_q\,[3]_q\,[4]_q}{q^{10}F_{5,q}\,(n)}, \tag{14.11}$$

Using Lemma 14.6 and $[n]_{q_n}=\frac{1-q_n^n}{1-q_n}$, we can give the following corollary.

Corollary 14.7. *Let* $0<q_n\leq 1$ *and* $(q_n)\to 1$ *as* $n\to\infty$. *Then we have*

$$\lim_{n\to\infty}[n]_{q_n}M_n^{q_n}(V_{10};(x,y))=2x+1,$$
$$\lim_{n\to\infty}[n]_{q_n}M_n^{q_n}(V_{20};(x,y))=2x(x+1),$$
$$\lim_{n\to\infty}[n]_{q_n}M_n^{q_n}(V_{11};(x,y))=0,$$
$$\lim_{n\to\infty}[n]_{q_n}^2M_n^{q_n}(V_{40};(x,y))=12x^2(x+1)^2.$$

Let us denote the space of functions having continuous partial derivatives up to order 2 on D, by $C^2(D)$. We now give a Voronovskaja type theorem for M_n^q operators.

Theorem 14.8. *Let* $0<q_n\leq 1$ *and* $(q_n)\to 1$ *as* $n\to\infty$. *Suppose that* $f\in C^2(D)$. *Then for each* (x,y) *in* D, *we have*

$$\lim_{n\to\infty}[n]_{q_n}\left(M_n^{q_n}(f;(x,y))-f(x,y)\right)$$
$$=(2x+1)f_x(x,y)+(2y+1)f_y(x,y)$$
$$+2x\,(x+1)\,f_{xx}(x,y)+2y\,(y+1)\,f_{yy}(x,y).$$

Proof. Let (x,y) be a fixed point in D and $f\in C^2(D)$. By Taylor formula for f we get

$$f(s,t)=f(x,y)+f_x(x,y)V_{10}(s,t)+f_y(x,y)V_{01}(s,t)$$
$$+\frac{1}{2}f_{xx}(x,y)V_{20}(s,t)\,(s-t)^2+f_{xy}(x,y)V_{11}(s,t)$$
$$+\frac{1}{2}f_{yy}(x,y)V_{02}(s,t)$$
$$+\psi((s,t);(x,y))\sqrt{(V_{40}+V_{04})\,(s,t)}. \tag{14.12}$$

From the symmetric properties $M_n^{q_n}, M_n^{q_n}(e_{ij};(x,y) = M_n^{q_n}\left(e_{ij};(y,x)\right)$, Lemma 14.6, $[n]_{q_n} = \frac{1-q_n^n}{1-q_n}$ and applying the operators $M_n^{q_n}$ to Eq. (14.12), we write

$$\begin{aligned}
&\lim_{n\to\infty} [n]_{q_n}\left[M_n^{q_n}(f;(x,y)) - f(x,y)\right] \\
&\quad = f_x(x,y)\lim_{n\to\infty}\left\{[n]_{q_n} M_n^{q_n}(V_{10};(x,y))\right\} \\
&\qquad + \lim_{n\to\infty} f_y(x,y)\left\{[n]_{q_n} M_n^{q_n}(V_{01};(x,y))\right\} \\
&\qquad + \frac{1}{2} f_{xx}(x,y)\lim_{n\to\infty}\left\{[n]_{q_n} M_n^{q_n}(V_{20};(x,y))\right\} \\
&\qquad + f_{xy}(x,y)\lim_{n\to\infty}\left\{[n]_{q_n} M_n^{q_n}(V_{11};(x,y))\right\} \\
&\qquad + \frac{1}{2} f_{yy}(x,y)\lim_{n\to\infty}\left\{[n]_{q_n} M_n^{q_n}(V_{02};(x,y))\right\} \\
&\qquad + \lim_{n\to\infty}\left\{[n]_{q_n} M_n^{q_n}\left(\psi(s,t)\sqrt{(V_{40}+V_{04})};(x,y)\right)\right\}. \qquad (14.13)
\end{aligned}$$

Coupled with $\lim_{n\to\infty} M_n^{q_n}\left(\psi^2;(x,y)\right) = 0$ and Corollary 14.7 yield

$$\lim_{n\to\infty} [n]_{q_n} M_n^{q_n}\left(\psi\sqrt{V_{4,0}+V_{0,4}};(x,y)\right) = 0. \qquad (14.14)$$

The proof of the theorem is completed by combining Corollary 14.7, (14.12)–(14.14).

14.4 Rate of Convergence

Full modulus of continuity of $f \in C(D_a)$ is denoted by $w(f;\delta)$ and defined as follows:

$$\begin{aligned}
w(f;\delta) := \max\Big\{ & |f(s,t) - f(x,y)| : (s,t),(x,y) \in D_a \\
& \text{and } \sqrt{(s-x)^2 + (t-y)^2} \le \delta \Big\}. \qquad (14.15)
\end{aligned}$$

Partial modulus of continuity with respect to x and y is given by

$$w^1(f;\delta) := \max_{0\le y\le a}\ \max_{|x_1-x_2|\le\delta} |f((x_1,y)) - f((x_2,y))| \qquad (14.16)$$

and

$$w^2(f;\delta) := \max_{0\le x\le a}\ \max_{|y_1-y_2|\le\delta} |f((y,y_1)) - f((y,y_2))|, \qquad (14.17)$$

respectively. We shall need some well-known properties of full and partial modulus of continuity in [11]:

$$w(f;\lambda\delta) \le (1+\lambda)\,w(f;\delta)$$

for any $\lambda \ge 0$ and $\lim\limits_{\delta\to 0} w(f;\delta) = 0$.

The following theorem gives us the rate of convergence of the sequence of linear positive operators (M_n^q) to f, by means of partial and full modulus of continuity.

Theorem 14.9. *Let $0 < q < 1$. For $f \in C(D_a)$, the following inequalities are satisfied*

$$\left\|M_n^q(f) - f\right\|_{C(D_a)} \le 4\sum_{i=1}^{2} w^i\left(f;\frac{3\,(a+1)}{q^3\sqrt{F_{0,q}(n-2)}}\right)$$

and

$$\left\|M_n^q\,(f) - f\right\|_{C(D_a)} \le 4w\left(f;\frac{4\,(a+1)}{q^3\sqrt{F_{0,q}(n-2)}}\right).$$

Proof. Let $(x,y) \in D_a$. We write

$$\begin{aligned}
&\left|M_n^q\,(f;(x,y)) - f(x,y)\right| \\
&\le [n-1]_q^2 \sum_{k,l\in\mathbb{N}_0} b_{n,k,l}^q(x,y) \int_0^{\infty/A}\int_0^{\infty/A} b_{n,k,l}^q(s,t)\,|f(s,t) - f(x,y)|\, d_q s d_q t \\
&\le [n-1]_q^2 \sum_{k,l\in\mathbb{N}_0} b_{n,k,l}^q(x,y) \int_0^{\infty/A}\int_0^{\infty/A} b_{n,k,l}^q(s,t) \\
&\quad\times \left|1 + \frac{|s-x|}{\delta_{n,1}}\right| w\,(f,\delta_{n,1})\, d_q s d_q t \\
&\quad + [n-1]_q^2 \sum_{k,l\in\mathbb{N}_0} b_{n,k,l}^q(x,y) \int_0^{\infty/A}\int_0^{\infty/A} b_{n,k,l}^q(s,t) \\
&\quad\times \left|1 + \frac{|t-y|}{\delta_{n,2}}\right| w\,(f,\delta_{n,2})\, d_q s d_q t. \qquad (14.18)
\end{aligned}$$

Using the inequality $(a+b)^p \le 2^{p-1}\,(a^p + b^p)$, for $p \ge 1$, (14.16), (14.17), Cauchy–Schwarz inequality and (14.10) in (14.18), we obtain

$$\begin{aligned}&\left|M_n^q(f;(x,y))-f(x,y)\right|\\&\leq 2\left\{1+\frac{1}{\delta_{n,1}^2}M_n^q(V_{20};(x,y))\right\}w(f,\delta_{n,1})\\&\quad+2\left\{1+\frac{1}{\delta_{n,2}^2}M_n^q(V_{02};(x,y))\right\}w(f,\delta_{n,2}).\end{aligned}\tag{14.19}$$

From (14.10), one can write the following inequality:

$$M_n^q(V_{20};(x,y))\leq\frac{8(x+1)^2}{q^6F_{0,q}(n-2)}.\tag{14.20}$$

By (14.19) and (14.20), we obtain

$$\left|M_n^q(f;(x,y))-f(x,y)\right|\leq 4w(f,\delta_{n,1})+4w(f,\delta_{n,2}),\tag{14.21}$$

where $\delta_{n,1}=\frac{8(x+1)^2}{q^6F_{0,q}(n-2)}$ and $\delta_{n,2}=\frac{8(y+1)^2}{q^6F_{0,q}(n-2)}$. Using the above inequalities and (14.15), we get

$$\begin{aligned}&\left|M_n^q(f;(x,y))-f(x,y)\right|\\&\leq[n-1]_q^2\sum_{k,l\in\mathbb{N}_0}b_{n,k,l}^q(x,y)\\&\quad\times\left\{\int_0^{\infty/A}\int_0^{\infty/A}b_{n,k,l}^q(t,s)\left|1+\frac{\sqrt{(s-x)^2+(t-y)^2}}{\delta_n}\right|w(f,\delta_n)\,d_qsd_qt\right\}\\&\leq 4w(f,\delta_n),\end{aligned}\tag{14.22}$$

where

$$\delta_n=\frac{8\left\{(x+1)^2+(y+1)^2\right\}}{q^6F_{0,q}(n-2)}.$$

Finally, taking supremum on $(x,y)\in D_a$ at (14.21) and (14.22), we get the desired results.

The class of the functions f in $C(D_a)$ satisfying the following relation

$$w(f;\delta)\leq M\delta^\alpha,\ \text{for all }\delta\geq 0,$$

is called a Lipschitz class and denoted by $Lip_M(\alpha)$ for $M>0$ and $0<\alpha\leq 1$. The following Corollary is routine and its proof is omitted.

Corollary 14.10. *Let* $f \in C(D_a)$. *If* $f \in Lip_M(\alpha)$, *for* $0 < \alpha \leq 1$, *then the following inequality*

$$\left\| M_n^q(f) - f \right\|_{C(D_a)} \leq 4M\delta_n^\alpha$$

holds, where $M > 0$ *and* δ_n *are given in the above theorem.*

References

1. O. Agratini, On a class of linear positive bivariate operators of King type. Stud. Univ. Babes-Bolyai Math. **LI**(4), 13–22 (2006)
2. A. Altın, O. Doğru, F. Taşdelen, The generalization of Meyer-König and Zeller operators by generating functions. J. Math. Anal. Appl. **312**(1), 181–194 (2005)
3. F. Altomare, M. Campiti, *Korovkin-Type Approximation Theory and Its Applications*. de Gruyter Studies in Mathematics, vol. 17 (Walter de Gruyter, Berlin, 1994)
4. A. Aral, T. Acar, Tuncer Voronovskaja type result for q-derivative of q-Baskakov operators. J. Appl. Funct. Anal. **7**(4), 321–331 (2012)
5. A. Aral, V. Gupta, Generalized q-Baskakov operators. Math. Slovaca **61**(4), 619–634 (2011)
6. A. Aral, V. Gupta, R.P. Agarwal, *Applications of q-Calculus in Operator Theory* (Springer, New York, 2013)
7. V.A. Baskakov, On a construction of converging sequences of linear positive operators, in *Studies of Modern Problems of Constructive Theory of Functions* Fizmatgiz, Moscow, 314–318 (1961)
8. M. Becker, Global approximation theorems for Szasz-Mirakyan and Baskakov operators in polynomial weight spaces. Indiana Univ. Math. J. **27**(1), 127–142 (1978)
9. F. Cao, Y. An, Lp approximation by multivariate Baskakov-Durrmeyer operator. J. Inequal. Appl., 7 pp. (2011). Art. ID 158219
10. F. Cao, Y. Liu, Approximation theorems by positive linear operators in weighted spaces. Positivity **15**(1), 87–103 (2011)
11. R.A. De Vore, G.G. Lorentz, *Constructive Approximation* (Springer, Berlin, 1993)
12. A.D. Gadzhiev, Positive linear operators in weighted spaces of functions of several variables. Izv. Akad. Nauk Azerbaidzhan. SSR Ser. Fiz.-Tekhn. Mat. Nauk **14**, 32–37 (1980)
13. V. Gupta, R.P. Agarwal, *Convergence Estimates in Approximation Theory* (Springer, Cham, 2014)
14. V. Gupta, A. Aral, Ali Some approximation properties of q-Baskakov-Durrmeyer operators. Appl. Math. Comput. **218**(3), 783–788 (2011)
15. P. Gupta, V. Gupta, Rate of convergence on Baskakov-Szasz type operators. Fasc. Math. **31**, 37–44 (2001)
16. V. Gupta, M.K. Gupta, V. Vasishtha, Simultaneous approximations by summation integral type operators. Nonlinear Funct. Anal. Appl. **8**(3), 399–412 (2003)
17. V. Gupta, D.K. Verma, P.N. Agrawal, Simultaneous approximation by certain Baskakov-Durrmeyer-Stancu operators. J. Egyptian Math. Soc. **20**(3), 183–187 (2012)
18. M. Heilmann, Direct and converse results for operators of Baskakov-Durrmeyer type. Approx. Theory Appl. **5**(1), 105–127 (1989)
19. V.N. Mishra, P. Patel, The Durrmeyer type modification of the q-Baskakov type operators with two parameter α and β. Numer. Algorithms **67**(4), 753–769 (2014)
20. G. Păltineanu, Classes of interpolation sets with respect to a space of cotinuous functions, in *Order Structures in Functional analysis*, vol. 3, 122–161 (Academiei, Bucharest, 1992) [in Romanian]

21. A. Sahai, G. Prasad, On the rate of convergence for modified Szász–Mirakyan operators on functions of bounded variation. Publ. Inst. Math. Nouv. Sér. (N.S.) **53**(67), 73–80 (1993)
22. D.D. Stancu, A new class of uniform approximating polynomial operators in two and several variables, in *Proceedings of the Conference on the Constructive Theory of Functions (Approximation Theory)* Budapest, 1969, pp. 443–455
23. F. Taşdelen, A. Olgun, G. Başcanbaz-Tunca, Approximation of functions of two variables by certain linear positive operators. Proc. Indian Acad. Sci. Math. Sci. **117**(3), 387–399 (2007)
24. V.I. Volkov, On the convergence of sequences of linear positive operators in the space of continuous functions of two variables (Russian). Dokl. Akad. Nauk SSSR (N.S.) **115**, 17–19 (1957)
25. Z. Walczak, Approximation properties of certain linear positive operators of functions of two variables. Acta Comment. Univ. Tartu. Math. **6**, 17–27 (2002)
26. Z. Walczak, Approximation of functions of two variables by some linear positive operators. Acta. Math. Univ. Comenianae **LXXIV**(1), 37–48 (2005)

Chapter 15
Blow Up of Solutions of Second Order Semilinear Parabolic Equations Under Robin Boundary Conditions

Jamila Kalantarova and Mustafa Polat

Abstract We considered initial boundary value problems for semilinear heat equation with a nonlinear source term under the Robin boundary conditions. Sufficient conditions for the finite-time blow up of solutions with negative initial energy and arbitrary positive initial energy are obtained.

15.1 Introduction

We study the following initial boundary value problem

$$u_t - \Delta u = f(u) + h(x,t), \quad x \in \Omega, \quad t > 0, \tag{15.1}$$

$$\frac{\partial u}{\partial n} + \nu u = 0, \quad x \in \partial\Omega, \quad t > 0, \tag{15.2}$$

$$u(x,0) = u_0(x), \quad x \in \Omega, \tag{15.3}$$

where Ω is a bounded domain in $\mathbb{R}^N$ with smooth boundary $\partial\Omega$, ν is a given non-zero number, $u_0, h, f(\cdot)$ are given sufficiently smooth functions for which the problem (15.1)–(15.3) has a local in time classical solution (for results on local solvability of the problem (15.1)–(15.3), we refer to [3, 5]). We assume also that

$$h \in L^2(0,\infty; L^2(\Omega)) \cap L^\infty(0,\infty; L^2(\Omega)), \tag{15.4}$$

and the nonlinear term satisfies the condition

$$f(s)s \geq 2(1+\alpha)F(s), \quad F(s) = \int_0^s f(\tau)\,d\tau, \text{ for all } s \in \mathbb{R}. \tag{15.5}$$

J. Kalantarova (✉) • M. Polat
Department of Mathematics, Yeditepe University, 34750 Istanbul, Turkey
e-mail: jkalantarova@gmail.com; mpolat@yeditepe.edu.tr

G.A. Anastassiou, O. Duman (eds.), *Computational Analysis*, Springer Proceedings in Mathematics & Statistics 155, DOI 10.1007/978-3-319-28443-9_15

Our main goal is to show that under these conditions solutions of the problem (15.1)–(15.3) corresponding to a wide class of initial data blow-up in a finite time. The problem of blow-up of solutions of nonlinear parabolic equations under Robin boundary conditions have been considered in [1, 6, 7]. In [6, 7], using the energy method the authors established blow-up of solutions and obtained a lower bound of blow-up time for the solutions of the problem (15.1)–(15.3) with $h(x,t) \equiv 0$, essentially employing positiveness of the coefficient ν and the initial function $u_0(x)$. In [1] similar results are obtained for a quasilinear parabolic equation.

In this note, by using the concavity method of Levine [4], we will derive sufficient conditions for the finite-time blow-up of solutions of the problem (15.1)–(15.3) regardless of the sign of ν and the initial functions $u_0(x)$ under the Robin boundary conditions. We also showed that solutions of initial boundary value problems for the semilinear heat equations under Robin boundary conditions with $\nu > 0$ may blow up in a finite time even when the initial energy is arbitrary positive number.

In what follows we use the following inequalities:

(a) The Cauchy inequality "with ε"

$$ab \le \varepsilon a^2 + \frac{1}{4\varepsilon} b^2, \tag{15.6}$$

which is valid for each $a, b \ge 0$ and $\varepsilon > 0$, and

(b) the following version of the Poincaré inequality (see, e.g., [2] Chap. I)

$$\int_{\partial\Omega} v^2 d\sigma \le \epsilon \int_{\Omega} |\nabla v|^2 dx + C(\epsilon) \int_{\Omega} v^2 dx, \tag{15.7}$$

which is valid for each function u from the Sobolev space $H^1(\Omega)$, where $\Omega \in \mathbb{R}^N$ is a bounded domain with boundary $\partial\Omega$, $\epsilon > 0$ is an arbitrary positive number, $C(\varepsilon) > 0$ depends on ϵ.

Our main result will be established by using the concavity method based on the following lemma.

Lemma 15.1 (see [4]). *Suppose that $\Psi(t)$ is a positive, twice differentiable function satisfying the inequality*

$$\Psi''(t)\Psi(t) - (1+\alpha_1)[\Psi'(t)]^2 \ge 0, \quad \textit{for all } t > 0,$$

with

$$\Psi(0) > 0, \quad \Psi'(0) > 0.$$

Then there exist $t_1 > 0$ such that $\Psi(t)$ tends to infinity as

$$t \to t_1^- \le t_2 = \frac{\Psi(0)}{\alpha_1 \Psi'(0)}.$$

In what follows we will use the following abbreviations $(\cdot,\cdot)$ and $\|\cdot\|$ for the inner product and the norm of $L^2(\Omega)$.

15.2 Blow-Up of Solutions

First we find sufficient conditions of finite-time blow-up of solutions of the problem (15.1)–(15.3) when $\nu \in \mathbb{R} \setminus \{0\}$ and the corresponding initial energy is negative.

Theorem 15.2. *Suppose that the conditions* (15.4) *and* (15.5) *are satisfied and the initial function* u_0 *satisfies*

$$\begin{aligned} &- \|\nabla u_0\|^2 - \nu \int_{\partial\Omega} u_0^2(x) d\sigma + 2 \int_{\Omega} F(u_0(x)) dx \\ &\geq \left(4 + \frac{4}{\alpha}\right) H_1 + \frac{H_2}{4\alpha(\alpha+1)} \\ &+ \frac{1}{2}\left(\frac{\alpha+2}{\alpha+1} + |\nu| C(|\nu|^{-1}) + 1\right) \|u_0\|^2, \end{aligned} \tag{15.8}$$

where $C(|\nu|^{-1})$ *is the constant in the Poincaré inequality* (15.7) *with* $\epsilon = |\nu|^{-1}$,

$$H_1 := \int_0^\infty \|h(t)\|^2 dt \text{ and } H_2 := \sup_{t \in \mathbb{R}+} \|h(t)\|^2 .$$

Then there exists $t_1 \leq t_2 := \dfrac{1}{2\alpha}$ *such that*

$$\lim_{t \to t_1^-} \int_0^t \|u(s)\|^2 ds = \infty.$$

Suppose that $u(x,t)$ is a local strong solution of the problem (15.1)–(15.3). It is clear that the function $v(x,t) = e^{-mt} u(x,t)$, $m > 0$ satisfies the equation

$$mv + v_t = \Delta v + e^{-mt} f(e^{mt} v) + e^{-mt} h(x,t), \; x \in \Omega, t > 0, \tag{15.9}$$

the boundary condition

$$\frac{\partial v}{\partial n} + \nu v = 0, \qquad x \in \partial\Omega, \quad t > 0, \tag{15.10}$$

and the initial condition

$$v(x,0) = u_0(x), \; x \in \Omega. \tag{15.11}$$

Our aim now is to find sufficient conditions for blow-up of solutions of the problem (15.9)–(15.11). It is obvious that blow-up of v implies the blow-up of u.

Proof. Assume $v(x,t)$ is a solution of the problem (15.9)–(15.11). Multiplying the Eq. (15.9) by v_t and integrating over Ω and using (15.10) we obtain

$$\frac{d}{dt}\left[\frac{m}{2}\|v\|^2 + \frac{1}{2}\|\nabla v\|^2 + \frac{\nu}{2}\int\limits_{\partial\Omega} v^2 d\sigma\right] + \|v_t\|^2$$
$$= e^{-mt}\int\limits_{\Omega} f(e^{mt}v)v_t dx + e^{-mt}\int\limits_{\Omega} hv_t dx. \tag{15.12}$$

It is easy to see that

$$\frac{d}{dt}F(e^{mt}v) = f(e^{mt}v)(e^{mt}v_t + me^{mt}v).$$

Plugging the expression

$$e^{-mt}f(e^{mt})v_t = e^{-2mt}\frac{d}{dt}F(e^{mt}v) - me^{-mt}f(e^{mt}v)v$$

into (15.12) we obtain

$$\frac{d}{dt}\left[\frac{m}{2}\|v\|^2 + \frac{1}{2}\|\nabla v\|^2 + \frac{\nu}{2}\int\limits_{\partial\Omega} v^2 d\sigma\right]$$
$$+ \|v_t\|^2 - e^{-2mt}\frac{d}{dt}\int\limits_{\Omega} F(e^{mt}v)dx$$
$$+ me^{-mt}\int\limits_{\Omega} f(e^{mt}v)v dx = e^{-mt}\int\limits_{\Omega} hv_t dx. \tag{15.13}$$

Since

$$e^{-2mt}\frac{d}{dt}\int\limits_{\Omega} F(e^{mt}v)dx = \frac{d}{dt}\left[e^{-2mt}\int\limits_{\Omega} F(e^{mt}v)dx\right]$$
$$+ 2me^{-2mt}\int\limits_{\Omega} F(e^{mt}v)dx,$$

we have

$$\frac{d}{dt}\left[\frac{m}{2}\|v\|^2+\frac{1}{2}\|\nabla v\|^2+\frac{\nu}{2}\int\limits_{\partial\Omega} v^2 d\sigma - e^{-2mt}\int\limits_{\Omega} F(e^{mt}v)dx\right]$$
$$+\ \|v_t\|^2 + me^{-mt}\int\limits_{\Omega} f(e^{mt}v)v dx - 2me^{-2mt}\int\limits_{\Omega} F(e^{mt}v)dx$$
$$= e^{-mt}\int\limits_{\Omega} hv_t dx. \tag{15.14}$$

By using the condition (15.5) we see that

$$e^{-mt}f(e^{mt}v)v = e^{-2mt}f(e^{mt}v)e^{mt}v \geq 2(\alpha+1)e^{-2mt}F(e^{mt}v). \tag{15.15}$$

Employing (15.15) and the Cauchy inequality with ε we deduce from (15.14) the following inequality

$$\frac{d}{dt}\left[\frac{m}{2}\|v\|^2+\frac{1}{2}\|\nabla v\|^2+\frac{\nu}{2}\int_{\partial\Omega} v^2 d\sigma - e^{-2mt}\int_{\Omega} F(e^{mt}v)dx\right]$$
$$+\|v_t\|^2 + 2m\alpha e^{-2mt}\int_{\Omega} F(e^{mt}v)dx$$
$$\leq \varepsilon_1\|v_t\|^2 + \frac{1}{4\varepsilon_1}\|h\|^2 e^{-2mt}.$$

From this inequality we obtain

$$\frac{d}{dt}\left[-\frac{m}{2}\|v\|^2-\frac{1}{2}\|\nabla v\|^2-\frac{\nu}{2}\int_{\partial\Omega} v^2 d\sigma + e^{-2mt}\int_{\Omega} F(e^{mt}v)dx\right]$$
$$\geq 2m\alpha\left[-\frac{m}{2}\|v\|^2-\frac{1}{2}\|\nabla v\|^2-\frac{\nu}{2}\int_{\partial\Omega} v^2 d\sigma + e^{-2mt}\int_{\Omega} F(e^{mt}v)dx\right]$$
$$+m\alpha\left[m\|v\|^2+\|\nabla v\|^2+\nu\int_{\partial\Omega} v^2 d\sigma\right]$$
$$+(1-\varepsilon_1)\|v_t\|^2 - \frac{1}{4\varepsilon_1}\|h\|^2 e^{-2mt}.$$

So we obtained the inequalities

$$\frac{d}{dt}\mathscr{E}(t) \geq 2m\alpha\mathscr{E}(t) + m\alpha\left[m\|v\|^2+\|\nabla v\|^2+\nu\int\limits_{\partial\Omega} v^2 d\sigma\right]$$
$$+(1-\varepsilon_1)\|v_t\|^2 - \frac{1}{4\varepsilon_1}\|h\|^2 e^{-2mt}, \tag{15.16}$$

where

$$\mathscr{E}(t) = -\frac{m}{2}\|v\|^2 - \frac{1}{2}\|\nabla v\|^2 - \frac{\nu}{2}\int_{\partial\Omega} v^2 d\sigma + e^{-2mt}\int_{\Omega} F(e^{mt}v)dx. \tag{15.17}$$

By using Poincare inequality (15.7) with $\epsilon = |\nu|^{-1}$ in (15.16) we get

$$\frac{d}{dt}\mathscr{E}(t) \geq 2m\alpha\mathscr{E}(t) + (1-\varepsilon_1)\|v_t\|^2$$
$$+ m\alpha(m - |\nu|C(|\nu|^{-1}))\|v\|^2 - \frac{1}{4\varepsilon_1}\|h\|^2 e^{-2mt}.$$

First choosing $m \geq |\nu|C(|\nu|^{-1})$, and then integrating the obtained differential inequality we obtain the estimate

$$\mathscr{E}(t) \geq e^{2m\alpha t}\mathscr{E}(0) + (1-\varepsilon_1)e^{2m\alpha t}\int_0^t \|v_s(s)\|^2 e^{-2ms} ds$$
$$- \frac{1}{4\varepsilon_1}e^{2m\alpha t}\int_0^t \|h(s)\|^2 e^{-2m(\alpha+1)s} ds. \tag{15.18}$$

Let us consider the function

$$\Psi(t) = \int_0^t \|v(s)\|^2 ds + c_0,$$

where c_0 is a nonnegative parameter to be specified below. It is clear that

$$\Psi'(t) = \|v(t)\|^2 = 2\int_0^t (v(s), v_s(s))ds + \|u_0\|^2$$

and $\Psi''(t) = 2(v, v_t)$. By using Eq. (15.9) and the inequality (15.15) we obtain the following lower estimate for the function $\Psi''(t)$:

$$\Psi''(t) = 2\int_{\Omega} v\left[-mv + \Delta v + e^{-mt}f(e^{mt}v) + e^{-mt}h(x,t)\right] dx$$
$$\geq -2m\|v\|^2 - 2\|\nabla v\|^2 - 2\nu\int_{\partial\Omega} v^2 d\sigma$$
$$+ 4(\alpha+1)e^{-2mt}\int_{\Omega} F(e^{mt}v)dx + 2e^{-mt}(h, v)$$

$$= 4(\alpha + 1)\left[-\frac{m}{2}\|v\|^2 - \frac{1}{2}\|\nabla v\|^2 - \frac{\nu}{2}\int_{\partial\Omega} v^2 d\sigma \right.$$
$$\left. + e^{-2mt}\int_{\Omega} F(e^{mt}v)dx\right]$$
$$+ 2m\alpha\|v\|^2 + 2\alpha\|\nabla v\|^2$$
$$+ 2\alpha\nu\int_{\partial\Omega} v^2 d\sigma + 2e^{-mt}(h, v). \tag{15.19}$$

Employing the inequality

$$2\alpha\nu\int_{\partial\Omega} v^2 d\sigma \geq -2\alpha\|\nabla v\|^2 - 2\alpha C(|\nu|^{-1})\|v\|^2$$

which follows from (15.7) and the inequality

$$2e^{-mt}(h, v) \geq -\epsilon_2\|v\|^2 - \frac{1}{\epsilon_2}e^{-2mt}\|h\|^2$$

with some $\epsilon_2 > 0$, we deduce from (15.19) the estimate

$$\Psi''(t) \geq 4(\alpha + 1)\mathscr{E}(t) + \left(2\alpha m - 2\alpha|\nu|C(|\nu|^{-1}) - \epsilon_2\right)\|v\|^2$$
$$-\frac{1}{\epsilon_2}e^{-2mt}\|h(t)\|^2.$$

Finally we choose here $\epsilon_2 = 2\alpha$ and $m = |\nu|C(|\nu|^{-1}) + 1$ (remember that already it is assumed that $m \geq |\nu|C(|\nu|^{-1})$) we get

$$\Psi''(t) \geq 4(\alpha + 1)\mathscr{E}(t) - \frac{1}{2\alpha}e^{-2mt}\|h(t)\|^2.$$

Thus employing the estimate (15.18) of $\mathscr{E}(t)$, from the last inequality, we obtain the following estimate

$$\Psi''(t) \geq 4(\alpha + 1)(1 - \epsilon_1)\left[\int_0^t \|v_s(s)\|^2 ds + c_0\right]$$
$$+ 4(\alpha + 1)\left[\mathscr{E}(0) - \frac{1}{\epsilon_1}\int_0^t \|h(s)\|^2 ds\right]$$
$$-\frac{1}{2\alpha}e^{-2mt}\|h(t)\|^2 - 4(\alpha + 1)(1 - \epsilon_1)c_0.$$

Let us choose $\epsilon_1 = \dfrac{\alpha}{2(\alpha+1)}$. Then

$$4(\alpha+1)\left(1-\frac{\varepsilon_1}{2}\right) = 4\left(\frac{\alpha}{2}+1\right),$$

and we note that if

$$\mathscr{E}(0) \geq \left(2+\frac{2}{\alpha}\right)\int_0^\infty \|h(t)\|^2\,dt + \frac{1}{8\alpha(\alpha+1)}\sup_{t\in\mathbb{R}+}\|h(t)\|^2 + \frac{\alpha+2}{2(\alpha+1)}c_0,$$

then

$$\Psi''(t) \geq 4\left(\frac{\alpha}{2}+1\right)\left[\int_0^t \|v_s(s)\|^2\,ds + c_0\right].$$

Therefore

$$\begin{aligned}\Psi''(t)\Psi(t) &- (\alpha_1+1)(\Psi'^2)\\ &\geq 4\left(\frac{\alpha}{2}+1\right)\left[\int_0^t \|v_s(s)\|^2\,ds + c_0\int_0^t \|v(s)\|^2\,ds + c_0\right]\\ &- 4\left(\frac{\alpha}{2}+1\right)\left[\int_0^t (v(s), v_s(s))ds + \frac{1}{2}\|u_0\|^2\right]^2. \end{aligned} \tag{15.20}$$

Finally we choose $c_0 = \frac{1}{2}\|u_0\|^2$. Then due to the Schwarz inequality we deduce from (15.20) the desired inequality $\Psi''(t)\Psi(t) - (\frac{\alpha}{2}+1)(\Psi'(t))^2 \geq 0$. The statement of the theorem follows from Lemma 15.1.

Remark 15.3. If $h(x,t) \equiv h(x) \in L^2$, $\nu > 0$ and in addition to (15.5)

$$F(s) \geq D_0|s|^p - D_1, \quad \forall s \in \mathbb{R} \tag{15.21}$$

holds for some $D_0 > 0$, $D_1 \geq 0$, and $p > 2$, then blow-up result can be obtained employing the standard energy equalities for solutions of the problem (15.1)–(15.3):

$$\frac{1}{2}\frac{d}{dt}\|u(t)\|^2 = \|\nabla u\|^2 - \nu\int_{\partial\Omega} u^2 d\sigma + (f(u), u) + (h, u) \tag{15.22}$$

and

$$\frac{d}{dt}\mathscr{E}_1(t) = \|u_t\|^2, \tag{15.23}$$

where

$$\mathscr{E}_1(t) \equiv -\frac{1}{2}\|\nabla u\|^2 - \frac{\nu}{2}\int_{\partial\Omega} u^2 d\sigma + (f(u), 1) + (h, u).$$

In fact the following proposition holds:

Proposition 15.4. *If the nonlinear term $f(\cdot)$ satisfies the conditions* (15.5), (15.21), $\nu > 0$ *and the initial function satisfies the condition*

$$\mathscr{E}_1(0) - 4\alpha D_1|\Omega| \geq 0, \tag{15.24}$$

then the solution of the problem (15.1)–(15.3) *blows up in a finite time.*

Proof. Utilizing the conditions (15.5) and (15.21) we obtain from (15.22) the inequality

$$\begin{aligned}\frac{d}{dt}\|u(t)\|^2 &\geq -2\|\nabla u(t)\|^2 - 2\nu\int_{\partial\Omega} u^2 d\sigma + 4(\alpha+1)(F(u), 1)\\ &\quad + 2(h, u)\\ &= 4\mathscr{E}(t) + 4\alpha(F(u), 1)\\ &\geq 4\mathscr{E}_1(t) + 4\alpha D_0\int_\Omega |u(x,t)|^p dx - 4\alpha D_1|\Omega|. \end{aligned}\tag{15.25}$$

Due to the inequality $E(t) \geq E(0)$ which follows from (15.23), and the condition (15.24) we obtain from (15.25) the following differential inequality for $\Psi(t) \equiv \|u(t)\|^2$:

$$\Psi'(t) \geq K_0[\Psi(t)]^{\frac{p}{2}},$$

where

$$K_0 \equiv |\Omega|^{-\frac{p-2}{2}}(4\alpha D_0|\Omega|)^{-\frac{p}{2}}. \tag{15.26}$$

Integrating (15.26) we see that

$$\Psi(t) \to \infty \text{ as } t \to (p-2)[2K_0]^{-1}[\Psi(0)]^{\frac{2-p}{2}}.$$

Remark 15.5. We would like also to note that a result on blow-up of solutions to a class of semilinear parabolic equations under the Robin boundary condition can be obtained by using the so-called method of eigenfunctions. In fact the following proposition holds true:

Proposition 15.6. *Suppose that $u_0(x) \geq 0$, $\forall x \in \Omega$, $\nu > 0$, the source term $h(x,t) \equiv h(x) \in L^2(\Omega)$ depends only on $x \in \Omega$, the nonlinear term is a convex, continuous function that satisfies also the conditions:*

$$f(u) - \lambda_1 u - h_0 > 0, \quad \forall u \geq \alpha_0 > 0,$$

with

$$\int_{\alpha_0}^{\infty} \frac{d\eta}{f(\eta) - \lambda_1 \eta - h_0} < \infty, \tag{15.27}$$

where $h_0 = \int_\Omega h(x)\psi_1(x)dx$, $\alpha_0 = \int_\Omega u_0(x)\psi_1(x)$, $\lambda_1 > 0$ *is the eigenvalue corresponding to the normalized principal eigenfunction* $\psi_1(x)$ *of the problem*

$$-\Delta\psi = \lambda\psi, \quad x \in \Omega; \ \frac{\partial\psi}{\partial n} + \nu\psi = 0, \quad x \in \partial\Omega.$$

Then the solution of the problem (15.1)–(15.3) *blows up in a finite time.*

In fact multiplying the equation (15.1) by ψ_1, and then integrating it over Ω and using the boundary condition (15.2) we get

$$\int_\Omega u_t\psi_1 dx + \lambda_1 \int_\Omega u\psi_1 dx = \int_\Omega f(u)\psi_1 dx + \int_\Omega hudx. \tag{15.28}$$

Due to the Jensen inequality for integrals we have (we refer to [2])

$$\int_\Omega f(u)\psi_1 dx \geq f\left(\int_\Omega u\psi_1 dx\right).$$

Thus from (15.28) we get the following differential inequality for the function $E(t) = \int_\Omega u(x,t)\psi_1(x)dx$:

$$E'(t) \geq f(E(t)) - \lambda_1 E(t) - h_0.$$

Integrating this inequality and using the condition (15.27) we obtain the desired result.

15.3 Blow Up When the Initial Energy is Positive

In this short section we will use again a concavity technique to prove the finite-time blow-up of solutions to the following problem

$$u_t - \Delta u = f(u), \quad x \in \Omega \quad t > 0, \tag{15.29}$$

$$\frac{\partial u}{\partial n} + \nu u = 0, \quad x \in \partial\Omega \quad t > 0, \tag{15.30}$$

$$u(x,0) = u_0(x), \quad x \in \Omega. \tag{15.31}$$

Theorem 15.7. *Assume that*

$$f(s)s - 2(1+\alpha)F(s) \geq -D_0, \quad \forall s \in R, \tag{15.32}$$

where $\alpha > 0$, $\nu > 0$, *and* D_0 *are given numbers, and* $F(s) = \int_0^s f(\tau)d\tau$. *Suppose also that*

$$\mu(\alpha,\nu)\|u_0\|^2 > 4(1+\alpha)\left[\frac{1}{2}\|\nabla u_0\|^2 + \frac{\nu}{2}\int_{\partial\Omega} u_0^2 d\sigma - (F(u_0),1)\right] + D_1|\Omega|, \tag{15.33}$$

where $\mu(\alpha,\nu) = \frac{\alpha(1+\nu)}{a_0}$, *and* a_0 *is a constant of the Poincaré inequality*

$$\int_\Omega u^2 dx \leq a_0\left(\int_{\partial\Omega} u^2 dx + \int_\Omega |\nabla u|^2 dx\right). \tag{15.34}$$

Then the corresponding local solution of the problem (15.29)–(15.31) *blows up in a finite time.*

Proof. Consider the function

$$\Psi(t) = \int_0^t \|u(\tau)\|^2 d\tau.$$

Employing Eq. (15.29) and the condition (15.32) we obtain

$$\begin{aligned}\Psi''(t) &= 2(u,u_t) = 2(u, \Delta u + f(u))\\ &\geq -2\|\nabla u\|^2 - 2\nu\int_{\partial\Omega} u^2 d\sigma + 4(1+\alpha)(F(u),1) - 2|\Omega|D_1. \end{aligned} \tag{15.35}$$

Multiplication of (15.29) by u_t in $L^2(\Omega)$ gives

$$\|u_t\|^2 + \frac{d}{dt}\left[\frac{1}{2}\|\nabla u\|^2 + \frac{\nu}{2}\int_{\partial\Omega} u^2 d\sigma - (F(u),1)\right] = 0$$

Integrating the last equality over the interval $(0,t)$ we obtain

$$E(t) := \frac{1}{2}\|\nabla u\|^2 + \frac{\nu}{2}\int_{\partial\Omega} u^2 d\sigma - (F(u),1) = E(0) - \int_0^t \|u_\tau(\tau)\|^2 d\tau. \tag{15.36}$$

Employing (15.36) in (15.35) we get

$$\begin{aligned}\Psi''(t) \geq\ & 4(1+\alpha)\left[-\frac{1}{2}\|\nabla u\|^2 - \frac{\nu}{2}\int_{\partial\Omega} u^2 d\sigma + (F(u),1)\right]\\ & - D_1|\Omega| + 2\alpha\|\nabla u\|^2 + \alpha\nu\int_{\partial\Omega} u^2 d\sigma\end{aligned}$$

$$= \alpha\|\nabla u\|^2 + \alpha\nu \int_{\partial\Omega} u^2 d\sigma - 4(1+\alpha)E(0)$$
$$+ 4(1+\alpha)\int_0^t \|u_\tau(\tau)\|^2 d\tau - D_1\,|\Omega|\,.$$

Using (15.34) for the last two terms in the mid-line we arrive

$$\Psi''^2 - M_0] + 4(1+\alpha)\int_0^t \|u_\tau(\tau)\|^2 d\tau, \tag{15.37}$$

where $M_0 = 4(1+\alpha)E(0) + D_1|\Omega|,\ \mu(\alpha,\nu) = \frac{\alpha(1+\nu)}{a_0}$. It follows from (15.37) that

$$\frac{d}{dt}\left(\Psi'(t) - M_1\right) \geq \mu(\alpha,\nu)\left(\Psi'(t) - M_1\right),$$

where $M_1 = \frac{M_0}{\mu(\alpha,\nu)}$. Using (15.32) we obtain the inequality:

$$\Psi'(t) \geq M_1 + e^{\mu(\alpha,\nu)t}\left(\Psi'(0) - M_1\right).$$

Hence $\Psi'(t) \to \infty$ as $t \to \infty$. Since (15.33) holds we deduce from (15.37) that

$$\Psi''(t) \geq 4(1+\alpha)\int_0^t \|u(\tau)\|^2 d\tau.$$

Therefore thanks to the equality

$$2\int_0^t (u, u_\tau)^2 d\tau + \|u_0\|^2 = \Psi'(t)$$

and the Schwarz inequality we have

$$\Psi''(t)\Psi(t) - (1+\alpha)\left(\left[\Psi'(t)\right]^2 - \|u_0\|^2\right) \geq 0. \tag{15.38}$$

It follows from (15.38) that

$$\Psi''(t)\Psi(t) - \left(1+\frac{\alpha}{2}\right)\left[\Psi'(t)\right]^2 \geq \frac{\alpha}{2}\left[\Psi'(t)\right]^2 - 2(1+\alpha)\Psi'(t)\|u_0\|^2 + \|u_0\|^4. \tag{15.39}$$

Since $\Psi'(t) \to \infty$ as $t \to \infty$, there exists $T_0 > 0$ such that

$$\frac{\alpha}{2}\left[\Psi'(t)\right]^2 - 2(1+\alpha)\Psi'(t)\|u_0\|^2 + \|u_0\|^4 \geq 0,\ \forall t \geq T_0.$$

Therefore we deduce from (15.38) the inequality

$$\Psi''(t)\Psi(t) - \left(1 + \frac{\alpha}{2}\right)\left[\Psi'(t)\right]^2 \geq 0, \quad \forall t \geq T_0.$$

Thus we can use the Levine's lemma and get the desired result.

Example 15.8. Consider the $1D$ version of the problem with $f(u) = u^3$ and $\Omega = (0, L)$, i.e. the following problem

$$\begin{cases} u_t - u_{xx} = u^3, \quad x \in (0, L), \\ -u_x(0,t) + \nu u(0,t) = 0, \quad u_x(L,t) + \nu u(L,t) = 0, \ t > 0, \\ u(x,0) = u_0(x), \end{cases}$$

where $u_0(x) = \lambda f(x),\ x \in [0, L],\ r > 0,\ \lambda$ will be specified below and $f : \mathbb{R} \to \mathbb{R}^+$ is a continuously differentiable function such that

$$0 < M_0 \leq f(X) \leq M_1, \ \forall X \in [0, L].$$

In this case

$$E(0) = \lambda^2 \left[\frac{1}{2}\int_0^L (f\prime(x))^2 dx + \nu f^2(L) + \nu f^2(0) - \frac{\lambda^2}{4}\int_0^L f^4(x) dx\right].$$

Let us choose here

$$\lambda^2 = 2\int_0^L (f\prime(x))^2 dx \Big/ \int_0^L f^4(x) dx.$$

Then

$$E(0) = \lambda^2 \nu \left[f^2(L) + f^2(0)\right]$$

It is clear that if M_0 is large enough then the initial energy $E(0)$ is large enough. Finally we see that in this case the condition (15.33) is satisfied when

$$L > \frac{8a_0(1+\alpha)\nu M_1^2}{\alpha(1+\nu)M_0^2}.$$

References

1. C. Enache, Blow-up phenomena for a class of quasilinear parabolic problems under Robin boundary condition. Appl. Math. Lett. **24**(3), 288–292 (2011)
2. O.A. Ladyzhenskaya, *The Boundary Value Problems of Mathematical Physics*. Applied Mathematical Sciences, vol. 49 (Springer, New York, 1985)

3. O.A. Ladyzhenskaja, V.A. Solonnikov, N.N. Ural'ceva, *Linear and Quasi-Linear Equations of Parabolic Type*. Translations of Mathematical Monographs, vol. 23 (American Mathematical Society, Providence, RI, 1968)
4. H.A. Levine, Some nonexistence and instability theorems for solutions of formally parabolic equations of the form $Pu_t = Au + F(u)$. Arch. Ration. Mech. Anal. **51**, 371–386 (1973)
5. C.V. Pao, *Nonlinear Parabolic and Elliptic Equations* (Plenum Press, New York, 1992)
6. L.A. Payne, G.A. Philippin, Blow-up in a class of nonlinear parabolic problems with time-dependent coefficients under Robin boundary conditions. Anal. Appl. **91**, 2245–2256 (2012)
7. L.E. Payne, P.W. Schaefer, Blow-up in parabolic problems under Robin boundary conditions. Appl. Anal. **87**(6), 699–707 (2008)

Chapter 16
Functional Inequalities in Fuzzy Normed Spaces

Jung Rye Lee, Reza Saadati, and Dong Yun Shin

Abstract In this paper, we investigate the following functional inequalities

$$\|f(x) + f(y) + f(z)\| \leq \|f(x + y + z)\|$$

and

$$\|f(x) + f(y) + 2f(z)\| \leq \left\|2f\left(\frac{x+y}{2} + z\right)\right\|$$

in fuzzy normed vector spaces, and prove the Hyers–Ulam stability of the above functional inequalities in fuzzy Banach spaces in the spirit of the Th. M. Rassias' stability approach.

16.1 Introduction and Preliminaries

The stability problem of functional equations originated from a question of Ulam [26] concerning the stability of group homomorphisms. Hyers [10] gave a first affirmative partial answer to the question of Ulam for Banach spaces. Hyers' Theorem was generalized by Aoki [1] for additive mappings and by Th. M. Rassias [24] for linear mappings by considering an unbounded Cauchy difference. A generalization of the Th. M. Rassias theorem was obtained by Găvruta [7] by replacing the unbounded Cauchy difference by a general control function in the spirit of the Th. M. Rassias' approach.

J.R. Lee (✉)
Department of Mathematics, Daejin University, Pocheon, South Korea
e-mail: jrlee@daejin.ac.kr

R. Saadati
Iran University of Science and Technology, Tehran, Iran
e-mail: rsaadati@eml.cc

D.Y. Shin
Department of Mathematics, University of Seoul, South Korea
e-mail: dyshin@uos.ac.kr

G.A. Anastassiou, O. Duman (eds.), *Computational Analysis*, Springer Proceedings in Mathematics & Statistics 155, DOI 10.1007/978-3-319-28443-9_16

During the last two decades a number of papers and research monographs have been published on various generalizations and applications of the Hyers–Ulam stability to a number of functional equations and mappings (see [11, 18–22]).

Gilányi [8] showed that if f satisfies the functional inequality

$$\|2f(x) + 2f(y) - f(x-y)\| \leq \|f(x+y)\| \tag{16.1}$$

then f satisfies the Jordan–von Neumann functional equation

$$2f(x) + 2f(y) = f(x+y) + f(x-y).$$

See also [25]. Fechner [5] and Gilányi [9] proved the Hyers–Ulam stability of the functional inequality (16.1). Park et al. [23] investigated the Cauchy additive functional inequality

$$\|f(x) + f(y) + f(z)\| \leq \|f(x+y+z)\| \tag{16.2}$$

and the Cauchy–Jensen additive functional inequality

$$\|f(x) + f(y) + 2f(z)\| \leq \left\| 2f\left(\frac{x+y}{2} + z\right) \right\| \tag{16.3}$$

and proved the Hyers–Ulam stability of the functional inequalities (16.2) and (16.3) in Banach spaces.

Katsaras [12] defined a fuzzy norm on a vector space to construct a fuzzy vector topological structure on the space. Some mathematicians have defined fuzzy norms on a vector space from various points of view [6, 14, 27]. In particular, Bag and Samanta [2], following Cheng and Mordeson [4], gave an idea of fuzzy norm in such a manner that the corresponding fuzzy metric is of Kramosil and Michalek type [13]. They established a decomposition theorem of a fuzzy norm into a family of crisp norms and investigated some properties of fuzzy normed spaces [3].

We use the definition of fuzzy normed spaces given in [2, 15, 17] to investigate a fuzzy version of the Hyers–Ulam stability for the Cauchy functional inequality (16.2) and the Cauchy–Jensen functional inequality (16.3) in the fuzzy normed vector space setting.

Definition 16.1 (See [2, 15–17]). Let X be a real vector space. A function $N : X \times \mathbf{R} \to [0, 1]$ is called a *fuzzy norm* on X if for all $x, y \in X$ and all $s, t \in \mathbf{R}$,

(N_1) $N(x, t) = 0$ for $t \leq 0$;
(N_2) $x = 0$ if and only if $N(x, t) = 1$ for all $t > 0$;
(N_3) $N(cx, t) = N\left(x, \frac{t}{|c|}\right)$ if $c \neq 0$;
(N_4) $N(x+y, s+t) \geq \min\{N(x, s), N(y, t)\}$;
(N_5) $N(x, \cdot)$ is a non-decreasing function of $\mathbf{R}$ and $\lim_{t\to\infty} N(x, t) = 1$;
(N_6) for $x \neq 0$, $N(x, \cdot)$ is continuous on $\mathbf{R}$.

The pair (X, N) is called a *fuzzy normed vector space.*

The properties of fuzzy normed vector spaces and examples of fuzzy norms are given in [15, 17].

Definition 16.2 ([2, 15–17]). Let (X, N) be a fuzzy normed vector space. A sequence $\{x_n\}$ in X is said to *be convergent* or *converge* if there exists an $x \in X$ such that $\lim_{n\to\infty} N(x_n - x, t) = 1$ for all $t > 0$. In this case, x is called the *limit* of the sequence $\{x_n\}$ and we denote it by $N - \lim_{n\to\infty} x_n = x$.

Definition 16.3 (See[2, 15, 17]). Let (X, N) be a fuzzy normed vector space. A sequence $\{x_n\}$ in X is called *Cauchy* if for each $\varepsilon > 0$ and each $t > 0$ there exists an $n_0 \in \mathbf{N}$ such that for all $n \geq n_0$ and all $p > 0$, we have $N(x_{n+p} - x_n, t) > 1 - \varepsilon$.

It is well known that every convergent sequence in a fuzzy normed vector space is Cauchy. If each Cauchy sequence is convergent, then the fuzzy norm is said to be *complete* and the fuzzy normed vector space is called a *fuzzy Banach space.*

We say that a mapping $f : X \to Y$ between fuzzy normed vector spaces X and Y is continuous at a point $x_0 \in X$ if for each sequence $\{x_n\}$ converging to x_0 in X, then the sequence $\{f(x_n)\}$ converges to $f(x_0)$. If $f : X \to Y$ is continuous at each $x \in X$, then $f : X \to Y$ is said to be *continuous* on X (see [3]).

In this paper, we investigate the functional inequalities (16.2) and (16.3) in fuzzy normed vector spaces, and prove the Hyers–Ulam stability of the functional inequalities (16.2) and (16.3) in fuzzy Banach spaces.

Throughout this paper, assume that X is a vector space and that (Y, N) is a fuzzy Banach space.

16.2 Hyers–Ulam Stability of Functional Inequalities in Fuzzy Normed Vector Spaces

In this section, we investigate the functional inequalities (16.2) and (16.3) in fuzzy normed vector spaces, and prove the Hyers–Ulam stability of the functional inequalities (16.2) and (16.3) in fuzzy Banach spaces.

Lemma 16.4. *Let (Z, N) be a fuzzy normed vector space. Let $f : X \to Z$ be a mapping such that*

$$N(f(x) + f(y) + f(z), t) \geq N\left(f(x + y + z), \frac{t}{2}\right) \tag{16.4}$$

for all $x, y, z \in X$ and all $t > 0$. Then f is Cauchy additive, i.e., $f(x+y) = f(x)+f(y)$ for all $x, y \in X$.

Proof. Letting $x = y = z = 0$ in (16.4), we get

$$N(3f(0), t) = N\left(f(0), \frac{t}{3}\right) \geq N\left(f(0), \frac{t}{2}\right)$$

for all $t > 0$. By (N_5) and (N_6), $N(f(0), t) = 1$ for all $t > 0$. It follows from (N_2) that $f(0) = 0$. Letting $y = -x$ and $z = 0$ in (16.4), we get

$$N(f(x) + f(-x), t) \geq N\left(f(0), \frac{t}{2}\right) = N\left(0, \frac{t}{2}\right) = 1$$

for all $t > 0$. It follows from (N_2) that $f(x) + f(-x) = 0$ for all $x \in X$. So

$$f(-x) = -f(x)$$

for all $x \in X$. Letting $z = -x - y$ in (16.4), we get

$$\begin{aligned} N(f(x) + f(y) - f(x+y), t) &= N(f(x) + f(y) + f(-x-y), t) \\ &\geq N\left(f(0), \frac{t}{2}\right) = N\left(0, \frac{t}{2}\right) = 1 \end{aligned}$$

for all $x, y \in X$ and all $t > 0$. By (N_2),

$$N(f(x) + f(y) - f(x+y), t) = 1$$

for all $x, y \in X$ and all $t > 0$. It follows from (N_2) that

$$f(x+y) = f(x) + f(y)$$

for all $x, y \in X$, as desired.

Lemma 16.5. *Let (Z, N) be a fuzzy normed vector space. Let $f : X \to Z$ be a mapping such that*

$$N(f(x) + f(y) + 2f(z), t) \geq N\left(2f\left(\frac{x+y}{2} + z\right), \frac{2}{3}t\right) \tag{16.5}$$

for all $x, y, z \in X$ and all $t > 0$. Then f is Cauchy additive, i.e.,

$$f(x+y) = f(x) + f(y)$$

for all $x, y \in X$.

Proof. Letting $x = y = z = 0$ in (16.5), we get

$$N(4f(0), t) = N\left(f(0), \frac{t}{4}\right) \geq N\left(2f(0), \frac{2}{3}t\right) = N\left(f(0), \frac{t}{3}\right)$$

for all $t > 0$. By (N_5) and (N_6), $N(f(0), t) = 1$ for all $t > 0$. It follows from (N_2) that $f(0) = 0$. Letting $y = -x$ and $z = 0$ in (16.5), we get

$$N(f(x) + f(-x), t) \geq N\left(2f(0), \frac{2}{3}t\right) = N\left(0, \frac{2}{3}t\right) = 1$$

for all $t > 0$. It follows from (N_2) that $f(x) + f(-x) = 0$ for all $x \in X$. So

$$f(-x) = -f(x)$$

for all $x \in X$. Letting $z = -\frac{x+y}{2}$ in (16.5), we get

$$\begin{aligned} N\left(f(x) + f(y) - 2f\left(\frac{x+y}{2}, t\right)\right) &= N\left(f(x) + f(y) + 2f\left(-\frac{x+y}{2}\right), t\right) \\ &\geq N\left(2f(0), \frac{2}{3}t\right) = N\left(0, \frac{2}{3}t\right) = 1 \end{aligned}$$

for all $x, y \in X$ and all $t > 0$. By (N_2),

$$N\left(f(x) + f(y) - 2f\left(\frac{x+y}{2}\right), t\right) = 1$$

for all $x, y \in X$ and all $t > 0$. It follows from (N_2) that

$$2f\left(\frac{x+y}{2}\right) = f(x) + f(y)$$

for all $x, y \in X$. Since $f(0) = 0$,

$$f(x+y) = 2f\left(\frac{x+y}{2}\right) = f(x) + f(y)$$

for all $x, y \in X$, as desired.

Now we prove the Hyers–Ulam stability of the Cauchy functional inequality (16.2) in fuzzy Banach spaces.

Theorem 16.6. *Let $\varphi : X^3 \to [0, \infty)$ be a function such that*

$$\tilde{\varphi}(x, y, z) := \sum_{n=0}^{\infty} 2^{-n} \varphi(2^n x, 2^n y, 2^n z) < \infty \tag{16.6}$$

for all $x, y, z \in X$. Let $f : X \to Y$ be an odd mapping such that

$$\lim_{t \to \infty} N(f(x) + f(y) + f(z), t\varphi(x, y, z)) = 1 \tag{16.7}$$

uniformly on X^3. *Then*

$$L(x) := N - \lim_{n\to\infty} \frac{f(2^n x)}{2^n}$$

exists for each $x \in X$ *and defines a Cauchy additive mapping* $L : X \to Y$ *such that if for some* $\delta > 0, \alpha > 0$

$$N(f(x) + f(y) + f(z), \delta\varphi(x, y, z)) \geq \alpha \tag{16.8}$$

for all $x, y, z \in X$, *then*

$$N\left(f(x) - L(x), \frac{\delta}{2}\tilde{\varphi}(x, x, -2x)\right) \geq \alpha$$

for all $x \in X$. *Furthermore, the additive mapping* $L : X \to Y$ *is a unique mapping such that*

$$\lim_{t\to\infty} N(f(x) - L(x), t\tilde{\varphi}(x, x, -2x)) = 1 \tag{16.9}$$

uniformly on X.

Proof. Since f is an odd mapping, $f(-x) = -f(x)$ for all $x \in X$ and $f(0) = 0$. Given $\varepsilon > 0$, by (16.7), we can find some $t_0 > 0$ such that

$$N(f(x) + f(y) + f(z), t\varphi(x, y, z)) \geq 1 - \varepsilon \tag{16.10}$$

for all $t \geq t_0$. By induction on n, we show that

$$N\left(2^n f(x) - f(2^n x), t\sum_{k=0}^{n-1} 2^{n-k-1}\varphi(2^k x, 2^k x, -2^{k+1}x)\right) \geq 1 - \varepsilon \tag{16.11}$$

for all $t \geq t_0$, all $x \in X$ and all $n \in N$. Letting $y = x$ and $z = -2x$ in (16.10), we get

$$N(2f(x) - f(2x), t\varphi(x, x, -2x)) \geq 1 - \varepsilon$$

for all $x \in X$ and all $t \geq t_0$. So we get (16.11) for $n = 1$. Assume that (16.11) holds for $n \in N$. Then

$$N\left(2^{n+1} f(x) - f(2^{n+1}x), t\sum_{k=0}^{n} 2^{n-k}\varphi(2^k x, 2^k x, -2^{k+1}x)\right)$$

$$\geq \min N\left(2^{n+1} f(x) - 2f(2^n x), t_0\sum_{k=0}^{n-1} 2^{n-k}\varphi(2^n x, 2^n x, -2^{n+1}x)\right),$$

$$N(2f(2^n x) - f(2^{n+1}x), t_0\varphi(2^n x, 2^n x, -2^{n+1}x))\Big\}$$

$$\geq \min\{1-\varepsilon, 1-\varepsilon\} = 1-\varepsilon.$$

This completes the induction argument. Letting $t = t_0$ and replacing n and x by p and $2^n x$ in (16.11), respectively, we get

$$N\left(\frac{f(2^n x)}{2^n} - \frac{f(2^{n+p}x)}{2^{n+p}}, \frac{t_0}{2^{n+p}}\sum_{k=0}^{p-1} 2^{p-k-1}\varphi(2^{n+k}x, 2^{n+k}x, -2^{n+k+1}x)\right)$$
$$\geq 1-\varepsilon \tag{16.12}$$

for all integers $n \geq 0, p > 0$. It follows from (16.6) and the equality

$$\sum_{k=0}^{p-1} 2^{-n-k-1}\varphi(2^{n+k}x, 2^{n+k}x, -2^{n+k+1}x)$$
$$= \frac{1}{2}\sum_{k=n}^{n+p-1} 2^{-k}\varphi(2^k x, 2^k x, -2^{k+1}x)$$

that for a given $\delta > 0$ there is an $n_0 \in N$ such that

$$\frac{t_0}{2}\sum_{k=n}^{n+p-1} 2^{-k}\varphi(2^k x, 2^k x, -2^{k+1}x) < \delta$$

for all $n \geq n_0$ and $p > 0$. Now we deduce from (16.12) that

$$N\left(\frac{f(2^n x)}{2^n} - \frac{f(2^{n+p}x)}{2^{n+p}}, \delta\right)$$
$$\geq N\left(\frac{f(2^n x)}{2^n} - \frac{f(2^{n+p}x)}{2^{n+p}}, \frac{t_0}{2^{n+p}}\sum_{k=0}^{p-1} 2^{p-k-1}\varphi(2^{n+k}x, 2^{n+k}x, -2^{n+k+1}x)\right)$$
$$\geq 1-\varepsilon$$

for each $n \geq n_0$ and all $p > 0$. Thus the sequence $\left\{\frac{f(2^n x)}{2^n}\right\}$ is Cauchy in Y. Since Y is a fuzzy Banach space, the sequence $\left\{\frac{f(2^n x)}{2^n}\right\}$ converges to some $L(x) \in Y$. So we can define a mapping $L : X \to Y$ by

$$L(x) := N - \lim_{n\to\infty} \frac{f(2^n x)}{2^n},$$

namely, for each $t > 0$ and $x \in X$,

$$\lim_{n\to\infty} N\left(\frac{f(2^n x)}{2^n} - L(x), t\right) = 1.$$

Let $x, y, z \in X$. Fix $t > 0$ and $0 < \varepsilon < 1$. Since

$$\lim_{n\to\infty} 2^{-n}\varphi(2^n x, 2^n y, 2^n z) = 0,$$

there is an $n_1 > n_0$ such that

$$t_0\varphi(2^n x, 2^n y, 2^n z) < \frac{2^n t}{4}$$

for all $n \geq n_1$. Hence for each $n \geq n_1$, we have

$$\begin{aligned}
&N(L(x) + L(y) + L(z), t) \geq \min \left\{N\left(L(x) - 2^{-n}f(2^n x), \frac{t}{16}\right),\right.\\
&N\left(L(y) - 2^{-n}f(2^n y), \frac{t}{16}\right), N\left(L(z) - 2^{-n}f(2^n z), \frac{t}{16}\right),\\
&N\left(L(x+y+z) - 2^{-n}f(2^n(x+y+z)), \frac{t}{16}\right),\\
&N\left(f(2^n(x+y+z)) - f(2^n x) - f(2^n y) - f(2^n z), \frac{2^n t}{4}\right),\\
&\left.N\left(L(x+y+z), \frac{t}{2}\right)\right\}.
\end{aligned}$$

The first four terms on the right-hand side of the above inequality tend to 1 as $n \to \infty$, and the fifth term is greater than

$$N(f(2^n(x+y+z)) - f(2^n x) - f(2^n y) - f(2^n z), t_0\varphi(2^n x, 2^n y, 2^n z)),$$

which is greater than or equal to $1 - \varepsilon$. Thus

$$N(L(x) + L(y) + L(z), t) \geq \min\left\{N\left(L(x+y+z), \frac{t}{2}\right), 1 - \varepsilon\right\}$$

for all $t > 0$ and $0 < \varepsilon < 1$. So

$$N(L(x) + L(y) + L(z), t) \geq N\left(L(x+y+z), \frac{t}{2}\right)$$

for all $t > 0$, or

$$N(L(x) + L(y) + L(z), t) \geq 1 - \varepsilon$$

for all $t > 0$. For the former case, the mapping $L : X \to Y$ is Cauchy additive, by Lemma 16.4. For the latter case,

$$N(L(x) + L(y) + L(z), t) = 1$$

for all $t > 0$. So $N(3L(x), t) = 1$ for all $t > 0$ and for all $x \in X$. By (N_2), $L(x) = 0$ for all $x \in X$. Thus the mapping $L : X \to Y$ is Cauchy additive, i.e.,

$$L(x + y) = L(x) + L(y)$$

for all $x, y \in X$. Now let for some positive δ and α (16.8) hold. Let

$$\varphi_n(x, y, z) := \sum_{k=0}^{n-1} 2^{-k-1}\varphi(2^k x, 2^k y, 2^k z)$$

for all $x, y, z \in X$. Let $x \in X$. By the same reasoning as in the beginning of the proof, one can deduce from (16.8) that

$$N\left(2^n f(x) - f(2^n x), \delta \sum_{k=0}^{n-1} 2^{n-k-1}\varphi(2^k x, 2^k x, -2^{k+1}x)\right) \geq \alpha \tag{16.13}$$

for all positive integers n. Let $t > 0$. We have

$$\begin{aligned} &N(f(x) - L(x), \delta\varphi_n(x, x, -2x) + t) \\ &\geq \min\left\{N\left(f(x) - \frac{f(2^n x)}{2^n}, \delta\varphi_n(x, x, -2x)\right)\right. \\ &\left. N\left(\frac{f(2^n x)}{2^n} - L(x), t\right)\right\} \end{aligned} \tag{16.14}$$

Combining (16.13) and (16.14) and the fact that

$$\lim_{n\to\infty} N\left(\frac{f(2^n x)}{2^n} - L(x), t\right) = 1,$$

we observe that

$$N(f(x) - L(x), \delta\varphi_n(x, x, -2x) + t) \geq \alpha$$

for large enough $n \in N$. Thanks to the continuity of the function $N(f(x) - L(x), \cdot)$, we see that

$$N\left(f(x) - L(x), \frac{\delta}{2}\tilde{\varphi}(x, x, -2x) + t\right) \geq \alpha.$$

Letting $t \to 0$, we conclude that

$$N\left(f(x) - L(x), \frac{\delta}{2}\tilde{\varphi}(x, x, -2x)\right) \geq \alpha.$$

To end the proof, it remains to prove the uniqueness assertion. Let T be another additive mapping satisfying (16.9). Fix $c > 0$. Given $\varepsilon > 0$, by (16.9) for L and T, we can find some $t_0 > 0$ such that

$$N\left(f(x) - L(x), \frac{t}{2}\tilde{\varphi}(x, x, -2x)\right) \geq 1 - \varepsilon,$$
$$N\left(f(x) - T(x), \frac{t}{2}\tilde{\varphi}(x, x, -2x)\right) \geq 1 - \varepsilon$$

for all $x \in X$ and all $t \geq t_0$. Fix some $x \in X$ and find some integer n_0 such that

$$t_0 \sum_{k=n}^{\infty} 2^{-k}\varphi(2^k x, 2^k x, -2^{k+1}x) < \frac{c}{2}$$

for all $n \geq n_0$. Since

$$\begin{aligned}
&\sum_{k=n}^{\infty} 2^{-k}\varphi(2^k x, 2^k x, -2^{k+1}x) \\
&= \frac{1}{2^n}\sum_{k=n}^{\infty} 2^{-(k-n)}\varphi(2^{k-n}(2^n x), 2^{k-n}(2^n x), 2^{k-n}(-2^{n+1}x)) \\
&= \frac{1}{2^n}\sum_{m=0}^{\infty} 2^{-m}\varphi(2^m(2^n x), 2^m(2^n x), 2^m(-2^{n+1}x)) \\
&= \frac{1}{2^n}\tilde{\varphi}(2^n x, 2^n x, -2^{n+1}x),
\end{aligned}$$

we have

$$\begin{aligned}
&N(L(x) - T(x), c) \\
&\geq \min\left\{N\left(\frac{f(2^n x)}{2^n} - L(x), \frac{c}{2}\right), N\left(T(x) - \frac{f(2^n x)}{2^n}, \frac{c}{2}\right)\right\} \\
&= \min\{N(f(2^n x) - L(2^n x), 2^{n-1}c), N(T(2^n x) - f(2^n x), 2^{n-1}c)\} \\
&\geq \min\left\{N\left(f(2^n x) - L(2^n x), 2^n t_0 \sum_{k=n}^{\infty} 2^{-k}\varphi(2^k x, 2^k x, -2^{k+1}x)\right),\right.
\end{aligned}$$

$$N\left(T(2^n x) - f(2^n x), 2^n t_0 \sum_{k=n}^{\infty} 2^{-k}\varphi(2^k x, 2^k x, -2^{k+1}x)\right)\Bigg\}$$
$$= \min\{N(f(2^n x) - L(2^n x), t_0\tilde{\varphi}(2^n x, 2^n x, -2^{n+1}x)),$$
$$N(T(2^n x) - f(2^n x), t_0\tilde{\varphi}(2^n x, 2^n x, -2^{n+1}x))\}$$
$$\geq 1 - \varepsilon.$$

It follows that

$$N(L(x) - T(x), c) = 1$$

for all $c > 0$. Thus $L(x) = T(x)$ for all $x \in X$.

Corollary 16.7. *Let $\theta \geq 0$ and let p be a real number with $0 < p < 1$. Let $f : X \to Y$ be an odd mapping such that*

$$\lim_{t\to\infty} N(f(x) + f(y) + f(z), t\theta(\|x\|^p + \|y\|^p + \|z\|^p)) = 1 \tag{16.15}$$

uniformly on X^3. Then

$$L(x) := N - \lim_{n\to\infty} \frac{f(2^n x)}{2^n}$$

exists for each $x \in X$ and defines a Cauchy additive mapping $L : X \to Y$ such that if for some $\delta > 0, \alpha > 0$

$$N(f(x) + f(y) + f(z), \delta\theta(\|x\|^p + \|y\|^p + \|z\|^p)) \geq \alpha$$

for all $x, y, z \in X$, then

$$N\left(f(x) - L(x), \frac{2 + 2^p}{2 - 2^p}\delta\theta\|x\|^p\right) \geq \alpha$$

for all $x \in X$. Furthermore, the additive mapping $L : X \to Y$ is a unique mapping such that

$$\lim_{t\to\infty} N\left(f(x) - L(x), \frac{2 + 2^p}{2 - 2^p} 2t\theta\|x\|^p\right) = 1$$

uniformly on X.

Proof. Define

$$\varphi(x, y, z) := \theta(\|x\|^p + \|y\|^p + \|z\|^p)$$

and apply Theorem 16.6 to get the result.

Similarly, we can obtain the following. We will omit the proofs.

Theorem 16.8. *Let $\varphi : X^3 \to [0, \infty)$ be a function such that*

$$\tilde{\varphi}(x, y, z) := \sum_{n=1}^{\infty} 2^n \varphi\left(\frac{x}{2^n}, \frac{y}{2^n}, \frac{z}{2^n}\right) < \infty \tag{16.16}$$

for all $x, y, z \in X$. Let $f : X \to Y$ be an odd mapping satisfying (16.7). Then

$$L(x) := N - \lim_{n\to\infty} 2^n f\left(\frac{x}{2^n}\right)$$

exists for each $x \in X$ and defines a Cauchy additive mapping $L : X \to Y$ such that if for some $\delta > 0, \alpha > 0$

$$N(f(x) + f(y) + f(z), \delta\varphi(x, y, z)) \geq \alpha$$

for all $x, y, z \in X$, then

$$N\left(f(x) - L(x), \frac{\delta}{2}\tilde{\varphi}(x, x, -2x)\right) \geq \alpha$$

for all $x \in X$. Furthermore, the additive mapping $L : X \to Y$ is a unique mapping such that

$$\lim_{t\to\infty} N(f(x) - L(x), t\tilde{\varphi}(x, x, -2x)) = 1$$

uniformly on X.

Corollary 16.9. *Let $\theta \geq 0$ and let p be a real number with $p > 1$. Let $f : X \to Y$ be an odd mapping satisfying (16.15). Then*

$$L(x) := N - \lim_{n\to\infty} 2^n f\left(\frac{x}{2^n}\right)$$

exists for each $x \in X$ and defines a Cauchy additive mapping $L : X \to Y$ such that if for some $\delta > 0, \alpha > 0$

$$N(f(x) + f(y) + f(z), \delta\theta(\|x\|^p + \|y\|^p + \|z\|^p)) \geq \alpha$$

for all $x, y, z \in X$, then

$$N\left(f(x) - L(x), \frac{2^p + 2}{2^p - 2}\delta\theta\|x\|^p\right) \geq \alpha$$

for all $x \in X$. Furthermore, the additive mapping $L : X \to Y$ is a unique mapping such that

$$\lim_{t\to\infty} N\left(f(x) - L(x), \frac{2^p + 2}{2^p - 2}2t\theta \|x\|^p\right) = 1$$

uniformly on X.

Proof. Define $\varphi(x, y, z) := \theta(\|x\|^p + \|y\|^p + \|z\|^p)$ and apply Theorem 16.8 to get the result.

Finally, we prove the Hyers–Ulam stability of the Cauchy–Jensen functional inequality (16.3) in fuzzy Banach spaces.

Theorem 16.10. *Let $\varphi : X^3 \to [0, \infty)$ be a function satisfying (16.6). Let $f : X \to Y$ be an odd mapping such that*

$$\lim_{t\to\infty} N(f(x) + f(y) + 2f(z), t\varphi(x, y, z)) = 1 \tag{16.17}$$

uniformly on X^3. Then

$$L(x) := N - \lim_{n\to\infty} \frac{f(2^n x)}{2^n}$$

exists for each $x \in X$ and defines a Cauchy additive mapping $L : X \to Y$ such that if for some $\delta > 0, \alpha > 0$

$$N(f(x) + f(y) + 2f(z), \delta\varphi(x, y, z)) \geq \alpha \tag{16.18}$$

for all $x, y, z \in X$, then

$$N\left(f(x) - L(x), \frac{\delta}{2}\tilde{\varphi}(0, -2x, x)\right) \geq \alpha$$

for all $x \in X$. Furthermore, the additive mapping $L : X \to Y$ is a unique mapping such that

$$\lim_{t\to\infty} N(f(x) - L(x), t\tilde{\varphi}(0, -2x, x)) = 1 \tag{16.19}$$

uniformly on X.

Proof. Since f is an odd mapping, $f(-x) = -f(x)$ for all $x \in X$ and $f(0) = 0$. Given $\varepsilon > 0$, by (16.17), we can find some $t_0 > 0$ such that

$$N(f(x) + f(y) + 2f(z), t\varphi(x, y, z)) \geq 1 - \varepsilon \tag{16.20}$$

for all $t \geq t_0$. By induction on n, we show that

$$N\left(2^n f(x) - f(2^n x), t\sum_{k=0}^{n-1} 2^{n-k-1}\varphi(0, -2^{k+1}x, 2^k x)\right) \geq 1 - \varepsilon \tag{16.21}$$

for all $t \geq t_0$, all $x \in X$, and all $n \in N$. Letting $x = 0, y = -2x$ and $z = x$ in (16.20), we get

$$N(2f(x) - f(2x), t\varphi(0, -2x, x)) \geq 1 - \varepsilon$$

for all $x \in X$ and all $t \geq t_0$. So we get (16.21) for $n = 1$. Assume that (16.21) holds for $n \in N$. Then

$$\begin{aligned} &N\left(2^{n+1} f(x) - f(2^{n+1}x), t\sum_{k=0}^{n} 2^{n-k}\varphi(0, -2^{k+1}x, 2^k x)\right) \\ &\geq \min\left\{N\left(2^{n+1} f(x) - 2f(2^n x), t_0\sum_{k=0}^{n-1} 2^{n-k}\varphi(0, -2^{k+1}x, 2^k x)\right),\right. \\ &\qquad \left. N(2f(2^n x) - f(2^{n+1}x), t_0\varphi(0, -2^{n+1}x, 2^n x))\right\} \\ &\geq \min\{1 - \varepsilon, 1 - \varepsilon\} = 1 - \varepsilon. \end{aligned}$$

This completes the induction argument. Letting $t = t_0$ and replacing n and x by p and $2^n x$ in (16.21), respectively, we get

$$\begin{aligned} &N\left(\frac{f(2^n x)}{2^n} - \frac{f(2^{n+p}x)}{2^{n+p}}, \frac{t_0}{2^{n+p}}\sum_{k=0}^{p-1} 2^{p-k-1}\varphi(0, -2^{n+k+1}x, 2^{n+k}x)\right) \\ &\geq 1 - \varepsilon \end{aligned} \tag{16.22}$$

for all integers $n \geq 0, p > 0$. It follows from (16.6) and the equality

$$\sum_{k=0}^{p-1} 2^{-n-k-1}\varphi(0, -2^{n+k+1}x, 2^{n+k}x) = \frac{1}{2}\sum_{k=n}^{n+p-1} 2^{-k}\varphi(0, -2^{k+1}x, 2^k x)$$

that for a given $\delta > 0$ there is an $n_0 \in \mathbf{N}$ such that

$$\frac{t_0}{2}\sum_{k=n}^{n+p-1} 2^{-k}\varphi(0, -2^{k+1}x, 2^k x) < \delta$$

for all $n \geq n_0$ and $p > 0$. Now we deduce from (16.22) that

$$\begin{aligned}
&N\left(\frac{f(2^n x)}{2^n} - \frac{f(2^{n+p}x)}{2^{n+p}}, \delta\right) \\
&\geq N\left(\frac{f(2^n x)}{2^n} - \frac{f(2^{n+p}x)}{2^{n+p}}, \frac{t_0}{2^{n+p}} \sum_{k=0}^{p-1} 2^{p-k-1}\varphi(0, -2^{n+k+1}x, 2^{n+k}x)\right) \\
&\geq 1 - \varepsilon
\end{aligned}$$

for each $n \geq n_0$ and all $p > 0$. Thus the sequence $\{\frac{f(2^n x)}{2^n}\}$ is Cauchy in Y. Since Y is a fuzzy Banach space, the sequence $\{\frac{f(2^n x)}{2^n}\}$ converges to some $L(x) \in Y$. So we can define a mapping $L : X \to Y$ by

$$L(x) := N - \lim_{n\to\infty} \frac{f(2^n x)}{2^n},$$

namely, for each $t > 0$ and $x \in X$,

$$\lim_{n\to\infty} N\left(\frac{f(2^n x)}{2^n} - L(x), t\right) = 1.$$

Let $x, y, z \in X$. Fix $t > 0$ and $0 < \varepsilon < 1$. Since

$$\lim_{n\to\infty} 2^{-n}\varphi(2^n x, 2^n y, 2^n z) = 0,$$

there is an $n_1 > n_0$ such that

$$t_0\varphi(2^n x, 2^n y, 2^n z) < \frac{2^n t}{12}$$

for all $n \geq n_1$. Hence for each $n \geq n_1$, we have

$$\begin{aligned}
N(L(x) + L(y) + 2L(z), t) \geq \min\Bigg\{&N\left(L(x) - 2^{-n}f(2^n x), \frac{t}{16}\right), \\
&N\left(L(y) - 2^{-n}f(2^n y), \frac{t}{16}\right), N\left(2L(z) - 2^{-n}f(2^n z), \frac{t}{16}\right), \\
&N\left(2L\left(\frac{x+y}{2} + z\right) - 2^{-n+1}f\left(2^n\left(\frac{x+y}{2} + z\right)\right), \frac{t}{16}\right), \\
&N\left(2f\left(2^n\left(\frac{x+y}{2} + z\right)\right) - f(2^n x) - f(2^n y) - f(2^n z), \frac{2^n t}{12}\right), \\
&N\left(2L\left(\frac{x+y}{2} + z\right), \frac{2}{3}t\right)\Bigg\}.
\end{aligned}$$

The first four terms on the right-hand side of the above inequality tend to 1 as $n \to \infty$, and the fifth term is greater than

$$N\left(2f\left(2^n\left(\frac{x+y}{2}+z\right)\right)-f(2^nx)-f(2^ny)-2f(2^nz), t_0\varphi(2^nx, 2^ny, 2^nz)\right),$$

which is greater than or equal to $1-\varepsilon$. Thus

$$N(L(x)+L(y)+2L(z), t) \geq \min\left\{N\left(2L\left(\frac{x+y}{2}+z\right), \frac{2}{3}t\right), 1-\varepsilon\right\}$$

for all $t > 0$ and $0 < \varepsilon < 1$. So

$$N(L(x)+L(y)+2L(z), t) \geq N\left(2L\left(\frac{x+y}{2}+z\right), \frac{2}{3}t\right)$$

for all $t > 0$, or

$$N(L(x)+L(y)+2L(z), t) \geq 1-\varepsilon$$

for all $t > 0$. For the former case, the mapping $L : X \to Y$ is Cauchy additive, by Lemma 16.5. For the latter case,

$$N(L(x)+L(y)+2L(z), t) = 1$$

for all $t > 0$. So $N(4L(x), t) = 1$ for all $t > 0$ and for all $x \in X$. By (N_2), $L(x) = 0$ for all $x \in X$. Thus the mapping $L : X \to Y$ is Cauchy additive, i.e.,

$$L(x+y) = L(x) + L(y)$$

for all $x, y \in X$. Now let for some positive δ and α (16.18) hold. Let

$$\varphi_n(x, y, z) := \sum_{k=0}^{n-1} 2^{-k-1}\varphi(2^kx, 2^ky, 2^kz)$$

for all $x, y, z \in X$. Let $x \in X$. By the same reasoning as in the beginning of the proof, one can deduce from (16.18) that

$$N\left(2^nf(x) - f(2^nx), \delta\sum_{k=0}^{n-1} 2^{n-k-1}\varphi(0, -2^{k+1}x, 2^kx)\right) \geq \alpha \tag{16.23}$$

for all positive integers n. Let $t > 0$. We have

$$N(f(x) - L(x), \delta\varphi_n(0, -2x, x) + t) \leq \min\left\{N\left(f(x) - \frac{f(2^n x)}{2^n}, \delta\varphi_n(0, -2x, x)\right)\right.$$
$$\left. N\left(\frac{f(2^n x)}{2^n} - L(x), t\right)\right\} \tag{16.24}$$

Combining (16.23) and (16.24) and the fact that $\lim_{n\to\infty} N\left(\frac{f(2^n x)}{2^n} - L(x), t\right) = 1$, we observe that

$$N(f(x) - L(x), \delta\varphi_n(0, -2x, x) + t) \geq \alpha$$

for large enough $n \in N$. Thanks to the continuity of the function $N(f(x) - L(x), \cdot)$, we see that

$$N\left(f(x) - L(x), \frac{\delta}{2}\tilde{\varphi}(0, -2x, x) + t\right) \geq \alpha.$$

Letting $t \to 0$, we conclude that

$$N\left(f(x) - L(x), \frac{\delta}{2}\tilde{\varphi}(0, -2x, x)\right) \geq \alpha.$$

To end the proof, it remains to prove the uniqueness assertion. Let T be another additive mapping satisfying (16.19). Fix $c > 0$. Given $\varepsilon > 0$, by (16.19) for L and T, we can find some $t_0 > 0$ such that

$$N\left(f(x) - L(x), \frac{t}{2}\tilde{\varphi}(0, -2x, x)\right) \geq 1 - \varepsilon,$$
$$N\left(f(x) - T(x), \frac{t}{2}\tilde{\varphi}(0, -2x, x)\right) \geq 1 - \varepsilon$$

for all $x \in X$ and all $t \geq t_0$. Fix some $x \in X$ and find some integer n_0 such that

$$t_0 \sum_{k=n}^{\infty} 2^{-k}\varphi(0, -2^{k+1}x, 2^k x) < \frac{c}{2}$$

for all $n \geq n_0$. Since

$$\begin{aligned}\sum_{k=n}^{\infty} 2^{-k}\varphi(0, -2^k x, 2^k x) &= \frac{1}{2^n}\sum_{k=n}^{\infty} 2^{-(k-n)}\varphi(0, 2^{k-n}(-2^{n+1}x), 2^{k-n}(2^n x)) \\ &= \frac{1}{2^n}\sum_{m=0}^{\infty} 2^{-m}\varphi(0, 2^m(-2^{n+1}x), 2^m(2^n x)) \\ &= \frac{1}{2^n}\tilde{\varphi}(0, -2^{n+1}x, 2^n x),\end{aligned}$$

we have

$$\begin{aligned}
&N(L(x)-T(x),c)\\
&\quad\geq \min\left\{N\left(\frac{f(2^nx)}{2^n}-L(x),\frac{c}{2}\right),N\left(T(x)-\frac{f(2^nx)}{2^n},\frac{c}{2}\right)\right\}\\
&\quad=\min\{N(f(2^nx)-L(2^nx),2^{n-1}c),N(T(2^nx)-f(2^nx),2^{n-1}c)\}\\
&\quad\geq \min\left\{N\left(f(2^nx)-L(2^nx),2^nt_0\sum_{k=n}^{\infty}2^{-k}\varphi(0,-2^{k+1}x,2^kx)\right),\right.\\
&\qquad\left.N\left(T(2^nx)-f(2^nx),2^nt_0\sum_{k=n}^{\infty}2^{-k}\varphi(0,-2^{k+1}x,2^kx)\right)\right\}\\
&\quad=\min\{N(f(2^nx)-L(2^nx),t_0\tilde{\varphi}(0,-2^{n+1}x,2^nx)),\\
&\qquad\qquad N(T(2^nx)-f(2^nx),t_0\tilde{\varphi}(0,-2^{n+1}x,2^nx))\}\\
&\quad\geq 1-\varepsilon.
\end{aligned}$$

It follows that

$$N(L(x)-T(x),c)=1$$

for all $c>0$. Thus $L(x)=T(x)$ for all $x\in X$.

Corollary 16.11. *Let $\theta\geq 0$ and let p be a real number with $0<p<1$. Let $f:X\to Y$ be an odd mapping such that*

$$\lim_{t\to\infty}N(f(x)+f(y)+2f(z),t\theta(\|x\|^p+\|y\|^p+\|z\|^p))=1 \tag{16.25}$$

uniformly on X^3. Then

$$L(x):=N-\lim_{n\to\infty}\frac{f(2^nx)}{2^n}$$

exists for each $x\in X$ and defines a Cauchy additive mapping $L:X\to Y$ such that if for some $\delta>0,\alpha>0$

$$N(f(x)+f(y)+2f(z),\delta\theta(\|x\|^p+\|y\|^p+\|z\|^p))\geq\alpha$$

for all $x,y,z\in X$, then

$$N\left(f(x)-L(x),\frac{1+2^p}{2-2^p}\delta\theta\|x\|^p\right)\geq\alpha$$

for all $x\in X$.

Furthermore, the additive mapping $L : X \to Y$ is a unique mapping such that

$$\lim_{t\to\infty} N\left(f(x) - L(x), \frac{1+2^p}{2-2^p} 2t\theta \|x\|^p\right) = 1$$

uniformly on X.

Proof. Define

$$\varphi(x, y, z) := \theta(\|x\|^p + \|y\|^p + \|z\|^p)$$

and apply Theorem 16.10 to get the result.

Similarly, we can obtain the following results. We will omit the proofs.

Theorem 16.12. *Let $\varphi : X^3 \to [0, \infty)$ be a function satisfying (16.16). Let $f : X \to Y$ be an odd mapping satisfying (16.17). Then*

$$L(x) := N - \lim_{n\to\infty} 2^n f\left(\frac{x}{2^n}\right)$$

exists for each $x \in X$ and defines a Cauchy additive mapping $L : X \to Y$ such that if for some $\delta > 0, \alpha > 0$

$$N(f(x) + f(y) + 2f(z), \delta\varphi(x, y, z)) \geq \alpha$$

for all $x, y, z \in X$, then

$$N\left(f(x) - L(x), \frac{\delta}{2}\tilde{\varphi}(0, -2x, x)\right) \geq \alpha$$

for all $x \in X$.

Furthermore, the additive mapping $L : X \to Y$ is a unique mapping such that

$$\lim_{t\to\infty} N(f(x) - L(x), t\tilde{\varphi}(0, -2x, x)) = 1$$

uniformly on X.

Corollary 16.13. *Let $\theta \geq 0$ and let p be a real number with $p > 1$. Let $f : X \to Y$ be an odd mapping satisfying (16.25). Then*

$$L(x) := N - \lim_{n\to\infty} 2^n f\left(\frac{x}{2^n}\right)$$

exists for each $x \in X$ and defines a Cauchy additive mapping $L : X \to Y$ such that if for some $\delta > 0, \alpha > 0$

$$N(f(x) + f(y) + 2f(z), \delta\theta(\|x\|^p + \|y\|^p + \|z\|^p)) \geq \alpha$$

for all $x, y, z \in X$, *then*

$$N\left(f(x) - L(x), \frac{2^p + 1}{2^p - 2}\delta\theta \|x\|^p\right) \geq \alpha$$

for all $x \in X$.

Furthermore, the additive mapping $L : X \to Y$ *is a unique mapping such that*

$$\lim_{t\to\infty} N\left(f(x) - L(x), \frac{2^p + 1}{2^p - 2}2t\theta \|x\|^p\right) = 1$$

uniformly on X.

Proof. Define

$$\varphi(x, y, z) := \theta(\|x\|^p + \|y\|^p + \|z\|^p)$$

and apply Theorem 16.12 to get the result.

References

1. T. Aoki, On the stability of the linear transformation in Banach spaces. J. Math. Soc. Jpn. **2**, 64–66 (1950)
2. T. Bag, S.K. Samanta, Finite dimensional fuzzy normed linear spaces. J. Fuzzy Math. **11**, 687–705 (2003)
3. T. Bag, S.K. Samanta, Fuzzy bounded linear operators. Fuzzy Sets Syst. **151**, 513–547 (2005)
4. S.C. Cheng, J.M. Mordeson, Fuzzy linear operators and fuzzy normed linear spaces. Bull. Calcutta Math. Soc. **86**, 429–436 (1994)
5. W. Fechner, Stability of a functional inequalities associated with the Jordan-von Neumann functional equation. Aequationes Math. **71**, 149–161 (2006)
6. C. Felbin, Finite dimensional fuzzy normed linear spaces. Fuzzy Sets Syst. **48**, 239–248 (1992)
7. G. Găvruta, A generalization of the Hyers-Ulam-Rassias stability of approximately additive mappings. J. Math. Anal. Appl. **184**, 431–436 (1994)
8. A. Gilányi, Eine zur Parallelogrammgleichung äquivalente Ungleichung. Aequationes Math. **62**, 303–309 (2001)
9. A. Gilányi, On a problem by K. Nikodem. Math. Inequal. Appl. **5**, 707–710 (2002)
10. D.H. Hyers, On the stability of the linear functional equation. Proc. Natl. Acad. Sci. U.S.A. **27**, 222–224 (1941)
11. S. Jung, *Hyers-Ulam-Rassias Stability of Functional Equations in Mathematical Analysis* (Hadronic Press, Palm Harbor, FL, 2001)
12. A.K. Katsaras, Fuzzy topological vector spaces II. Fuzzy Sets Syst. **12**, 143–154 (1984)
13. I. Kramosil, J. Michalek, Fuzzy metric and statistical metric spaces. Kybernetica **11**, 326–334 (1975)
14. S.V. Krishna, K.K.M. Sarma, Separation of fuzzy normed linear spaces. Fuzzy Sets Syst. **63**, 207–217 (1994)
15. A.K. Mirmostafaee, M.S. Moslehian, Fuzzy versions of Hyers-Ulam-Rassias theorem. Fuzzy Sets Syst. **159**, 720–729 (2008)

16. A.K. Mirmostafaee, M.S. Moslehian, Fuzzy approximately cubic mappings. Inform. Sci. **178**, 3791–3798 (2008)
17. A.K. Mirmostafaee, M. Mirzavaziri, M.S. Moslehian, Fuzzy stability of the Jensen functional equation. Fuzzy Sets Syst. **159**, 730–738 (2008)
18. C. Park, On the stability of the linear mapping in Banach modules. J. Math. Anal. Appl. **275**, 711–720 (2002)
19. C. Park, Modified Trif's functional equations in Banach modules over a C^*-algebra and approximate algebra homomorphisms. J. Math. Anal. Appl. **278**, 93–108 (2003)
20. C. Park, On an approximate automorphism on a C^*-algebra. Proc. Am. Math. Soc. **132**, 1739–1745 (2004)
21. C. Park, Lie $*$-homomorphisms between Lie C^*-algebras and Lie $*$-derivations on Lie C^*-algebras. J. Math. Anal. Appl. **293**, 419–434 (2004)
22. C. Park, Homomorphisms between Poisson JC^*-algebras. Bull. Braz. Math. Soc. **36**, 79–97 (2005)
23. C. Park, Y. Cho, M. Han, Stability of functional inequalities associated with Jordan-von Neumann type additive functional equations. J. Inequal. Appl. **2007** (2007). Art. ID 41820, pp. 13
24. T.M. Rassias, On the stability of the linear mapping in Banach spaces. Proc. Am. Math. Soc. **72**, 297–300 (1978)
25. J. Rätz, On inequalities associated with the Jordan-von Neumann functional equation. Aequationes Math. **66**, 191–200 (2003)
26. S.M. Ulam *A Collection of the Mathematical Problems* (Interscience Publishers, New York, 1960)
27. J.Z. Xiao, X.H. Zhu, Fuzzy normed spaces of operators and its completeness. Fuzzy Sets Syst. **133**, 389–399 (2003)

Chapter 17
Principal Vectors of Matrix-Valued Difference Operators

Yelda Aygar and Murat Olgun

Abstract In this paper, we investigate the principal vectors corresponding to the eigenvalues and the spectral singularities of matrix-valued difference operator and get some properties of these vectors.

17.1 Introduction

Spectral theory of self-adjoint difference operators is well-known in literature [1–3, 5]. Spectral analysis of nonselfadjoint Sturm–Liouville and difference operators with continuous and point spectrum was investigated in [10]. In [10], the author proved that the spectrum of a nonselfadjoint Sturm–Liouville operator consists of continuous spectrum, eigenvalues and spectral singularities. The spectral singularities are poles of the kernel of the resolvent and are also imbedded in the continuous spectrum, but they are not eigenvalues. The effect of spectral singularities in the spectral expansion of Sturm–Liouville operators in terms of the principal vectors was considered in [9]. Some problems of spectral theory of difference and differential operators with matrix coefficients were also investigated in [4, 6, 11, 13, 14]. Furthermore, a lot of mathematicians studied about principal vectors of differential and difference operators with scalar coefficients, also principal vectors of differential operators with matrix coefficients [7, 8, 12]. But principal vectors of matrix difference operators have not been investigated yet. Let us introduce the Hilbert space $\ell_2(\mathbb{N}, \mathbb{C}^m)$ consisting of all vector sequences $y = \{y_n\}_{n\in\mathbb{N}}$, ($y_n \in \mathbb{C}^m$), such that $\|y_n\|^2_{\mathbb{C}^m} < \infty$ with the inner product $\langle y, z\rangle = \sum_{n=1}^{\infty}(y_n, z_n)_{\mathbb{C}^m}$, where $\mathbb{C}^m$ is m-dimensional ($m < \infty$) Euclidean space, $\|.\|_{\mathbb{C}^m}$ and $\langle ., .\rangle_{\mathbb{C}^m}$ denote norm and inner product in $\mathbb{C}^m$, respectively. Let L denote the difference operator of second-order generated in $\ell_2(\mathbb{N}, \mathbb{C}^m)$ by matrix difference expression

$$(ly)_n := A_{n-1}y_{n-1} + B_n y_n + A_n y_{n+1}, \quad n \in \mathbb{N},$$

Y. Aygar • M. Olgun (✉)
Department of Mathematics, Faculty of Science, Ankara University, Turkey
e-mail: yaygar@science.ankara.edu.tr; Murat.Olgun@ankara.edu.tr

G.A. Anastassiou, O. Duman (eds.), *Computational Analysis*, Springer Proceedings in Mathematics & Statistics 155, DOI 10.1007/978-3-319-28443-9_17

and the boundary condition $y_0 = 0$, where A_n, $n \in \mathbb{N} \cup \{0\}$ and B_n, $n \in \mathbb{N}$ are linear operators (matrices) acting in $\mathbb{C}^m$. Throughout the paper, we will assume that $\det A_n \neq 0$, $A_n = A_n^*$, $(n \in \mathbb{N} \cup \{0\})$ and $B_n = B_n^*$, where $*$ denotes the adjoint operator. In [14], it is proved that the operator L has a finite number of eigenvalues and spectral singularities with finite multiplicities under the condition

$$\sup_{n \in \mathbb{N}} \{e^{\varepsilon n} (\|I - A_n\| + \|B_n\|)\} < \infty, \quad \varepsilon > 0, \tag{17.1}$$

where I denotes identity matrix and $\|.\|$ shows the matrix norm in $\mathbb{C}^m$. The aim of this paper is to extend some results of paper [14] by using principal vectors corresponding to the eigenvalues and spectral singularities of the operator L. The paper is organized as follows: Sect. 17.2 contains some information about Jost solution and spectral properties of L which are given in [14]. We will get main results by using these information in the next section. In Sect. 17.3, we get principal vectors corresponding to eigenvalues and spectral singularities of L, and give some properties of these vectors.

17.2 Jost Solution and Spectral Properties of *L*

Related to the operator L, consider the equation

$$(ly)_n = \lambda y_n, \quad n \in \mathbb{N}. \tag{17.2}$$

Assume (17.1). Then the Jost solution of (17.2) is given in [14], as

$$F_n(z) = T_n e^{inz} \left[I + \sum_{m=1}^{\infty} K_{nm} e^{imz} \right], \quad n \in \mathbb{N} \cup \{0\}, \tag{17.3}$$

for $\lambda = 2\cos z$ and $z \in \overline{\mathbb{C}}_+ := \{z \in \mathbb{C} : \operatorname{Im} z \geq 0\}$. Also we can write T_n and K_{nm} in terms of A_n and B_n as

$$T_n = \prod_{p=n}^{\infty} A_p^{-1}, \quad K_{n1} = -\sum_{p=n+1}^{\infty} T_p^{-1} B_p T_p,$$

$$K_{n2} = \sum_{p=n+1}^{\infty} T_p^{-1} (I - A_p^2) T_p - \sum_{p=n+1}^{\infty} T_p^{-1} B_p T_p K_{p1}, \tag{17.4}$$

$$K_{n,m+2} = K_{n+1,m} + \sum_{p=n-1}^{\infty} T_p^{-1} (I - A_p^2) T_p K_{p+1,m} - \sum_{p=n-1}^{\infty} T_p^{-1} B_p T_p K_{p,m+1},$$

where $n \in \mathbb{N} \cup \{0\}$ and $m \in \mathbb{N}$. Moreover, K_{nm} satisfies

$$\|K_{nm}\| \leq C \sum_{p=n+\lfloor \frac{m}{2} \rfloor} \left(\|I - A_p\| + \|B_p\|\right), \tag{17.5}$$

where $\lfloor \frac{m}{2} \rfloor$ is the integer part of $\frac{m}{2}$ and $C > 0$ is a constant. T_n and K_{nm}, $n \in \mathbb{N} \cup \{0\}$, and $m \in \mathbb{N}$ are absolutely convergent. Analogously to the Sturm–Liouville equation, the solution $F(z) := \{F_n(z)\}_{n \in \mathbb{N} \cup \{0\}}$ and the function

$$F_0(z) = T_0 \left[I + \sum_{m=1}^{\infty} K_{0m} e^{imz} \right]$$

are called the Jost solution and Jost function of (17.2), respectively [14]. Also in [14], the author found asymptotic behavior and analytical properties of F, and showed $\sigma_c(L) = [-2, 2]$, where $\sigma_c(L)$ is continuous spectrum of L. Let us define $f(z) = \det F_0(z)$, $z \in \overline{\mathbb{C}}_+$. Then the function f is analytic in $\mathbb{C}_+$, continuous in $\overline{\mathbb{C}}_+$ and $f(z) = f(z + 2\pi)$. If we define the semi-strips

$$P_0 = \{z \in \mathbb{C}_+ : 0 \leq \mathrm{Re} z \leq 2\pi\}$$

and $P = P_0 \cup [0, 2\pi]$, we get

$$\begin{aligned} \sigma_d(L) &= \{\lambda : \lambda = 2\cos z, \ z \in P_0, \quad f(z) = 0\} \\ \sigma_{ss}(L) &= \{\lambda : \lambda = 2\cos z, \ z \in [0, 2\pi], \quad f(z) = 0\} \setminus \{0\}, \end{aligned} \tag{17.6}$$

by using the definition of eigenvalues and the spectral singularities of nonselfadjoint operators [14], where $\sigma_d(L)$ and $\sigma_{ss}(L)$ denote the set of eigenvalues and spectral singularities of L, respectively.

Definition 17.1. The multiplicity of a zero of F in P is called the multiplicity of the corresponding eigenvalue or spectral singularity of L.

It has been shown in [14] that the operator L has a finite number of zeros in P with finite multiplicities under the condition (17.1).

17.3 Principal Functions of L

Let us define the functions

$$E_n(\lambda) := F_n\left(\arccos \frac{\lambda}{2}\right), \quad n \in \mathbb{N}$$

and

$$B(\lambda) := f\left(\arccos\frac{\lambda}{2}\right).$$

Using (17.3) and

$$\arccos\frac{\lambda}{2} = -i\ln\left(\frac{\lambda+\sqrt{\lambda^2-4}}{2}\right),$$

we get that

$$E_n(\lambda) = T_n\left(\frac{\lambda+\sqrt{\lambda^2-4}}{2}\right)^n\left(I+\sum_{m=1}^{\infty}K_{nm}\left(\frac{\lambda+\sqrt{\lambda^2-4}}{2}\right)^m\right)$$

for $n\in\mathbb{N}$. Since $\lambda = 2\cos z$ maps the semi-strip P_0 to the domain $\Lambda := \mathbb{C}\backslash[-2,2]$, the functions $E_n(\lambda)$ and $B(\lambda)$ are analytic in Λ, and continuous up to the interval $[-2,2]$. Using (17.4), we can write

$$\sigma_{\mathrm{d}}(L) = \{\lambda\in\Lambda : B(\lambda)=0\}$$
$$\sigma_{\mathrm{ss}}(L) = \{\lambda\in[-2,2] : B(\lambda)=0\}\backslash\{0\}.$$

Therefore the function B has a finite number of zeros in Λ and $[-2,2]$, and each of them is of a finite multiplicity. Let $\lambda_1,\lambda_2,\ldots,\lambda_s$ and $\lambda_{s+1},\lambda_{s+2},\ldots,\lambda_\nu$ denote the zeros of B in Λ and in $[-2,2]$ with multiplicities $m_1,m_2,\ldots,m_s$ and $m_{s+1},m_{s+2},\ldots,m_\nu$, respectively.

Definition 17.2. Let λ_0 be an eigenvalue of L. If the vectors $y^{(0)},y^{(1)},\ldots,y^{(s)}$ satisfy the equation

$$\begin{cases}(ly^{(0)})_n-\lambda_0 y_n^{(0)}=0\\ (ly^{(k)})_n-\lambda_0 y_n^{(k)}-y_n^{(k-1)}=0\quad k=1,2,\ldots,s;\quad n\in\mathbb{N},\end{cases}\tag{17.7}$$

then the vector $y^{(0)}$ is called the eigenvector corresponding to the eigenvalue $\lambda=\lambda_0$ of L. The vectors $y^{(1)},y^{(2)},\ldots,y^{(s)}$ are called the associated vectors corresponding to λ_0. The eigenvector and the associated vectors corresponding to λ_0 are called the principal vectors of the eigenvalue $\lambda=\lambda_0$. The principal vectors of the spectral singularities of L are defined similarly.

Now, we define the matrix-functions for $\lambda = 2\cos z$, $z\in P$

$$U_n^{(k)}(\lambda_j) = \frac{1}{k!}\left\{\frac{\mathrm{d}^k}{\mathrm{d}\lambda^k}E_n(\lambda)\right\}_{\lambda=\lambda_j},$$
$$(k=0,1,\ldots,m_j-1;\ j=1,2,\ldots,s)$$

and

$$U_n^{(k)}(\lambda_j) = \frac{1}{k!}\left\{\frac{\mathrm{d}^k}{\mathrm{d}\lambda^k}E_n(\lambda)\right\}_{\lambda=\lambda_j},$$
$$(k = 0, 1, \ldots, m_j - 1;\ j = s+1, s+2, \ldots, \nu).$$

If $y(\lambda) = \{y_n(\lambda)\}_{n\in\mathbb{N}}$ is a solution of (17.2), then

$$\left(\frac{\mathrm{d}^k}{\mathrm{d}\lambda^k}\right)y(\lambda) = \left\{\left(\frac{\mathrm{d}^k}{\mathrm{d}\lambda^k}\right)y_n(\lambda)\right\}_{n\in\mathbb{N}}$$

satisfies

$$\begin{aligned}A_{n-1}\frac{\mathrm{d}^k}{\mathrm{d}\lambda^k}y_{n-1}(\lambda) + B_n\frac{\mathrm{d}^k}{\mathrm{d}\lambda^k}y_n(\lambda) + A_n\frac{\mathrm{d}^k}{\mathrm{d}\lambda^k}y_{n+1}(\lambda)\\ = \lambda\frac{\mathrm{d}^k}{\mathrm{d}\lambda^k}y_n(\lambda) + k\frac{\mathrm{d}^{k-1}}{\mathrm{d}\lambda^{k-1}}y_n(\lambda).\end{aligned} \tag{17.8}$$

Using (17.8) and the definition of $U_n^{(k)}(\lambda_j)$ for $k = 0, 1, \ldots, m_j - 1; j = 1, 2, \ldots, s, s+1, \ldots, \nu$, we obtain

$$\begin{cases}\left(lU^{(0)}(\lambda_j)\right)_n - \lambda_j U_n^{(0)}(\lambda_j) = 0,\\ \left(lU^{(k)}(\lambda_j)\right)_n - \lambda_j U_n^{(k)}(\lambda_j) - U_n^{(k-1)}(\lambda_j) = 0.\end{cases}$$

Last equations show that $U^{(k)}(\lambda_j)$ are the principal vectors of eigenvalues of L for $k = 0, 1, \ldots, m_j - 1; j = 1, 2, \ldots, s$ and $U^{(k)}(\lambda_j)$ are the principal vectors of spectral singularities of L for $k = 0, 1, \ldots, m_j - 1; j = s+1, s+2, \ldots, \nu$.

Theorem 17.3. *Assume* (17.1). *Then for* $k = 0, 1, \ldots, m_j - 1$ *and* $j = 1, 2, \ldots, s$, $U^{(k)}(\lambda_j) \in \ell_2(\mathbb{N}, \mathbb{C}^m)$, *but* $U^{(k)}(\lambda_j) \notin \ell_2(\mathbb{N}, \mathbb{C}^m)$ *for* $k = 0, 1, \ldots, m_j - 1$ *and* $j = s+1, s+2, \ldots, \nu$.

Proof. Since $E_n(\lambda) = F_n(\arccos\frac{\lambda}{2})$, we can write

$$\left\{\frac{\mathrm{d}^k}{\mathrm{d}\lambda^k}E_n(\lambda)\right\}_{\lambda=\lambda_j} = \sum_{m=0}^{k} C_m\left\{\frac{\mathrm{d}^m}{\mathrm{d}\lambda^m}F_n(z)\right\}_{z=z_j}, \quad n\in\mathbb{N},$$

where $\lambda_j = 2\cos z_j$, $z_j \in P, j = 1, 2, \ldots, \nu$ and C_m is a constant depending on λ_j. Using the definition of $F_n(z)$, we get

$$\left\{\frac{\mathrm{d}^p}{\mathrm{d}\lambda^p}F_n(z)\right\}_{z=z_j} = T_n e^{inz_j}\left\{(in)^p + \sum_{m=1}^{\infty}(i(n+m))^p K_{nm}e^{imz_j}\right\}$$

for all $n \in \mathbb{N}$ and $j = 1, 2, \ldots, \nu$. Consider the principal vectors $U_n^{(k)}(\lambda_j), k = 0, 1, \ldots, m_j - 1$ and $j = 1, 2, \ldots, s$ corresponding to the eigenvalues $\lambda_j = 2\cos z_j$ of L, then we get

$$\left\{\frac{\mathrm{d}^{\mathrm{k}}}{\mathrm{d}\lambda^k} E_n(\lambda)\right\}_{\lambda=\lambda_j} = \sum_{p=0}^{k} C_p \left\{T_n e^{inz_j}\left[(in)^p + \sum_{m=1}^{\infty}(i(n+m))^p K_{nm} e^{imz_j}\right]\right\}$$

and

$$U^{(k)}(\lambda_j) = \frac{1}{k!}\left\{\sum_{p=0}^{k} C_p \left[T_n e^{inz_j}\left((in)^p + \sum_{m=1}^{\infty}(i(n+m))^p K_{nm} e^{imz_j}\right)\right]\right\} \tag{17.9}$$

for $k = 0, 1, \ldots, m_j - 1$ and $j = 1, 2, \ldots, s$. Since $\mathrm{Im} z_j > 0$ for $j = 0, 1, \ldots, s$, we obtain

$$\begin{aligned}
&\sum_{n=1}^{\infty}\left\|\frac{1}{(k!)}\sum_{p=0}^{k} C_p T_n e^{inz_j}(in)^p\right\|^2 \\
&\leq \frac{1}{(k!)^2}\left\{\sum_{n=1}^{\infty}\sum_{p=0}^{k}|C_p|\|T_n\| e^{-n\mathrm{Im}z_j} n^p\right\}^2 \\
&\leq H\left\{\sum_{n=1}^{\infty} e^{-n\mathrm{Im}z_j}\left(1 + n + n^2 + \cdots + n^k\right)\right\}^2 \\
&\leq H(p+1)^2\left(\sum_{n=1}^{\infty} e^{-n\mathrm{Im}z_j} n^p\right)^2 < \infty,
\end{aligned} \tag{17.10}$$

where H is a constant. Now, we define the function

$$g_n(z) = \frac{1}{k!}\sum_{p=0}^{k} C_p T_n e^{inz_j}\sum_{m=1}^{\infty}(i(n+m))^p K_{nm} e^{imz_j} \tag{17.11}$$

for $j = 1, 2, \ldots, s$. Using (17.5), we can also write

$$\begin{aligned}
\|g_n(z)\| &\leq \sum_{p=0}^{k}|C_p|\|T_n\| e^{-n\,\mathrm{Im}z_j}\sum_{m=1}^{\infty}|n+m|^p\|K_{nm}\| e^{-m\mathrm{Im}z_j} \\
&\leq \|T_n\| e^{-n\mathrm{Im}z_j}|C_0|\sum_{m=1}^{\infty}\|K_{nm}\| e^{-m\mathrm{Im}z_j}
\end{aligned}$$

$$+\|T_n\|e^{-n\mathrm{Im}z_j}|C_1|\sum_{m=1}^{\infty}(n+m)\|K_{nm}\|e^{-m\mathrm{Im}z_j}$$

$$+\cdots+\|T_n\|e^{-n\mathrm{Im}z_j}|C_k|\sum_{m=1}^{\infty}(n+m)^k\|K_{nm}\|e^{-m\mathrm{Im}z_j}$$

$$\leq \widetilde{C}\|T_n\|e^{-n\mathrm{Im}z_j}\left(\sum_{m=1}^{\infty}\sum_{p=0}^{k}\|K_{nm}\|e^{-m\mathrm{Im}z_j}(n+m)^p\right)$$

$$\leq \backsim Ce^{-n\mathrm{Im}z_j},$$

where $\widetilde{C}=\max\{|C_0|,|C_1|,\ldots,|C_k|\}$ and

$$\backsim C=\widetilde{C}\|T_n\|\left(\sum_{m=1}^{\infty}\sum_{p=0}^{k}\|K_{nm}\|e^{-m\mathrm{Im}z_j}(n+m)^p\right).$$

Then, we get for $j=1,2,\ldots,s$ that

$$\sum_{n=1}^{\infty}\|g_n(z)\|^2\leq\sum_{n=1}^{\infty}\backsim C^2e^{-2n\mathrm{Im}z_j}<\infty. \tag{17.12}$$

It follows from (17.10) and (17.12) that $U_n^{(k)}(\lambda_j)\in\ell_2(\mathbb{N},\mathbb{C}^m)$ for $k=0,1,\ldots,m_j-1$ and $j=1,2,\ldots,s$. Now, we will use the equation like (17.9) for the principal vectors corresponding to the spectral singularities of L for $\lambda_j=2\cos z_j$ and $j=s+1,s+2,\ldots,\nu$. Then we have

$$U_n^{(k)}(\lambda_j)=\frac{1}{k!}\left\{\sum_{p=0}^{k}C_p\|T_n\|e^{inz_j}\left((in)^p+\sum_{m=1}^{\infty}(in+im)^p\|K_{nm}\|e^{imz_j}\right)\right\} \tag{17.13}$$

for $k=0,1,\ldots,m_j-1$ and $j=s+1,s+2,\ldots,\nu$. Since $\mathrm{Im}z_j=0$ for the spectral singularities, $j=s+1,s+2,\ldots,\nu$ of L, we get that

$$\frac{1}{k!}\sum_{n=1}^{\infty}\left\|\sum_{p=0}^{k}C_pT_ne^{inz_j}(in)^p\right\|^2=\infty. \tag{17.14}$$

Also, if we define the function t as

$$t_n(z):=\sum_{p=0}^{k}C_p(in+im)^pKnme^{imz_j},$$

then using (17.1) and (17.5), we obtain

$$\begin{aligned}
\|t_n(z)\| &\le \sum_{p=0}^{k} |C_p| \sum_{m=1}^{\infty} |n+m|^p \|K_{nm}\| \\
&\le \sum_{p=0}^{k} |C_p| \sum_{m=1}^{\infty} C(n+m)^p \sum_{k=n+\lfloor \frac{m}{2} \rfloor} (\|I-A_k\| + \|B_k\|) \\
&= \sum_{p=0}^{k} |C_p| \sum_{m=1}^{\infty} C(n+m)^p \\
&\quad \times \sum_{k=n+\lfloor \frac{m}{2} \rfloor} \exp(-\varepsilon k) \exp(\varepsilon k)(\|I-A_k\| + \|B_k\|) \\
&\le C_1 \sum_{p=0}^{k} \sum_{m=1}^{\infty} (n+m)^p \exp\left(-\frac{\varepsilon}{4}(n+m)\right) \\
&= C_1 \exp\left(-\frac{\varepsilon}{4} n\right) \sum_{m=1}^{\infty} \sum_{p=0}^{k} (n+m)^p \exp\left(-\frac{\varepsilon}{4} m\right) \\
&= A \exp\left(-\frac{\varepsilon}{4} n\right),
\end{aligned}$$

where

$$C_1 = C\widetilde{C} \sum_{k=n+\lfloor \frac{m}{2} \rfloor} \exp(\varepsilon k)(\|I-A_k\| + \|B_k\|)$$

and

$$A = C_1 \sum_{m=1}^{\infty} \sum_{p=0}^{k} (n+m)^p \exp\left(-\frac{\varepsilon}{4} m\right).$$

Therefore, we can write

$$\begin{aligned}
&\frac{1}{k!} \sum_{n=1}^{\infty} \left\| T_n e^{inz_j} \sum_{p=0}^{k} \sum_{m=1}^{\infty} (in+im)^p Knm e^{imz_j} \right\|^2 \\
&\le \frac{1}{k!} \sum_{n=1}^{\infty} \|T_n\|^2 C_1 \exp\left(-\frac{\varepsilon}{2} m\right).
\end{aligned} \tag{17.15}$$

Since

$$\frac{1}{k!}\sum_{n=1}^{\infty}\|T_n\|^2 C_1 \exp\left(-\frac{\varepsilon}{2}m\right) < \infty,$$

we find

$$U_n^{(k)}(\lambda_j) \notin \ell_2(\mathbb{N}, \mathbb{C}^m)$$

for $k = 0, 1, \ldots, m_j - 1$ and $j = s + 1, s + 2, \ldots, \nu$ by using (17.14) and (17.15). This completes the proof.

Let us introduce Hilbert spaces,

$$H_k(\mathbb{N}, \mathbb{C}^m) := \left\{ y = \{y_n\}_{n\in\mathbb{N}} : \quad \sum_{n\in\mathbb{N}} (1 + |n|)^{2k} \|y_n\|^2 < \infty \right\}$$

and

$$H_{-k}(\mathbb{N}, \mathbb{C}^m) := \left\{ u = \{u_n\}_{n\in\mathbb{N}} : \quad \sum_{n\in\mathbb{N}} (1 + |n|)^{-2k} \|u_n\|^2 < \infty \right\}$$

for $k = 0, 1, 2, \ldots$ with the norms

$$\|y\|_k^2 = \sum_{n\in\mathbb{N}} (1 + n)^{2k} \|y_n\|^2$$

and

$$\|u\|_k^2 = \sum_{n\in\mathbb{N}} (1 + n)^{-2k} \|u_n\|^2,$$

respectively. Therefore, $H_0(\mathbb{N}, \mathbb{C}^m) = \ell_2(\mathbb{N}, \mathbb{C}^m)$ and

$$\begin{aligned} H_{k+1}(\mathbb{N}) &\subsetneqq H_k(\mathbb{N}, \mathbb{C}^m) \subsetneqq \ell_2(\mathbb{N}, \mathbb{C}^m) \\ &\subsetneqq H_{-k}(\mathbb{N}, \mathbb{C}^m) \subsetneqq H_{-(k+1)}(\mathbb{N}, \mathbb{C}^m), \quad k = 1, 2, \ldots \end{aligned}$$

Theorem 17.4. *$U_n^{(k)}(\lambda_j) \in H_{k+1}(\mathbb{N}, \mathbb{C}^m)$ for $k = 0, 1, \ldots, m_j - 1$ and $j = s + 1$, $s + 2, \ldots, \nu$.*

Proof. Using (17.13), we obtain

$$\sum_{n=1}^{\infty}(1+n)^{-2(k+1)}\left\|\frac{1}{k!}\sum_{p=0}^{k}C_pT_ne^{inz_j}(in)^p\right\|^2$$
$$\leq K_1\sum_{n=1}^{\infty}(1+n)^{-2(k+1)}(1+n+n^2+\cdots+n^k)^2,$$

where $K_1=\left(\widetilde{C}\|T_n\|\frac{1}{k!}\right)^2$. Since

$$(1+n+n^2+\cdots+n^k)^2<(k+1)^2(n+1)^{2k},$$

we can write

$$\sum_{n=1}^{\infty}(1+n)^{-2(k+1)}\left\|\frac{1}{k!}\sum_{p=0}^{k}C_pT_ne^{inz_j}(in)^p\right\|^2$$
$$\leq K_1(k+1)^2\sum_{n=1}^{\infty}(1+n)^{-2}<\infty \tag{17.16}$$

Also, we easily get

$$\sum_{n=1}^{\infty}(1+n)^{-2(k+1)}\left\|\frac{1}{k!}\sum_{p=0}^{k}C_pT_ne^{inz_j}\sum_{m=1}^{\infty}(im+im)^pK_{nm}e^{imz_j}\right\|^2$$
$$\leq D\sum_{n=1}^{\infty}(1+n)^{-2(k+1)}\exp\left\{-\frac{\varepsilon}{2}n\right\}<\infty, \tag{17.17}$$

where D is a constant. It follows from (17.16) and (17.17) that

$$U_n^{(k)}(\lambda_j)\in H_{k+1}(\mathbb{N},\mathbb{C}^m)$$

for $k=0,1,\ldots,m_j-1$ and $j=s+1,s+2,\ldots,\nu$.

If we choose $m_0=\max\{m_{s+1},m_{s+2},\ldots,m_\nu\}$, then the following result can be given using Theorem 17.4.

Remark 17.5. $U_n^{(k)}(\lambda_j)\in H_{m_0}(\mathbb{N},\mathbb{C}^m)$ for $k=0,1,\ldots,m_j-1$ and $j=s+1,s+2,\ldots,\nu$.

References

1. R.P. Agarwal, *Difference Equations and Inequalities*. Monographs and Textbooks in Pure and Applied Mathematics, 2nd edn., vol. 228 (Marcel Dekker, New York, 2000)
2. R.P. Agarwal, P.J.Y. Wong, *Advanced Topics in Difference Equations*. Mathematics and its Applications, vol. 404 (Kluwer Academic Publishers Group, Dordrecht, 1997)
3. N.I. Akhiezer, *The Classical Moment Problem and Some Related Questions in Analysis*. Translated by N. Kemmer (Hafner Publishing Co., New York, 1965)
4. Y. Aygar, E. Bairamov, Jost solution and the spectral properties of the matrix-valued difference operators. Appl. Math. Comput. **218**(19), 9676–9681 (2012)
5. E. Biramov, Y. Aygar, M. Olgun. Jost solution and the spectrum of the discrete Dirac systems. Bound. Value Probl., 11 pp. (2010). Art. ID 306571
6. S. Clark, F. Gesztesy, W. Renger, Trace formulas and Borg-type theorems for matrix-valued Jacobi and Dirac finite difference operators. J. Differ. Equ. **219**(1), 144–182 (2005)
7. C. Coskun, M. Olgun, Principal functions of nonselfadjoint matrix Sturm-Liouville equations. J. Comput. Appl. Math. **235**(16), 4834–4838 (2011)
8. E. Kir, Spectrum and principal functions of the nonselfadjoint Sturm-Liouville operators with a singular potential. Appl. Math. Lett. **18**(11), 1247–1255 (2005)
9. V.E. Lyance, A differential operator with spectral singularities I, II. Am. Math. Soc. Transl. **60**(2), 185–225 (1967)
10. M.A. Naimark, Investigation of the spectrum and the expansion in eigenfunctions of a nonselfadjoint differential operator of the second order on a semi-axis. Am. Math. Soc. Transl. **16**(2), 103–193 (1960)
11. M. Olgun, C. Coskun, Nonselfadjoint matrix Sturm-Liouville operators with spectral singularities. Appl. Math. Comput. **216**(8), 2271–2275 (2010)
12. M. Olgun, T. Koprubasi, Y. Aygar, Principal functions of nonselfadjoint difference operator with spectral parameter in boundary conditions. Abstr. Appl. Anal., 10 pp. (2011). Art. ID 608329
13. V.P. Serebrjakov, An inverse problem of the scattering theory for difference equations with matrix coefficients. Dokl. Akad. Nauk SSSR **250**(3), 562–565 (1980)
14. S. Yardimci, A note on the spectral singularities of nonselfadjoint matrix-valued difference operators. J. Comput. Appl. Math. **234**(10), 3039–3042 (2010)

Chapter 18
Some Extensions of Preinvexity for Stochastic Processes

Nurgül Okur Bekar, Hande Günay Akdemir, and İmdat İşcan

Abstract In this paper, we introduce some important extensions of preinvexity for stochastic processes, and investigate mutual relation of main preinvex stochastic processes. Besides, we obtain a Kuhn-type result and well-known Hermite–Hadamard integral type inequality for strongly preinvex stochastic processes.

18.1 Introduction

The well-known Hermite–Hadamard integral inequality

$$f\left(\frac{a+b}{2}\right) \leq \frac{1}{b-a}\int_a^b f(x)dx \leq \frac{f(a)+f(b)}{2} \tag{18.1}$$

is used to provide estimations of the mean value of a continuous convex function $f : [a, b] \to \mathbb{R}$. In probabilistic point of view, (18.1) gives a lower bound and an upper bound for $E[f(X)]$ where X is uniformly distributed over the interval $[a, b]$ [5]. In recent years, there has been an extensive interest in providing inequalities involving variety of convexity extensions.

A stochastic process $\{X(t) : t \in I\}$ is a parameterized collection of random variables defined on a common probability space $(\Omega, \Im, P)$. Its parameter t is considered to be time. Then $X(t)$, which can also be shown as $X(t, \omega)$ for $\omega \in \Omega$, is considered to be state or position of the process at time t. For any fixed outcome ω of sample space Ω, the deterministic mapping $t \to X(t, \omega)$ denotes a realization, trajectory, or sample path of the process. For any particular $t \in I$ the mapping

N.O. Bekar (✉)
Departments of Statistics and Mathematics, Giresun University, Turkey
e-mail: nrgokur@gmail.com

H.G. Akdemir • İ. İşcan
Department of Mathematics, Giresun University, Turkey
e-mail: hande.akdemir@giresun.edu.tr; imdat.iscan@giresun.edu.tr

G.A. Anastassiou, O. Duman (eds.), *Computational Analysis*, Springer Proceedings in Mathematics & Statistics 155, DOI 10.1007/978-3-319-28443-9_18

depends ω alone, i.e., then we obtain a random variable. It can be said that $X(t, \omega)$ changes in time in a random manner. We restrict our attention to continuous time stochastic processes, i.e., index set is $I = [0, \infty)$.

There are various ways to define stochastic monotonicity and convexity for stochastic processes, and it is of great importance in optimization, especially in optimal designs, and also useful for numerical approximations when there exist probabilistic quantities [15]. Temporal and spatiotemporal stochastic convexity was defined in [16] and [17], respectively, for discrete time stochastic processes with illustrative examples. Convexity notions in sample path sense can also be found in [4], and the references therein. Time stochastic s-convexity was taken into account in [6] by using order preserving functions of majorizations.

In [13] Nikodem proposed convex stochastic processes and gave some properties which are also known for classical convex functions. Kotrys [8] extended the classical Hermite–Hadamard inequality to convex stochastic processes. Strongly convex stochastic processes was also proposed by Kotrys in [9]. Bekar et al. [2] studied on strongly GA-convex functions and stochastic processes.

Jensen-convex, λ-convex, Wright-convex stochastic processes were introduced in [18, 19]. Two of the significant generalizations of convexity are invex and preinvex functions introduced by Ben-Israel and Mond [3] and Hanson [7], respectively. Akdemir et al. [1] considered preinvex stochastic processes which is a class of the generalized convex stochastic processes, and provided related well-known Hermite–Hadamard integral inequality for the preinvex stochastic process of $X(t, \omega)$ with respect to η, as follows:

$$X\left(\frac{2u + \eta(v, u)}{2}, \cdot\right)$$
$$\leq \frac{1}{\eta(v, u)} \int_{u}^{u+\eta(v,u)} X(t, \cdot)dt \leq \frac{X(u, \cdot) + X(v, \cdot)}{2} \ (a.e.), \tag{18.2}$$

where a given vector function $\eta : I \times I \to \mathbb{R}^n$, I is a non-empty closed subset of $\mathbb{R}^n$.

Our goal in this paper are to establish an important extension of preinvex stochastic processes, to investigate some characteristics of them. The remainder of this article is organized as follows. Section 18.2 contains brief basic definitions which will be required for our further considerations. In Sect. 18.3, we propose mutual relation of main preinvex stochastic processes. Also, we present strongly preinvex stochastic processes, and obtain a Kuhn type result and Hermite–Hadamard type inequality for strongly preinvex stochastic processes in Sect. 18.4.

18.2 Preliminary Discussions

In this section we recall some basic definitions and notions about invex sets, preinvex, invex and strongly preinvex functions, additionally on continuity concepts and differentiability for stochastic processes, mean-square integral of a stochastic process.

Definition 18.1. A non-empty closed subset I of $\mathbb{R}^n$ is said to be invex set with respect to the given vector function $\eta : I \times I \to \mathbb{R}^n$(or η-invex, or η-connected set) if $u + \lambda\eta(v,u) \in I$ for all $u, v \in I$ and $\lambda \in [0,1]$.

Clearly, any convex set is an invex set with respect to $\eta(v,u) = v - u$. Geometrically, endpoints belonging to the set and line segment joining the endpoints are contained in a convex set. Convex sets cannot be disconnected, but invex sets can be disconnected. Definition 18.1 essentially says that there is a path starting from a point u which is contained in I. We do not require that the point v should be the one of endpoints of the path [14].

Definition 18.2. Let $I \subseteq \mathbb{R}^n$ be an invex set with respect to $\eta : I \times I \to \mathbb{R}^n$. Then the function (not necessarily differentiable) $f : I \to \mathbb{R}$ is said to be preinvex with respect to η if

$$f(u + \lambda\eta(v,u)) \leq (1-\lambda)f(u) + \lambda f(v) \tag{18.3}$$

for each $u, v \in I$ and $\lambda \in [0,1]$.

Any convex function is preinvex with respect to $\eta(v,u) = v - u$, but the converse is not necessarily true.

Definition 18.3. For a differentiable function $f : \mathbb{R}^n \to \mathbb{R}$ is said to be invex if there exists a vector function $\eta : \mathbb{R}^n \times \mathbb{R}^n \longrightarrow \mathbb{R}^n$ such that

$$f(x) - f(u) \geq [\nabla f(u)]^T \eta(x,u) \tag{18.4}$$

for all $x, u \in \mathbb{R}^n$.

Any differentiable preinvex function is also an invex function [3]. An invex function may not be preinvex, $f(x) = \exp(x)$ is a counterexample, it is invex with respect to $\eta(x,u) = -1$, but not preinvex with respect to same η.

Mohan and Neogy [12] proved that an invex function is also preinvex under the following Condition C.

Condition C. *Let $\eta : I \times I \to R^n$. It is told that the function η satisfies Condition C if*

(C1) $\eta(u, u + \lambda\eta(v,u)) = -\lambda\eta(v,u)$,
(C2) $\eta(v, u + \lambda\eta(v,u)) = (1-\lambda)\eta(v,u)$,

for all $u, v \in I$ and $\lambda \in [0,1]$.

Additionally, note that from Condition C, we have

$$\eta(u + \lambda_2\eta(v,u), u + \lambda_1\eta(v,u)) = (\lambda_2 - \lambda_1)\eta(v,u) \tag{18.5}$$

for all $u, v \in I$ and $\lambda_1, \lambda_2 \in [0,1]$. See [11] for more detail on preinvex and invex functions.

Definition 18.4. Let $I \subseteq \mathbb{R}^n$ be an invex set with respect to $\eta : I \times I \to \mathbb{R}^n$. Then the function $f : I \to \mathbb{R}$ is called strongly preinvex with modulus $c > 0$ on I if

$$f(u + \lambda\eta(v, u)) \leq (1 - \lambda)f(u) + \lambda f(v) - c\lambda(1 - \lambda)\eta^2(v, u) \tag{18.6}$$

for all $u, v \in I$ and $\lambda \in [0, 1]$.

Obviously, every strongly preinvex function is preinvex.

Throughout this paper, we assume that $I \subseteq [0, \infty)$ is a η-invex interval and the function η satisfies Condition C unless stated otherwise.

18.3 Mutual Relation of Main Preinvex Stochastic Processes

In this section, we give inter-preinvex stochastic processes relation. Let's remember concepts related to preinvexity for stochastic processes in [1, 9].

Definition 18.5 (See [9]). A real-valued stochastic process $\{X(t)|t \in I\}$ is said to be

(i) *continuous in probability* in I if

$$P - \lim_{t\to t_0} X(t, \cdot) = X(t_0, \cdot)$$

(where $P - \lim$ denotes limit in probability), or equivalently

$$\lim_{t\to t_0} P\{|X(t, \cdot) - X(t_0, \cdot)| > \varepsilon\} = 0$$

for any arbitrary small enough $\varepsilon > 0$ and all $t_0 \in I$.

(ii) *mean-square continuous* (or *continuous in quadratic mean*) in I if

$$\lim_{t\to t_0} E[(X(t) - X(t_0))^2] = 0$$

such that $E[X(t)^2] < \infty$, for all $t_0 \in I$.

(iii) *mean-square differentiable* in I if it is mean square continuous and there exists a process $X'(t, \cdot)$("speed" of the process) such that

$$\lim_{t\to t_0} E\left[\left(\frac{X(t) - X(t_0)}{t - t_0} - X'(t_0)\right)^2\right] = 0.$$

Different types of continuity concepts can be defined for stochastic processes. Surely (everywhere) and almost surely (almost everywhere or sample path) convergences are rarely used in applications due to restrictive requirement, that is, as $t \to t_0, X(t, \omega)$ has to approach $X(t_0, \omega)$ for each outcome $\omega \in S \subseteq \Omega$. For further reading on stochastic calculus, reader may refer to [20].

Definition 18.6 (See [1]). Let $X : I \times \Omega \to \mathbb{R}$ be a stochastic process (not necessarily mean-square differentiable) on η. $X(t, \cdot)$ is called preinvex with respect to η if

$$X(u + \lambda\eta(v, u), \cdot) \leq (1 - \lambda)X(u, \cdot) + \lambda X(v, \cdot)(a.e.) \tag{18.7}$$

for all $u, v \in I$ and $\lambda \in [0, 1]$.

For a preinvex stochastic process, the inequality (18.7) holds almost everywhere on Ω, i.e., almost every sample path of X will be a preinvex function. In the inequality (18.7), if λ is fixed number in $(0, 1)$, then X is called λ-preinvex stochastic process, and mid-preinvex (Jensen-preinvex) stochastic process for $\lambda = \frac{1}{2}$.

If we choose $\eta(v, u) = v - u$, then preinvex $X(t, \cdot)$ is also a convex stochastic process, that is, class of convex stochastic processes is contained by the class of preinvex stochastic processes. Now, let us denote by

- P—the set of all preinvex stochastic processes,
- P_λ—the set of all λ-preinvex stochastic processes,
- $P_{\frac{1}{2}}$—the set of all mid-preinvex stochastic processes,
- $P^C_{\frac{1}{2}}$—the set of all mid-preinvex stochastic processes by assuming the function η satisfies Condition C.

Theorem 18.7. $P \subset P_\lambda \subset P^C_{\frac{1}{2}}$.

Proof. If $X \in P$, then X satisfies (18.7) for a fixed $\lambda \in (0, 1)$, so the first inclusion can be proven trivially. As regards the second one, let $\lambda \in (0, 1)$ be a fixed number and $X \in P_\lambda$. If we take

$$A = u + \frac{\lambda + 1}{2}\eta(v, u)$$

and

$$B = u + \frac{\lambda}{2}\eta(v, u)$$

for $u, v \in (a, b)$, then using Condition C, we get

$$\begin{aligned} A + \lambda\eta(B, A) &= u + \frac{\lambda + 1}{2}\eta(v, u) \\ &\quad + \lambda\eta\left(u + \frac{\lambda}{2}\eta(v, u), u + \frac{\lambda + 1}{2}\eta(v, u)\right) \\ &= u + \frac{\lambda + 1}{2}\eta(v, u) + \lambda\left(-\frac{1}{2}\eta(v, u)\right) \\ &= u + \frac{1}{2}\eta(v, u). \end{aligned}$$

Thus using definition of λ-preinvexity, we get

$$
\begin{aligned}
X\left(u+\frac{1}{2}\eta(v,u),\cdot\right) &= X(A+\lambda\eta(B,A),\cdot) \\
&\le (1-\lambda)X(A,\cdot)+\lambda X(B,\cdot) \\
&= (1-\lambda)X\left(u+\frac{\lambda+1}{2}\eta(v,u),\cdot\right) \\
&\quad +\lambda X\left(u+\frac{\lambda}{2}\eta(v,u),\cdot\right) \\
&\le (1-\lambda)\left(\left(1-\frac{\lambda+1}{2}\right)X(u,\cdot)+\frac{\lambda+1}{2}X(v,\cdot)\right) \\
&\quad +\lambda\left(\left(1-\frac{\lambda}{2}\right)X(u,\cdot)+\frac{\lambda}{2}X(v,\cdot)\right) \\
&= \frac{X(u,\cdot)+X(v,\cdot)}{2} \quad (a.e.)
\end{aligned}
$$

which ends the proof.

18.4 Strongly Preinvex Stochastic Processes and Related Well-Known Results for Them

In this section we propose an important extension of preinvexity for stochastic processes which is called strongly preinvexity. Moreover, we obtain a Kuhn type result and Hermite–Hadamard inequality for these processes.

Definition 18.8. Let $C:\Omega\to\mathbb{R}$ denote a positive random variable, $X:I\times\Omega\to\mathbb{R}$ be a stochastic process. $X:I\times\Omega\to\mathbb{R}$ is called strongly preinvex with modulus $C(\cdot)$, if the following inequality is satisfied

$$
\begin{aligned}
&X(u+\lambda\eta(v,u),\cdot) \\
&\quad \le (1-\lambda)X(u,\cdot)+\lambda X(v,\cdot)-C(\cdot)\lambda(1-\lambda)\eta^2(v,u) \;\; (a.e.) \qquad (18.8)
\end{aligned}
$$

for all $u,v\in I$ and $\lambda\in[0,1]$.

Note that, if (18.8) holds for a fixed number $\lambda\in(0,1)$, then we describe that the process is strongly λ-preinvex with modulus $C(\cdot)$. Assuming that (18.8) holds only for $\lambda=\frac{1}{2}$, then $X:I\times\Omega\to\mathbb{R}$ is called strongly mid-preinvex with modulus $C(\cdot)$:

$$
X\left(\frac{2u+\eta(v,u)}{2},\cdot\right)\le\frac{X(u,\cdot)+X(v,\cdot)}{2}-\frac{C(\cdot)}{4}\eta^2(v,u) \;\; (a.e.).
$$

Obviously, for a strongly preinvex stochastic process, it can be immediately found that X is λ-preinvex for a fixed number $\lambda \in (0,1)$, and also mid-preinvex for $\lambda = \frac{1}{2}$ from Definition 18.8.

Lemma 18.9. *Let $[u, u+\eta(v,u)] \subset I$. A stochastic process $X : I \times \Omega \to \mathbb{R}$ is strongly λ-preinvex (strongly preinvex, respectively) with modulus $C(\cdot)$ if and only if the stochastic process*

$$Y : [u, u+\eta(v,u)] \times \Omega \to \mathbb{R}$$

defined by

$$Y(t,\cdot) = X(t,\cdot) - C(\cdot)\eta^2(t,u)$$

is λ-preinvex (preinvex, respectively).

Proof. In the first part of the proof, let's assume that X is strongly λ-preinvex and $Y(t,\cdot) = X(t,\cdot) - C(\cdot)\eta^2(t,u)$. Then for $t = u + \lambda\eta(v,u)$, we get

$$\begin{aligned}
&Y(u+\lambda\eta(v,u),\cdot)\\
&\quad= X(u+\lambda\eta(v,u),\cdot) - C(\cdot)\eta^2(u+\lambda\eta(v,u),u)\\
&\quad= X(u+\lambda\eta(v,u),\cdot) - C(\cdot)\lambda^2\eta^2(v,u)\\
&\quad\leq (1-\lambda)X(u,\cdot) + \lambda X(v,\cdot)\\
&\qquad - C(\cdot)\lambda(1-\lambda)\eta^2(v,u) - C(\cdot)\lambda^2\eta^2(v,u)\\
&\quad= (1-\lambda)X(u,\cdot) + \lambda X(v,\cdot) - C(\cdot)\lambda\eta^2(v,u)\\
&\quad= \lambda\left[X(v,\cdot) - C(\cdot)\eta^2(v,u)\right] + (1-\lambda)\left[X(u,\cdot) - C(\cdot)\eta^2(u,u)\right]\\
&\quad= \lambda Y(v,\cdot) + (1-\lambda)Y(u,\cdot)(a.e.),
\end{aligned}$$

and so $Y(t,\cdot)$ is λ-preinvex. The proof of the second part is similar, so we omit it.

18.4.1 A Kuhn-Type Result for Strongly Preinvex Stochastic Processes

The classical result due to Kuhn states in [10] that if $f : I \to \mathbb{R}$ fulfills for some fixed $\lambda \in (0,1)$ and for all $u, v \in I$, f is a λ-convex function then f is also mid-convex. Furthermore, Skowronski proved in [18] that a λ-convex stochastic process is also mid-convex.

Now, we prove the counterparts of these facts for strongly λ-preinvex stochastic processes in following theorem.

Theorem 18.10. *Let $\lambda \in (0,1)$ be a fixed number and $X : I \times \Omega \to \mathbb{R}$ be a strongly λ-preinvex stochastic process with modulus $C(\cdot)$. Then X is strongly mid-preinvex stochastic process with modulus $C(\cdot)$.*

Proof. Assume that X is strongly λ-preinvex stochastic process with modulus $C(.)$, then Lemma 18.9 yields that the process

$$Y(t,\cdot) = X(t,\cdot) - C(\cdot)\eta^2(t,u)$$

is λ-preinvex. By Theorem 18.7, Y is also mid-preinvex stochastic process by assuming the function η satisfies Condition C, which means that

$$Y\left(\frac{2u+\eta(v,u)}{2},\cdot\right) \le \frac{Y(u,\cdot)+Y(v,\cdot)}{2} \quad (a.e.).$$

Considering the definition of $Y(t,\cdot)$ for $t = \dfrac{2u+\eta(v,u)}{2}$, we have

$$X\left(\frac{2u+\eta(v,u)}{2},\cdot\right) - C(\cdot)\eta^2\left(\frac{2u+\eta(v,u)}{2},u\right)$$
$$\le \frac{X(u,\cdot) - C(\cdot)\eta^2(u,u) + X(v,\cdot) - C(\cdot)\eta^2(v,u)}{2} \quad (a.e.).$$

Finally, using Condition C, and $\eta(u,u) = 0$ for all $u \in I$, we can easily obtain that X is strongly mid-preinvex stochastic process with modulus $C(\cdot)$.

18.4.2 Hermite–Hadamard Inequality for Strongly Preinvex Stochastic Processes

Now, we would like to prove Hermite–Hadamard type inequality for strongly preinvex stochastic processes. Let's start some essential definitions.

Definition 18.11 (See [9]). Let $X : I \times \Omega \to \mathbb{R}$ be a stochastic process with $E[X(t)^2] < \infty$ for all $t \in I$. Let

$$[a,b] \subset I, a = t_0 < t_1 < \cdots < t_n = b$$

be a partition of $[a,b]$ and $\Theta_k \in [t_{k-1}, t_k]$ arbitrarily for $k = 1,\ldots,n$. A random variable $Y : \Omega \to \mathbb{R}$ is called mean-square integral of the process $X(t)$ on $[a,b]$ if the following identity holds:

$$\lim_{n\to\infty} E\left[\left(\sum_{k=1}^{n} X(\Theta_k)(t_k - t_{k-1}) - Y\right)^2\right] = 0. \tag{18.9}$$

Then we can write

$$\int_a^b X(t,\cdot)dt = Y(\cdot)\ (a.e.).$$

Mean square integral operator is increasing, that is,

$$\int_a^b X(t,\cdot)dt \le \int_a^b Z(t,\cdot)dt \quad (a.e.)$$

where $X(t,\cdot) \le Z(t,\cdot)$ $(a.e.)$ in $[a,b]$.

Definition 18.12 (See [1]). Let $X : I \times \Omega \to \mathbb{R}$ be a mean square differentiable stochastic process. Then $X(t,\cdot)$ is called invex with respect to η if

$$X(t,\cdot) - X(t_0,\cdot) \ge X'(t_0,\cdot)\eta(t,t_0)(a.e.) \tag{18.10}$$

for all $t, t_0 \in I$.

If X is a mean square differentiable stochastic process, then it is also mean square continuous by Definition 18.5. Mean square continuity guarantees continuity in probability, as $t \to t_0$, for any small enough $\varepsilon > 0$ and all $t_0 \in I$.

For a preinvex stochastic process, from now on, let us assume that η is skew-symmetric, i.e., $\eta(t,t_0) = -\eta(t_0,t)$ for all $t, t_0 \in intI$, and $\eta(t,t_0) \ge 0$ for such $t \ge t_0$.

Now, in order to establish our argument, we consider the following lemma.

Lemma 18.13. *If stochastic process $X : I \times \Omega \to \mathbb{R}$ has the form*

$$X(t,\cdot) = C(\cdot)\eta^2(t,u)$$

where $C(\cdot)$ is random variable such that $E[C^2] < \infty$, and $[u, u+\eta(v,u)] \subset I$, then

$$\int_u^{u+\eta(v,u)} X(t,\cdot)\,dt = C(\cdot)\frac{\eta^3(v,u)}{3}\ (a.e.). \tag{18.11}$$

Proof. By dividing the interval $[u, u+\eta(v,u)]$ into n subintervals and choosing endpoints of the subintervals as points in the partition, we yield

$$E\left(\sum_{k=1}^{n} X(\Theta_k)\cdot(t_k - t_{k-1}) - C\frac{\eta^3(v,u)}{3}\right)^2$$
$$= E\left(\sum_{k=1}^{n} C\eta^2(\Theta_k,u)\cdot(t_k - t_{k-1}) - C\frac{\eta^3(v,u)}{3}\right)^2$$

$$= E[C^2]\left(\sum_{k=1}^{n}\eta^2\left(u+\frac{k}{n}\eta(v,u),u\right)(t_k-t_{k-1})-\frac{\eta^3(v,u)}{3}\right)^2$$

$$= E[C^2]\left(\frac{n(n+1)(2n+1)}{6n^3}\eta^3(v,u)-\frac{\eta^3(v,u)}{3}\right)^2.$$

If $n \to \infty$, then the above expression tends to zero, because of the definition of the Riemann integral. This completes the proof.

Theorem 18.14. *Let the stochastic process $X : I \times \Omega \to \mathbb{R}$ be a strongly preinvex with modulus $C(\cdot)$ and mean-square continuous in the interval I. Then we obtain Hermite–Hadamard type inequality for any $u, v \in I$, as follows:*

$$X\left(\frac{2u+\eta(v,u)}{2},\cdot\right)+\frac{C(\cdot)}{12}\eta^2(v,u)$$

$$\leq \frac{1}{\eta(v,u)}\int_u^{u+\eta(v,u)} X(t,\cdot)\,dt$$

$$\leq \frac{X(u,\cdot)+X(v,\cdot)}{2}-\frac{C(\cdot)}{6}\eta^2(v,u) \quad (a.e.).$$

Proof. The process X is strongly preinvex with modulus $C(.)$, the process

$$Y(t,.) = X(t,.) - C(.)\eta^2(t,u)$$

is preinvex by Lemma 18.9. Using (18.2), we get

$$Y\left(\frac{2u+\eta(v,u)}{2},\cdot\right) \leq \frac{1}{\eta(v,u)}\int_u^{u+\eta(v,u)} Y(t,\cdot)\,dt$$

$$\leq \frac{Y(u,\cdot)+Y(v,\cdot)}{2} \quad (a.e.).$$

Considering the definition of $Y(t,.)$ for $t = \dfrac{2u+\eta(v,u)}{2}$, we have

$$X\left(\frac{2u+\eta(v,u)}{2},\cdot\right)-C(\cdot)\eta^2\left(\frac{2u+\eta(v,u)}{2},u\right)$$

$$\leq \frac{1}{\eta(v,u)}\int_u^{u+\eta(v,u)} (X(t,\cdot)-C(\cdot)\eta^2(t,u))\,dt$$

$$\leq \frac{X(u,\cdot)-C(\cdot)\eta^2(u,u)+X(v,\cdot)-C(\cdot)\eta^2(v,u)}{2}.$$

Furthermore, using Lemma 18.13 and taking into account Condition C and $\eta(u,u)=0$ for all $u \in I$, we get

$$X\left(\frac{2u+\eta(v,u)}{2},.\right)-\frac{C(\cdot)}{4}\eta^2(v,u)$$

$$\leq \frac{1}{\eta(v,u)}\int\limits_{u}^{u+\eta(v,u)} X(t,\cdot)\,dt-\frac{C(\cdot)}{3}\eta^2(v,u)$$

$$\leq \frac{X(u,\cdot)+X(v,\cdot)}{2}-\frac{C(\cdot)}{2}\eta^2(v,u).$$

Consequently, after some rearrangement, we obtain Hermite–Hadamard type inequality for the process X.

18.5 Conclusion

In this paper, we propose strongly preinvex stochastic processes. We also obtain a Kuhn-type result and Hermite–Hadamard type inequality for strongly preinvex stochastic processes under some suitable conditions. Preinvexity concepts are particularly interesting from optimization viewpoint, since it provides a broader setting to study the mathematical programming problems.

As special cases, one can obtain several new and correct versions of the previously known results for various classes of these stochastic processes. Applying this type inequalities for stochastic processes is another promising direction for future research.

References

1. H.G. Akdemir, N.O. Bekar, I. Iscan, On preinvexity for stochastic processes. Stat.: J. Turkish Stat. Assoc. **7**(1), 15–22 (2014)
2. N.O. Bekar, H.G. Akdemir, I. Iscan, On strongly GA-convex and functions and stochastic processes, in *AIP Conference Proceedings*, ed. by A. Ashyralyev, E. Malkowsky, vol. 1611(1) (AIP Publishers, Melville, 2014), pp. 363–368
3. A. Ben-Israel, B. Mond, What is invexity? J. Aust. Math. Soc. **28**(1), 1–9 (1986)
4. C.S. Chang, X.L. Chao, M. Pinedo, J.G. Shanthikumar, Stochastic convexity for multidimensional processes and its applications. IEEE Trans. Autom. Control **36**, 1341–1355 (1991)
5. J. De la Cal, J. Carcamo, Multidimensional Hermite-Hadamard inequalities and the convex order. J. Math. Anal. Appl. **324**, 248–261 (2006)
6. M. Denuit, Time stochastic s-convexity of claim processes. Insurance Math. Econ. **26**(2–3), 203–211 (2000)
7. M.A. Hanson, On sufficiency of the Kuhn-Tucker conditions. J. Math. Anal. Appl. **80**(2), 545–550 (1981)

8. D. Kotrys, Hermite-Hadamard inequality for convex stochastic processes. Aequationes Math. **83**, 143–151 (2012)
9. D. Kotrys, Remarks on strongly convex stochastic processes. Aequationes Math. **86**, 91–98 (2013)
10. N. Kuhn, A note on t-convex functions, in *General Inequalities 4*. International Schriftenreihe Numerical Mathematics, vol. 71 (Birkhauser, Basel, 1984), pp. 269–276
11. S.K. Mishra, G. Giorgi, *Invexity and Optimization, Nonconvex Optimization and Its Applications*, vol. 88 (Springer, Berlin, 2008)
12. S.R. Mohan, S.K. Neogy, On invex sets and preinvex functions. J. Math. Anal. Appl. **189**(3), 901–908 (1995)
13. K. Nikodem, On convex stochastic processes. Aequationes Math. **20**, 184–197 (1980)
14. M.A. Noor, On Hadamard integral inequalities involving two log-preinvex functions. J. Inequal. Pure Appl. Math. **8** (3), 1–6 (2007)
15. M. Shaked, J.G. Shanthikumar, Stochastic convexity and its applications. Adv. Appl. Probab. **20**, 427–446 (1988)
16. M. Shaked, J.G. Shanthikumar, Temporal stochastic convexity and concavity. Stochastic Process. Appl. **27**, 1–20 (1988)
17. J. G. Shanthikumar, D.D. Yao, Spatiotemporal convexity of stochastic processes and applications. Probab. Eng. Inform. Sci. **6**, 1–16 (1992)
18. A. Skowronski, On some properties of J-convex stochastic processes. Aequationes Math. **44**, 249–258 (1992)
19. A. Skowronski, On Wright-convex stochastic processes. Ann. Math. Silesianae **9**, 29–32 (1995)
20. J.J. Shynk, *Probability, Random Variables, and Random Processes: Theory and Signal Processing Applications* (Wiley, New York, 2013)
21. M. Tomar, E. Set, N.O. Bekar, On Hermite-Hadamard-type inequalities for strongly-log convex stochastic processes. J. Global Eng. Stud. **1**(2), 53–61 (2014)

Chapter 19
On One Boundary-Value Problem with Two Nonlocal Conditions for a Parabolic Equation

Olga Danilkina

Abstract This work is concerned with a boundary-value problem for a parabolic equation with nonlocal integral conditions of the second kind. Existence and uniqueness of a generalized solution are proved.

19.1 Introduction

In recent years, nonlocal problems for PDEs have received a great deal of attention as a convenient way of description of different physical phenomena. These problems arise in a wide variety of applications, including heat conduction, processes in liquid plasma, dynamics of ground waters, thermo-elasticity and some technological processes.

In this paper, our main interest lies in the field of nonlocal problems with integral conditions that generalizes the discrete case. We mention the first papers in this area [6, 14] devoted to problems for parabolic equations. Then these results were extended [2, 7–11, 13, 16, 26, 28, 29]. For papers related to nonlocal problems for other evolution equations, we refer the reader to [1, 3–5, 12, 17, 19–25, 27].

In [25], the author studied two problems for the hyperbolic equation

$$u_{tt} - u_{xx} + c(x,t)u = f(x,t)$$

with the initial condition

$$u(x,0) = \varphi(x), \quad u_t(x,0) = \psi(x)$$

and two types of nonlocal conditions.

O. Danilkina (✉)
Department of Mathematics, The University of Dodoma, Tanzania
e-mail: olga.danilkina@gmail.com; olga.danilkina@udom.ac.tz

G.A. Anastassiou, O. Duman (eds.), *Computational Analysis*, Springer Proceedings in Mathematics & Statistics 155, DOI 10.1007/978-3-319-28443-9_19

Case 1 :

$$\int_0^l K_i(x)u(x,t)\,dx = 0, \quad i = 1,2.$$

Case 2 :

$$u_x(0,t) - \int_0^l K_1(x,t)u(x,t)\,dx = 0,$$

$$u_x(l,t) - \int_0^l K_2(x,t)u(x,t)\,dx = 0.$$

Motivated by the ideas of Pulkina [25], in this paper we extend the results of Pulkina [25] to a special class of boundary-value problems with nonlocal integral conditions for parabolic equations. The proof of the main result is based on the method of energy estimates and the Faedo–Galerkin approximations.

19.2 Preliminaries

In the cylinder $Q_T = \{(x,t)\colon x \in (0,l),\ t \in (0,T)\}$ we consider the problem for the equation

$$u_t = u_{xx} + c(x,t)u \tag{19.1}$$

with the initial condition

$$u(x,0) = \varphi(x) \tag{19.2}$$

and the nonlocal conditions

$$u_x(0,t) = \int_0^l K_1(x,t)u_x(x,t)\,dx + \int_0^l M_1(x,t)u(x,t)\,dx, \tag{19.3}$$

$$u_x(l,t) = \int_0^l K_2(x,t)u_x(x,t)\,dx + \int_0^l M_2(x,t)u(x,t)\,dx. \tag{19.4}$$

In this paper, we shall assume that the following assumptions are satisfied.

(*A*1) $c(x,t) \in C(\overline{Q}_T)$, $\varphi(x) \in C^1[0,l]$;
(*A*2) $K_1(x,t)$, $K_2(x,t)$, $M_1(x,t)$, $M_2(x,t) \in C^1(\overline{Q}_T)$.

We note that presence of partial derivatives on the right-hand side of the nonlocal conditions (19.3), (19.4) can cause difficulties in constructing of a priori estimates. Therefore, to avoid this we integrate by parts in (19.3), (19.4) and obtain

$$u_x(0,t) = K_1(l,t)u(l,t) - K_1(0,t)u(0,t) + \int_0^l R_1(x,t)u(x,t)\,dx, \tag{19.5}$$

$$u_x(l,t) = K_2(l,t)u(l,t) - K_2(0,t)u(0,t) + \int_0^l R_2(x,t)u(x,t)\,dx, \tag{19.6}$$

where $R_1(x,t) = M_1(x,t) - (K_1(x,t))_x$, $R_2(x,t) = M_2(x,t) - (K_2(x,t))_x$.

Let $W_2^{1,0}(Q_T)$ be the usual Sobolev space. We define the space $V_2(Q_T)$ which consists of elements of $W_2^{1,0}(Q_T)$ with the norm

$$|u|^2 = ess \sup_{0\le t\le T} \int_0^l u^2(x,t)\,dt + \int_{Q_T} u_x^2(x,t)\,dx\,dt.$$

Definition 19.1. A function $u(x,t) \in V_2(Q_T)$ is said to be a generalized solution to the problem (19.1), (19.2), (19.5), (19.6) provided for any function $\eta(x,t) \in W_2^1(Q_T)$, $\eta(x,T) = 0$, the following integral identity holds:

$$\begin{aligned}
&\int_{Q_T} (-u\eta_t + u_x\eta_x - cu\eta)\,dx\,dt \\
&= \int_0^l \varphi(x)\eta(x,0)\,dx + \int_0^T (K_1(0,t)\eta(0,t) - K_2(0,t)\eta(l,t))\,u(0,t)\,dt \\
&+ \int_0^T (K_2(l,t)\eta(l,t) - K_1(l,t)\eta(0,t))\,u(l,t)\,dt \\
&+ \int_{Q_T} (R_1(x,t)\eta(l,t) - R_2(0,t)\eta(0,t))\,u(x,t)\,dx\,dt.
\end{aligned} \tag{19.7}$$

Lemma 19.2. *Let a function $u(x,t)$ be a solution to the problem* (19.1), (19.2), (19.5), (19.6). *Then the following identity holds:*

$$\frac{1}{2}\int\limits_0^l u^2(x,\tau)\,dx + \int\limits_{Q_T} u_x^2\,dx\,dt = \frac{1}{2}\int_0^l \varphi^2(x)\,dx + \int\limits_{Q_T} cu^2\,dx\,dt$$

$$+\int\limits_0^\tau K_2(l,t)u^2(l,t)\,dt - \int\limits_0^\tau K_1(l,t)u(0,t)u(l,t)\,dt$$

$$+\int\limits_0^\tau K_1(0,t)u^2(0,t)\,dt - \int\limits_0^\tau K_2(0,t)u(0,t)u(l,t)\,dt$$

$$+\int\limits_{Q_T} R_1(x,t)u(x,t)\,dx\,u(l,t)\,dt$$

$$-\int\limits_{Q_T} R_2(x,t)u(x,t)\,dx\,u(0,t)\,dt$$

for a.e. $\tau \in [0,T]$.

Proof. Let a function $u(x,t) \in W_2^1(Q_\tau)$ and satisfy the integral identity (19.7) for all functions $\eta(x,t) \in W_2^1(Q_T)$, $\eta(x,T) = 0$. For an arbitrary $\tau \in [0,T]$, we take

$$\eta(x,t) = \begin{cases} u(x,t), & 0 < t < \tau, \\ 0, & \tau \le t < T. \end{cases}$$

After integration by parts in (19.7) we obtain

$$\frac{1}{2}\int\limits_0^l u^2(x,\tau)\,dx + \int\limits_{Q_\tau} u_x^2\,dx\,dt = \frac{1}{2}\int\limits_0^l \varphi^2(x)\,dx + \int\limits_{Q_\tau} cu^2\,dx\,dt$$

$$+\int\limits_0^\tau K_2(l,t)u^2(l,t)\,dt - \int\limits_0^\tau K_1(l,t)u(0,t)u(l,t)\,dt$$

$$+\int\limits_0^\tau K_1(0,t)u^2(0,t)\,dt - \int\limits_0^\tau K_2(0,t)u(0,t)u(l,t)\,dt$$

$$+\int\limits_{Q_\tau} R_1(x,t)u(x,t)\,dx\,u(l,t)\,dt$$

$$-\int\limits_{Q_\tau} R_2(x,t)u(x,t)\,dx\,u(0,t)\,dt. \qquad (19.8)$$

We shall prove that a function $u(x,t) \in V_2(Q_T)$ also satisfies (19.8). To this aim consider a sequence $v^m(x,t) \in W_2^1(Q_T)$ which satisfies the identity (19.7) and hence, (19.8), that is

$$\begin{aligned}\frac{1}{2}\int_0^l (v^m)^2(x,\tau)\,dx + \int_{Q_\tau} (v^2)_x^2\,dx\,dt = \frac{1}{2}\int_0^l \varphi^2(x)\,dx + \int_{Q_\tau} c(v^m)^2\,dx\,dt \\ + \int_0^\tau K_2(l,t)(v^m)^2(l,t)\,dt \\ - \int_0^\tau K_1(l,t)v^m(0,t)v^m(l,t)\,dt \\ + \int_0^\tau K_1(0,t)(v^m)^2(0,t)\,dt \\ - \int_0^\tau K_2(0,t)v^m(0,t)v^m(l,t)\,dt \\ + \int_{Q_\tau} R_1(x,t)v^m(x,t)\,dx\,v^m(l,t)\,dt \\ - \int_{Q_\tau} R_2(x,t)v^m(x,t)\,dx\,v^m(0,t)\,dt.\end{aligned} \tag{19.9}$$

Note that $W_2^1(Q_\tau)$ is dense in $V_2^{1,0}(Q_T)$ [15] and hence, in $V_2(Q_T)$. Therefore, there exists a function $u^* \in V_2(Q_T)$ such that $|v^m - u^*|_{Q_T} \to 0$ as $m \to \infty$:

$$ess \sup_{0\le t\le T} \int_0^l (v^m - u^*)^2\,dx + \int_{Q_T} (v^m - u^*)_x^2\,dx\,dt \to 0.$$

It implies that $v^m(x,t) \to u^*$ strongly in $L_2(0,l)$ and $v_x^m(x,t) \to u_x^*$ strongly in $L_2(Q_T)$. We also note that $v^m \to u^*$ in $L_2(Q_T)$. Our next aim is to estimate terms on the right-hand side of (19.9). The assumptions $(A1)$, $(A2)$ imply that there exist positive numbers c_1, k_2 such that $|c(x,t)| \le c_1$, $|K_2(x,t)| \le k_2$. Applying ε-inequality [18]

$$v^2\big|_{x=0,\,x=l} \le \int_0^l \left(\varepsilon v_x^2 + C(\varepsilon)v^2\right)\,dx$$

we obtain

$$\begin{aligned}\left|\int_0^\tau K_2(l,t)(v^m(l,t))^2\,dt\right| &\le k_2\int_0^\tau (v^m(l,t))^2\,dt\\ &\le k_2\varepsilon||v_x^m||^2 + k_2C_\varepsilon||v^m||^2\\ &\le H_1\left(||v_x^m||^2 + ||v^m||^2\right),\end{aligned}\tag{19.10}$$

where $H_1 = \max\{k_2\varepsilon,\ k_2C_\varepsilon\}$. Similarly, we derive the estimates

$$\left|\int_0^\tau K_1(0,t)(v^m(0,t))^2\,dt\right| \le H_2\left(||v_x^m||^2 + ||v^m||^2\right),\tag{19.11}$$

$$\left|\int_0^\tau (K_1(l,t) + K_2(0,t))v^m(0,t)v^m(l,t)\,dt\right| \le H_3\left(||v_x^m||^2 + ||v^m||^2\right).\tag{19.12}$$

To obtain an estimate for the term

$$\int_{Q_\tau} R_1(x,t)v^m(x,t)\,dx\,v^m(l,t)\,dt,$$

we use Young's inequality and the Cauchy–Schwartz inequality and then

$$\begin{aligned}&\left|\int_{Q_\tau} R_1(x,t)v^m(x,t)\,dx\,v^m(l,t)\,dt\right|\\ &\quad\le \frac{l}{2}\int_0^\tau (v^m(l,t))^2\,dt + \frac{r_1}{2}\int_{Q_\tau}(v^m(x,t))^2\,dx\,dt.\end{aligned}$$

Similarly,

$$\left|\int_{Q_\tau} R_1(x,t)v^m(x,t)\,dx\,v^m(l,t)\,dt\right| \le H_4\left(||v_x^m||^2 + ||v^m||^2\right)\tag{19.13}$$

and

$$\left|\int_{Q_\tau} R_2(x,t)v^m(x,t)\,dx\,v^m(0,t)\,dt\right| \le H_5\left(||v_x^m||^2 + ||v^m||^2\right),\tag{19.14}$$

where the constants H_i, $i = 2, 3, 4, 5$ do not depend on m. Furthermore, we note that

$$\left| \int_{Q_\tau} c(v^m)^2 \, dx \, dt \right| \le c_1 ||v^m||^2. \tag{19.15}$$

Therefore, using strong convergence $v^m(x,t) \to u^*$ and the estimates (19.10)–(19.15) we pass to the limit as $m \to \infty$ in and obtain (19.8) for $u(x,t) \in V_2(Q_T)$.

Lemma 19.3. *Let a function $u(x,t)$ be a solution to the problem* (19.1), (19.2), (19.5), (19.6). *Then there exists $H > 0$ such that $|u|_{Q_T} \le H$.*

Proof. By Lemma 19.2, the solution $u(x,t)$ satisfies the integral identity (19.8). We shall estimate the right-hand side of (19.8). Note that for $\varepsilon_1, \varepsilon_2 > 0$

$$u^2(0,t) \le \int_0^l \left(\varepsilon_1 u_x^2 + C(\varepsilon_1) u^2\right) dx, \; u^2(l,t) \le \int_0^l \left(\varepsilon_2 u_x^2 + C(\varepsilon_2) u^2\right) dx$$

and hence,

$$|u(0,t)u(l,t)| \le \frac{1}{2} \int_0^l \left((\varepsilon_1 + \varepsilon_2) u_x^2 + (C(\varepsilon_1) + C(\varepsilon_2)) u^2\right) dx.$$

Therefore,

$$\left| \int_0^\tau K_2(l,t) u^2(l,t) \, dt \right| \le k_2 \varepsilon \int_0^\tau \int_0^l u_x^2 \, dx \, dt + k_2 C_\varepsilon \int_0^\tau \int_0^l u^2 \, dx \, dt, \tag{19.16}$$

$$\left| \int_0^\tau (K_1(l,t) + K_2(0,t)) \, u(0,t) u(l,t) \right| \le \frac{(k_1 + k_2)}{2} (\varepsilon_1 + \varepsilon_2) \int_{Q_\tau} u_x^2 \, dx \, dt + \\ + \frac{(k_1 + k_2)}{2} (C_{\varepsilon_1} + C_{\varepsilon_2}) \int_{Q_\tau} u^2 \, dx \, dt.$$

Moreover,

$$\left|\int\limits_{Q_\tau} R_1(x,t)u(x,t)\,dx\,u(l,t)\,dt\right| \leq \frac{1}{2}(r_1 + lC_{\varepsilon_2})\int\limits_{Q_\tau}(u(x,t))^2\,dx\,dt + \frac{\varepsilon_2 l}{2}\int\limits_{Q_\tau}(u_x(x,t))^2\,dx\,dt$$

and

$$\left|\int\limits_{Q_\tau} R_2(x,t)u(x,t)\,dx\,u(0,t)\,dt\right| \leq \frac{1}{2}(r_2 + lC_{\varepsilon_1})\int\limits_{Q_\tau}(u(x,t))^2\,dx\,dt + \frac{\varepsilon_1 l}{2}\int\limits_{Q_\tau}(u_x(x,t))^2\,dx\,dt. \tag{19.17}$$

From the estimates (19.16)–(19.17) and the integral identity (19.8) it follows that

$$\int\limits_0^l u^2(x,\tau)\,dx + \int\limits_{Q_\tau} u_x^2\,dx\,dt \leq \frac{1}{2}\int\limits_0^l \varphi^2(x)\,dx + P\int\limits_{Q_\tau} u^2\,dx\,dt.$$

In particular,

$$\int\limits_0^l u^2(x,\tau)\,dx \leq \frac{1}{2}\int\limits_0^l \varphi^2(x)\,dx + P\int\limits_{Q_\tau} u^2\,dx\,dt. \tag{19.18}$$

By Gronwall's lemma we conclude that

$$\int\limits_{Q_\tau} u^2(x,t)\,dx\,dt \leq H_1\int\limits_0^l \varphi^2(x)\,dx, \tag{19.19}$$

and hence,

$$\int\limits_{Q_\tau} u_x^2(x,t)\,dx\,dt \le H_2 \int\limits_0^l \varphi^2(x)\,dx. \tag{19.20}$$

Therefore, from (19.19) and (19.20) we obtain

$$|u|_{Q_T} \le H.$$

19.3 The Main Result

In this section we shall prove existence and uniqueness theorem for the problem (19.1), (19.2), (19.5), (19.6).

Theorem 19.4. *Let the conditions* (A1)–(A2) *hold and*

(A3) $K_1(\xi_1, 0)$, $K_2(\xi_i, 0) = 0$, $i = 1, 2$, $\xi_1 = 0$, $\xi_2 = l$, $K_1(l,t) = K_2(0,t)$,
(A4) $R_1^2 + R_2^2 \le \frac{1}{2}$.

Then there exists a unique generalized solution to the problem (19.1), (19.2), (19.5), (19.6).

Proof. The proof of the theorem is organized as follows. First, to prove the existence part we construct a sequence of Faedo–Galerkin approximations and show its convergence to the solution of the problem. Second, we prove uniqueness of the generalized solution. Let a system of functions $\{\varphi_i(x)\} \in C^1[0,l]$ be complete in W_2^1 and

$$\left(\varphi_i, \varphi_j\right)_{L_2(0,l)} = \begin{cases} 1, & i = j, \\ 0, & i \ne j. \end{cases}$$

We define for each $N \in \mathbb{N}$ the approximate solution in the following form

$$u^N(x,t) = \sum_{k=1}^{N} c_k^N(t)\varphi_k(x),$$

where the functions $c_k(t)$ are unknown for the moment. We shall consider $c_k(t)$ which are solutions to the Cauchy problem

$$\int\limits_0^l u_t^N \varphi_i\,dx + \int\limits_0^l u_x^N \varphi_i'\,dx - \int\limits_0^l c(x,t)u^N \varphi_i\,dx$$
$$= K_1(0,t)u^N(0,t)\varphi_i(0) - K_1(l,t)u^N(l,t)\varphi_i(0)$$

$$+\int_0^l R_1(x,t)u^N\,dx\varphi_i(0)-\int_0^l R_2(x,t)u^N\,dx\varphi_i(l)$$

$$-K_2(0,t)u^N(0,t)\varphi_i(l)+K_2(l,t)u^N(l,t)\varphi_i(l), \tag{19.21}$$

$$c_i^N(0)=(\varphi,\,\varphi_i)\,, \tag{19.22}$$

$i=\overline{1,N}$. We write the Cauchy problem (19.21)–(19.22) such that

$$\frac{d}{dt}c_i^N(t)+\sum_{k=1}^N c_k^N(t)A_{k,i}(t)=0,\;\; i=\overline{1,N}, \tag{19.23}$$

where

$$A_{k,i}(t)=\int_0^l \varphi_k'(x)\varphi_i'(x)\,dx-\int_0^l c(x,t)\varphi_k(x)\varphi_i(x)\,dx$$

$$-\varphi_i(0)\left(K_1(l,t)\varphi_k(l)-K_1(0,t)\varphi_k(0)+\int_0^l R_1(x,t)\varphi_k(x)\,dx\right)$$

$$+\varphi_i(l)\left(K_2(l,t)\varphi_k(l)-K_2(0,t)\varphi_k(0)+\int_0^l R_2(x,t)\varphi_k(x)\,dx\right).$$

We estimate the coefficients $A_{k,i}$ as follows:

$$|A_{k,i}(t)|\le\frac{1}{2}\int_0^l \varphi_k'^2(x)\,dx+\frac{1}{2}\int_0^l \varphi_i'^2(x)\,dx$$

$$+\frac{\overline{c}}{2}\int_0^l \varphi_k^2(x)\,dx+\frac{\overline{c}}{2}\int_0^l \varphi_i^2(x)\,dx$$

$$+|\varphi_i(0)|\left(k_1\left(\varphi_k(l)|+|\varphi_k(0)|\right)+\frac{r_1}{2}+\frac{1}{2}\int_0^l \varphi_k^2(x)\,dx\right)$$

$$+\,|\varphi_i(0)|\left(k_2\left(\varphi_k(l)|+|\varphi_k(0)|\right)+\frac{r_2}{2}+\frac{1}{2}\int_0^l \varphi_k^2(x)\,dx\right).$$

The assumptions $(A1)$–$(A2)$ imply that $A_{k,i}$ are bounded. Therefore, the Cauchy problem has a unique solution $c_k^N \in C^1(0,T)$ and all approximations $u^N(x,t)$ are defined. The next aim is to show that the sequence $\{u^N(x,t)\}$ converges to the solution to the problem (19.1), (19.2), (19.5), (19.6). To this aim we multiply each (19.21) by $c_i^N(t)$, sum it up from $i = 0$ to $i = N$ and integrate the result with respect to t from 0 to $t_1 < T$. Thus we obtain

$$\frac{1}{2}\int\limits_0^l \left(u^N(x,t_1)\right)^2 dx + \int\limits_{Q_{t_1}} \left(u_x^N\right)^2 dx\,dt = \frac{1}{2}\int\limits_0^l \varphi^2(x)\,dx + \int\limits_{Q_{t_1}} c(u^N)^2\,dx\,dt$$

$$+ \int\limits_0^{t_1} K_2(l,t)(u^N(l,t))^2\,dt$$

$$- \int\limits_0^{t_1} K_1(l,t)u^N(0,t)u^N(l,t)\,dt$$

$$+ \int\limits_0^{t_1} K_1(0,t)(u^N(0,t))^2\,dt$$

$$- \int\limits_0^{t_1} K_2(0,t)u^N(0,t)u^N(l,t)\,dt$$

$$+ \int\limits_{Q_{t_1}} R_1(x,t)u^N(x,t)\,dx\,u^N(l,t)\,dt$$

$$- \int\limits_{Q_{t_1}} R_2(x,t)u^N(x,t)\,dx\,u^N(0,t)\,dt.$$

Therefore, from Lemmas 19.2 and 19.3 it follows that $\left|u^N\right|_{Q_T} \le \text{Const}$. It implies that there exists a subsequence of $\{u^N(x,t)\}$ which converges weakly in $L_2(0,l)$ and uniformly with respect to $t \in [0,T]$ to some function $u(x,t)$ [18]. We shall prove that this function $u(x,t)$ satisfies the integral identity (19.7) from the definition of a generalized solution to the problem (19.1), (19.2), (19.5), (19.6). To this end, we multiply each (19.21) by a smooth function $d_i(t)$, $d_i(T) = 0$, sum it up from $i = 1$ to $i = N$, integrate with respect to t from 0 to T and denote $\Phi^{N'} = \sum\limits_{i=1}^{N'} d_i(t)\varphi_i(x)$. As a result we obtain

$$\int_0^T \left(-(u^N, \Phi_t^{N'}) + (u_x^N, \Phi_x^{N'}) - (cu^N, \Phi^{N'})\right)$$

$$= \int_0^l \varphi(x)\Phi^{N'}(x,0)\,dx + \int_0^T \int_0^l R_1(x,t)u^N(x,t)\,dx\,\Phi^{N'}(0,t)\,dt$$

$$- \int_0^T \int_0^l R_2(x,t)u^N(x,t)\,dx\,\Phi^{N'}(l,t)\,dt - \int_0^T K_2(0,t)u^N(0,t)\Phi^{N'}(l,t)\,dt$$

$$+ \int_0^T K_2(l,t)u^N(l,t)\Phi^{N'}(l,t)\,dt - \int_0^T K_1(l,t)u^N(l,t)\Phi^{N'}(0,t)\,dt$$

$$- \int_0^T K_1(0,t)u^N(l,t)\Phi^{N'}(0,t)\,dt. \tag{19.24}$$

Since $\Phi^{N'}, \Phi_t^{N'}, \Phi_x^{N'} \in L_2(Q_T)$, the subsequence $\{u^{N_m}(x,t)\}$ converges weakly in $L_2(Q_T)$, so it is possible to pass to the limit in (19.24) as $m \to \infty$ for any fixed $\Phi^{N'}$. Thus, for any $u(x,t) \in V_2(Q_T)$ the following identity holds

$$\int_0^T \left(-(u, \Phi_t^{N'}) + (u_x, \Phi_x^{N'}) - (cu, \Phi^{N'})\right)$$

$$= \int_0^l \varphi(x)\Phi^{N'}(x,0)\,dx + \int_0^T \int_0^l R_1(x,t)u(x,t)\,dx\,\Phi^{N'}(0,t)\,dt$$

$$- \int_0^T \int_0^l R_2(x,t)u(x,t)\,dx\,\Phi^{N'}(l,t)\,dt - \int_0^T K_2(0,t)u(0,t)\Phi^{N'}(l,t)\,dt$$

$$+ \int_0^T K_2(l,t)u(l,t)\Phi^{N'}(l,t)\,dt - \int_0^T K_1(l,t)u(l,t)\Phi^{N'}(0,t)\,dt$$

$$+ \int_0^T K_1(0,t)u(l,t)\Phi^{N'}(0,t)\,dt. \tag{19.25}$$

Denote $\Phi = \bigcup_{N'=1}^{\infty} \Phi^{N'}$. The set Φ is dense in $W_2^1(Q_T)$ and hence, there exists a function $\Phi(x,t) \in W_2^1(Q_T)$ that is the limit of the sequence $\Phi^{N'}$. Finally, we conclude that the relation (19.25) holds for all functions $\Phi(x,t) \in W_2^1(Q_T)$ and therefore, there exists the solution $u(x,t) \in V_2(Q_T)$ to the problem (19.1), (19.2), (19.5), (19.6) in sense of Definition 19.1. Assume that there exist two different generalized solutions $u_1(x,t)$, $u_2(x,t) \in V_2(Q_T)$ to the problem (19.1), (19.2), (19.5), (19.6). Then

$$u = u_1 - u_2 \in V_2(Q_T)$$

satisfies the following identity

$$\begin{aligned}
&\int\limits_{Q_T} (-u\eta_t + u_x\eta_x - cu\eta)\, dx\, dt \\
&\quad = \int\limits_{Q_T} (R_1(x,t)\eta(l,t) - R_2(0,t)\eta(0,t))\, u(x,t)\, dx\, dt \\
&\quad + \int\limits_0^T (K_1(0,t)\eta(0,t) - K_2(0,t)\eta(l,t))\, u(0,t)\, dt \\
&\quad + \int\limits_0^T (K_2(l,t)\eta(l,t) - K_1(l,t)\eta(0,t))\, u(l,t)\, dt.
\end{aligned} \tag{19.26}$$

We take

$$\eta(x,t) = \begin{cases} 0, & b \le t \le T, \\ \int\limits_b^t u(x,\tau)\, d\tau, & 0 \le t \le b, \end{cases}$$

where $b \in [0,T]$ is arbitrary. Note that

$$\eta(x,t) \in W_2^1(Q_T), \eta(x,T) = 0$$

and since $\eta_{xt} = u_x$, so $\eta_{xt} \in L_2(Q_T)$. We substitute $\eta(x,t)$ into (19.26) and express u, u_x in terms of η. Then (19.26) becomes

$$\begin{aligned}
&\int_{Q_b} \left(-\eta_t^2 + \eta_{tx}\eta_x - c\eta\eta_t\right) dx\, dt \\
&\quad = \int_{Q_b} (R_1(x,t)\eta(l,t) - R_2(0,t)\eta(0,t))\, \eta_t(x,t)\, dx\, dt \\
&\quad + \int_0^b (K_1(0,t)\eta(0,t) - K_2(0,t)\eta(l,t))\, \eta_t(0,t)\, dt \\
&\quad + \int_0^T (K_2(l,t)\eta(l,t) - K_1(l,t)\eta(0,t))\, \eta_t(l,t)\, dt.
\end{aligned} \tag{19.27}$$

Integrating by parts in (19.27) we obtain

$$\frac{1}{2}\int_0^l \eta_x^2(x,0)\,dx + \int_{Q_b} \eta_t^2\,dx\,dt$$
$$= \frac{1}{2}\left(-\int_0^l c(x,0)\eta^2(x,0)\,dx + \int_{Q_b} c_t\eta^2\,dt\right)$$
$$+ \int_{Q_b} \left(-R_1(x,t)\eta(l,t) + R_2(0,t)\eta(0,t)\right)\eta_t(x,t)\,dx\,dt$$
$$+ \frac{1}{2}K_1(0,0)\eta^2(0,0) + \frac{1}{2}\int_0^b (K_1(0,t))_t\eta^2(0,t)\,dt$$
$$+ \frac{1}{2}K_2(l,0)\eta^2(l,0) + \frac{1}{2}\int_0^b (K_2(l,t))_t\eta^2(l,t)\,dt$$
$$- K_2(0,0)\eta(l,0)\eta(0,0) - \int_0^b K_2(0,t)\eta_t(l,t)\eta(0,t)\,dt$$
$$+ \int_0^b K_1(l,t)\eta(l,t)_t\eta(0,t)\,dt - \int_0^b (K_2(0,t))_t\eta(l,t)\eta(0,t)\,dt. \quad (19.28)$$

Under the assumptions $(A1)$–$(A2)$ from (19.28) we obtain the following estimate

$$\int_0^l \eta_x^2(x,0)\,dx + \int_{Q_b} \eta_t^2\,dx\,dt \le (c_1 b + P_1 b^2)\int_{Q_b} \eta_t^2\,dx\,dt + P_2\int_{Q_b} \eta_x^2\,dx\,dt,$$

where

$$P_1 = 2 + c_2 + \frac{k_3 + 2k_4}{l},\ P_2 = 2(2l^2 + k_3 l + 3k_4 l).$$

Since $b > 0$ is an arbitrary, so let b be such that $1 - c_1 b - Cb^2 > 0$ and in particular,

$$1 - c_1 b - Cb^2 \ge \frac{1}{2}.$$

That is,

$$b \in [0, \nu]),\ \nu = \frac{-c_1 + \sqrt{c_1^2 + 2C}}{2C}.$$

Then for all $b \in [0, \nu]$

$$\int_0^l \eta_x^2(x,0)\,dx + \frac{1}{2}\int_{Q_b} \eta_t^2\,dx\,dt \le P_2 \int_{Q_b} \eta_x^2\,dx\,dt. \tag{19.29}$$

Define the function $u(x,t) = \int_0^t u(x,\tau)\,d\tau$. Then $\eta(x,t) = y(x,t) - y(x,b)$ for $t \in [0,b]$ and (19.29) can be represented as

$$\int_0^l y_x^2(x,b)\,dx + \frac{1}{2}\int_{Q_b} y_t^2\,dx\,dt \le P_2 \int_{Q_b} (y(x,t) - y(x,b))_x^2\,dx\,dt,$$

which implies that

$$\begin{aligned}\int_0^l y_x^2(x,b)\,dx &\le P_2 \int_{Q_b} (y(x,t) - y(x,b))_x^2\,dx\,dt \\ &\le 2P_2 \int_{Q_b} y_x^2(x,t)\,dx\,dt + 2P_2 b \int_0^l y_x^2(x,b)\,dx.\end{aligned}$$

In particular, for $b \le \dfrac{1}{4P_2}$ we obtain

$$\int_0^l y_x^2(x,b)\,dx \le 4P_2 \int_{Q_b} y_x^2(x,t)\,dx\,dt. \tag{19.30}$$

The estimate (19.30) is valid for all $b \in [0, b_1]$, where $b_1 = \min\left\{\dfrac{1}{4P_2}, \nu\right\}$. From (19.30), Gronwall's lemma and the condition $y_x(x,0) = 0$ we obtain that $y_x^2(x,b) = 0$ for all $b \in [0,b_1]$. And hence, $\eta_x(x,t) = 0$, $t \in [0,b_1]$. Then from (19.30) it follows that $\eta_t(x,t) = u(x,t) = 0$, $t \in [0,b_1]$. Likewise, we repeat the above arguments and obtain $u(x,t) = 0$ for all $t \in [b_1, 2b_1]$ and so on. Finally, we conclude that $u(x,t) = 0$ in Q_T that in turn implies uniqueness of the solution to the problem (19.1), (19.2), (19.5), (19.6).

References

1. A.M. Abdrakhmanov, Solvability of a boundary-value problem with an integral boundary condition of the second kind for equations of odd order. Math. Notes **88**(1), 151–159 (2010)
2. A. Ashyralyev, A. Sarsenbi, Well-posedness of a parabolic equation with nonlocal boundary condition. Bound. Value Probl. **2015**, 38 (2015)
3. G. Avalishvili, M. Avalishvili, D. Gordeziani, On integral nonlocal boundary value problems for some partial differential equations. Bull. Georgian Natl. Acad. Sci. **5**(1), 31–37 (2011)
4. A. Bouziani, Solution forte d'un problem mixte avec condition non locales pour une classe d'equations hyperboliques. Bull. Cl. Sci. Acad. R. Belg. **8**, 53–70 (1997)
5. A. Bouziani, Initial-boundary value problems for a class of pseudoparabolic equations with integral boundary conditions. J. Math. Anal. Appl. **291**, 371–386 (2004)
6. J.R. Cannon, The solution of the heat equation subject to the specification of energy. Quart. Appl. Math. **21**(2), 155–160 (1963)
7. J.R. Cannon, Y. Lin, An inverse problem of finding a parameter in a semi-linear heat equation. J. Math. Anal. Appl. **145**, 470–484 (1990)
8. J.R. Cannon, Y. Lin, S. Wang, Determination of a control parameter in a parabolic partial differential equation. J. Austral. Math. Soc. Ser. B **33**, 149–163 (1991)
9. O. Danilkina, On a nonlocal problem with the second kind integral condition for a parabolic equation. Int. J. Part. Differ. Equ. Appl. **2**(4), 62–67 (2014)
10. O. Danilkina, On a certain nonlocal problem for a heat equation. Glob. J. Math. Anal. **2**(4), 235–242 (2014)
11. O. Danilkina, On one non-classical problem for a multidimensional parabolic equation. J. Concr. Appl. Math. **4**(13), 266–278 (2015)
12. D.G. Gordeziani, G.A. Avalishvili, Solutions of nonlocal problems for one-dimensional oscillations of the medium. Mat. Modelir. **12**(1), 94–103 (2000)
13. B. Jumarhon, S. McKee, On the heat equation with nonlinear and nonlocal boundary conditions. J. Math. Anal. Appl. **190**, 806–820 (1995)
14. L.I. Kamynin, On a boundary problem in the theory of heat conduction with a non-classical boundary conditions. Zh. Vychisl. Math. Math. Fiz. **4**(6), 1006–1024 (1964)
15. A.N. Kolmogorov, S.V. Fomin, *Elements of the Theory of Functions and Functional Analysis* (Dover Publications, New York, 1999)
16. A.I. Kozhanov, Parabolic equations with an unknown time-dependent coefficient. Comput. Math. Math. Phys. **12**, 2085–2101 (2005)
17. A.I. Kozhanov, L.S. Pulkina, On the solvability of boundary value problems with a nonlocal boundary condition of integral form for multidimensional hyperbolic equations. Differ. Equ. **42**(9), 1233–1246 (2006)
18. O.A. Ladyzhenskaya, *Boundary-Value Problems of Mathematical Physics* (Nauka, Moscow, 1973)
19. A. Merad, A. Bouziani, Numerical solutions of the hyperbolic equation with purely integral condition by using Laplace transform method. Palest. J. Math. **4**(1), 30–36 (2015)
20. S. Mesloub, A nonlinear nonlocal mixed problem for a second order pseudoparabolic equation. J. Math. Anal. Appl. **316**, 189–209 (2006)
21. L.S. Pulkina, A mixed problem with integral condition for the hyperbolic equation. Math. Notes **74**(3), 411–421 (2003)
22. L.S. Pulkina, A nonlocal problem with integral conditions for a hyperbolic equation. Differ. Equ. **40**(7), 947–953 (2004)
23. L.S. Pulkina, Initial-boundary value problem with a nonlocal boundary condition for a multidimensional hyperbolic equation. Differ. Equ. **44**(8), 1119–1125 (2008)
24. L.S. Pulkina, A nonlocal problem with integral conditions for the hyperbolic equation. Nanosyst.: Phys. Chem. Math. **2**(4), 61–70 (2011)
25. L.S. Pulkina, Boundary value problems for a hyperbolic equation with nonlocal conditions of the I and II kind. Russ. Math. (Iz. VUZ) **56**(4), 62–69 (2012)

26. J.M. Rassias, E.T. Karimov, Boundary-value problems with non-local initial condition for parabolic equations with parameter. Eur. J. Pure Appl. Math. **3**(6), 948–957 (2010)
27. A.A. Samarskii, On certain problems of the modern theory of differential equations. Differ. Equ. **16**(11), 1221–1228 (1980)
28. R.V. Shamin, Nonlocal parabolic problems with the support of nonlocal terms inside a domain. Funct. Differ. Equ. **10**(1–2), 307–314 (2003)
29. M. Tadi, M. Radenkovic, A numerical method for 1-D parabolic equation with nonlocal boundary conditions. Int. J. Comput. Math. **2014**, Article ID 923693, 9 (2014)

Chapter 20
Neighborhoods of Analytic Functions Associated with Fractional Derivative

Osman Altıntaş

Abstract In this paper we define a differential operator and introduce the subclasses $F^q_{n,p}(\lambda, \alpha, \delta)$ and $K^q_{n,p}(\lambda, \alpha, \delta, \mu)$ of functions which are analytic and p-valent in the open unit disk. Also we derive coefficient bounds, distortion inequalities, associated inclusion relation for (n, ε)-neighborhoods of the classes, which are defined by means of a certain non-homogeneous differential equation.

20.1 Introduction and Definitions

Let $F(n, p)$ denote the class of functions $f(z)$ normalized by

$$f(z) = z^p - \sum_{k=n+p}^{\infty} a_k z^k \qquad (a_k \geq 0, n, p \in \mathbb{N} = \{1, 2, 3, \ldots\}), \tag{20.1}$$

which are analytic and p-valent in the unit disk

$$U = \{z : z \in \mathbb{C} \text{ and } |z| < 1\}.$$

The fractional derivative is defined as follows (see [8, 9]).

Definition 20.1. The fractional derivative of order δ is defined by

$$D_z^{\delta} f(z) = \frac{1}{\Gamma(1-\delta)} \frac{d}{dz} \int_0^z \frac{f(\xi)}{(z-\xi)^{\delta}} d\xi \quad (0 \leq \delta < 1),$$

where $f(z)$ is an analytic function in a simply connected region of the z-plane containing the origin and the multiplicity of $(z-\xi)^{\delta}$ is removed by requiring $\log(z-\xi)$ to be real when $z-\xi > 0$.

O. Altıntaş (✉)
Departments of Mathematics Education & Secondary Education, Institute of Education Sciences, University of Başkent, Ankara, Turkey
e-mail: oaltinta@baskent.edu.tr

G.A. Anastassiou, O. Duman (eds.), *Computational Analysis*, Springer Proceedings in Mathematics & Statistics 155, DOI 10.1007/978-3-319-28443-9_20

Definition 20.2. Under the hypotheses of Definition 20.2, the fractional derivative of order $(n+\delta)$ is defined by

$$D_z^{n+\delta} f(z) = \frac{d^n}{dz^n} D_z^{\delta} f(z) \quad \left(0 \le \delta < 1,\ n \in \mathbb{N}_0 = \mathbb{N} \cup \{0\}\right).$$

Following the earlier investigation by Goodman [7] and Rusheweyh [10], we define the (n,ε)-neighborhoods of a function $f^{(q+\delta)}(z)$ when $f \in F(n,p)$ by

$$N_{n,p}^{\varepsilon}\left(f^{(q+\delta)}, g^{(q+\delta)}\right) = \left\{ \begin{array}{l} g \in F(n,p) : g(z) = z^p - \sum_{k=n+p}^{\infty} b_k z^k \text{ and} \\ \sum\limits_{k=n+p}^{\infty} \dfrac{\Gamma(k+1)}{\Gamma(k-q-\delta+1)} k \left|a_k - b_k\right| \le \varepsilon \end{array} \right\}$$

$$\left(p > q+\delta,\ p \in \mathbb{N},\ q \in \mathbb{N}_0 = \mathbb{N} \cup \{0\},\ 0 \le \delta < 1,\ z \in U\right). \tag{20.2}$$

So that obviously,

$$N_{n,p}^{\varepsilon}\left(h^{(q+\delta)}, g^{(q+\delta)}\right) = \left\{ \begin{array}{l} g \in F(n,p) : g(z) = z^p - \sum\limits_{k=n+p}^{\infty} b_k z^k \\ \text{and } \sum\limits_{k=n+p}^{\infty} \dfrac{\Gamma(k+1)}{\Gamma(k-q-\delta+1)} k \left|b_k\right| \le \varepsilon \end{array} \right\} \tag{20.3}$$

where $h(z) = z^p$ $(p \in \mathbb{N},\ q \in \mathbb{N}_0)$.

We also let $F_{n,p}^{q}(\lambda,\alpha,\delta)$ denote the subclass of $F(n,p)$ consisting of functions $f(z)$ which satisfy the inequality

$$Re\left\{ \frac{\Gamma(p+1)}{\Gamma(p-q-\delta+1)} z^{q+\delta-p} \left[\lambda z D_z^{1+q+\delta} f(z) + (1-\lambda) D_z^{q+\delta} f(z)\right] \right\} > \alpha \tag{20.4}$$

where $0 \le \lambda \le 1, 0 \le \alpha < 1, 0 \le \delta < 1, q < n+p, q+\alpha+\delta < p, z \in U$.

$$f^{(q+\delta)}(z) = \frac{\Gamma(p+1)}{\Gamma(p-q-\delta+1)} z^{p-q-\delta} - \sum_{k=n+p}^{\infty} \frac{\Gamma(k+1)}{\Gamma(k-q-\delta+1)} a_k z^{k-q-\delta} \ (p > q+\delta). \tag{20.5}$$

The various special cases of the class $F_{n,p}^{q}(\lambda,\alpha,\delta)$ were considered by many earlier researchers. For example, we have the following relationships with the class which were studied in the earlier works:

$$F_{n,p}^{q}(\lambda,\alpha,0) = T_p^n(q,\lambda,\alpha) \quad \text{(see [2])}$$

$$F_{1,1}^{0}(\lambda,\alpha,0) = T_1^1(0,\lambda,\alpha) \quad (0 \le \lambda \le 1,\ 0 \le \alpha < 1) \text{ (see [12])}$$

$$F_{n,p}^{0}(1,\alpha,0) = T_p^n(0,1,\alpha) \quad (0 \le \alpha < p) \text{ (see [9])}$$

$$F_{n,1}^{0}(1,\alpha,0) = T_1^n(0,1,\alpha) \quad (0 \le \alpha < 1) \text{ (see [11])}.$$

The neighborhoods of certain subclasses of analytic functions were studied by Altıntaş in [1] and by Altıntaş et al. in [3–6].

Finally, $K_{n,p}^{q}(\lambda,\alpha,\delta,\mu)$ denote the subclass of the general class $F(n,p)$ consisting of functions $f\in F(n,p)$ satisfying the following non-homogeneous Cauchy–Euler differential equation:

$$z^2\frac{d^{2+q+\delta}w}{dz^{2+q+\delta}}+2(1+\mu)z\frac{d^{1+q+\delta}w}{dz^{1+q+\delta}}+\mu(1+\mu)\frac{d^{q+\delta}w}{dz^{q+\delta}}$$
$$=(p-q-\delta+\mu)(p-q-\delta+\mu+1)\frac{d^{q+\delta}g}{dz^{q+\delta}}, \tag{20.6}$$

where $w=f(z)\in F(n,p), g=g(z)\in F_{n,p}^{q}(\lambda,\alpha,\delta)$ and $\mu>q-p+\delta$.

The main object of the present investigation is to derive coefficient bounds, distortion inequalities, and associated inclusion relation for the (n,ε)-neighborhoods of functions $f\in F(n,p)$ in both classes $T_{n,p}^{q}(\lambda,\alpha,\delta)$ and $K_{n,p}^{q}(\lambda,\alpha,\delta,\mu)$.

20.2 Coefficient Bounds and Distortion Inequalities

We begin the following Lemmas.

Lemma 20.3. *Let the function $f(z)\in F(n,p)$ be defined by* (20.1). *Then $f(z)$ is in the class $F_{n,p}^{q}(\lambda,\alpha,\delta)$ if only if*

$$\sum_{k=n+p}^{\infty}\frac{\Gamma(k+1)}{\Gamma(k-q-\delta+1)}\left[1+\lambda(k-q-\delta-1)\right]a_k$$
$$\le\frac{\Gamma(p+1)}{\Gamma(p-q-\delta+1)}\left[1+\lambda(p-q-\delta-1)\right]-\alpha \tag{20.7}$$
$$0\le\lambda\le1,\ 0\le\delta<1,\ 0\le\alpha<\frac{\Gamma(p+1)}{\Gamma(p-q-\delta+1)}\left[1+\lambda(p-q-\delta-1)\right],$$
$$p\in\mathbb{N},\ q\in\mathbb{N}_0.$$

The result is sharp for the function $f(z)$ given by

$$f(z)=z^p-\frac{\{\Gamma(p+1)\left[1+\lambda(p-q-\delta-1)\right]-\alpha\}\,\Gamma(n+p-q-\delta+1)}{\Gamma(p-q-\delta+1)\left[1+\lambda(n+p-q-\delta-1)\right]\Gamma(n+p+1)}z^{n+p}.$$

Proof. Using (20.4) we have

$$Re\left\{\begin{array}{l}\dfrac{\Gamma(p+1)}{\Gamma(p-q-\delta+1)}\left[1+\lambda(p-q-\delta-1)\right]z^{p-q-\delta}\\ -\displaystyle\sum_{k=n+p}^{\infty}\frac{\Gamma(k+1)}{\Gamma(k-q-\delta+1)}\left[1+\lambda(k-q-\delta-1)\right]a_kz^{k-q-\delta}\end{array}\right\}>\alpha.$$

By letting $z \to 1^-$ along the real axis, we arrive easily at the inequality (20.7). Conversely, suppose that the inequality (20.7) holds true and let

$$z \in \partial u = \{z : z \in \mathbb{C} \text{ and } |z| = 1\}.$$

Then we find

$$\begin{aligned}
&\left| \frac{\Gamma(p+1)}{\Gamma(p-q-\delta+1)} z^{q+\delta-p} \left[\lambda z D_z^{1+q+\delta} f(z) + (1-\lambda) D_z^{q+\delta} f(z) \right] \right. \\
&\quad \left. - \frac{\Gamma(p+1)}{\Gamma(p-q-\delta+1)} [1+\lambda(p-q-\delta-1)] \right| \\
&\leq \frac{\Gamma(p+1)}{\Gamma(p-q-\delta+1)} [1+\lambda(p-q-\delta-1)] - \alpha.
\end{aligned}$$

Hence by the maximum modulus theorem we have $f(z) \in F_{n,p}^q(\lambda, \alpha, \delta)$.

Lemma 20.4. *Let the function* $f(z)$ *given by* (20.1) *be in the class* $F_{n,p}^q(\lambda, \alpha, \delta)$. *Then*

$$\begin{aligned}
&\sum_{k=n+p}^{\infty} \frac{\Gamma(k+1)}{\Gamma(k-q-\delta+1)} a_k \\
&\quad \leq \frac{\Gamma(p+1)[1+\lambda(p-q-\delta-1)] - \alpha}{\Gamma(p-q-\delta+1)[1+\lambda(n+p-q-\delta-1)]}
\end{aligned} \tag{20.8}$$

and

$$\begin{aligned}
&\sum_{k=n+p}^{\infty} \frac{\Gamma(k+1)}{\Gamma(k-q-\delta+1)} k a_k \\
&\quad \leq \frac{\{\Gamma(p+1)[1+\lambda(p-q-\delta-1)] - \alpha\}(n+p)}{\Gamma(p-q-\delta+1)[1+\lambda(n+p-q-\delta-1)]}.
\end{aligned} \tag{20.9}$$

Proof. By using Lemma 20.3, we find from (20.7) that

$$\begin{aligned}
1+\lambda(n+p-q-\delta-1) &\sum_{k=n+p}^{\infty} \frac{\Gamma(k+1)}{\Gamma(k-q-\delta+1)} a_k \\
&\leq \frac{\Gamma(k+1)}{\Gamma(k-q-\delta+1)} [1+\lambda(k-q-\delta-1)] a_k \\
&\leq \frac{\Gamma(p+1)}{\Gamma(p-q-\delta+1)} [1+\lambda(p-q-\delta-1)] - \alpha,
\end{aligned}$$

which is the first assertion (20.8). For the proof of second assertion, by appealing (20.7), we also have

$$\lambda \sum \frac{\Gamma(k+1)}{\Gamma(k-q-\delta+1)} k a_k$$
$$+ [1+\lambda(-q-\delta-1)] \sum \frac{\Gamma(k+1)}{\Gamma(k-q-\delta+1)} a_k$$
$$\leq \frac{\Gamma(p+1)}{\Gamma(p-q-\delta+1)} [1+\lambda(p-q-\delta-1)] - \alpha. \tag{20.10}$$

By using (20.8) in (20.10), we can get the assertion (20.9) of Lemma 20.4.

The distortion inequalities for functions in the classes

$$F^q_{n,p}(\lambda, \alpha, \delta) \text{ and } K^q_{n,p}(\lambda, \alpha, \delta, \mu)$$

are given by the below theorem.

Theorem 20.5. *Let a function $f \in F(n,p)$ be in the class $F^q_{n,p}(\lambda, \alpha, \delta)$ then*

$$|f(z)| \leq |z|^p + \frac{\{\Gamma(p+1)[1+\lambda(p-q-\delta-1)] - \alpha\} \Gamma(n+p-q-\delta+1)}{\Gamma(p-q-\delta+1)[1+\lambda(n+p-q-\delta-1)] \Gamma(n+p+1)} |z|^{n+p} \tag{20.11}$$

and

$$|f(z)| \geq |z|^p - \frac{\{\Gamma(p+1)[1+\lambda(p-q-\delta-1)] - \alpha\} \Gamma(n+p-q-\delta+1)}{\Gamma(p-q-\delta+1)[1+\lambda(n+p-q-\delta-1)] \Gamma(n+p+1)} |z|^{n+p}. \tag{20.12}$$

Proof. Assume that a function $f(z) \in F(n,p)$ is in the class $F^q_{n,p}(\lambda, \alpha, \delta)$ then we have

$$|f(z)| \leq |z|^p + \sum_{k=n+p}^{\infty} a_k z^k$$

and using (20.8) in Lemma 20.4, we have

$$|f(z)| \leq |z|^p + \frac{\{\Gamma(p+1)[1+\lambda(p-q-\delta-1)] - \alpha\} \Gamma(n+p-q-\delta+1)}{\Gamma(p-q-\delta+1)[1+\lambda(n+p-q-\delta-1)] \Gamma(n+p+1)} |z|^{n+p}.$$

The assertion (20.12) of Theorem 20.5 can be proven by similarly.

Theorem 20.6. *Let a function $f \in F(n,p)$ be in the class $K^q_{n,p}(\lambda,\alpha,\delta,\mu)$ then*

$$|f(z)| \le |z|^p + \frac{\{\Gamma(p+1)[1+\lambda(p-q-\delta-1)]-\alpha\}\,\Gamma(n+p-q-\delta+1)}{\Gamma(p-q-\delta+1)[1+\lambda(n+p-q-\delta-1)]\,\Gamma(n+p+1)} \cdot \frac{(p-q-\delta+\mu)(p-q-\delta+\mu+1)}{(n+p-q-\delta+\mu)} \tag{20.13}$$

and

$$|f(z)| \ge |z|^p - \frac{\{\Gamma(p+1)[1+\lambda(p-q-\delta-1)]-\alpha\}\,\Gamma(n+p-q-\delta+1)}{\Gamma(p-q-\delta+1)[1+\lambda(n+p-q-\delta-1)]\,\Gamma(n+p+1)} \cdot \frac{(p-q-\delta+\mu)(p-q-\delta+\mu+1)}{(n+p-q-\delta+\mu)}\,|z|^{n+p}. \tag{20.14}$$

Proof. Suppose that $f(z)$ is in the class $K^q_{n,p}(\lambda,\alpha,\delta,\mu)$. Also let the function $g(z) \in F^q_{n,p}(\lambda,\alpha,\delta)$ occurring in the non-homogeneous differential equation (20.6) be given as in the definition (20.2) with $b_k \ge 0 \quad (k = n+p, n+p+1, \ldots)$. Then we have from (20.6) that

$$a_k = \frac{(p-q-\delta+\mu)(p-q-\delta+\mu+1)}{(k-q-\delta+\mu)(k-q-\delta+\mu+1)} b_k, \; (k = n+p, n+p+1, \ldots). \tag{20.15}$$

So that

$$f(z) = z^p - \sum_{k=n+p}^{\infty} a_k z^k = z^p - \sum_{k=n+p}^{\infty} \frac{(p-q-\delta+\mu)(p-q-\delta+\mu+1)}{(k-q-\delta+\mu)(k-q-\delta+\mu+1)} b_k z^k$$

and

$$|f(z)| \le |z|^p + |z|^{n+p} \sum_{k=n+p}^{\infty} \frac{(p-q-\delta+\mu)(p-q-\delta+\mu+1)}{(k-q-\delta+\mu)(k-q-\delta+\mu+1)} b_k. \tag{20.16}$$

Since $g(z) \in F^q_{n,p}(\lambda,\alpha,\delta)$, the first assertion (20.8) of Lemma 20.4 yields the following inequality:

$$|b_k| \le \frac{\{\Gamma(p+1)[1+\lambda(p-q-\delta-1)]-\alpha\}\,\Gamma(n+p-q-\delta+1)}{\Gamma(p-q-\delta+1)[1+\lambda(n+p-q-\delta-1)]\,\Gamma(n+p+1)}. \tag{20.17}$$

We have from (20.16) and (20.17),

$$|f(z)| \leq |z|^p + |z|^{n+p}$$
$$\cdot \frac{\{\Gamma(p+1)[1+\lambda(p-q-\delta-1)]-\alpha\}\Gamma(n+p-q-\delta+1)}{\Gamma(p-q-\delta+1)[1+\lambda(n+p-q-\delta-1)]\Gamma(n+p+1)}$$
$$\cdot (p-q-\delta+\mu)(p-q-\delta+\mu+1)$$
$$\cdot \sum_{k=n+p}^{\infty} \frac{1}{(k-q-\delta+\mu)(k-q-\delta+\mu+1)} \tag{20.18}$$

and also note the following identity that

$$\sum_{k=n+p}^{\infty} \frac{1}{(k-q-\delta+\mu)(k-q-\delta+\mu+1)}$$
$$= \sum_{k=n+p}^{\infty} \frac{1}{(k-q-\delta+\mu)} - \frac{1}{(k-q-\delta+\mu+1)}$$
$$= \frac{1}{(n+p-q-\delta+\mu)} \tag{20.19}$$

where $\mu \in \mathbb{R}\backslash\{-n-p,-n-p-1,\ldots\}$. Using (20.19) in (20.18) we get the assertion (20.13) of Theorem 20.6. The assertion (20.14) of Theorem 20.6 can be proven by similarly.

20.3 Neighborhoods for the Classes $F_{n,p}^{q}(\lambda,\alpha,\delta)$ and $K_{n,p}^{q}(\lambda,\alpha,\delta,\mu)$

In this section, we determine inclusion relations for the classes $F_{n,p}^{q}(\lambda,\alpha,\delta)$ and $K_{n,p}^{q}(\lambda,\alpha,\delta,\mu)$ involving the (n,ε)-neighborhoods defined by (20.2) and (20.3).

Theorem 20.7. *Let a function $f \in F(n,p)$ be in the class $F_{n,p}^{q}(\lambda,\alpha,\delta)$ then*

$$F_{n,p}^{q}(\lambda,\alpha,\delta) \subset N_{n,p}^{\varepsilon}\left(h^{(q+\delta)}, g^{(q+\delta)}\right) \tag{20.20}$$

where $h(z)$ is given by (20.3) *and ε is the given by*

$$\varepsilon = \frac{\{\Gamma(p+1)[1+\lambda(p-q-\delta-1)]-\alpha\}(n+p)}{\Gamma(p-q-\delta+1)[1+\lambda(n+p-q-\delta-1)]}.$$

Proof. Assertion (20.20) would follow easily from the definition (20.3) with $g(z)$ replaced by $f(z)$ and the assertion (20.9) of Lemma 20.4.

Theorem 20.8. *Let a function $f \in F(n,p)$ be in the class $K_{n,p}^{q}(\lambda,\alpha,\delta,\mu)$ then*

$$K_{n,p}^{q}(\lambda,\alpha,\delta,\mu) \subset N_{n,p}^{\varepsilon}\left(g^{(q+\delta)}, f^{(q+\delta)}\right)$$

where $g(z)$ is given by (20.6) and ε is the given by

$$\varepsilon = \frac{\{\Gamma(p+1)[1+\lambda(p-q-\delta-1)]-\alpha\}(n+p)[n+(p-q-\delta+\mu)(p-q-\delta+\mu+2)]}{\Gamma(p-q-\delta+1)[1+\lambda(n+p-q-\delta-1)](n+p-q-\delta+\mu)}.$$

Proof. Suppose that $f \in K_{n,p}^{q}(\lambda,\alpha,\delta,\mu)$. Then, upon substituting from (20.15) into the following inequality:

$$\sum_{k=n+p}^{\infty} \frac{\Gamma(k+1)}{\Gamma(k-q-\delta+1)} k|b_k - a_k| \le \sum_{k=n+p}^{\infty} \frac{\Gamma(k+1)}{\Gamma(k-q-\delta+1)} kb_k + \sum_{k=n+p}^{\infty} \frac{\Gamma(k+1)}{\Gamma(k-q-\delta+1)} ka_k,$$

where $a_k \ge 0$ and $b_k \ge 0$, we obtain

$$\sum_{k=n+p}^{\infty} \frac{\Gamma(k+1)}{\Gamma(k-q-\delta+1)} k|b_k - a_k| \le \sum_{k=n+p}^{\infty} \frac{\Gamma(k+1)}{\Gamma(k-q-\delta+1)} kb_k + \sum_{k=n+p}^{\infty} \frac{\Gamma(k+1)(p-q-\delta+\mu)(p-q-\delta+\mu+1)}{\Gamma(k-q-\delta+1)(k-q-\delta+\mu)(k-q-\delta+\mu+1)} kb_k. \tag{20.21}$$

Since $g(z) \in F_{n,p}^{q}(\lambda,\alpha,\delta)$, the second assertion (20.9) of Lemma 20.4. yields that

$$\frac{\Gamma(k+1)}{\Gamma(k-q-\delta+1)} kb_k \le \frac{\{\Gamma(p+1)[1+\lambda(p-q-\delta-1)]-\alpha\}(n+p)}{\Gamma(p-q-\delta+1)[1+\lambda(n+p-q-\delta-1)]}. \tag{20.22}$$

Finally, by making use of (20.9) as well as (20.22) on the right-hand side of (20.21), we find that

$$\sum_{k=n+p}^{\infty} \frac{\Gamma(k+1)}{\Gamma(k-q-\delta+1)} k|b_k - a_k| \le \frac{\{\Gamma(p+1)[1+\lambda(p-q-\delta-1)]-\alpha\}(n+p)}{\Gamma(p-q-\delta+1)[1+\lambda(n+p-q-\delta-1)]}$$

$$\cdot \left(1 + \sum_{k=n+p}^{\infty} \frac{(p-q-\delta+\mu)(p-q-\delta+\mu+1)}{(k-q-\delta+\mu)(k-q-\delta+\mu+1)} \right),$$

which by virtue of the identity (20.19) yields that

$$\sum_{k=n+p}^{\infty} \frac{\Gamma(k+1)}{\Gamma(k-q-\delta+1)} k\,|b_k - a_k|$$
$$\leq \frac{\{\Gamma(p+1)[1+\lambda(p-q-\delta-1)]-\alpha\}(n+p)}{\Gamma(p-q-\delta+1)[1+\lambda(n+p-q-\delta-1)]}$$
$$\cdot \left(\frac{n+(p-q-\delta+\mu)(p-q-\delta+\mu+2)}{(n+p-q-\delta+\mu)} \right) = \varepsilon.$$

Thus, by definition (20.2) with $g(z)$ interchanged by $f(z)$,

$$f \in N_{n,p}^{\varepsilon}\left(g^{(q+\delta)}, f^{(q+\delta)}\right).$$

This completes the proof of Theorem 20.8.

References

1. O. Altıntaş, Neighborhoods of certain p-valently analytic functions with negative coefficients. Appl. Math. Comput. **187**(1), 47–53 (2007)
2. O. Altıntaş, H. Irmak, On a family of multivalently analytic functions with negative coefficients. Hacet. J. Math. Stat. **26**, 83–94 (1997)
3. O. Altintas, S. Owa, Neighborhoods of certain analytic functions with negative coefficients. Int. J. Math. Math. Sci. **19**(4), 797–800 (1996)
4. O. Altintaş, Ö. Őzkan, H.M. Srivastava, Neighborhoods of a class of analytic functions with negative coefficients. Appl. Math. Lett. **13**(3), 63–67 (2000)
5. O. Altintaş, Ö. Özkan, H.M. Srivastava, Neighborhoods of a certain family of multivalent functions with negative coefficients. Comput. Math. Appl. **47**(10–11), 1667–1672 (2004)
6. O. Altıntaş, H. Irmak, H.M. Srivastava, Neighborhoods for certain subclasses of multivalently analytic functions defined by using a differential operator. Comput. Math. Appl. **55**(3), 331–338 (2008)
7. A.W. Goodman, *Univalent Functions* (Springer, New York, 1983)
8. S. Owa, On the distortion theorems I. Kyungpook Math. J. **18**, 53–59 (1978)
9. S. Owa, *Current Topics in Analytic Function Theory* (World Scientific, Singapore, 1992), pp. 234–251
10. S. Ruscheweyh, Neighborhoods of univalent functions. Proc. Am. Math. Soc. **81**(4), 521–527 (1981)
11. H. Silverman, Univalent functions with negative coefficients. Proc. Am. Math. Soc. **51**, 109–116 (1975)
12. S.R. Swamy, S.S. Bloosnurmath, Certain classes of analytic functions with negative coefficients, Indian J. Math. **27**, 89–98 (1985)

References

Chapter 21
Spectrums of Solvable Pantograph Type Delay Differential Operators for First Order

Zameddin I. Ismailov and Pembe Ipek

Abstract Based on Vishik's method on the description of solvable extensions of a densely defined operator all solvable extensions of the minimal operator generated by some delay differential-operator expression for first order in the Hilbert space of vector-functions at finite interval are described. Later on, the structure of spectrum of these extensions is surveyed.

21.1 Introduction

The first work in the area of extension of linear densely defined operator in a Hilbert space was studied by Neumann. In his paper [16], the self-adjoint extensions of the linear densely defined having equal and nonzero deficiency indexes symmetric operator in any Hilbert space have been described. But in the years of 1949 and 1952, the boundedly (compact, regular and normal) invertible extensions of any linear operator with regular point zero in a Hilbert space have been established in works by Vishik [14, 15]. Lastly, these results have been generalized to the nonlinear operators and complete additive Hausdorff topological spaces in abstract terms in works by Otelbayev, Kokebaev, and Shynybekov [6–8, 11]. In monograph [1] Dezin gave a general method for the description of regular extensions for some classes of linear differential operators in the Hilbert space of vector-functions in finite interval.

In 1985, all boundedly solvable extensions of a minimal operator generated by linear parabolic and hyperbolic type differential expression for first order with self-adjoint operator coefficient in the Hilbert space vector-functions at finite interval in terms of boundary values were given, respectively, by Pivtorak [12] and Ismailov [5].

In considered works the operator coefficients in differential expressions are self-adjoint or bounded constant operators. Unfortunately, since delay type differential expression has not been expressed with remarkable coefficient, theories mentioned above are not applicable to this theory. On the other hand, in noted above works spectral investigations have not been done.

Z.I. Ismailov • P. Ipek (✉)
Department of Mathematics, Karadeniz Technical University, 61080 Trabzon, Turkey
e-mail: zameddin.ismailov@gmail.com; pembeipek@hotmail.com

G.A. Anastassiou, O. Duman (eds.), *Computational Analysis*, Springer Proceedings in Mathematics & Statistics 155, DOI 10.1007/978-3-319-28443-9_21

Note that the general theory of delay differential equations is given in many books (see, for example, [3, 4]). Applications of this theory can be found in economy, biology, control theory, electrodynamics, chemistry, ecology, epidemiology, tumor growth, neural networks, etc. (see [2, 10, 13]).

Let's remember that an operator $S : D(S) \subset H \to H$ in Hilbert space H is called solvable, if S is one-to-one, $SD(S) = H$ and $S^{-1} \in L(H)$.

The main goal of this work is to describe all solvable extensions of the minimal operator generated by some delay differential expression for first order with operator coefficients in the Hilbert space of vector-functions at finite interval and investigate the structure of spectrum of these extensions. Finally, will be given some applications.

21.2 Representation of Solvable Extensions

In Hilbert space of vector-functions $L^2(H, (0, 1))$ consider the following pantograph type delay differential-operator expression in form

$$l(u) = u'(t) + A(t)u(\alpha t),\ 0 < \alpha < 1 \tag{21.1}$$

where

(1) H is a separable Hilbert space with inner product $(\cdot, \cdot)_H$ and norm $\|\cdot\|_H$;
(2) operator-function $A(\cdot) : [0, 1] \to L(H)$ satisfies the condition $\|A(t)\|_H \in L_1(0, 1)$;

On the other hand, in the space $L^2(H, (0, 1))$ the following differential expression corresponding to (21.1) will be considered

$$m(u) = u'(t) \tag{21.2}$$

By standard methods it can be defined minimal operator $L_0(M_0)$ and maximal operator $L(M)$ corresponding to (21.1) ((21.2)) in $L^2(H, (0, 1))$.

Now define an operator P_α in form

$$P_\alpha u(t) = u(\alpha t),\ u \in L^2(H, (0, 1)),$$
$$P_\alpha : L^2(H, (0, 1)) \longrightarrow L^2(H, (0, 1))$$

Then (21.1) can be written in form

$$l(u) = u'(t) + A(t)P_\alpha u(t),$$

In this case it is clear that

$$\int_0^1 \|P_\alpha u(t)\|_H^2 dt = \int_0^1 \|u(\alpha t)\|_H^2 dt = \frac{1}{\alpha}\int_0^\alpha \|u(x)\|_H^2 dx \le \frac{1}{\alpha}\|u\|_{L^2}^2,$$

that is,

$$\|P_\alpha\| \le \left(\frac{1}{\alpha}\right)^{1/2}$$

Define an operator A_α in $L^2(H,(0,1))$ in form

$$A_\alpha(t) = A(t)P_\alpha,\ t \in [0,1]$$

In this case the following proposition is true.

Lemma 21.1. *For any fixed* $s \in [0,1]$ *the operator*

$$\exp\left(-\int_s^t A_\alpha(x)dx\right),\ t \in [0,1]$$

is a linear bounded in $L^2(H,(0,1))$.

Proof. Indeed, in this case

$$\begin{aligned}
&\left\|\exp\left(-\int_s^t A_\alpha(x)dx\right)u(t)\right\|_{L^2(H,(0,1))}^2 \\
&\le \int_0^1 \left\|\exp\left(-\int_s^t A_\alpha(x)dx\right)\right\|_{L^2(H,(0,1))}^2 \|u(t)\|_H^2 dt \\
&\le \int_0^1 \exp\left(2\left\|\int_s^t A_\alpha(x)dx\right\|\right)\|u(t)\|_H^2 dt \\
&\le \int_0^1 \exp\left(2\int_s^t \|A_\alpha(x)\|dx\right)\|u(t)\|_H^2 dt,
\end{aligned}$$

which gives

$$\begin{aligned}
&\left\| \exp\left(-\int_s^t A_\alpha(x)dx \right) u(t) \right\|^2_{L^2(H,(0,1))} \\
&\leq \int_0^1 \exp\left(2\int_0^1 \|A_\alpha(x)\|dx \right) \|u(t)\|_H^2 dt \\
&= \exp\left(2\int_0^1 \|A(x)P_\alpha\|dx \right) \|u\|^2_{L^2(H,(0,1))} \\
&\leq \exp\left(2(\alpha)^{-1/2}\|A(x)\|_{L_1(0,1)}\right) \|u\|^2_{L^2(H,(0,1))}.
\end{aligned}$$

Then for any fixed $s \in [0,1]$ it is obtained that

$$\left\| \exp\left(-\int_s^t A_\alpha(x)dx \right) \right\| \leq \exp\left((\alpha)^{-1/2}\|A(t)\|_{L_1(0,1)}\right)$$

for $t \in [0,1]$.

Now let $U(t,s),\ t,s \in [0,1]$, be the family of evolution operators corresponding to the homogeneous differential equation

$$\begin{cases} U_t(t,s)f + A_\alpha(t)U(t,s)f = 0,\ t,s \in [0,1] \\ U(s,s)f = f,\ f \in H. \end{cases}$$

The operator $U(t,s),\ t,s \in [0,1]$ is a linear continuous, boundedly invertible in H and

$$U^{-1}(t,s) = U(s,t),\ s,t \in [0,1]$$

(for more detailed analysis of this concept, see [9]). Let us introduce the operator

$$Uz(t) := U(t,0)z(t),\ U : L^2(H,(0,1)) \to L^2(H,(0,1)).$$

In this case it is easy to see that for the differentiable vector-function

$$z \in L^2(H,(0,1)),\ z : [0,1] \to H$$

satisfies the following relation:

$$l(Uz) = (Uz)'(t) + A_\alpha(t)Uz(t) = Uz'(t) + (U_t' + A_\alpha(t)U)z(t) = Um(z).$$

From this, $U^{-1}l(Uz) = m(z)$. Hence it is clear that if $\widetilde{L}$ is some extension of the minimal operator L_0, that is $L_0 \subset \widetilde{L} \subset L$, then

$$U^{-1}L_0U = M_0,\ M_0 \subset U^{-1}LU = \widetilde{M} \subset M,\ U^{-1}LU = M.$$

For example, prove the validity of the last relation. It is known that

$$D(M_0) = \overset{o}{W}{}_2^1(H,(0,1)),\ D(M) = W_2^1(H,(0,1)).$$

If $u \in D(M)$, then $l(Uz) = Um(z) \in L^2(H,(0,1))$, that is $Uu \in D(L)$. From the last relation $M \subset U^{-1}LU$. On the contrary, if a vector-function $u \in D(L)$, then

$$m(U^{-1}v) = U^{-1}l(v) \in L^2(H,(0,1)),$$

that is, $U^{-1}v \in D(M)$. From last relation it is obtained $U^{-1}L \subset MU^{-1}$, that is, $U^{-1}LU \subset M$. Hence, $U^{-1}LU = M$.

Before of all prove the following claim.

Theorem 21.2. *$KerL_0 = \{0\}$ and $\overline{R(L_0)} \neq L^2(H,(0,1))$.*

Proof. It is sufficient to prove $KerM_0 = \{0\}$ and $\overline{R(M_0)} \neq L^2(H,(0,1))$. Consider the boundary value problem in form

$$\begin{cases} M_0u = u'(t) = 0,\ u \in D(M_0) \\ u(0) = u(1) = 0. \end{cases}$$

Then $u(t) = f,\ f \in H$ and $u(1) = u(0) = f = 0$. Hence $KerM_0 = \{0\}$. Now for any $f \in L^2(H,(0,1))$ consider the following differential equation in a form

$$M_0u(t) = f(t),$$

that is,

$$\begin{cases} u'(t) = f(t), \\ u(0) = u(1) = 0 \end{cases}$$

From this for the general solution

$$u(t) = f_0 + \int_0^t f(s)ds,$$

we have $f_0 = 0$ and $\int_0^1 f(s)ds = 0$. Consequently, for $h \in H$ and arbitrary $f \in L^2(H,(0,1))$ we have

$$(h,f)_{L^2(H,(0,1))} = \int_0^1 (h,f(t))_H dt = \left(h, \int_0^1 f(t)dt \right) = 0.$$

The last equation means that

$$H \perp R(L_0),$$

and from this

$$dimcoker\overline{R(L_0)} \geq dimH > 0$$

Now prove the following assertion on the description of solvable extension of minimal operator L_0.

Theorem 21.3. *Each solvable extension* $\widetilde{L}$ *of the minimal operator* L_0 *in* $L^2(H,(0,1))$ *is generated by the pantograph differential-operator expression* (21.1) *and boundary condition*

$$(K + E)u(0) = KU(0,1)u(1), \tag{21.3}$$

where $K \in L(H)$ *and* E *is an identity operator in* H. *The operator* K *is determined uniquely by the extension* $\widetilde{L}$, *i.e.* $\widetilde{L} = L_K$.

On the contrary, the restriction of the maximal operator L *to the linear manifold of vector-functions satisfy the condition* (21.3) *for some bounded operator* $K \in L(H)$ *is a solvable extension of the minimal operator* L_0 *in the* $L^2(H,(0,1))$.

Proof. Firstly, all solvable extensions $\widetilde{M}$ of the minimal operator M_0 in $L^2(H,(0,1))$ in terms of boundary values are described. Consider the following so-called Cauchy extension M_c

$$M_c u = u'(t), \ u(0) = 0,$$

$$M_c : D(M_c) = \{u \in W_2^1(H,(0,1)) : u(0) = 0\}$$
$$\subset L^2(H,(0,1)) \to L^2(H,(0,1))$$

of the minimal operator M_0. It is clear that M_c is a solvable extension of M_0 and

$$M_c^{-1} f(t) = \int_0^t f(x)dx, f \in L^2(H,(0,1)),$$

$$M_c^{-1} : L^2(H,(0,1)) \to L^2(H,(0,1)).$$

Now assume that $\widetilde{M}$ is a solvable extension of the minimal operator M_0 in $L^2(H,(0,1))$. In this case it is known that the domain of $\widetilde{M}$ can be written in direct sum in form

$$D(\widetilde{M}) = D(M_0) \oplus (M_c^{-1} + K)V,$$

where $V = KerM = H, K \in L(H)$ [14, 15]. Therefore for each $u(t) \in D(\widetilde{M})$ it is true that

$$u(t) = u_0(t) + M_c^{-1}f + Kf, u_0 \in D(M_0), f \in H$$

That is, $u(t) = u_0(t) + tf + Kf$, $u_0 \in D(M_0)$, $f \in H$. Hence $u(0) = Kf$, $u(1) = f + Kf = (K + E)f$. Hence $u(0) = Kf$, $u(1) = f + Kf = (K + E)f$ and from these relations it is obtained that

$$(K + E)u(0) = Ku(1) \tag{21.4}$$

On the other hand, uniqueness of operator $K \in L(H)$ follows from [14]. Therefore, $\widetilde{M} = M_K$. This consequently, the validity of necessary part of this assertion it is clear.

On the contrary, if M_K is an operator generated by differential expression (21.2) and boundary condition (21.4), then M_K is bounded, boundedly invertible and

$$M_K^{-1} : L^2(H, (0, 1)) \to L^2(H, (0, 1)),$$

$$M_K^{-1}f(t) = \int_0^t f(x)dx + K \int_0^1 f(x)dx, \; f \in L^2(H, (0, 1)).$$

Consequently, all solvable extensions of the minimal operator M_0 in $L^2(H, (0, 1))$ are generated by differential expression (2.2) and boundary condition (2.4) with any linear bounded operator K. Now consider the general case. For this in $L^2(H, (0, 1))$ introduce an operator in the form

$$U : L^2(H, (0, 1)) \to L^2(H, (0, 1)),$$
$$(Uz)(t) := U(t, 0)z(t), z \in L^2(H, (0, 1))$$

From the properties of the family of evolution operators $U(t, s), t, s \in [0, 1]$ it is implied that an operator U is linear bounded and has a bounded inverse and

$$(U^{-1}z)(t) = U(0, t)z(t).$$

On the other hand, from the relations

$$U^{-1}L_0U = M_0, U^{-1}\widetilde{L}U = \widetilde{M}, U^{-1}LU = M$$

it is implied that an operator U is a one-to-one between sets of solvable extensions of minimal operators L_0 and M_0 in $L^2(H, (0, 1))$. The extension $\widetilde{L}$ of the minimal operator L_0 is solvable in $L^2(H,(0,1))$ if and only if the operator $\widetilde{M} = U^{-1}\widetilde{L}U$ is an extension of the minimal M_0 in $L^2(H,(0,1))$. Then, $u \in D(\widetilde{L})$ if and only if

$$(K+E)U(0,0)u(0) = KU(0,1)u(1),$$

that is,

$$(K+E)u(0) = KU(0,1)u(1).$$

This proves the validity of the claims of theorem.

Corollary 21.4. *In particular the resolvent operator $R_\lambda(L_K)$, $\lambda \in \rho(L_K)$ of any solvable extension L_K of the minimal operator L_0, generated by pantograph type delay differential expression*

$$l(u) = u'(t) + A(t)u(\alpha t),\ 0 < \alpha < 1,$$

with boundary condition

$$(K+E)u(0) = KU(0,1)u(1),$$

in $L^2(H,(0,1))$ is in form

$$R_\lambda(L_K)f(t)$$
$$= U(t,0)\left[\left(E + K(1-e^{\lambda})\right)^{-1} K \int_0^1 e^{\lambda(1-s)}U(0,s)f(s)ds + \int_0^t e^{\lambda(t-s)}U(0,s)f(s)ds\right],$$

$f \in L^2(H,(0,1))$

Corollary 21.5. *Assume that for any $t \in (0,1)$*

$$A(t) = A = const.$$

In this case, all solvable extensions of minimal operator L_0 generated by the following differential expression

$$l(u) = u'(t) + Au(\alpha t),\ 0 < \alpha < 1,$$

and boundary condition

$$(K+E)u(0) = K\sum_{n=0}^{\infty} \frac{(-1)^n}{n!} A^n u(\alpha^n),\ K \in B(H),$$

in the Hilbert $L^2(H,(0,1))$ and vice versa.

Corollary 21.6. *All solvable extensions L_K of the minimal operator L_0 generated by pantograph type differential expression $l(u) = u'(t) + u(\alpha t)$, $0 < \alpha < 1$ are described with boundary condition*

$$\begin{aligned}(K+E)u(0) &= K\left[u(1) - \frac{u(\alpha)}{1!} - \frac{u(\alpha^2)}{2!} + \cdots\right] \\ &= K\sum_{n=0}^{\infty}\frac{(-1)^n}{n!}u(\alpha^n)\end{aligned}$$

in the Hilbert space $L^2(H,(0,1))$.

21.3 Spectrum of Solvable Extensions

In this section the structure of spectrum of solvable extensions of minimal operator L_0 in $L^2(H,(0,1))$ will be investigated. Firstly, prove the following fact.

Theorem 21.7. *If $\widetilde{L}$ is a solvable extension of a minimal operator L_0 and $\widetilde{M} = U^{-1}\widetilde{L}U$ is corresponding solvable extension of a minimal operator M_0, then for the spectrum of these extensions is true $\sigma(\widetilde{L}) = \sigma(\widetilde{M})$.*

Proof. Consider a problem to the spectrum for a solvable extension L_K of a minimal operator L_0 generated by pantograph differential-operator expression (21.1), that is,

$$L_K u = \lambda u + f,\ \lambda \in \mathbb{C},\ f \in L^2(H,(0,1)).$$

From this it is obtained that

$$(L_K - \lambda E)u = f$$

or $(UM_KU^{-1} - \lambda E)u = f$. Hence $U(M_K - \lambda)(U^{-1}u) = f$. Therefore, the validity of the theorem is clear.

Now prove the following result for the spectrum of solvable extension.

Theorem 21.8. *If L_K is a solvable extension of the minimal operator L_0 in the space $L^2(H,(0,1))$, then spectrum of L_K has the form*

$$\begin{aligned}\sigma(L_K) = \Bigg\{\lambda \in \mathbb{C} : \lambda = \ln\left|\frac{\mu+1}{\mu}\right| \\ + i\arg\left(\frac{\mu+1}{\mu}\right) + 2n\pi i,\ \mu \in \sigma(K)\backslash\{0,-1\},\ n \in \mathbb{Z}\Bigg\}.\end{aligned}$$

Proof. Firstly, the spectrum of the solvable extension $M_K = U^{-1}L_KU$ of the minimal operator M_0 in $L^2(H,(0,1))$ will be investigated. For this consider the following problem for the spectrum, that is, $M_K u = \lambda u + f, \lambda \in \mathbb{C}, f \in L^2(H,(0,1))$. Then

$$u' = \lambda u + f, (K+E)u(0) = Ku(1),\ \lambda \in \mathbb{C},\ f \in L^2(H,(0,1)),\ K \in L(H).$$

It is clear that a general solution of the above differential equation in $L^2(H,(0,1))$ has the form

$$u_\lambda(t) = \exp(\lambda t)f_0 + \int_0^t \exp(\lambda(t-s))f(s)ds,\ f_0 \in H.$$

Therefore, from the boundary condition $(K+E)u_\lambda(0) = Ku_\lambda(1)$ it is obtained that

$$(E + K(1-\exp(\lambda)))f_0 = K\int_0^1 \exp(\lambda(1-s))f(s)ds. \tag{21.5}$$

For $\lambda_m = 2m\pi i, m \in \mathbb{Z}$ from the last relation it is established that

$$f_0^{(m)} = K\int_0^1 \exp(\lambda_m(1-s))f(s)ds,\ m \in \mathbb{Z}.$$

Consequently, in this case the resolvent operator of M_K is in the form

$$R_{\lambda_m}(M_K)f(t) = K\exp(\lambda_m t)\int_0^1 \exp(\lambda_m(1-s))f(s)ds$$
$$+ \int_0^t \exp(\lambda_m(t-s))f(s)ds,\ f \in L^2(H,(0,1)),\ m \in \mathbb{Z}.$$

On the other hand, it is clear that $R_{\lambda_m}(M_K) \in B((L^2(H,(0,1)), m \in \mathbb{Z}$. If $\lambda \neq 2m\pi i$, $m \in \mathbb{Z},\ \lambda \in \mathbb{C}$, then using Eq. (21.5) we have

$$\left(K - \frac{1}{\exp(\lambda)-1}E\right)f_0 = \frac{1}{1-\exp(\lambda)}K\int_0^1 \exp(\lambda(1-s))f(s)ds, f_0 \in H,\ f \in (L^2(H,(0,1)).$$

Therefore, $\lambda \in \sigma(M_K)$ if and only if

$$\mu = \frac{1}{\exp(\lambda)-1} \in \sigma(K).$$

In this case since $\mu \neq 0$ and $\mu \neq -1$,

$$\exp(\lambda) = \frac{\mu + 1}{\mu}, \ \mu \in \sigma(K),$$

and

$$\lambda_n = \ln\left|\frac{\mu + 1}{\mu}\right| + i \arg\left(\frac{\mu + 1}{\mu}\right) + 2n\pi i, \ n \in \mathbb{Z}$$

Later on, using the last relation and Theorem 21.7 the validity of the claim of theorem is proved.

21.4 Applications

Example 21.9. Let

$$(H, \|\cdot\|_H) = (\mathbb{C}, |\cdot|), \ A(t) = \frac{1}{\sqrt{t}}, \ t \in (0, 1)$$

By Theorem 21.3, all solvable extensions L_k of minimal operator L_0 generated by

$$l(u) = u'(t) + \frac{1}{\sqrt{t}} u(\alpha t), 0 < \alpha < t,$$

in $L^2(0, 1)$ are described with $l(\cdot)$ and boundary condition

$$(k + 1)u(0) = k \exp\left(-\int\limits_0^1 \frac{1}{\sqrt{t}} P_\alpha dt\right) u(1), \ k \in \mathbb{C}.$$

In addition, the resolvent operators of these extensions are in the form

$$\begin{aligned} R_\lambda(L_k)f(t) = \exp&\left(-\int\limits_0^t \frac{1}{\sqrt{x}} P_\alpha dx\right) \\ \times&\left[\left(1 + k(1 - \exp(\lambda))^{-1}\right) k \int\limits_0^1 \exp\left(\lambda(1 - s) + \int\limits_0^s \frac{1}{\sqrt{x}} P_\alpha dx\right) f(s) ds \right. \\ &\left. + \int\limits_0^t \exp\left(\lambda(t - s) + \int\limits_0^s \frac{1}{\sqrt{x}} P_\alpha dx\right) f(s) ds\right], \end{aligned}$$

$\lambda \in \rho(L_k), f \in L^2(0,1)$ and for $k \neq 0, -1$ spectrum of this extension L_k is in the form,

$$\sigma(L_k) = \{\lambda \in \mathbb{C} : \lambda = \ln\left|\frac{k+1}{k}\right| + i\arg\left(\frac{k+1}{k}\right) + 2n\pi i,\ n \in \mathbb{Z}\}$$

Example 21.10. All solvable extensions of minimal operator generated by differential expression

$$l(u) = \frac{\partial u(t,x)}{\partial t} + t^{\lambda} \sin x u(\alpha t, x),$$
$$x \in (-1,1),\ 0 < t < 1,\ \lambda < -\frac{1}{2},\ 0 < \alpha < 1,$$

in the Hilbert space $L^2((-1,1) \times (0,1))$ are described by this $l(\cdot)$ and boundary condition

$$(K+E)u(0,x) = KU(0,1)u(1,x),$$

where $K \in B(L^2(-1,1))$ and $U(t,s), t, s \in [0,1]$ is a solution of operator equation

$$U_t'(t,s)f + t^{\lambda}\sin(\cdot)P_{\alpha}U(t,s)f = 0,$$
$$t, s \in [0,1],\ U(s,s)f = f, f \in L^2(-1,1),$$

where $P_{\alpha}u(t) = u(\alpha t),\ P_{\alpha} : L^2(0,1) \to L^2(0,1)$.

References

1. A.A. Dezin, *General Problems in the Theory of Boundary Value Problems* (Nauka, Moscow, 1980) [in Russian]
2. L. Edelstein-Keshet, *Mathematical Models in Biology* (McGraw-Hill, New York, 1988)
3. S. Hal, *An Introduction to Delay Differential Equations with Applications to the Life Sciences* (Springer, New York, 2011)
4. J.K. Hale, S.M. Verduyn Lunel, *Introduction to Functional Differential Equations* (Springer-Verlog, New York, 1993)
5. Z.I. Ismailov, Description of all regular operators (of solvable extensions) for first-order differential equations in a Hilbert space. Ukr. Math. J. **37**(4), 285–287 (1985)
6. B.K. Kokebaev, M. Otelbaev, A.N. Shynybekov, On the theory of contraction and extension of operators. I. Izv. Akad. Nauk Kazakh. SSR Ser. Fiz.-Mat. **5**, 24–26 (1982) [in Russian]
7. B.K. Kokebaev, M. Otelbaev, A.N. Shynybekov, On the theory of contraction and extension of operators. II. Izv. Akad. Nauk Kazakh. SSR Ser. Fiz.-Mat. **110**, 24–26 (1983) [in Russian]
8. B.K. Kokebaev, M. Otelbaev, A.N. Shynybekov, On questions of extension and restriction of operator. Sov. Math. Dokl. **28**(1), 259–262 (1983)
9. S.G. Krein, *Linear Differential Equations in Banach Space*, vol. 29. Translations of Mathematical Monographs (American Mathematical Society, Providence, RI, 1971)

10. J.R. Ockendon, A.B. Tayler, The dynamics of a current collection system for an electric locomotive. Proc. R. Soc. Lond. Ser. A **322**, 447–468 (1971)
11. M. Otelbaev, A.N. Shynybekov, Well-posed problems of Bitsadze-Samarskii type. Sov. Math. Dokl. **26**(1), 157–161 (1983)
12. N.I. Pivtorak, Solvable boundary value problems for an operator-differential equations for parabolic type. Akad. Nauk Ukr. SSR Inst. Mat. Kiev **9**, 104–107 (1985) [in Russian]
13. H. Smith, *Applied Delay Differential Equations* (Springer, New York, 2009)
14. M.I. Vishik, On linear boundary problems for differential equations. Doklady Akad. Nauk SSSR (N.S.) **65**, 785–788 (1949)
15. M.I. Vishik, On general boundary problems for elliptic differential equations. Am. Math. Soc. Transl. II **24**, 107–172 (1963)
16. J. von Neumann, Allgemeine eigenwerttheorie Hermitescher funktionaloperatoren. Math. Ann. **102**, 49–131 (1929–1930)

Chapter 22
Non-commutative Geometry and Applications to Physical Systems

Slimane Zaim

Abstract We obtain exact solutions of the 2D Schrödinger equation with the central potentials $V(r) = ar^2 + br^{-2} + cr^{-4}$ and $V(r) = ar^{-1} + br^{-2}$ in a non-commutative space up to the first order of noncommutativity parameter using the power-series expansion method similar to the 2D Schrödinger equation with the singular even-power and inverse-power potentials, respectively, in commutative space. We derive the exact non-commutative energy levels and show that the energy is shifted to m levels, as in the Zeeman effect.

22.1 Introduction

Non-commutative quantum mechanics is motivated by the natural extension of the usual quantum mechanical commutation relations between position and momentum, by imposing further commutation relations between position coordinates themselves. As in usual quantum mechanics, the non-commutativity of position coordinates immediately implies a set of uncertainty relations between position coordinates analogous to the Heisenberg uncertainty relations between position and momentum; namely:

$$[x^\mu, x^\nu]_* = i\theta^{\mu\nu}, \tag{22.1}$$

where $\theta^{\mu\nu}$ are the non-commutativity parameters of dimension of area that signify the smallest area in space that can be probed in principle. We use the symbol $*$ in Eq. (22.1) to denote the product of the non-commutative structure. This idea is similar to the physical meaning of the Plank constant in the relation $\left[x_i, p_j\right] = i\hbar\delta_{ij}$, which is known as the smallest phase-space in quantum mechanics.

S. Zaim (✉)
Département des Sciences de la Mtière, Faculté des Sciences, Université Hadj Lakhdar - Batna, Algeria
e-mail: zaim69slimane@yahoo.com

G.A. Anastassiou, O. Duman (eds.), *Computational Analysis*, Springer Proceedings in Mathematics & Statistics 155, DOI 10.1007/978-3-319-28443-9_22

Our motivation is to study the effect of non-commutativity on the level of quantum mechanics when space non-commutativity is accounted for. One can study the physical consequences of this theory by making detailed analytical estimates for measurable physical quantities and compare the results with experimental data to find an upper bound on the θ parameter. The most obvious natural phenomena to use in hunting for non-commutative effects are simple quantum mechanics systems with central potential, such as the hydrogen atom [1–3]. In the non-commutative space one expects the degeneracy of the initial spectral line to be lifted, thus one may say that non-commutativity plays the role of magnetic field.

It has recently been shown that the non-inertial motion of the atom also induces corrections to the Lamb shift [4–7]. However, all the aforementioned studies are concerned with flat space-time. Therefore, it remains interesting to see what happens if the atom with central potential is placed in a non-commutative space rather than a flat one.

Studies involving both exact and approximate solutions to the Schrödinger equation with central potentials have received much attention in the literature [8–40]. However two-dimensional Schrödinger equation which characterises the relative planar motion of the electron and proton, by a single particle with a reduced mass, can also be very fruitful. In fact this problem initially originated as purely theoretical construction in which a 2D scenario plays the role of a toy model for higher dimensional systems which are harder to deal with. But later this 2D problem proved to have important applications in real physical situations such as semiconductors. Subsequently this 2D problem received much attention with the growing of the semiconductor technology and which led to the development of 2D structures. The Runge–Lenz vector in the 2D case was first defined in [41], and the solutions to the Schrödinger equation was obtained for 2D atomic physics problems in [42].

Furthermore, it is clear from the Bohr model quantisation of angular momentum that the latter is quantised in units of $\hbar$, $L = l\hbar$, where l is an integer. In the non-commutative space we can understand the maximum value $+l\hbar$ and minimum value $-l\hbar$, by the model we use in this work, which depends on the interaction between the noncommutativity and the dipole moment M of the form $\theta \cdot M$. In the two-dimensional space the non-commutative parameter takes two values $|\theta|$ and $-|\theta|$. Then the possible values of the dipole moment M in the θ direction are $\pm M_\theta$ ($M_\theta = \pm\mu_B \theta l$) for each value of l, which reflects the physical reality of the phenomenon.

In this work we present an important contribution to the non-commutative approach to the Schrödinger equation with central potentials. Many interesting quantum mechanical problems have been studied in non-commutative space and the effect of the non-commutativity on observables was analysed [43–52].

Our goal is to solve the Schrödinger equation with singular even-power and inverse-power potentials induced by the non-commutativity of space. We thus find the exact non-commutative energy levels and that the non-commutativity effects are similar to the Zeeman splitting in commutative space. In this work, we apply the power-series expansion method to study the solutions of the Schrödinger equation in two dimensions for perturbation operator of pseudo-harmonic and the Kratzer potentials. Instead of considering pseudo-harmonic and Kratzer potentials as perturbation operators it suffices to write them in non-commutative space which leads to the same formulation as that of the perturbation operators. Similar applications in atomic interaction potentials may be found in condensed matter physics such as the quantum Hall effect and fractional statistics.

This paper is organised as follows. In Sect. 22.2, we derive the deformed 2D Schrödinger equation for a central potentials $V(r) = ar^2 + br^{-2} + cr^{-4}$ and $V(r) = ar^{-1} + br^{-2}$ in non-commutative space. We exactly solve the deformed Schrödinger equation in closed form [39] and obtain the exact non-commutative energy levels. Finally, Sect. 22.3 is devoted to a discussion.

22.2 Non-commutative Schrödinger Equation

In this section we study the exact solutions of the Schrödinger equation for the potentials $V(r) = ar^2 + br^{-2} + cr^{-4}$ and $V(r) = ar^{-1} + br^{-2}$ in the non-commutative space. The non-commutative model specified by Eq. (22.1) is defined by a star-product, where the normal product between two functions is replaced by the $\star$-product:

$$(\varphi \star \psi)(x) = \varphi(x) \exp\left(\frac{i}{2}\theta^{\mu\nu} \overleftarrow{\partial}_{\mu} \overrightarrow{\partial}_{\nu}\right) \psi(y)_{x=y}\,.$$

In a canonical non-commutative space-space type, the non-commutative quantum mechanics is described by the following equation:

$$H(p,x) \star \psi(x) = E\psi.$$

This equation reduces to the usual one described by Mezincescu [53]:

$$H(\hat{p},\hat{x})\,\psi(x) = E\psi,$$

where

$$\hat{x}_i = x_i - \frac{\theta_{ij}}{2} p_j, \qquad \hat{p}_i = p_i.$$

22.2.1 Exact Solution with the Potential $V(r) = ar^2 + br^{-2} + cr^{-4}$

We can write the deformed potential $V(r) = ar^2 + br^{-2} + cr^{-4}$ in non-commutative space up to $\mathcal{O}(\Theta^2)$ as:

$$V(\hat{r}) = \frac{a}{2}\theta L_z + ar^2 + br^{-2} + \left(c + \frac{b}{4}\theta L_z\right) r^{-4} + \frac{c}{2}\theta L_z r^{-6},$$

which is similar to the singular even-power potential which was studied in [39].

The Schrödinger equation in a 2D non-commutative space in the presence of the potential $V(\hat{r})$ can be cast into:

$$\left(-\frac{1}{r}\frac{\partial}{\partial r}r\frac{\partial}{\partial r} - \frac{1}{r^2}\frac{\partial^2}{\partial \varphi^2} + V(\hat{r})\right)\psi(\hat{r}) = E\psi(\hat{r}). \tag{22.2}$$

The solution to Eq. (22.2) in polar coordinates $(\hat{r}, \hat{\varphi})$ takes the separable form [39]:

$$\psi(\hat{r}) = r^{-1/2} R_m(\hat{r})\, e^{im\varphi}.$$

Then Eq. (22.2) reduces to the radial equation up to $\mathcal{O}(\Theta^2)$:

$$\frac{d^2 R_m(\hat{r})}{dr^2} + \left[\tilde{E} + \tilde{V}(r) - \frac{m^2 - 1/4}{r^2}\right] R_m(\hat{r}) = 0, \tag{22.3}$$

where

$$\tilde{V}(r) = ar^2 + br^{-2} + \tilde{c}r^{-4} + \tilde{d}r^{-6},$$

and

$$\tilde{E} = E - \frac{a}{2}\theta m, \quad \tilde{c} = c + \frac{b}{4}\theta m \text{ and } \tilde{d} = \frac{c}{2}\theta m, \quad \tilde{d} > 0.$$

Equation (22.3) is similar to the radial Schrödinger equation with singular even-power potential [39]. To solve Eq. (22.2), we write the radial functions as [6, 12]:

$$R_m(\hat{r}) = e^{\tilde{p}_{|m|}(r)} \sum_{n=0} \tilde{a}_n r^{2n+\tilde{\nu}}, \tag{22.4}$$

where

$$\tilde{p}_{|m|}(r) = \frac{\alpha}{2}r^2 + \frac{\tilde{\beta}}{2}r^{-2}.$$

Substituting Eq. (22.4) into Eq. (22.3) and equating the coefficients of $r^{n+\nu}$ to zero, we obtain:

$$\tilde{A}_n\tilde{a}_n + \tilde{B}_{n+1}\tilde{a}_{n+1} + \tilde{C}_{n+2}\tilde{a}_{n+2} = 0, \tag{22.5}$$

where

$$\begin{aligned}
\tilde{A}_n &= \tilde{E} + \alpha\,(1 + 2\tilde{\nu} + 4n)\\
\tilde{B}_{n+1} &= -b - 2\alpha\tilde{\beta} - \left(m^2 - \frac{1}{4}\right) + (\tilde{\nu} + 2n)\,(\tilde{\nu} - 1 + 2n)\\
\tilde{C}_n &= \tilde{\beta}\,(3 - 2\tilde{\nu} - 4n) - \tilde{c},
\end{aligned}$$

and

$$\alpha^2 = a, \tilde{\beta}^2 = \left|\tilde{d}\right|.$$

We can choose α and $\tilde{\beta}$ such that [39]:

$$\alpha = -\sqrt{a}, \tilde{\beta} = \sqrt{|\tilde{d}|}.$$

If $a_0 \neq 0$, then one obtains $C_0 = 0$, a condition that forbids the existence of the s energy levels ($|m| = 2l + 1$ in $2D$, where l is the eigenvalue of the angular momentum and the eigenvalues of L_z are denoted by m, there are $(2l + 1)$ values of m for a given l. However, as $l = n$, we see that there are $(2n + 1)$ values of m for a given energy). This is in fact a particularity of the non-commutative Schrödinger equation solution, which is not present in the ordinary Schrödinger framework [39]. Then we obtain:

$$\tilde{\nu} = \left(\frac{3}{2} + \frac{\tilde{c}}{2\tilde{\beta}}\right) = \left(\frac{3}{2} + \tilde{\gamma}\right),$$

where

$$\tilde{\gamma} = \frac{\tilde{c}}{2\tilde{\beta}}.$$

However if $a_n \neq 0$, with $a_{n+1} = a_{n+2} = \cdots = 0$, then $\tilde{A}_n = 0$, from which one obtains the non-commutative energy eigenvalues exact up to $\mathcal{O}\left(\Theta^2\right)$:

$$\tilde{E}_{n,m} = \sqrt{a}\,(4 + 2\tilde{\gamma} + 4n) + \frac{a}{2}\theta m, \qquad |m| = 1, 2, 3, \ldots. \tag{22.6}$$

We have thus shown that the degeneracy with respect to the angular quantum number m is removed since θ may take two values being positive or negative, $(\theta = \pm|\theta|)$ and that non-commutativity here acts like a Lamb shift.

Now, we discuss the corresponding exact solution for $n = 1$. From Eq. (22.6) the non-commutative energy splitting of the energy levels up to $\mathcal{O}(\Theta^2)$ is:

$$\begin{aligned}\tilde{E}_{1,m} &= \sqrt{a}\,(8+\tilde{\gamma}) + \frac{a}{2}\theta m \\ &= \sqrt{a}\left(8 + \frac{c}{\sqrt{|\tilde{d}|}}\right) + \frac{a}{4}\left(2 + \frac{b}{\sqrt{a}}\right)\theta m \\ &= \sqrt{a}\left(8+\tilde{\lambda}\right) + \frac{a}{4}(2+\delta)\,\theta m,\end{aligned}$$

where

$$\tilde{\lambda} = \frac{c}{\sqrt{|\tilde{d}|}} \text{ and } \delta = \frac{b}{\sqrt{a}}.$$

We have shown that the non-commutative energy splitting is similar to the Zeeman effects and removes the degeneracy with respect to m. Furthermore we can say that the displacement of the energy levels is actually induced by the space non-commutativity which plays the role of a magnetic field. The corresponding eigenfunction is:

$$\psi_1(\hat{r}) = \left(\tilde{a}_0 + \tilde{a}_1 r^2\right) r^{\tilde{\nu}-1/2} e^{-\frac{1}{2}\left(\sqrt{a}r^2 + \sqrt{|\tilde{d}|}r^{-2}\right)} e^{im\varphi},$$

where $\tilde{a}_0$ and $\tilde{a}_1$ can be calculated from Eq. (22.5) and the normalisation condition. Following this method, we can obtain a class of exact solutions.

22.2.2 Exact Solution with the Potential $V(r) = ar^{-1} + br^{-2}$

The deformed potential $V(r) = ar^{-1} + br^{-2}$ in non-commutative space up to $\mathcal{O}(\Theta^2)$ is:

$$V(\hat{r}) = ar^{-1} + br^{-2} + \tilde{c}r^{-3} + \tilde{d}r^{-4}, \tag{22.7}$$

where

$$\tilde{c} = \frac{a}{2}\theta m \text{ and } \tilde{d} = \frac{b}{4}\theta m, \quad \tilde{d} \succ 0$$

where the third term is the dipôle–dipôle interaction created by the noncommutativity, the second term is similar to the interaction between an ion and a neutral atom created by the non-commutativity. These interactions show us that the effect of space non-commutativity on the interaction of a single-electron atom, for example, is similar to that of a charged ion interacting with the atom on the one hand and on the other hand interacting with the electron to create a dipole and with the nucleus to create a second dipole.

The approach of the potential in Eq. (22.7) is similar to that for the inverse-power potential in a commutative space. Thus we can take as solutions the eigenfunctions from [40]:

$$R_m(\hat{r}) = h_m(\hat{r})\, e^{f(\hat{r})}, \quad m = 1, 2, 3, \ldots$$

where

$$f(\hat{r}) = Ar^{-1} + Br + C\log r, \quad A \prec 0 \text{ and } B \prec 0,$$

and

$$h_m(\hat{r}) = \prod_{j=1}^{m}\left(r - \tilde{\sigma}_j^m\right) = \sum_{j=1}^{m} \tilde{a}_j r^j.$$

Then the radial Schrödinger Eq. (22.6) reduces to the following equation:

$$\left[f'' + f'^2 + \frac{h_m'' + 2h_m' f'}{h_m} + E - V(\hat{r}) - \frac{m^2 - 1/4}{r^2}\right] R_m(\hat{r}) = 0.$$

We arrive at the equation [40]:

$$f'' + f'^2 + \frac{h_m'' + 2h_m' f'}{h_m} = -E + V(\hat{r}) - \frac{m^2 - 1/4}{r^2}. \tag{22.8}$$

Now using the fact that:

$$f'' + f'^2 = B^2 + \frac{2BC}{r} - \frac{2AB}{r^2} + \frac{2A - 2AC}{r^3} + \frac{A^2}{r^4}, \tag{22.9}$$

and

$$h_m' = \sum_{j=1}^{m} j\tilde{a}_j r^{j-1} \text{ and } h_m'' = \sum_{j=1}^{m} j(j-1)\,\tilde{a}_j r^{j-2}, \tag{22.10}$$

where

$$a_m = 1,$$
$$a_{m-1} = -\sum_{j=1}^{m} \tilde{\sigma}_j^m, \tag{22.11}$$
$$a_{m-2} = -\sum_{j\prec i}^{m} \tilde{\sigma}_j^m \tilde{\sigma}_i^m,$$

and so on, then Eqs. (22.8)–(22.10) lead to an algebraic equation where we equate equivalent coefficients of r^s between both sides of the equation, taking into account Eq. (22.11), we find:

$$A^2 = \tilde{d}, \quad \tilde{E} = -B^2 \tag{22.12}$$
$$2A(1 - C) = \tilde{c} \tag{22.13}$$
$$a = 2B(C + m), \tag{22.14}$$

and

$$\lambda = b + m^2 - 1/4 = C(C + 2m - 1) + m(m - 1) - 2B\left(A - \sum_{j=1}^{m} \sigma_j^m\right), \tag{22.15}$$

and

$$m\sqrt{\tilde{d}} + (m + 1 + C)\sum_{j=1}^{m} \sigma_j^m + B\sum_{j=1}^{m} \left(\sigma_j^m\right)^2 = 0,$$

$$(m-1)\sqrt{\tilde{d}}\sum_{j=1}^{m} \sigma_j^m + 2(m - 1 + C)\sum_{j\prec i}^{m} \sigma_j^m \sigma_i^m + B\sum_{j\prec i}^{m} \sigma_j^m \sigma_i^m \sum_{l\prec k}^{m} \left(\sigma_l^m + \sigma_k^m\right) = 0,$$

$$(m-2)\sqrt{\tilde{d}}\sum_{j\prec i}^{m} \sigma_j^m \sigma_i^m + 3(m - 2 + C)\sum_{j\prec i\prec k}^{m} \sigma_j^m \sigma_i^m \sigma_k^m$$
$$+B\sum_{j\prec i\prec k}^{m} \sigma_j^m \sigma_i^m \sigma_k^m \sum_{l\prec q\prec s}^{m} \left(\sigma_l^m + \sigma_q^m + \sigma_s^m\right) = 0,$$

Moreover, multiplying Eq. (22.15) by B and using Eqs. (22.12)–(22.14) we find the following algebraic equation for B as:

$$4\left(A - \sum_{j=1}^{m} \sigma_j^m\right) B^2 + 2\tilde{\omega}B - a(\tilde{\nu} + 2m) = 0, \tag{22.16}$$

where

$$\tilde{\omega} = \lambda + m\left(2m + \tilde{\nu}\right) - m\left(m - 1\right),$$

and

$$\tilde{\nu} = C - 1 = \frac{\tilde{c}}{2\sqrt{\tilde{d}}}.$$

Equation (22.16) is solved by:

$$\begin{aligned} B_{\pm} &= \frac{\tilde{\omega} \pm \sqrt{\tilde{\omega}^2 + 4\left(A - \sum_{j=1}^{m} \sigma_j^m\right) a\left(\tilde{\nu} + 2m\right)}}{4\left(A - \sum_{j=1}^{m} \sigma_j^m\right)} \\ &= \tilde{\omega}\frac{1 \pm \sqrt{1 + 4\left(A - \sum_{j=1}^{m} \sigma_j^m\right) \frac{a(\tilde{\nu}+2m)}{\tilde{\omega}^2}}}{4\left(A - \sum_{j=1}^{m} \sigma_j^m\right)}. \end{aligned}$$

So the non-commutative energy spectrum up to $\mathcal{O}\left(\Theta^2\right)$ is given by:

$$\tilde{E} = -\tilde{\omega}^2\frac{\left(1 \pm \sqrt{1 + 4\left(A - \sum_{j=1}^{m} \sigma_j^m\right) \frac{a(\tilde{\nu}+2m)}{\tilde{\omega}^2}}\right)^2}{16\left(A - \sum_{j=1}^{m} \sigma_j^m\right)^2},$$

where

$$\tilde{\omega}^2 = \theta m\frac{a^2}{4b}\left(m^2 + \frac{2m}{\tilde{\nu}}\left(\lambda + m\left(m + 1\right)\right)\right) + \left(\lambda + m\left(m + 1\right)\right)^2.$$

We have thus shown that the non-commutativity effects are manifested in energy levels, so that they are split into m levels, similarly to the effects of the magnetic field. Thus we can say that the non-commutativity plays the role of the magnetic field. It is also found that if the limit $\theta \to 0$ is taken, then we recover the results of the commutative case [40].

22.3 Conclusions

In this paper we started from a quantum particle with the central potentials

$$V\left(r\right) = ar^2 + br^{-2} + cr^{-4} \text{ and } V\left(r\right) = ar^{-1} + br^{-2}$$

in a canonical non-commutative space. Using the Moyal product method, we have derived the deformed Schrödinger equation, we showed that it is similar to the

Schrödinger equation with singular even-power and inverse-power potentials in commutative space. Using the power-series expansion method we solved it exactly and we found that the non-commutative energy is shifted to m levels. The non-commutativity acts here like a Lamb shift. This proves that the non-commutativity has an effect similar to the Zeeman effects, where the non-commutativity leads the role of the magnetic field. This method is simple in producing the exact bound-state solutions for central potentials as perturbation operators in the non-commutativity parameter.

References

1. M. Chaichian, M.M. Sheikh-Jabbari, A. Tureanu, Hydrogen atom spectrum and the Lamb shift in noncommutative QED. Phys. Rev. Lett. **86**, 2716 (2001)
2. T.C. Adorno, M.C. Baldiotti, M. Chaichian, D.M. Gitman, A. Tureanu, Phys. Lett. B **682**, 235–239 (2009)
3. H. Motavalli, A.R. Akbarieh, Klein–Gordon equation for the Coulomb potential in noncommutative space. Mod. Phys. Lett. A **25**, 2523–2528 (2010)
4. J. Audretsch, R. Müller, Radiative energy shifts of an accelerated two-level system. Phys. Rev. A **52**, 629 (1995)
5. R. Passante, Radiative level shifts of an accelerated hydrogen atom and the Unruh effect in quantum electrodynamics. Phys. Rev. A **57**, 1590 (1998)
6. L. Rizzuto, Casimir-Polder interaction between an accelerated two-level system and an infinite plate. Phys. Rev. A **76**, 062114 (2007)
7. Z. Zhu, H. Yu, Position-dependent energy-level shifts of an accelerated atom in the presence of a boundary. Phys. Rev. A **82**, 042108 (2010)
8. A. Share, S.N. Behra, Pramana J. Phys. **14**, 327 (1980)
9. D. Amin, Phys. Today **35**, 35 (1982); Phys. Rev. Lett. **36**, 323 (1976)
10. S. Coleman, *Aspects of Symmetry: Selected Erice Lectures* (Cambridge University Press, Cambridge, 1988)
11. H. Hashimoto, Int. J. Electron. **46**, 125 (1979); Opt. Commun. **32**, 383 (1980)
12. C.E. Reid, J. Mol. Spectrosc. **36**, 183 (1970)
13. R.S. Kaushal, Quantum mechanics of noncentral harmonic and anharmonic potentials in two-dimensions. Ann. Phys. **206**, 90–105 (1991)
14. R.S. Kaushal, D. Parashar, On the quantum bound states for the potential $V(r) = ar^2 + br^{-4} + cr^{-6}$. Phys. Lett. A **170**, 335–338 (1992)
15. R.S. Kaushal, An exact solution of the Schrödinger wave equation for a sextic potential. Phys. Lett. A **142**, 57–58 (1989)
16. S.K. Bose, N. Varma, Comment on: "An exact solution of the Schrödinger wave equation for a sextic potential" [Phys. Lett. A **142**, 57–58 (1989)] by R.S. Kaushal. Phys. Lett. A **147**, 85–86 (1990)
17. S.K. Bose, Exact bound states for the central fraction power singular potential $V(r) = \sigma r^{2/3} + \beta r^{-2/3} + \gamma r^{-4/3}$. Nuovo Cimento B **109**(11), 1217–1220 (1994)
18. A. Voros, J. Phys. A **32**, 5993 (1999)
19. Y.P. Varshni, The first three bound states for the potential $V(r) = ar^2 + br^{-4} + cr^{-6}$. Phys. Lett. A **183**, 9–13 (1993)
20. S. Özcelik, M. Simsek, Exact solutions of the radial Schrödinger equation for inverse-power potentials. Phys. Lett. A **152**, 145–150 (1991)
21. M. Simsek, S. Özcelik, Bound state solutions of the Schrödinger equation for reducible potentials: general Laurent series and four-parameter exponential-type potentials. Phys. Lett. A **186**, 35–40 (1994)

22. M. Simsek, Negative-energy levels of the Dirac equation in N dimensions. Phys. Lett. A **259**, 215–219 (1999)
23. S.-H. Dong, Z.-Q. Ma, Exact solutions to the Schrödinger equation for the potential $V(r) = ar^2 + br^{-4} + cr^{-6}$ in two dimensions. J. Phys. A **31**, 9855–9859 (1998)
24. S.-H. Dong, Z.-Q. Ma, G. Esposito, Exact solutions of the Schrödinger equation with inverse-power potential. Found. Phys. Lett. **12**, 465–474 (1999)
25. M. Znojil, Singular anharmonicities and the analytic continued fractions. J. Math. Phys. **30**, 23–27 (1989)
26. M. Znojil, Singular anharmonicities and the analytic continued fractions. II. The potentials $V(r) = ar^2 + br^{-4} + cr^{-6}$. J. Math. Phys. **31**, 108–112 (1990)
27. M. Znojil, Elementary bound states for the power-law potentials. J. Phys. A **15**, 2111–2122 (1982)
28. V. de Alfaro, T. Regge, *Potential Scattering* (North Holland, Amsterdam, 1965)
29. S. Fubini, R. Stroffolini, Nuovo Cimento **37**, 1812 (1965)
30. F. Calogero, *Variable Phase Approach to Potential Scattering* (Academic, New York, 1967)
31. R.G. Newton, *Scattering Theory of Waves and Particles* (McGraw Hill, New York, 1967)
32. W.M. Frank, D.J. Land, R.M. Spector, Rev. Mod. Phys. **43**, 36 (1971)
33. R. Stroffolini, Nuovo Cimento A **2**, 793 (1991)
34. G. Esposito, Scattering from singular potentials in quantum mechanics. J. Phys. A **31**, 9493–9504 (1998)
35. G. Esposito, Complex parameters in quantum mechanics. Found. Phys. Lett. **11**, 535–547 (1998)
36. A.O. Barut, Magnetic resonances between massive and massless spin-12 particles with magnetic moments. J. Math. Phys. **21**, 568–570 (1980)
37. B.H. Bransden, C.J. Joachain, *Physics of Atoms and Molecules* (Longman, London, 1983)
38. G.C. Maitland, M.M. Higby, E.B. Smith, V.A. Wakoham, *Intermolecular Forces* (Oxford University Press, Oxford, 1987)
39. S.-H. Dong, Levinson's theorem for the nonlocal interaction in one dimension. Int. J. Theor. Phys. **39**, 1529–1541 (2000)
40. S.-H. Dong, Z.-Q. Ma, G. Esposito, Exact solutions of the Schrödinger equation with inverse-power potential. Found. Phys. Lett. **12**, 465–474 (1999)
41. X.L. Yang, M. Lieber, F.T. Chan, The Runge–Lenz vector for the two-dimensional hydrogen atom. Am. J. Phys. **59**, 231 (1991)
42. X.L. Yang, S.H. Guo, F.T. Chan, K.W. Wong, W.Y. Ching, Analytic solution of a two-dimensional hydrogen atom. I. Nonrelativistic theory. Phys. Rev. A **43**(3), 1186–1196 (1991)
43. G.V. Dunne, J. Jackiw, C. Trugenberger, "Topological" (Chern-Simons) quantum mechanics. Phys. Rev. D **41**(3), 661–666 (1990)
44. H. Falomir, J. Gamboa, M. Loewe, F. Mendez, Phys. Rev. D **66**, 045018 (2002)
45. F.S. Bemfica, H.O. Girotti, The noncommutative degenerate electron gas. J. Phys. A **38**, L539–L547 (2005)
46. D.V. Vassilevich, Quantum aspects of a noncommutative supersymmetric kink. J. High Energy Phys. **093**, 19 pp. (2008)
47. F.G. Scholtz, B. Chakraborty, S. Gangopadhyay, A. Hazra, Phys. Rev. D **71**, 085005 (2005)
48. F.G. Scholtz, L. Gouba, A. Hafver, C.M. Rohwer, J. Phys. A **42**, 175303 (2009)
49. P.A. Horvathy, M.S. Plyushchay, Nucl. Phys. B **714**, 269 (2005)
50. B. Basu, S. Ghosh, S. Dhar, Noncommutative geometry and geometric phases. Europhys. Lett. **76**, 395–401 (2006)
51. S. Dhar, B. Basu, S. Ghosh, Spin Hall effect for anyons. Phys. Lett. A **371**, 406–409 (2007)
52. B.K. Pal, B. Roy, B. Basu, Quantum dot with spin–orbit interaction in noncommutative phase space and analog Landau levels. Phys. Lett. A **374**, 4369–4374 (2010)
53. L. Mezincescu, Star operation in quantum mechanics (2000). arXiv:hep-th/0007046

Chapter 23
The (s,t)-Generalized Jacobsthal Matrix Sequences

Şükran Uygun and Kemal Uslu

Abstract In this study, we consider sequences named (s,t)-Jacobsthal, (s,t)-Jacobsthal–Lucas and defined generalized (s,t)-Jacobsthal integer sequences. After that, by using these sequences, we define generalized (s,t)-Jacobsthal matrix sequence in which it generalizes (s,t)-Jacobsthal matrix sequence, (s,t)-Jacobsthal–Lucas matrix sequence at the same time. Finally we investigate some properties of the sequence and present some important relationship among (s,t)-Jacobsthal matrix sequence, (s,t)-Jacobsthal–Lucas matrix sequence and generalized (s,t)-Jacobsthal matrix sequence.

23.1 Introduction

We can find a great deal of study on the different integer sequences in [1, 2, 8, 9, 11]. Many properties of these sequences were deduced directly from elementary matrix algebra. For example, Köken and Bozkurt [7] defined a Jacobsthal matrix of the type nxn and using this matrix derived a lot of properties on Jacobsthal numbers. Of course the most known integer sequence is made of Fibonacci numbers which are very important because of golden section. So the authors are interested in Fibonacci matrix sequences. Civciv and Turkmen, in [3, 4], defined (s,t)-Fibonacci and (s,t)-Lucas matrix sequences by using (s,t)-Fibonacci and (s,t)-Lucas sequences.

Jacobsthal and Jacobsthal–Lucas numbers are defined for $n \geq 1$ by recurrence relations

$$j_{n+1} = j_n + 2j_{n-1},\ j_0 = 0,\ j_1 = 1$$

Ş. Uygun (✉)
Department of Mathematics, Gaziantep University, Turkey
e-mail: suygun@gantep.edu.tr

K. Uslu
Department of Mathematics, Selcuk University, Konya, Turkey
e-mail: kuslu@selcuk.edu.tr

G.A. Anastassiou, O. Duman (eds.), *Computational Analysis*, Springer Proceedings in Mathematics & Statistics 155, DOI 10.1007/978-3-319-28443-9_23

and

$$c_{n+1} = c_n + 2c_{n-1},\ c_0 = 2,\ c_1 = 1,$$

respectively. Particular cases of these numbers were investigated earlier by Horadam [5, 6]. Also Uslu and Uygun, in [10], defined (s,t)-Jacobsthal and (s,t)-Jacobsthal–Lucas matrix sequences by using (s,t)-Jacobsthal and (s,t)-Jacobsthal–Lucas integer sequences.

In this study, we firstly define (s,t)-Jacobsthal and (s,t)-Jacobsthal–Lucas integer sequences, then by using these sequences, we also define (s,t)-Jacobsthal and (s,t)-Jacobsthal–Lucas matrix sequences. After that, by using them, we establish generalized (s,t)-Jacobsthal integer and matrix sequences. In the last of the study, we investigate the relationships among each matrix sequences.

Additionally, in [10], the (s,t)-Jacobsthal and (s,t)-Jacobsthal–Lucas integer sequences are defined recurrently by

$$\hat{j}_n(s,t) = s\hat{j}_{n-1}(s,t) + 2t\hat{j}_{n-2}(s,t), \quad (\hat{j}_0(s,t) = 0, \hat{j}_1(s,t) = 1)$$
$$\hat{c}_n(s,t) = s\hat{c}_{n-1}(s,t) + 2t\hat{c}_{n-2}(s,t), \quad (\hat{c}_0(s,t) = 2, \quad \hat{c}_1(s,t) = s),$$

where $s > 0,\ t \neq 0,\ s^2 + 8t > 0,\ n \geq 1$ any integer.

Particular cases of previous definition are

- If $s = 1,\ t = 1/2$ and $\hat{j}_0(1, 1/2) = 0,\ \hat{j}_1(1, 1/2) = 1$, then we have the classic Fibonacci sequence.
- If $s = 1,\ t = 1/2$ and $\hat{c}_0(1, 1/2) = 2,\ \hat{c}_1(1, 1/2) = 1$, then we have the classic Lucas sequence.
- If $s = t = 1$ and $\hat{j}_0(1, 1) = 0,\ \hat{j}_1(1, 1) = 1$, then we have the classic Jacobsthal sequence.
- If $s = t = 1$ and $\hat{c}_0(1, 1) = 2,\ \hat{c}_1(1, 1) = 1$, then we have the classic Jacobsthal–Lucas sequence.

Furthermore, in the following proposition, (s,t)-Jacobsthal $\{J_n(s,t)\}_{n\in\mathbb{N}}$ and (s,t)-Jacobsthal–Lucas $\{C_n(s,t)\}_{n\in\mathbb{N}}$ matrix sequences are defined by carrying to matrix theory (s,t)-Jacobsthal and (s,t)-Jacobsthal–Lucas integer sequences.

Proposition 23.1. *Let us consider $s > 0,\ t \neq 0$ and $s^2 + 8t > 0,\ n \geq 1$ any integer, the following properties are hold:*

1. $J_{n+1}(s,t) = sJ_n(s,t) + 2tJ_{n-1}(s,t)$ *with initial conditions*

$$J_0(s,t) = \begin{pmatrix} 1 & 0 \\ 0 & 1 \end{pmatrix} \text{ and } J_1(s,t) = \begin{pmatrix} s & 2 \\ t & 0 \end{pmatrix},$$

2. $C_{n+1}(s,t) = sC_n(s,t) + 2tC_{n-1}(s,t)$ *with initial conditions*

$$C_0(s,t) = \begin{pmatrix} s & 4 \\ 2t & -s \end{pmatrix} \text{ and } C_1(s,t) = \begin{pmatrix} s^2 + 4t & 2s \\ st & 4t \end{pmatrix},$$

3.

$$J_n(s,t) = \begin{pmatrix} \hat{j}_{n+1} & 2\hat{j}_n \\ t\hat{j}_n & 2t\hat{j}_{n-1} \end{pmatrix} \text{ and } C_n(s,t) = \begin{pmatrix} \hat{c}_{n+1} & 2\hat{c}_n \\ t\hat{c}_n & 2t\hat{c}_{n-1} \end{pmatrix},$$

4. $\hat{j}_n = \dfrac{r_1^n - r_2^n}{r_1 - r_2}$, $c_n = r_1^n + r_2^n$, *where*

$$r_1 = \frac{s + \sqrt{s^2 + 8t}}{2} \text{ and } r_2 = \frac{s - \sqrt{s^2 + 8t}}{2},$$

5. For $m, n \in \mathbb{Z}^+$,

$$J_{n+m}(s,t) = J_n(s,t) J_m(s,t),$$
$$J_m(s,t) C_{n+1}(s,t) = C_{n+1}(s,t) J_m(s,t).$$

Throughout this paper, we will use the notation J_n instead of $J_n(s,t)$ and C_n instead of $C_n(s,t)$.

Since there are certainly some new developments over these numbers and matrices, we have aimed to define the generalized (s,t)-Jacobsthal integer sequence and generalized (s,t)-Jacobsthal matrix sequence which are new generalizations of them. Then, of course, it needs to investigate the relationship among this generalized (s,t)-Jacobsthal number sequence, (s,t)-Jacobsthal and (s,t)-Jacobsthal–Lucas integer sequences. Also, we investigate the relationships among the generalized (s,t)-Jacobsthal matrix sequence, (s,t)-Jacobsthal and (s,t)-Jacobsthal–Lucas matrix sequences.

23.2 Main Results

Firstly, let us first consider the following definition of generalized (s,t)-Jacobsthal number sequence which will be needed for the definition of generalized (s,t)-Jacobsthal matrix sequence and relationships among them.

Definition 23.2. For any integer, $n \geq 0$, let $a, b \in \mathbb{R}$ and $s^2 + 8t > 0$, $s > 0$, $t \neq 0$. Then the generalized (s,t)-Jacobsthal integer sequence $\{G_n(s,t)\}_{n\in\mathbb{N}}$ is defined by the following equation:

$$G_{n+2}(s,t) = sG_{n+1}(s,t) + 2tG_n(s,t) \tag{23.1}$$

with initial conditions $G_0(s,t) = a$, $G_1(s,t) = bs$.

As previously, we will use G_n instead of $G_n(s,t)$.

Definition 23.3. For any integer, $n \geq 0$, let $a, b \in \mathbb{R}$, $s > 0$, $t \neq 0$ and $s^2 + 8t > 0$. Then the generalized (s,t)-Jacobsthal matrix sequence $(\Re_n(s,t))_{n\in\mathbb{N}}$ is defined by the following equation:

$$\Re_{n+2}(s,t) = s\Re_{n+1}(s,t) + 2t\Re_n(s,t) \tag{23.2}$$

with initial conditions

$$\Re_0(s,t) = \begin{pmatrix} bs & 2a \\ at & (b-a)s \end{pmatrix} \text{ and } \Re_1(s,t) = \begin{pmatrix} bs^2 + 2at & 2bs \\ bst & 2at \end{pmatrix}.$$

By considering definitions 23.2 and 23.3, we obtain the following equalities:

- For $a = b = 1$, $\begin{array}{l} G_n = \hat{j}_{n+1}, \\ \Re_n = J_{n+1}. \end{array}$
- For $a = 2, b = 1$, $\begin{array}{l} G_n = \hat{c}_n, \\ \Re_n = C_n. \end{array}$

The following result gives us the nth general term of the matrix sequence given in (23.2).

Theorem 23.4. *For any integer $n \geq 1$, we have*

$$\Re_n = \begin{pmatrix} G_{n+1} & 2G_n \\ tG_n & 2tG_{n-1} \end{pmatrix}. \tag{23.3}$$

Proof. Let us consider $n = 1$ in (23.3). Then we clearly have $G_0 = a$, $G_1 = bs$, $G_2 = bs^2 + 2at$ and then

$$\Re_1 = \begin{pmatrix} G_2 & 2G_1 \\ tG_1 & 2tG_0 \end{pmatrix} = \begin{pmatrix} bs^2 + 2at & 2bs \\ bst & 2at \end{pmatrix}.$$

As a next step of that, for $n = 2$, we also get

$$\Re_2 = \begin{pmatrix} G_3 & 2G_2 \\ tG_2 & 2tG_1 \end{pmatrix} = \begin{pmatrix} bs^3 + 2ast + 2bst & 2bs^2 + 4at \\ bs^2t + 2at^2 & 2bst \end{pmatrix}.$$

By iterating this procedure and considering induction steps, let us assume that the equality in (23.3) holds for all $n = k \in \mathbb{Z}^+$. To end up the proof, we have to show that the case also holds for $n = k + 1$. Therefore, we get

$$\begin{aligned} \Re_{k+1} &= s\Re_k + 2t\Re_{k-1} \\ &= s\begin{pmatrix} G_{k+1} & 2G_k \\ tG_k & 2tG_{k-1} \end{pmatrix} + 2t\begin{pmatrix} G_k & 2G_{k-1} \\ tG_{k-1} & 2tG_{k-2} \end{pmatrix} \end{aligned}$$

$$= \begin{pmatrix} sG_{k+1} + 2tG_k & 2sG_k + 4tG_{k-1} \\ stG_k + 2t^2G_{k-1} & 2stG_{k-1} + 4t^2G_{k-2} \end{pmatrix}$$

$$= \begin{pmatrix} G_{k+2} & 2G_{k+1} \\ tG_{k+1} & 2tG_k \end{pmatrix}.$$

Hence the result.

Corollary 23.5. *In the above theorem, if we choose suitable values on s, t, a and b, then some special matrix sequences are obtained. For example, by taking $a = 1$, $b = 1$, we obtain the (s,t)-Jacobsthal matrix:*

$$\Re_n = \begin{pmatrix} \hat{j}_{n+1} & 2\hat{j}_n \\ \hat{j}_n & 2\hat{j}_{n-1} \end{pmatrix},$$

where $\hat{j}_n$ is nth (s,t)-Jacobsthal number and by taking $a = 2$, $b = 1$, we obtain the (s,t)-Jacobsthal–Lucas matrix:

$$\Re_n = \begin{pmatrix} \hat{c}_{n+1} & 2\hat{c}_n \\ \hat{c}_n & 2\hat{c}_{n-1} \end{pmatrix},$$

where $\hat{c}_n$ is nth (s,t)-Jacobsthal–Lucas number.

Let us consider the following theorem which will be needed for the results in this section. In fact, by this theorem, it will be given a relationship among the sequences $\{G_n(s,t)\}_{n\in\mathbb{N}}$, $\left\{\hat{j}_n(s,t)\right\}_{n\in\mathbb{N}}$ and $\{\hat{c}_n(s,t)\}_{n\in\mathbb{N}}$.

Theorem 23.6. *For any integer, $n \geq 1$, we have*

1. $G_n = bs\hat{j}_n + 2at\hat{j}_{n-1}$,
2. $G_{n+1} + 2tG_{n-1} = bs\hat{c}_n + 2at\hat{c}_{n-1}$.

Proof. To prove the existence of these equalities, we need to consider the sequence given in (23.1) with its initial conditions.

1. If we consider the initial conditions $G_1 = bs, G_2 = bs^2 + 2at$, then it can be clearly written as

$$G_1 = bs = (bs)\hat{j}_1 + (2at)\hat{j}_0$$

and

$$G_2 = bs^2 + 2at = (bs)\hat{j}_2 + (2at)\hat{j}_1,$$

By keeping the (s,t)-Jacobsthal sequence and using same technique, we get $(bs)\hat{j}_3 + (2at)\hat{j}_2$, which gives G_3 in the statement of proposition. So, by iterating this above all progresses, we obtain the general term as the form of $bs\hat{j}_n + 2at\hat{j}_{n-1}$ that implies G_n, as required.

2. The replacing of (s,t)-Jacobsthal–Lucas initial conditions $\hat{c}_0$ and $\hat{c}_1$ in place of (s,t)-Jacobsthal's initial conditions in $\hat{j}_0$ and $\hat{j}_1$ above, we then get the equality

$$G_{n+1} + 2tG_{n-1} = bs\hat{c}_n + 2at\hat{c}_{n-1}.$$

We will usually reveal the required relationships among

$$\{\Re_n(s,t)\}_{n\in\mathbb{N}}, \ \{J_n(s,t)\}_{n\in\mathbb{N}} \text{ and } \{C_n(s,t)\}_{n\in\mathbb{N}}$$

in the rest of this paper.

Theorem 23.7. *For any integer, $n \geq 1$, we have*

1. $\Re_n = bsJ_n + 2atJ_{n-1}$,
2. $\Re_{n+1} + 2t\Re_{n-1} = bsC_n + 2atC_{n-1}$

Proof. 1. If we consider the initial condition for

$$\Re_1 = \begin{pmatrix} bs^2 + 2at & 2bs \\ bst & 2at \end{pmatrix},$$

it can be clearly written as

$$\Re_1 = bs\begin{pmatrix} s & 2 \\ t & 0 \end{pmatrix} + 2at\begin{pmatrix} 1 & 0 \\ 0 & 1 \end{pmatrix} = \begin{pmatrix} bs^2 + 2at & 2bs \\ bst & 2at \end{pmatrix}.$$

If we apply the same idea to

$$\Re_2 = \begin{pmatrix} bs^3 + 2ast + 2bst & 2bs^2 + 4at \\ bs^2t + 2at^2 & 2bst \end{pmatrix},$$

then

$$\Re_2 = bs\begin{pmatrix} bs^2 + 2t & 2s \\ st & 2t \end{pmatrix} + 2at\begin{pmatrix} s & 2 \\ t & 0 \end{pmatrix}$$
$$= \begin{pmatrix} bs^3 + 2ast + 2bst & 2bst + 4at \\ bs^2t + 2at^2 & 2bst \end{pmatrix}.$$

In fact these above re-written conditions contain the initial conditions J_0 and J_1 of (s,t)-Jacobsthal sequence. Therefore, by replacing these conditions on these new $\Re_1$ and $\Re_2$, then we obtain

$$\Re_1 = (bs)\,J_1 + (2at)\;J_0 \text{ and } \Re_2 = (bs)\;J_2 + (2at)\;J_1,$$

respectively. By keeping the (s,t)-Jacobsthal sequence and using the same technique, we get

$$(bs)\,J_3 + (at)\;J_2,$$

which gives $\Re_3$ in the statement of definition 23.3. So, by iterating this above all progresses, we obtain the general term as the form of $bsJ_n + 2atJ_{n-1}$ that implies $\Re_n$, as required.

2. Replacing of (s,t)-Jacobsthal–Lucas initial conditions C_0 and C_1 in place of (s,t)-Jacobsthal's initial conditions in $J_0$and J_1 above, we then get the equality

$$\Re_{n+1} + 2t\Re_{n-1} = bsC_n + 2atC_{n-1}.$$

Theorem 23.8. *For $m,n \in N$, we have*

$$\Re_{m+n} = J_n\Re_m \tag{23.4}$$

Proof. To prove the equation in (23.4), let us follow the induction steps on m. For $m = 0$,

$$\begin{aligned} J_n\Re_0 &= \begin{pmatrix} \hat{j}_{n+1} & 2\hat{j}_n \\ t\hat{j}_n & 2t\hat{j}_{n-1} \end{pmatrix} \begin{pmatrix} bs & 2a \\ at & (b-a)\,s \end{pmatrix} \\ &= \begin{pmatrix} bs\hat{j}_{n+1} + 2at\hat{j}_n & 2(a\hat{j}_{n+1} + (b-a)st\hat{j}_n) \\ t(a\hat{j}_{n+1} + (b-a)st\hat{j}_n) & 2t(a\hat{j}_n + (b-a)s\hat{j}_{n-1}) \end{pmatrix} \\ &= \Re_n. \end{aligned}$$

Now, assuming that it is true for all positive integers m, that is,

$$\Re_{m+n} = J_n\Re_m.$$

Therefore, we have to show that the case also holds for $m+1$. If we use the property of (23.2), then

$$\begin{aligned} J_n\Re_{m+1} &= J_n(s\Re_m + 2t\Re_{m-1}) = sJ_n\Re_m + 2tJ_n\Re_{m-1} \\ &= s\Re_{m+n} + 2t\Re_{m+n-1} = \Re_{m+n+1}, \end{aligned}$$

hence the result.

Theorem 23.9. *Binet Formula enables us to state (s,t)-generalized Jacobsthal numbers. It can be clearly obtained from the roots r_1 and r_2 of characteristic equation of* (23.1) *as the form $x^2 = sx + 2t$, where*

$$r_1 = \frac{s + \sqrt{s^2 + 8t}}{2}, \quad r_2 = \frac{s - \sqrt{s^2 + 8t}}{2}.$$

Then, the Binet Formula for the nth (s,t)-generalized Jacobsthal number is given by

$$G_n = \frac{Xr_1^n - Yr_2^n}{r_1 - r_2}, \tag{23.5}$$

where $X = bs + \frac{2at}{r_1}$ and $Y = bs + \frac{2at}{r_2}$.

$$\begin{aligned} G_n &= bs\hat{j}_n + 2at\hat{j}_{n-1} \\ &= G_n(s,t) = bs\frac{r_1^n - r_2^n}{r_1 - r_2} + 2at\frac{r_1^{n-1} - r_2^{n-1}}{r_1 - r_2} \\ &= \frac{\left(bs + \frac{2at}{r_1}\right) r_1^n - \left(bs + \frac{2at}{r_2}\right) r_2^n}{r_1 - r_2} \\ &= \frac{Xr_1^n - Yr_2^n}{r_1 - r_2} \end{aligned}$$

Theorem 23.10. *For $a, b \in \mathbb{R}$, $n \in \mathbb{N}$, $s > 0$, $t \neq 0$ and $s^2 + 8t > 0$, we have*

$$\Re_1 C_n = \Re_{n+2} + 2t\Re_n \tag{23.6}$$

Proof. We get

$$\begin{aligned} \Re_1 C_n &= \begin{pmatrix} bs^2 + 2at & 2bs \\ bst & 2at \end{pmatrix} \begin{pmatrix} \hat{c}_{n+1} & 2\,\hat{c}_n \\ t\,\hat{c}_n & 2t\,\hat{c}_{n-1} \end{pmatrix} \\ &= \begin{pmatrix} bs\,\hat{c}_{n+2} + 2at\,\hat{c}_{n+1} & 2(bs\,\hat{c}_{n+1} + 2at\,\hat{c}_n) \\ t(bs\,\hat{c}_{n+1} + 2at\,\hat{c}_n) & 2t(bs\,\hat{c}_n + 2at\,\hat{c}_{n-1}) \end{pmatrix} \\ &= \begin{pmatrix} G_{n+3} & 2G_{n+2} \\ tG_{n+2} & 2tG_{n+1} \end{pmatrix} + 2t \begin{pmatrix} G_{n+1} & 2G_n \\ tG_n & 2tG_{n-1} \end{pmatrix} \\ &= \Re_{n+2} + 2t\Re_n. \end{aligned}$$

Theorem 23.11. *For $m, n \in \mathbb{N}$, the following equalities hold:*

1. $C_n\Re_1 = \Re_1 C_n$,
2. $J_m\Re_{n+1} = \Re_{n+1}J_m$,
3. $C_m\Re_{n+1} = \Re_{n+1}C_m$.

Proof. Firstly, we can easily prove (1) by the equality of two sides of multiplying of matrices

$$\mathfrak{R}_1.C_n = \begin{pmatrix} bs^2+2at & 2bs \\ bst & 2at \end{pmatrix} \cdot \begin{pmatrix} \hat{c}_{n+1} & 2\,\hat{c}_n \\ t\,\hat{c}_n & 2t\,\hat{c}_{n-1} \end{pmatrix}$$

$$= \begin{pmatrix} (bs^2+2at)\,\hat{c}_{n+1}+2bst\,\hat{c}_n & 2((bs^2+2at)\,\hat{c}_n+2bst\,\hat{c}_{n-1}) \\ t(bs\,\hat{c}_{n+1}+2at\,\hat{c}_n) & 2t(bs\,\hat{c}_n+2at\,\hat{c}_{n-1}) \end{pmatrix}$$

$$= \begin{pmatrix} bs(s\hat{c}_{n+1}+2t\,\hat{c}_n)+2at\,\hat{c}_{n+1} & 2(bs(s\,\hat{c}_n+2t\,\hat{c}_{n-1})+2at\,\hat{c}_n) \\ t(bs\,\hat{c}_{n+1}+2at\,\hat{c}_{n)} & 2t(bs\,\hat{c}_n+2at\,\hat{c}_{n-1}) \end{pmatrix},$$

$$C_n.\mathfrak{R}_1 = \begin{pmatrix} \hat{c}_{n+1} & 2\,\hat{c}_n \\ t\,\hat{c}_n & 2t\,\hat{c}_{n-1} \end{pmatrix} \cdot \begin{pmatrix} bs^2+2at & 2bs \\ bst & 2at \end{pmatrix}$$

$$\times \begin{pmatrix} (bs^2+2at)\,\hat{c}_{n+1}+2bst\,\hat{c}_n & 2(bs\,\hat{c}_{n+1}+2at\,\hat{c}_n) \\ t((bs^2+2at)\hat{c}_n+2bst\hat{c}_{n-1}) & 2t(bs\,\hat{c}_n+2at\,\hat{c}_{n-1}) \end{pmatrix}$$

$$\times \begin{pmatrix} bs(s\hat{c}_{n+1}+2t\,\hat{c}_n)+2at\,\hat{c}_{n+1} & 2(bs(s\,\hat{c}_n+2t\,\hat{c}_{n-1})+2at\,\hat{c}_n) \\ t(bs\,\hat{c}_{n+1}+2at\,\hat{c}_n) & 2t(bs\,\hat{c}_n+2at\,\hat{c}_{n-1}) \end{pmatrix}.$$

We can prove (2) from

$$J_m\mathfrak{R}_{n+1} = J_m\mathfrak{R}_1J_n = J_m(bsJ_1+2atJ_0)J_n = bsJ_{m+n+1}+2atJ_{m+n}$$
$$= (bsJ_1+2atJ_0)J_{m+n} = \mathfrak{R}_1J_nJ_m = \mathfrak{R}_{n+1}J_m.$$

We can prove (3) from

$$C_m\mathfrak{R}_{n+1} = C_m\mathfrak{R}_1J_n = \mathfrak{R}_1C_mJ_n = \mathfrak{R}_1J_nC_m = \mathfrak{R}_{n+1}C_m.$$

In the following result, as the same approximation with Theorem 23.10, we will depict that there are also some other relations between $\{\mathfrak{R}_n(s,t)\}_{n\in\mathbb{N}}$ and $\{J_n(s,t)\}_{n\in\mathbb{N}}$.

Theorem 23.12. *For $m, n \in \mathbb{N}$, the following equality holds:*

$$\mathfrak{R}_{n+1}^m = \mathfrak{R}_1^m J_{mn}.$$

Proof. To prove the equation, let us follow induction steps on m. For $m = 1$, the proof is clear by (23.4). Now, assuming that it is true for all positive integers m, that is,

$$\mathfrak{R}_{n+1}^m = \mathfrak{R}_1^m J_{mn}.$$

Therefore, we have to show that it is true for $m+1$. If we multiply this mth step by $\Re_{n+1}$ on both sides, then

$$\Re_{n+1}^{m+1} = \Re_1^m J_{mn} \Re_{n+1}.$$

By considering Eq. (23.4), we can write

$$\Re_{n+1}^{m+1} = \Re_1^{m+1} J_n J_{mn}.$$

and then, from Proposition 23.1, we obtain

$$\Re_{n+1}^{m+1} = \Re_1^{m+1} J_{(m+1)n},$$

which ends up the induction and the proof of theorem.

Corollary 23.13. *For $n \geq 0$, in the equation given in the above theorem*

- *by taking $m = 1$, we obtain*

$$\Re_{n+1}^2 = \Re_1^2 J_{2n} = \Re_1 \Re_{2n+1}.$$

- *by taking $m = 2$, $a = 1$ and $b = 1$, we obtain*

$$J_{n+2}^2 = J_{n+1}^2 J_1^2 = J_1^2 J_{2n,} J_1^2 = J_{2n+4}.$$

- *by taking $m = 2$, $a = 2$ and $b = 1$, we obtain*

$$C_{n+1}^2 = C_1^2 J_{2n}.$$

Proposition 23.14. *For $a, b \in \mathbb{R}$, $n \in \mathbb{N}$, $s > 0$, $t \neq 0$ and $s^2 + 8t > 0$, we have*

1. $G_{n+2}^2 + 2tG_{n+1}^2 = (b^2s^2 + 2a^2t)\hat{j}_{2n+3} + 2ast(2b - a)\hat{j}_{2n+2}$,
2. $G_{n+2}^2 + 2tG_{n+1}^2 = bG_{2n+4} + 2(a - b)\, tG_{2n+2}$,
3. $G_{2n} = \hat{j}_n G_{n+1} + 2t\hat{j}_{n-1} G_n$.

Proof. For the proof of (1), we use the corollary

$$\begin{aligned}
\Re_{n+1}^2 &= \begin{pmatrix} G_{n+2} & 2G_{n+1} \\ tG_{n+1} & 2tG_n \end{pmatrix}^2 \\
&= \Re_1^2 J_{2n} \\
&= \begin{pmatrix} bs^2 + 2at & 2bs \\ bst & 2at \end{pmatrix}^2 \begin{pmatrix} \hat{j}_{2n+1} & 2\hat{j}_{2n} \\ t\hat{j}_{2n} & 2t\hat{j}_{2n-1} \end{pmatrix}.
\end{aligned}$$

The desired result is easily seen from the equality of the matrices of elements of the indices $(1, 1)$.

For the proof of (2), we use the corollary

$$\begin{aligned}\mathfrak{R}_{n+1}^2 &= \begin{pmatrix} G_{n+2} & 2G_{n+1} \\ tG_{n+1} & 2tG_n \end{pmatrix}^2 \\ &= \mathfrak{R}_1 \mathfrak{R}_{2n+1} \\ &= \begin{pmatrix} bs^2 + 2at & 2bs \\ bst & 2at \end{pmatrix} \begin{pmatrix} G_{2n+2} & 2G_{2n+1} \\ tG_{2n+1} & 2tG_{2n} \end{pmatrix}.\end{aligned}$$

The desired result is easily seen from the equality of the matrices of elements of the indices $(1, 1)$.

For the proof of (3), we use the corollary

$$\begin{aligned}\mathfrak{R}_{2n+1} &= \begin{pmatrix} G_{2n+2} & 2G_{2n+1} \\ tG_{2n+1} & 2tG_{2n} \end{pmatrix} \\ &= \mathfrak{R}_{n+1} J_n \\ &= \begin{pmatrix} G_{n+2} & 2G_{n+1} \\ tG_{n+1} & 2tG_n \end{pmatrix} \begin{pmatrix} \hat{j}_{n+1} & 2\hat{j}_n \\ t\hat{j}_n & 2t\hat{j}_{n-1} \end{pmatrix}.\end{aligned}$$

The desired result is easily seen from the equality of the matrices of elements of the indices $(2, 2)$.

23.3 Conclusion

In the present paper we introduce the generalized (s, t)-Jacobsthal integer sequences, whose entries are numbers G_n satisfying the recurrence formula

$$G_{n+2}(s,t) = sG_{n+1}(s,t) + 2tG_n(s,t), \quad s > 0,\ t \neq 0, s^2 + 8t > 0, n \geq 1,$$

and initial conditions

$$G_0(s,t) = a,\ G_1(s,t) = bs.$$

In the case $a = b = 1$, we define the (s, t)-Jacobsthal integer sequence, whose entries are $\hat{j}_n$ (s, t)-Jacobsthal numbers satisfying known recursive formula and initial conditions

$$j_0(s,t) = 0,\ j_1(s,t) = 1.$$

After that by using these integer sequences we introduce the generalized (s,t)-Jacobsthal matrix sequences, whose entries are matrices $\Re_k$ satisfying the recurrence formula

$$\Re_{k+1} = s\Re_k + 2t\Re_{k-1},$$

and initial conditions

$$\Re_0(s,t) = \begin{pmatrix} bs & 2a \\ at & (b-a)s \end{pmatrix} \text{ and } \Re_1(s,t) = \begin{pmatrix} bs^2 + 2at & 2bs \\ bst & 2at \end{pmatrix}$$

and give some properties of the generalized (s,t)-Jacobsthal matrix sequences.

References

1. Z. Cerin, Formulae for sums of Jacobsthal Lucas numbers. Int. Math. Forum **2**, 1969–1984 (2007)
2. Z. Cerin, Sums of squares and products of Jacobsthal numbers. J. Integer. Seq. **10**, 1–15 (2007)
3. H. Civciv, R. Turkmen, Notes on the (s,t) Lucas and Lucas matrix sequences. ARS Combinatoria **89**, 271–285 (2008)
4. H. Civciv, R. Turkmen, On the (s,t) Fibonacci and Fibonacci matrix sequences. ARS Combinatoria **87**, 161–173 (2008)
5. A.F. Horadam, Jacobsthal representation numbers. Fibonacci Q. **37**(2), 141–144 (1996)
6. A.F. Horadam, Jacobsthal representation polynomials. Fibonacci Q. **35**(2), 137–148 (1997)
7. F. Köken, D. Bozkurt, On the Jacobsthal numbers by matrix methods. Int. J. Contemp. Math. Sci. **3**(13), 605–614 (2008)
8. T. Koshy, *Fibonacci and Lucas Numbers with Applications* (Wiley, New York, 2001)
9. J.R. Silvester, Fibonacci properties by matrix methods. Math. Gaz. **63**(425), 188–191 (1979)
10. K. Uslu, Ş. Uygun, The (s,t) Jacobsthal and (s,t) Jacobsthal-Lucas matrix sequences. ARS Combinatoria **108**, 13–22 (2013)
11. E.W. Weisstein, Jacobsthal Number, Wolfram Mathworld, online, http://mathworld.wolfram.com/JacobsthalNumber.html. Retrieved 2007

Chapter 24
Domain Decomposition Approximation Approach for an Elliptic Partial Differential Equation

Tahira Nasreen Buttar and Naila Sajid

Abstract Domain Decomposition Methods present a strong and general class of techniques for the approximate solution of partial differential equations. A non-overlapping Domain Decomposition Method for the solution of Elliptic Partial Differential Equation is formulated. This DDM involves to find solution of Dirichlet and Neumann problem on each sub-domain, along with smoothing operation on the interfaces of the sub-domains. Analysis of this iterative non-overlapping scheme is made for an elliptic problem. At odd iteration levels, we enforce Dirichlet boundary value among sub-domain problems at their interfaces, whereas at even iterative levels are imposed Neumann boundary values. Fourier analysis is applied to show the fast convergence rate of this DDM in case of constant coefficient and four rectangular sub-domains.

24.1 Introduction

The idea of Domain Decomposition is very simple: it is to split a domain into finite number of sub-domains. DDMs are usually applied to boundary value problems (BVP). When the domain is decomposed into sub-domains, then the BVP is applicable on each sub-domain separately with its respective boundary conditions. The problems on each sub-domain are independent and matching conditions on the interfaces are imposed which makes this method suitable for parallel computing. There are techniques for splitting the domain into overlapping, such as Schwarz alternating method and Additive Schwarz method. Many DDMs can be analyzed and written as a special case of abstract additive Schwarz method. Other DDMs are with non-overlapping sub-domains. In this type each pair of adjacent sub-domains has intersection only on one interface, i.e. a curve or a line Γ. Methods of Schur complement, Fictitious domain, Neumann–Dirichlet, Neumann–Neumann, balancing domain decomposition (BDD) are non-overlapping DDMs.

T.N. Buttar (✉) • N. Sajid
Department of Mathematics, Lahore College for Women University, Pakistan
e-mail: abdullahbuttar@hotmail.com; sajid.naila@yahoo.com

G.A. Anastassiou, O. Duman (eds.), *Computational Analysis*, Springer Proceedings in Mathematics & Statistics 155, DOI 10.1007/978-3-319-28443-9_24

In engineering community, DDMs are much admired. These methods solve problems with deficiency of large RAM and save time of computation, as well as handle the complications which arise from usage of the finite element method. Now-a-days, large computer simulations are in concentrate of science and engineering. While solving a PDE numerically, we get a system of algebraic equations, which can be solved by direct (such as Gaussian elimination) or iterative methods which are based on minimization of the iterated form of $Ax - b$. Iterative schemes like Gauss Seidel and SOR are effective. To solve large system of algebraic equations efficiently, both of the above methods are not suitable. Parallelization of sequential codes in general can be very complicated in some cases. However, the most complicated part of parallelization of the FEM is the parallelization of the solver of system of algebraic equations. DDMs solve the problems by overcoming these deficiencies. A non-overlapping DDM to obtain the numerical solution of an Elliptic PDE problem is formulated. The proposed numerical scheme involves the solution of the problems with Dirichlet condition as well as problems with Neumann boundary conditions imposed on each sub-domain, by using smoothing process on the interfaces of the sub-domains. Emphasis is given to formulation and implementation of this method

We have developed a modified domain decomposition scheme for solving elliptic partial differential equation and applied it to solve Poisson equation for different choices of relaxation parameters and grid sizes. We have used this iterative DDM, for elliptic problems with non-overlapping decomposition of domain. The conditions on boundary vary at even iteration (where Neumann boundary conditions are applied) and at odd iterations (where we entertain Dirichlet boundary conditions) among sub-domains and are interchanged at their interfaces. Let us consider a domain D that is decomposed into a family of non-overlapping sub-domains $\{N_i, 1 \leq i \leq N\}$ with

$$\overline{D} = \bigcup_i \overline{D_i}, \;\; D_i \cap D_j = \emptyset, \;\; i \neq j.$$

If $\Gamma_i = \partial D_i - \partial D$, the interface Γ is defined so that Γ and Γ_i are open.

The conditions at Γ are phrased as transmission conditions, moreover, the function and their derivatives which are traced along with their independent linear combinations are termed n_i which is outward normal to D_i.

24.2 Derivation of the Modified Domain Decomposition Scheme

The BVP selected here is the following:

$$Lw = f \text{ in } D,$$
$$w = g \text{ on } \partial D,$$

where L is the operator defined by

$$Lw = -\sum_{i,j=1}^{d} \frac{\partial}{\partial x_i}\left(a_{ij}(x)\frac{\partial w}{\partial x_i}\right) + a_0(x)w.$$

We introduce a generalized version of the scheme given in [1].

Let D be a polygon which is convex in R^d, $d = 1, 2, 3, \ldots$, having a boundary ∂D. Given $f \in L^2(D)$ and $g \in H^{\frac{1}{2}}(\partial D)$, we find $w \in H^1(D)$ for which

$$Lw = f \text{ in } D, \; w = g \text{ on } \partial D.$$

A Domain Decomposition scheme is proposed for second order Elliptic problem. The initial domain D is sub-divided into sub-domains D_i, $i = 1, 2, 3, 4$, inside of each subdomain the differential equation is satisfied. At every interface between two adjacent sub-domains, the continuity of the solution and of its derivatives is imposed. DDM is quite easy to formulate and can be applied to randomly chosen decompositions. To ease the appropriate analysis, the domain D is divided into four non-overlapping sub-domains D_1, D_2, D_3, and D_4 such that

$$\begin{aligned}
&\overline{D} = \overline{D_1} \cup \overline{D_2} \cup \overline{D_3} \cup \overline{D_4},\\
&D_1 \cap D_2 = \emptyset,\; D_2 \cap D_3 = \emptyset,\; D_3 \cap D_4 = \emptyset,\\
&\partial D_i \cap \partial D \neq \emptyset,\; i = 1, 2, 3, 4
\end{aligned}$$

We denote the interfaces of the four sub-domains by

$$\begin{aligned}
\Gamma_1 &= \partial D_1 \cap \partial D_2,\\
\Gamma_2 &= \partial D_2 \cap \partial D_3,\\
\Gamma_3 &= \partial D_3 \cap \partial D_4.
\end{aligned}$$

As the earlier research suggests that subject to appropriate regularity conditions, the problem becomes equivalent to the four problems mentioned as under:

In Domain D_1:

$$\begin{aligned}
Lw_1 &= f \quad \text{in } D_1,\\
w_1 &= g \text{ on } \partial D_1 \cap \partial D,\\
w_1 &= w_2 \text{ on } \Gamma_1,\\
\frac{\partial w_1}{\partial v^1} + \frac{\partial w_2}{\partial v^2} &= 0 \text{ on } \Gamma_1.
\end{aligned}$$

In Domain D_2:

$$\begin{aligned} Lw_2 &= f \text{ in } D_2, \\ w_2 &= g \text{ on } \partial D_2 \cap \partial D, \\ w_2 &= \begin{cases} w_1, \text{ on } \Gamma_1 \\ w_3, \text{ on } \Gamma_2, \end{cases} \\ \frac{\partial w_2}{\partial v^2} + \frac{\partial w_1}{\partial v^2} &= 0 \text{ on } \Gamma_1, \\ \frac{\partial w_2}{\partial v^2} + \frac{\partial w_3}{\partial v^3} &= 0 \text{ on } \Gamma_2. \end{aligned}$$

In Domain D_3:

$$\begin{aligned} Lw_3 &= f \text{ in } D_3, \\ w_3 &= g \text{ on } \partial D_3 \cap \partial D, \\ w_3 &= \begin{cases} w_2, \text{ on } \Gamma_2 \\ w_4, \text{ on } \Gamma_3, \end{cases} \\ \frac{\partial w_3}{\partial v^3} + \frac{\partial w_2}{\partial v^2} &= 0 \text{ on } \Gamma_2, \\ \frac{\partial w_3}{\partial v^3} + \frac{\partial w_4}{\partial v^4} &= 0 \text{ on } \Gamma_3. \end{aligned}$$

In Domain D_4:

$$\begin{aligned} Lw_4 &= f \text{ in } D_4, \\ w_4 &= g \text{ on } \partial D_4 \cap \partial D, \\ w_4 &= w_3 \text{ on } \Gamma_3, \\ \frac{\partial w_4}{\partial v^4} + \frac{\partial w_3}{\partial v^3} &= 0 \text{ on } \Gamma_3, \end{aligned}$$

where for $n = 1, 2, 3, 4$, $w_n = w|_{D_n}$ and v^n is the unit vector in direction of outward normal to ∂D_n. For this Domain Decomposition scheme, we assume $w_n^{(0)} \in H^{(1)}(D_n)$ with $w_n^{(0)}|_{\partial D_n \cap \partial D} = g$, $n = 1, 2, 3, 4$. For $k = 0, 1, 2, 3, \ldots$, we get the sequence so constructed as (w_n^{k+1}) where $w_n^{k+1} \in H^{(1)}(D_n)$ with $w_n^{(0)}|_{\partial D_n \cap \partial D} = g$, $n = 1, 2, 3, 4$, satisfying:

$$\begin{aligned} Lw_1^{(2k+1)} &= f \text{ in } D_1, \\ w_1^{(2k+1)} &= \alpha w_1^{(2k)} + (1-\alpha) w_2^{(2k)} \text{ on } \Gamma_1, \end{aligned}$$

$$
\begin{aligned}
Lw_2^{(2k+1)} &= f \text{ in } D_2, \\
w_2^{(2k+1)} &= \begin{cases} \alpha w_2^{(2k)} + (1-\alpha) w_1^{(2k)}, \text{ on } \Gamma_1 \\ \alpha w_2^{(2k)} + (1-\alpha) w_3^{(2k)}, \text{ on } \Gamma_2, \end{cases} \\
Lw_3^{(2k+1)} &= f \text{ in } D_3, \\
w_3^{(2k+1)} &= \begin{cases} \alpha w_3^{(2k)} + (1-\alpha) w_2^{(2k)}, \text{ on } \Gamma_2 \\ \alpha w_3^{(2k)} + (1-\alpha) w_4^{(2k)}, \text{ on } \Gamma_3, \end{cases} \\
Lw_4^{(2k+1)} &= f \text{ in } D_4, \\
w_4^{(2k+1)} &= \alpha w_4^{(2k)} + (1-\alpha) w_3^{(2k)} \text{ on } \Gamma_3.
\end{aligned}
$$

While

$$
\begin{aligned}
Lw_1^{(2k+2)} &= f \text{ in } D_1, \\
\frac{\partial w_1^{(2k+2)}}{\partial v^1} &= \beta \frac{\partial w_1^{(2k+1)}}{\partial v^1} + (1-\beta) \frac{\partial w_2^{(2k+1)}}{\partial v^1} \text{ on } \Gamma_1, \\
Lw_2^{(2k+2)} &= f \text{ in } D_2, \\
\frac{\partial w_2^{(2k+2)}}{\partial v^2} &= \begin{cases} \beta \frac{\partial w_2^{(2k+1)}}{\partial v^2} + (1-\beta) \frac{\partial w_1^{(2k+1)}}{\partial v^2}, \text{ on } \Gamma_1 \\ \beta \frac{\partial w_2^{(2k+1)}}{\partial v^2} + (1-\beta) \frac{\partial w_3^{(2k+1)}}{\partial v^2}, \text{ on } \Gamma_2, \end{cases} \\
Lw_3^{(2k+2)} &= f \text{ in } D_3, \\
\frac{\partial w_3^{(2k+2)}}{\partial v^3} &= \begin{cases} \beta \frac{\partial w_3^{(2k+1)}}{\partial v^3} + (1-\beta) \frac{\partial w_2^{(2k+1)}}{\partial v^3}, \text{ on } \Gamma_2 \\ \beta \frac{\partial w_3^{(2k+1)}}{\partial v^3} + (1-\beta) \frac{\partial w_4^{(2k+1)}}{\partial v^3}, \text{ on } \Gamma_3, \end{cases} \\
Lw_4^{(2k+2)} &= f \text{ in } D_4, \\
\frac{\partial w_4^{(2k+2)}}{\partial v^4} &= \beta \frac{\partial w_4^{(2k+1)}}{\partial v^4} + (1-\beta) \frac{\partial w_3^{(2k+1)}}{\partial v^4} \text{ on } \Gamma_3,
\end{aligned}
$$

where $\alpha, \beta \in (0, 1)$ are the "relaxation parameters."The choice of "relaxation parameters" decides to assurance of convergence and to enhance its rate of convergence. The parameters α and β depend upon the sub-domains selected and the PDE under consideration. There is not any definite approach to estimate the most suitable values for α and β but $\left(\frac{1}{2}, \frac{1}{2}\right)$ is observed as suitable pair of values.

Continuity of the variables w and $\frac{\partial w}{\partial v}$ in this scheme is imposed on the interface alternately at each iterative stage. Using DDM, the applications for approximation and numerical solution will be investigated for convergence while optimum value of relaxation parameters α and β will be determined. Numerical result will be displayed.

24.3 Analysis of the Proposed Domain Decomposition Method

Here we discuss the formulated DD scheme for PDE problems on rectangular subdomains and analyze the error in each subdomain. We suppose the model problem:

$$-\bigtriangleup w + \lambda w = f \;\text{ on }\; D = [-x_1, x_2] \times [-1, 1],$$

$w = 0$ on boundary ∂D, where $x_1, x_2 > 0$ and $\lambda > 0$. We tear the domain D into four sub-domains

$$\begin{aligned}
D_1 &= \left[-x_1, \frac{-x_1}{2}\right] \times [-1, 1],\\
D_2 &= \left[\frac{-x_1}{2}, 0\right] \times [-1, 1],\\
D_3 &= \left[0, \frac{x_2}{2}\right] \times [-1, 1],\\
D_4 &= \left[\frac{x_2}{2}, x_2\right] \times [-1, 1],
\end{aligned}$$

so that the interfaces line Γ_i, $i = 1, 2, 3$ are $x = -\frac{x_1}{2}, 0, \frac{x_2}{2}$. If at jth iteration the solution of the DDM is defined by $w_i^{(j)}$ on each sub-domain D_i, then it is simple to prove that the error functions $E_i^{(j)}$ accordingly defined by

$$E_i^{(j)}(x, y) = w(x, y) - w_i^{(j)}(x, y) \text{ for } (x, y) \in D_i$$

satisfy, for $k = 0, 1, 2, \ldots$, the PDE problems:

In Domain D_1:

$$\begin{aligned}
-\bigtriangleup E_1^{(2k+1)} + \lambda E_1^{(2k+1)} &= 0 \;\text{ in }\; D_1,\\
E_1^{(2k+1)} &= c\left(\alpha, E_1^{(2k)}, E_2^{(2k)}\right) \;\text{ on }\; \Gamma_1,\\
E_1^{(2k+1)} &= 0 \text{ on } \partial D_1 \setminus \Gamma_1.
\end{aligned}$$

In Domain D_2:

$$\begin{aligned} -\triangle E_2^{(2k+1)} + \lambda E_2^{(2k+1)} &= 0 \text{ in } D_2, \\ E_2^{(2k+1)} &= c\left(\alpha, E_2^{(2k)}, E_1^{(2k)}\right) \text{ on } \Gamma_1, \\ E_2^{(2k+1)} &= 0 \text{ on } \partial D_2 \setminus \Gamma_1 \end{aligned}$$

and

$$\begin{aligned} -\triangle E_2^{(2k+1)} + \lambda E_2^{(2k+1)} &= 0 \text{ in } D_2, \\ E_2^{(2k+1)} &= c\left(\alpha, E_2^{(2k)}, E_3^{(2k)}\right) \text{ on } \Gamma_2, \\ E_2^{(2k+1)} &= 0 \text{ on } \partial D_2 \setminus \Gamma_2. \end{aligned}$$

In Domain D_3:

$$\begin{aligned} -\triangle E_3^{(2k+1)} + \lambda E_3^{(2k+1)} &= 0 \text{ in } D_3, \\ E_3^{(2k+1)} &= c\left(\alpha, E_3^{(2k)}, E_2^{(2k)}\right) \text{ on } \Gamma_2, \\ E_3^{(2k+1)} &= 0 \text{ on } \partial D_3 \setminus \Gamma_2 \end{aligned}$$

and

$$\begin{aligned} -\triangle E_3^{(2k+1)} + \lambda E_2^{(2k+1)} &= 0 \text{ in } D_2, \\ E_3^{(2k+1)} &= c\left(\alpha, E_3^{(2k)}, E_4^{(2k)}\right) \text{ on } \Gamma_2, \\ E_3^{(2k+1)} &= 0 \text{ on } \partial D_2 \setminus \Gamma_3. \end{aligned}$$

In Domain D_4:

$$\begin{aligned} -\triangle E_4^{(2k+1)} + \lambda E_4^{(2k+1)} &= 0 \text{ in } D_4, \\ E_4^{(2k+1)} &= c\left(\alpha, E_4^{(2k)}, E_3^{(2k)}\right) \text{ on } \Gamma_3, \\ E_4^{(2k+1)} &= 0 \text{ on } \partial D_4 \setminus \Gamma_3. \end{aligned}$$

In Domain D_1:

$$\begin{aligned} -\triangle E_1^{(2k+2)} + \lambda E_1^{(2k+2)} &= 0 \text{ in } D_1, \\ \frac{\partial E_1^{(2k+2)}}{\partial x} &= \left(\beta \frac{\partial E_1^{(2k+1)}}{\partial x}, (1-\beta)\frac{\partial E_2^{(2k+1)}}{\partial x}\right) \text{ on } \Gamma_1, \\ E_1^{(2k+2)} &= 0 \text{ on } \partial D_1 \setminus \Gamma_1. \end{aligned}$$

In Domain D_2:

$$\begin{aligned} - \triangle E_2^{(2k+2)} + \lambda E_2^{(2k+2)} &= 0 \text{ in } D_2, \\ \frac{\partial E_2^{(2k+2)}}{\partial x} &= \left(\beta \frac{\partial E_2^{(2k+1)}}{\partial x}, (1-\beta) \frac{\partial E_1^{(2k+1)}}{\partial x} \right) \text{ on } \Gamma_1, \\ E_2^{(2k+2)} &= 0 \text{ on } \partial D_2 \setminus \Gamma_1 \end{aligned}$$

and

$$\begin{aligned} - \triangle E_2^{(2k+2)} + \lambda E_1^{(2k+2)} &= 0 \text{ in } D_2, \\ \frac{\partial E_2^{(2k+2)}}{\partial x} &= \left(\beta \frac{\partial E_2^{(2k+1)}}{\partial x}, (1-\beta) \frac{\partial E_3^{(2k+1)}}{\partial x} \right) \text{ on } \Gamma_2, \\ E_2^{(2k+2)} &= 0 \text{ on } \partial D_2 \setminus \Gamma_2. \end{aligned}$$

In Domain D_3:

$$\begin{aligned} - \triangle E_3^{(2k+2)} + \lambda E_3^{(2k+2)} &= 0 \text{ in } D_3 \\ \frac{\partial E_3^{(2k+2)}}{\partial x} &= \left(\beta \frac{\partial E_3^{(2k+1)}}{\partial x}, (1-\beta) \frac{\partial E_2^{(2k+1)}}{\partial x} \right) \text{ on } \Gamma_2 \\ E_3^{(2k+2)} &= 0 \text{ on } \partial D_3 \setminus \Gamma_2 \end{aligned}$$

and

$$\begin{aligned} - \triangle E_3^{(2k+2)} + \lambda E_3^{(2k+2)} &= 0 \text{ in } D_2 \\ \frac{\partial E_3^{(2k+2)}}{\partial x} &= \left(\beta \frac{\partial E_3^{(2k+1)}}{\partial x}, (1-\beta) \frac{\partial E_4^{(2k+1)}}{\partial x} \right) \text{ on } \Gamma_3 \\ E_3^{(2k+2)} &= 0 \text{ on } \partial D_3 \setminus \Gamma_3. \end{aligned}$$

In Domain D_4:

$$\begin{aligned} - \triangle E_4^{(2k+2)} + \lambda E_4^{(2k+2)} &= 0 \text{ in } D_3, \\ \frac{\partial E_4^{(2k+2)}}{\partial x} &= \left(\beta \frac{\partial E_4^{(2k+1)}}{\partial x}, (1-\beta) \frac{\partial E_3^{(2k+1)}}{\partial x} \right) \text{ on } \Gamma_3, \\ E_4^{(2k+2)} &= 0 \text{ on } \partial D_4 \setminus \Gamma_3. \end{aligned}$$

Now we set $\lambda_i = \lambda + \left(\frac{1}{2}\iota\pi\right)$ and define the functions such as

$$V_i(y) = \sin\left(\frac{1}{2}\iota\pi(y+1)\right), \text{ where } y \in (-1, 1),$$

$$W_i(x) = \frac{\sinh(\sqrt{\lambda_i}(-x - x_1))}{\sinh(\sqrt{\lambda_i}(-\frac{x_1}{2}))}, \text{ where } x \in \left(-x_1, -\frac{x_1}{2}\right),$$

$$X_i(x) = \begin{cases} \dfrac{\sinh(\sqrt{\lambda_i}(x))}{\sinh(\sqrt{\lambda_i}(-\frac{x_1}{2}))}, & x \in \left(-\frac{x_1}{2}, 0\right) \\[2ex] \dfrac{\sinh(\sqrt{\lambda_i}(-\frac{x_1}{2} - x))}{\sinh(\sqrt{\lambda_i}(-\frac{x_1}{2}))}, & x = 0, \end{cases}$$

$$Y_i(x) = \begin{cases} \dfrac{\sinh(\sqrt{\lambda_i}(x))}{\sinh(\sqrt{\lambda_i}(\frac{x_2}{2} - x))}, & x \in \left(0, \frac{x_2}{2}\right) \\[2ex] \dfrac{\sinh(\sqrt{\lambda_i}(-\frac{x_1}{2} - x))}{\sinh(\sqrt{\lambda_i}(\frac{x_2}{2}))}, & x = \frac{x_2}{2}, \end{cases}$$

and

$$Z_i(x) = \frac{\sinh(\sqrt{\lambda_i}(x))}{\sinh(\sqrt{\lambda_i}(x_2 - x))}, \text{ where } x \in \left(\frac{x_2}{2}, x_2\right).$$

It can be easily verified that the above listed functions $V_i(y)$, $W_i(x)$, $X_i(x)$, $Y_i(x)$, and $Z_i(x)$ are solutions of the following problems, respectively:

$$V_i''(y) + \left(\frac{1}{2}\pi i^2\right) V_i(y) = 0,$$

$$y \in (-1, 1),\ V_i(-1) = V_i(1) = 0,$$

$$-W_i''(x) + \lambda_i W_i(x) = 0,$$

$$x \in \left(-x_1, \frac{x_1}{2}\right),\ V_i(-x_1) = 0,\ W_i\left(-\frac{x_1}{2}\right) = 1,$$

$$-X_i''(x) + \lambda_i X_i(x) = 0,$$

$$x \in \left(-\frac{x_1}{2}, 0\right),\ X_i\left(-\frac{x_1}{2}\right) = 1,\ X_i(0) = 1,$$

$$-Y_i''(x) + \lambda_i Y_i(x) = 0,$$

$$x \in \left(0, \frac{x_2}{2}\right),\ Y_i(0) = 1,\ Y_i\left(\frac{x_2}{2}\right) = 1,$$

and

$$-Z_i''(x) + \lambda_i Z_i(x) = 0,$$
$$x \in \left(\frac{x_2}{2}, x_2\right), \; Z_i\left(\frac{x_2}{2}\right) = 1, \; Z_i(x_2) = 0.$$

Now, for the jth iteration, the error $E_i^j(x, y)$ can be expanded in each sub-domain in terms of the V_i, W_i, X_i, Y_i, Z_i as follows: The error functions in D_1, D_2, D_3, and D_4 are defined respectively as follows:

$$E_1^j(x, y) = \sum_{i=1}^{\infty} a_i^{(j)} W_i(x) V_i(y) \text{ on } \Gamma_1, \text{ i.e., at } x = -\frac{x_1}{2},$$

$$E_2^j(x, y) = \begin{cases} \sum_{i=1}^{\infty} b_i^{(j)} X_i(x) V_i(y), \text{ on } \Gamma_1, \text{ i.e., at } x = -\frac{x_1}{2} \\ \sum_{i=1}^{\infty} \hat{b}_i^{(j)} X_i(x) V_i(y), \text{ on } \Gamma_2, \text{ i.e., at } x = 0 \end{cases}$$

$$E_3^j(x, y) = \begin{cases} \sum_{i=1}^{\infty} c_i^{(j)} Y_i(x) V_i(y), \text{ on } \Gamma_2, \text{ i.e., at } x = 0 \\ \sum_{i=1}^{\infty} \hat{c}_i^{(j)} Y_i(x) V_i(y), \text{ on } \Gamma_3, \text{ i.e., at } x = \frac{x_2}{2} \end{cases}$$

$$E_4^j(x, y) = \sum_{i=1}^{\infty} d_i^{(j)} Z_i(x) V_i(y) \text{ on } \Gamma_3, \text{ i.e., at } x = \frac{x_2}{2}.$$

The coefficients $a_i^{(2k+1)}$, $b_i^{(2k+1)}$, $c_i^{(2k+1)}$, and $d_i^{(2k+1)}$ of the series given in the following are same as coefficients for expansions $E_i^{(2k+1)}$ and $\frac{\partial E_i^{(2k+2)}}{\partial x}$ on Γ_i for $k = 0, 1, 2, 3, \ldots$:

$$a_i^{(2k+1)} = \int_{-1}^{1} \left[\alpha E_1^{(2k)}\left(-\frac{x_1}{2}, y\right) + (1-\alpha) E_2^{(2k)}\left(-\frac{x_1}{2}, y\right)\right] V_i(y) \mathrm{d}y,$$

$$b_i^{(2k+1)} = \int_{-1}^{1} \left[\alpha E_2^{(2k)}\left(-\frac{x_1}{2}, y\right) + (1-\alpha) E_1^{(2k)}\left(-\frac{x_1}{2}, y\right)\right] V_i(y) \mathrm{d}y,$$

$$\hat{b}_i^{(2k+1)} = \int_{-1}^{1} \left[\alpha E_2^{(2k)}(0, y) + (1-\alpha) E_3^{(2k)}(0, y)\right] V_i(y) \mathrm{d}y,$$

$$c_i^{(2k+1)} = \int_{-1}^{1} \left[\alpha E_3^{(2k)}(0,y) + (1-\alpha)E_2^{(2k)}(0,y)\right] V_i(y)\mathrm{d}y,$$

$$\hat{c}_i^{(2k+1)} = \int_{-1}^{1} \left[\alpha E_3^{(2k)}\left(\frac{x_2}{2},y\right) + (1-\alpha)E_4^{(2k)}\left(\frac{x_2}{2},y\right)\right] V_i(y)\mathrm{d}y,$$

$$d_i^{(2k+1)} = \int_{-1}^{1} \left[\alpha E_4^{(2k)}\left(\frac{x_2}{2},y\right) + (1-\alpha)E_3^{(2k)}\left(\frac{x_2}{2},y\right)\right] V_i(y)\mathrm{d}y.$$

As the $V_i's$ are orthogonal in $L^2(\Gamma_i. i = 1, 2, 3.)$, we get

$$\begin{aligned}
a_i^{(2k+1)} &= \left[\alpha a_i^{(2k)} W_i\left(-\frac{x_1}{2}\right) + (1-\alpha)b_i^{(2k)} X_i\left(-\frac{x_1}{2}\right)\right] \\
&= \left[\alpha a_i^{(2k)} + (1-\alpha)b_i^{(2k)}\right], \\
b_i^{(2k+1)} &= \left[\alpha b_i^{(2k)} X_i\left(-\frac{x_1}{2}\right) + (1-\alpha)a_i^{(2k)} W_i\left(-\frac{x_1}{2}\right)\right] \\
&= \left[\alpha b_i^{(2k)} + (1-\alpha)a_i^{(2k)}\right], \\
\hat{b}_i^{(2k+1)} &= \left[\alpha \hat{b}_i^{(2k)} X_i(0) + (1-\alpha)c_i^{(2k)} Y_i(0)\right] \\
&= \left[\alpha \hat{b}_i^{(2k)} + (1-\alpha)c_i^{(2k)}\right], \\
c_i^{(2k+1)} &= \left[\alpha c_i^{(2k)} Y_i(0) + (1-\alpha)\hat{b}_i^{(2k)} X_i(0)\right] \\
&= \left[\alpha c_i^{(2k)} + (1-\alpha)\hat{b}_i^{(2k)}\right], \\
\hat{c}_i^{(2k+1)} &= \left[\alpha \hat{c}_i^{(2k)} Y_i\left(\frac{x_2}{2}\right) + (1-\alpha)d_i^{(2k)} Z_i\left(\frac{x_2}{2}\right)\right] \\
&= \left[\alpha \hat{c}_i^{(2k)} + (1-\alpha)d_i^{(2k)}\right], \\
d_i^{(2k+1)} &= \left[\alpha d_i^{(2k)} Z_i\left(\frac{x_2}{2}\right) + (1-\alpha)\hat{c}_i^{(2k)} Y_i\right] \\
&= \left[\alpha d_i^{(2k)} + (1-\alpha)\hat{c}_i^{(2k)}\right].
\end{aligned}$$

Now for even iterations:

$$a_i^{(2k+2)} = -t_i\left(-\frac{x_1}{2}\right) \int_{-1}^{1} \left[\beta \frac{\partial E_1^{(2k+1)}(-\frac{x_1}{2},y)}{\partial x} + (1-\beta)\frac{\partial E_2^{(2k+1)}(-\frac{x_1}{2},y)}{\partial x}\right] V_i(y)\mathrm{d}y,$$

$$b_i^{(2k+2)} = -t_i\left(-\frac{x_1}{2}\right)\int_{-1}^{1}\left[\beta\frac{\partial E_2^{(2k+1)}(-\frac{x_1}{2},y)}{\partial x} + (1-\beta)\frac{\partial E_1^{(2k+1)}(-\frac{x_1}{2},y)}{\partial x}\right]V_i(y)\mathrm{d}y,$$

$$\hat{b}_i^{(2k+2)} = -t_i\left(-\frac{x_1}{2}\right)\int_{-1}^{1}\left[\beta\frac{\partial E_2^{(2k+1)}(0,y)}{\partial x} + (1-\beta)\frac{\partial E_3^{(2k+1)}(0,y)}{\partial x}\right]V_i(y)\mathrm{d}y,$$

$$c_i^{(2k+2)} = -t_i\left(-\frac{x_1}{2}\right)\int_{-1}^{1}\left[\beta\frac{\partial E_3^{(2k+1)}(0,y)}{\partial x} + (1-\beta)\frac{\partial E_2^{(2k+1)}(0,y)}{\partial x}\right]V_i(y)\mathrm{d}y,$$

$$\hat{c}_i^{(2k+2)} = -t_i\left(-\frac{x_1}{2}\right)\int_{-1}^{1}\left[\beta\frac{\partial E_3^{(2k+1)}(\frac{x_2}{2},y)}{\partial x} + (1-\beta)\frac{\partial E_4^{(2k+1)}(\frac{x_2}{2},y)}{\partial x}\right]V_i(y)\mathrm{d}y,$$

$$d_i^{(2k+2)} = -t_i\left(-\frac{x_1}{2}\right)\int_{-1}^{1}\left[\beta\frac{\partial E_4^{(2k+1)}(\frac{x_2}{2},y)}{\partial x} + (1-\beta)\frac{\partial E_3^{(2k+1)}(\frac{x_2}{2},y)}{\partial x}\right]V_i(y)\mathrm{d}y,$$

where $t_i(x) = \left[\tanh\sqrt{\lambda_i}(x)\right]/\sqrt{\lambda_i}$.

$$\frac{\partial E_1^{(j)}(-\frac{x_1}{2},y)}{\partial x} = \sum_{i=1}^{\infty} a_i^{(j)} W_i'\left(-\frac{x_1}{2}\right)V_i(y),$$

$$\frac{\partial E_2^{(j)}(-\frac{x_1}{2},y)}{\partial x} = \begin{cases}\sum_{i=1}^{\infty} b_i^{(j)} X_i'\left(-\frac{x_1}{2}\right)V_i(y),\\[2ex] \sum_{i=1}^{\infty} \hat{b}_i^{(j)} X_i'(0)\,V_i(y),\end{cases}$$

$$\frac{\partial E_3^{(j)}(0,y)}{\partial x} = \begin{cases}\sum_{i=1}^{\infty} c_i^{(j)} Y_i'(0)\,V_i(y),\\[2ex] \sum_{i=1}^{\infty} \hat{c}_i^{(j)} Y_i'\left(\frac{x_2}{2}\right)V_i(y),\end{cases}$$

$$\frac{\partial E_4^{(j)}(\frac{x_2}{2},y)}{\partial x} = \sum_{i=1}^{\infty} d_i^{(j)} Z_i'\left(\frac{x_2}{2}\right)V_i(y),$$

$$W_i(x) = \frac{\sinh\left(\sqrt{\lambda_i}(-x-x_1)\right)}{\sinh\left(\sqrt{\lambda_i}(-\frac{x_1}{2})\right)}.$$

Differentiating,

$$W_i'(x) = \frac{\cosh\left(\sqrt{\lambda_i}(-x-x_1)\right)}{\sinh\left(\sqrt{\lambda_i}(-\frac{x_1}{2})\right)}\left(-\sqrt{\lambda_i}\right),$$

$$W_i'\left(-\frac{x_1}{2}\right) = -\left(\sqrt{\lambda_i}\right)\coth\left(\sqrt{\lambda_i}\left(-\frac{x_1}{2}\right)\right).$$

Now

$$X_i(x) = \frac{\sinh\left(\sqrt{\lambda_i}(x)\right)}{\sinh\left(\sqrt{\lambda_i}(-\frac{x_1}{2})\right)},$$

$$X_i'(x) = \frac{\cosh\left(\sqrt{\lambda_i}(x)\right)}{\sinh\left(\sqrt{\lambda_i}(-\frac{x_1}{2})\right)}\left(\sqrt{\lambda_i}\right),$$

$$X_i'\left(-\frac{x_1}{2}\right) = \left(\sqrt{\lambda_i}\right)\coth\left(\sqrt{\lambda_i}\left(-\frac{x_1}{2}\right)\right).$$

In the same way,

$$X_i'(0) = -\left(\sqrt{\lambda_i}\right)\coth\left(\sqrt{\lambda_i}\left(-\frac{x_1}{2}\right)\right),$$

$$Y_i'(0) = -\left(\sqrt{\lambda_i}\right)\coth\left(\sqrt{\lambda_i}\left(\frac{x_2}{2}\right)\right),$$

$$Y_i'\left(\frac{x_2}{2}\right) = \left(\sqrt{\lambda_i}\right)\coth\left(\sqrt{\lambda_i}\left(\frac{x_2}{2}\right)\right),$$

$$Z_i'\left(\frac{x_2}{2}\right) = -\left(\sqrt{\lambda_i}\right)\coth\left(\sqrt{\lambda_i}\left(\frac{x_2}{2}\right)\right).$$

Using the orthogonally of the $V_i's$ in $L^2(\Gamma_1,\ i = 1, 2, 3.)$ we obtain that

$$a_i^{(2k+2)} = -t_i\left(-\frac{x_1}{2}\right)\left[\beta a_i^{(2k+1)} W_i'\left(-\frac{x_1}{2}\right) + (1-\beta) b_i^{(2k+1)} X_i'\left(-\frac{x_1}{2}\right)\right],$$

$$a_i^{(2k+2)} = \frac{\left[\tanh\left(\sqrt{\lambda_i}\left(-\frac{x_1}{2}\right)\right)\right]}{\sqrt{\lambda_i}}$$
$$\left[\beta a_i^{(2k+1)}\left(-\left(\sqrt{\lambda_i}\right)\coth\left(\sqrt{\lambda_i}\left(\frac{x_1}{2}\right)\right)\right)\right.$$
$$\left. + (1-\beta) b_i^{(2k+1)}\left(\sqrt{\lambda_i}\right)\coth\left(\sqrt{\lambda_i}\left(\frac{x_1}{2}\right)\right)\right],$$

$$a_i^{(2k+2)} = \left[\beta a_i^{(2k+1)} - (1-\beta) b_i^{(2k+1)}\right]$$

$$b_i^{(2k+2)} = t_i\left(-\frac{x_1}{2}\right)\left[\beta b_i^{(2k+1)} X_i'\left(-\frac{x_1}{2}\right)(1-\beta) a_i^{(2k+1)} W_i'\left(-\frac{x_1}{2}\right)\right],$$

$$b_i^{(2k+2)} = \left[\beta b_i^{(2k+1)} - (1-\beta) a_i^{(2k+1)}\right],$$

$$\hat{b}_i^{(2k+2)} = -t_i\left(-\frac{x_1}{2}\right)\left[\beta \hat{b}_i^{(2k+1)} X_i'(0) + (1-\beta) c_i^{(2k+1)} Y_i'(0)\right],$$

$$\hat{b}_i^{(2k+2)} = \left[\beta \hat{b}_i^{(2k+1)} p_i + (1-\beta) c_i^{(2k+1)}\right],$$

where $p_i = \dfrac{\left[\tanh\left(\sqrt{\lambda_i}\left(\frac{x_2}{2}\right)\right)\right]}{\left[\tanh\left(\sqrt{\lambda_i}\left(-\frac{x_1}{2}\right)\right)\right]}$ and

$$c_i^{(2k+2)} = -t_i\left(\frac{x_2}{2}\right)\left[\beta c_i^{(2k+1)} Y_I'(0) + (1-\beta)\hat{b}_i^{(2k+1)} X_I'(0)\right],$$

$$c_i^{(2k+2)} = \left[\beta c_i^{(2k+1)} + (1-\beta)\hat{b}_i^{(2k+1)} p_i^{-1}\right],$$

$$\hat{c}_i^{(2k+2)} = t_i\left(\frac{x_2}{2}\right)\left[\beta\hat{c}_i^{(2k+1)} Y_i'\left(\frac{x_2}{2}\right) + (1-\beta)d_i^{(2k+1)} Z_I'\left(\frac{x_2}{2}\right)\right],$$

$$\hat{c}_i^{(2k+2)} = \left[\beta\hat{c}_i^{(2k+1)} - (1-\beta)d_i^{(2k+1)}\right],$$

$$d_i^{(2k+2)} = -t_i\left(\frac{x_2}{2}\right)\left[\beta d_i^{(2k+1)} Z_I'\left(\frac{x_2}{2}\right) + (1-\beta)\hat{c}_i^{(2k+1)} Y_i'\left(\frac{x_2}{2}\right)\right],$$

$$d_i^{(2k+2)} = \left[\beta d_i^{(2k+1)} - (1-\beta)\hat{c}_i^{(2k+1)}\right].$$

Example 24.1. The following BVP of Poisson equation is considered: $\nabla^2 w = f$, on $|x| \leq 1$, $|y| \leq 1$, $w = g$, on the boundary ∂D.

Solution. Using the technique derived in Sect. 24.2, we decompose the domain in four sub-domains

$$D_1 = \left[-1, -\frac{1}{2}\right] \times [-1, -1],$$

$$D_2 = \left[-\frac{1}{2}, 0\right] \times [-1, -1],$$

$$D_3 = \left[0, \frac{1}{2}\right] \times [-1, 1],$$

$$D_4 = \left[\frac{1}{2}, 1\right] \times [-1, 1].$$

Initial guesses for the coefficients a_i, b_i, c_i, and d_i are taken along with appropriate choice of α and β. For $f = 1$ and $g = 0$ we have

Numerical result for the Example 24.1 with interfaces at $x = -\frac{1}{2}, 0, \frac{1}{2}$ and the total absolute errors are shown in Tables 24.1 and 24.2.

Table 24.1 Total error in whole domain $E = |E_1| + |E_2| + |E_3| + |E_4|$ $\alpha = 0.45$, $\beta = 0.5$, and grid size is $\frac{1}{20} \times \frac{1}{5}$

Iteration	Error
1.	9.94×10^{-4}
2.	1.50×10^{-4}
3.	2.8×10^{-5}

Table 24.2 Total error in whole domain $E = |E_1| + |E_2| + |E_3| + |E_4|$ $\alpha = 0.35$, $\beta = 0.65$, and grid size is $\frac{1}{40} \times \frac{1}{10}$

Iteration	Error
1.	6.7×10^{-5}
2.	1.8×10^{-5}
3.	5.0×10^{-6}

24.4 Conclusion

A new domain decomposition scheme is developed for solving elliptic partial differential equation and applied it to solve Poisson equation for different choices of relaxation parameters and grid sizes.

Reference

1. J.R. Rice, E.A. Vavalis, D. Yang, Analysis of a non-overlapping domain decomposition method for elliptic partial differential equations. J. Comput. Appl. Math. **87**, 11–19 (1997)

Chapter 25
Voronovskaja Type Approximation Theorem for q-Szasz–Schurer Operators

Tuba Vedi and Mehmet Ali Özarslan

Abstract In 2011, Özarslan (Miscolc Math Notes, 12:225–235, 2011) introduced the q-Szasz–Schurer operators and investigated their approximation properties. In the present paper, we state the Voronovskaja-type asymptotic formula for q-analogue of Szasz–Schurer operators.

25.1 Introduction

In the last decade, different types of Szasz operators were studied in [1, 6, 13], etc. On the other hand, q-analogue of Szasz operators were investigated in [4, 5, 7, 8, 12, 18–21, 23–25, 30]. Also, many researchers investigate q-analogue of Bernstein polynomials in [2, 3, 10, 17, 22, 27–29]. Particularly, Voronovskaja type theorem was stated in [9, 14, 15, 20]. In 2011, q-Szasz–Schurer operators were introduced by Özarslan [26] as

$$S_{n,q}(f;x;p) = \frac{1}{E_q\left([n+p]\,x\right)} \sum_{k=0}^{\infty} f\left(\frac{[k]}{q^{k-1}\,[n]}\right) q^{\frac{k(k-1)}{2}} \frac{[n+p]^k\, x^k}{[k]!}, \quad x \in [0,\infty),$$

where $0 < q < 1, f \in C\,[0,\infty)$ and $E_q\left([n+p]\,x\right)$ was given in [26].

Before giving our results, let us give some definitions related to q-calculus. Let $q \in \mathbb{R}$ and the function f be calculated at the q-integers $\frac{[k]}{[n]}$. Recall that the q-integer of $k \in \mathbb{R}$ is [16]

$$[k]_q = \begin{cases} \left(1-q^k\right)/\left(1-q\right), & q \neq 1 \\ k, & q = 1, \end{cases}$$

T. Vedi (✉) • M.A. Özarslan
Department of Mathematics, Eastern Mediterranean University, Famagusta, North Cyprus
e-mail: tuba.vedi@emu.edu.tr; mehmetali.ozarslan@emu.edu.tr

G.A. Anastassiou, O. Duman (eds.), *Computational Analysis*, Springer Proceedings in Mathematics & Statistics 155, DOI 10.1007/978-3-319-28443-9_25

the q-factorial is defined by

$$[k]_q! = \begin{cases} [k]_q [k-1]_q \dots [1]_q, & k = 1, 2, 3, \dots, \\ 1, & k = 0 \end{cases}$$

and q-binomial coefficients are defined by

$$\begin{bmatrix} n \\ k \end{bmatrix}_q = \frac{[n]_q!}{[n-k]_q! \, [k]_q!}$$

for $n \geq 0$, $k \geq 0$. On the other hand, q-integers satisfy the following well-known property

$$[n] = [s] + q^s [n-s]. \tag{25.1}$$

Firstly, let us state the following lemma and corollary which were given in [26]:

Lemma 25.1. *For fixed $p \in \mathbb{N}_0$ and $n \in \mathbb{N}$, we have the following relation for q-Szasz–Schurer operators*

$$S_{n,q}\left(t^{m+1}; x; p\right) = \frac{[n+p]_q}{[n]_q} x \sum_{j=0}^{m} \binom{m}{j} \frac{1}{q^j [n]_q^{m-j}} S_{n,q}\left(t^j; x; p\right). \tag{25.2}$$

Corollary 25.2. *Using* (25.2) *we have the following moments for the q-Szasz–Schurer operators:*

(*i*) $S_{n,q}(1; x; p) = 1,$

(*ii*) $S_{n,q}(t; x; p) = \dfrac{[n+p]_q}{[n]_q} x,$

(*iii*) $S_{n,q}\left(t^2; x; p\right) = \dfrac{1}{q} \left(\dfrac{[n+p]_q}{[n]_q} x \right)^2 + \dfrac{[n+p]_q}{[n]_q^2} x,$

(*iv*) $S_{n,q}\left(t^3; x; p\right) = \dfrac{1}{q^3} \left(\dfrac{[n+p]_q}{[n]_q} x \right)^3 + \dfrac{2q+1}{q^2} \dfrac{[n+p]_q^2}{[n]_q^3} x^2 + \dfrac{[n+p]_q}{[n]_q^3} x,$

(*v*) $S_{n,q}\left((t-x)^2; x; p\right) = \left[\left(\dfrac{[n+p]_q}{q [n]_q} - 2 \right) \dfrac{[n+p]_q}{[n]_q} + 1 \right] x^2 + \dfrac{[n+p]_q}{[n]_q^2} x.$

In this paper, we consider the following space:

$$E := \left\{ f \in C_2 [0, \infty) \text{ such that } \lim_{x \to \infty} \frac{|f(x)|}{1 + x^2} \text{ exists finitely} \right\}.$$

Also before, Özarslan gave the following corollary on the space E :

Corollary 25.3 (See [26]). *Let $q := q_n \in (0,1)$ and fix $p \in \mathbb{N}_0 := \mathbb{N} \cup \{0\}$. Then, for all $f \in E$, $\{S_{n,q}(f;x;p)\}$ converges uniformly to f on $[0,b]$ if and only if $\lim_{n\to\infty} q_n = 1$.*

25.2 Voronovskaja Type Theorem

Recently, Duman and Özarslan gave the Voronovskaja type theorem in [11] for Szasz–Mirakjan type operators. In this section, we state our main result for q-analogue of Szasz–Schurer operators. For proving our main theorem we need the following:

Lemma 25.4. *Using Lemma* 25.1, *for fixed $p \in \mathbb{N}_0$ and $n \in \mathbb{N}$, we get the following results:*

(*i*)

$$S_{n,q}\left(t^4;x;p\right) = \frac{1}{q^6}\left(\frac{[n+p]_q}{[n]_q}x\right)^4 + \left(\frac{3}{q^3\,[n]_q} + \frac{2q+1}{q^5\,[n]_q}\right)\left(\frac{[n+p]_q}{[n]_q}x\right)^3 + \left(\frac{3}{q\,[n]_q^2} + \frac{3}{q^2\,[n]_q^2} + \frac{1}{q^3\,[n]_q^2}\right)\left(\frac{[n+p]_q}{[n]_q}x\right)^2 + \frac{[n+p]_q}{[n]_q^3}x,$$

(*ii*)

$$\begin{aligned} S_{n,q}\left((t-x)^4;x;p\right) &= \left(\frac{[n+p]_q^4}{q^6\,[n]_q^4} - 4\frac{[n+p]_q^3}{q^3\,[n]_q^3} + 6\frac{[n+p]_q^2}{q\,[n]_q^2} - 4\frac{[n+p]_q}{[n]_q} + 1\right)x^4 \\ &\times \left(\left(\frac{3}{q^3} + \frac{2q+1}{q^5}\right)\frac{[n+p]_q^3}{[n]_q^4} - 4\frac{(2q+1)\,[n+p]_q^2}{q^2\,[n]_q^3} + 6\frac{[n+p]_q}{[n]_q^2}\right)x^3 \\ &+ \left(\left(\frac{3}{q} + \frac{3}{q^2} + \frac{1}{q^3}\right)\frac{[n+p]_q^2}{[n]_q^4} - 4\frac{[n+p]_q}{[n]_q^3}\right)x^2 + \frac{[n+p]_q}{[n]_q^3}x. \end{aligned}$$

Proof. By the help of Lemma 25.1 and Corollary 25.2 and linearity of the operators we can obtain the desired result.

Lemma 25.5. *Let (q_n) be a sequence with $0 < q_n < 1$, $\lim_{n\to\infty} q_n = 1$ and $\lim_{n\to\infty} q_n^n = A$. Then, for fixed $p \in \mathbb{N}_0$ and $n \in \mathbb{N}$ we have the following:*

(*i*) $\lim_{n\to\infty} [n]_{q_n} S_{n,q_n}((t-x);x;p) = pAx,$

(ii) $\lim_{n\to\infty}[n]_{q_n} S_{n,q_n}\left((t-x)^2;x;p\right)=x,$

(iii) $\lim_{n\to\infty}[n]_{q_n}^2 S_{n,q_n}\left((t-x)^4;x;p\right)=3(1-A)^2x^4-12(1-A)x^3+3x^2+x$

uniformly with respect to $x\in[0,c]$, $(c>0)$.

Proof. (i) By Corollary 25.2, we obtain the desired result directly with $0<q_n<1$, $\lim_{n\to\infty}q_n=1$ and $\lim_{n\to\infty}q_n^n=A$,

$$\begin{aligned}\lim_{n\to\infty}[n]_{q_n} S_{n,q_n}((t-x);x;p) &= \lim_{n\to\infty}[n]_{q_n}\left(\frac{[n+p]_{q_n}}{[n]_{q_n}}-1\right)x\\ &=\lim_{n\to\infty}\left([n+p]_{q_n}-[n]_{q_n}\right)x\\ &=\lim_{n\to\infty}\frac{q_n^n\left(1-q_n^p\right)}{1-q_n}x\\ &=pAx.\end{aligned}$$

(ii) If we take limit on both sides of $[n]_{q_n}S_{n,q_n}\left((t-x)^2;x;p\right)$ as n tends to infinity and taking into account that $0<q_n<1$, $\lim_{n\to\infty}q_n=1$ and $\lim_{n\to\infty}q_n^n=A$, and also using (25.1), we get

$$\begin{aligned}&\lim_{n\to\infty}[n]_{q_n}S_{n,q_n}\left((t-x)^2;x;p\right)\\ &=\lim_{n\to\infty}\left[\left(\frac{[n+p]_{q_n}}{q_n[n]_{q_n}}-2\right)\frac{[n]_{q_n}[n+p]_{q_n}}{[n]_{qn}}+[n]_{q_n}\right]x^2+\lim_{n\to\infty}\frac{[n]_{q_n}[n+p]_{q_n}}{[n]_{q_n}^2}x\\ &=\lim_{n\to\infty}\left(\frac{[n+p]_{q_n}^2-2q_n[n]_{q_n}[n+p]_{q_n}+q_n[n]_{q_n}^2}{q_n[n]_{q_n}}\right)x^2+x,\end{aligned}$$

which gives

$$\begin{aligned}&\lim_{n\to\infty}[n]_{q_n}S_{n,q_n}\left((t-x)^2;x;p\right)\\ &=\lim_{n\to\infty}\left(\frac{\left([p]_{q_n}+q_n^p[n]_{q_n}\right)^2-2q_n[n]_{q_n}\left([p]_{q_n}+q_n^p[n]_{q_n}\right)+q_n[n]_{q_n}^2}{q_n[n]_{q_n}}\right)x^2+x\\ &=\lim_{n\to\infty}\left(\frac{\left(q_n^{2p}-2q_n^{p+1}+q_n\right)[n]_{q_n}^2+[n]_{q_n}\left(2[p]_{q_n}q_n^p-2q_n[p]_{q_n}\right)+[p]_{q_n}^2}{q_n[n]_{q_n}}\right)x^2+x\\ &=\lim_{n\to\infty}\left(\frac{\left(q_n^p-1\right)^2}{q_n}\frac{1-q_n^n}{1-q_n}\right)x^2+x=\lim_{n\to\infty}\left(\frac{\left(q_n^p-1\right)[p]_{q_n}}{q_n}\right)x^2+x=x.\end{aligned}$$

(iii) Now, let us prove our last results of the lemma in the following way:

$$\lim_{n\to\infty} [n]_{q_n}^2 S_{n,q_n}\left((t-x)^4;x;p\right)$$
$$= \lim_{n\to\infty} [n]_{q_n}^2 \left[\left(\frac{[n+p]_{q_n}^4}{q_n^6 [n]_{q_n}^4} - 4\frac{[n+p]_{q_n}^3}{q_n^3 [n]_{q_n}^3} + 6\frac{[n+p]_{q_n}^2}{q_n [n]_{q_n}^2} - 4\frac{[n+p]_{q_n}}{[n]_{q_n}} + 1\right) x^4\right.$$
$$+ \left(\left(\frac{3}{q_n^3} + \frac{2q_n+1}{q_n^5}\right)\frac{[n+p]_{q_n}^3}{[n]_{q_n}^4} - 4\frac{(2q_n+1)[n+p]_{q_n}^2}{q_n^2 [n]_{q_n}^3} + 6\frac{[n+p]_{q_n}}{[n]_{q_n}^2}\right) x^3$$
$$\left. + \left(\left(\frac{3}{q_n} + \frac{3}{q_n^2} + \frac{1}{q_n^5}\right)\frac{[n+p]_{q_n}^2}{[n]_{q_n}^4} - 4\frac{[n+p]_{q_n}}{[n]_{q_n}^3}\right) x^2 + \frac{[n+p]_{q_n}}{[n]_{q_n}^3} x\right],$$

and hence

$$\lim_{n\to\infty} [n]_{q_n}^2 S_{n,q_n}\left((t-x)^4;x;p\right)$$
$$= \lim_{n\to\infty} [n]_{q_n}^2 \left(\frac{[n+p]_{q_n}^4}{q_n^6 [n]_{q_n}^4} - 4\frac{[n+p]_{q_n}^3}{q_n^3 [n]_{q_n}^3} + 6\frac{[n+p]_{q_n}^2}{q_n [n]_{q_n}^2} - 4\frac{[n+p]_{q_n}}{[n]_{q_n}} + 1\right) x^4$$
$$+ \lim_{n\to\infty} [n]_{q_n}^2 \left(\left(\frac{3}{q_n^3} + \frac{2q_n+1}{q_n^5}\right)\frac{[n+p]_{q_n}^3}{[n]_{q_n}^4} - 4\frac{(2q_n+1)[n+p]_{q_n}^2}{q^2 [n]_{q_n}^3} + 6\frac{[n+p]_{q_n}}{[n]_{q_n}^2}\right) x^3$$
$$+ \lim_{n\to\infty} [n]_{q_n}^2 \left(\left(\left(\frac{3}{q_n} + \frac{3}{q_n^2} + \frac{1}{q_n^3}\right)\frac{[n+p]_{q_n}^2}{[n]_{q_n}^4} - 4\frac{[n+p]_{q_n}}{[n]_{q_n}^3}\right) x^2 + \frac{[n+p]_{q_n}}{[n]_{q_n}^3}\right) x.$$

Now, again using (25.1) and re-arranging the above terms, we have

$$\lim_{n\to\infty} [n]_{q_n}^2 S_{n,q_n}\left((t-x)^4;x;p\right)$$
$$= \left[\lim_{n\to\infty}\left(\frac{q_n^{4p} - 4q_n^{3p+3} + 6q_n^{2p+5} - 4q_n^{p+6} + q_n^6}{q_n^6}\right)\left(\frac{1-q_n^n}{1-q_n}\right)_{q_n}^2\right.$$
$$+ \lim_{n\to\infty}\left(\frac{4[p]_{q_n} q_n^{3p} - 12q_n^{2p+3}[p]_{q_n} + 12q_n^{p+5}[p]_{q_n} - 4q_n^6[p]_{q_n}}{q_n^6}\right)\frac{1-q_n^n}{1-q_n}$$
$$+ \lim_{n\to\infty}\left(\frac{6[p]_{q_n}^2 q_n^{2p} - 12q_n^{p+3}[p]_{q_n}^2 + 6q_n^5[p]_{q_n}^2}{q_n^6}\right)$$
$$\left. + \lim_{n\to\infty}\left(\frac{4[p]_{q_n}^3 q_n^p - 4q_n^3[p]_{q_n}^3}{q_n^6}\right)\frac{1-q_n}{1-q_n^n} + \lim_{n\to\infty}\frac{[p]_{q_n}^4(1-q_n)}{q_n^6(1-q_n^n)}\right] x^4$$

$$+\left[\lim_{n\to\infty}\left(\frac{\frac{3q_n^2+2q_n+1}{q_n^3}3\left[p\right]_{q_n}^2 q_n^p-4\left(2q_n+1\right)q_n^{2p}+6q_n^{p+2}}{q_n^2}\right)\frac{1-q_n}{1-q_n^n}\right.$$

$$+\lim_{n\to\infty}\left(\frac{\frac{3q_n^2+2q_n+1}{q_n^3}3\left[p\right]_{q_n}q_n^{2p}-4\left(2q_n+1\right)2\left[p\right]_{q_n}q_n^p+6q_n^2\left[p\right]_{q_n}}{q_n^2}\right)$$

$$+\lim_{n\to\infty}\left(\frac{\frac{3q_n^2+2q_n+1}{q_n^3}3\left[p\right]_{q_n}^2 q_n^p-4\left(2q_n+1\right)\left[p\right]_{q_n}^2}{q_n^2}\right)\frac{1-q_n^n}{1-q_n}$$

$$\left.+\lim_{n\to\infty}\frac{\frac{3q_n^2+2q_n+1}{q_n^3}\left[p\right]_{q_n}^3}{q_n^2}\left(\frac{1-q_n}{1-q_n^n}\right)^2\right]x^3+3x^2+x.$$

Finally if $0 < q_n < 1$, $\lim_{n\to\infty} q_n = 1$ and $\lim_{n\to\infty} q_n^n = A$, then we obtain the following result:

$$\lim_{n\to\infty}\left[n\right]_{q_n}^2 S_{n,q_n}\left(\left(t-x\right)^4;x;p\right)$$
$$=3\left(1-A\right)^2x^4-12\left(1-A\right)x^3+3x^2+x,$$

whence the result.

Now, we can give the Voronovskaja type theorem for q-Szasz–Schurer operators.

Theorem 25.6. *Let (q_n) be a sequence with $0 < q_n < 1$, $\lim_{n\to\infty} q_n = 1$ and $\lim_{n\to\infty} q_n^n = A$. For any $f \in E[0,\infty)$ such that $f' \in E[0,\infty)$ and $f'' \in E[0,\infty)$. Then, we have the following relation:*

$$\lim_{n\to\infty}\left[n\right]_{q_n}\left(S_{n,q_n}\left(f\left(x\right);x;p\right)-f\left(x\right)\right)$$
$$=f'\left(x\right)pAx+f''\left(x\right)\left(\left(1-A\right)\frac{x^2}{2}+\frac{1}{2}x\right)$$

uniformly with respect to $x \in [0,c]$, $(c > 0)$.

Proof. From the Taylor expansion for f, we get

$$f\left(t\right)=f\left(x\right)+f'\left(x\right)\left(t-x\right)+\frac{1}{2}f''\left(x\right)\left(t-x\right)^2+\varepsilon\left(t,x\right)\left(t-x\right)^2,$$

where $\varepsilon(t,x) \to 0$ as $t \to x$. Using the linearity of the operators S_{n,q_n}, we get

$$S_{n,q_n}\left(f;x;p\right)-f\left(x\right)$$

$$
\begin{aligned}
&= f'(x)\left(\frac{[n+p]_{q_n}}{[n]_{q_n}}x - x\right) \\
&+ \frac{1}{2}f''(x)\left(\left[\left(\frac{[n+p]_{q_n}}{q_n[n]_{q_n}} - 2\right)\frac{[n+p]_{q_n}}{[n]_{q_n}} + 1\right]x^2 + \frac{[n+p]_{q_n}}{[n]_{q_n}^2}x\right) \\
&+ S_{n,q_n}\left(\varepsilon(t,x)(t-x)^2;x;p\right).
\end{aligned}
$$

Applying Cauchy–Schwarz inequality to the third term on the right-hand side, we get

$$
\begin{aligned}
&\lim_{n\to\infty}[n]_{q_n}S_{n,q_n}\left(\varepsilon(t,x)(t-x)^2;x;p\right) \\
&\leq \left(\lim_{n\to\infty}S_{n,q_n}\left(\varepsilon^2(t,x);x;p\right)\right)^{\frac{1}{2}}\left(\lim_{n\to\infty}[n]_{q_n}S_{n,q_n}\left((t-x)^4;x;p\right)\right)^{\frac{1}{2}}.
\end{aligned}
$$

Since $\varepsilon^2(t,x) \in E$, we have by Corollary 25.3 that

$$
\lim_{n\to\infty}[n]_{q_n}S_{n,q_n}\left(\varepsilon(t,x)(t-x)^2;x;p\right) = 0
$$

uniformly with respect to $x \in [0,c]$, $(c > 0)$. Then, we have by Lemma 25.5 (i) and (ii) that

$$
\begin{aligned}
&\lim_{n\to\infty}[n]_{q_n}\left(S_{n,q_n}(f(t);x;p) - f(x)\right) \\
&= f'(x)\lim_{n\to\infty}[n]_{q_n}\left(\frac{[n+p]_{q_n}}{[n]_{q_n}}x - x\right) \\
&+ \frac{1}{2}f''(x)\lim_{n\to\infty}[n]_{q_n}\left(\left[\left(\frac{[n+p]_{q_n}}{q_n[n]_{q_n}} - 2\right)\frac{[n+p]_{q_n}}{[n]_{q_n}} + 1\right]x^2 + \frac{[n+p]_{q_n}}{[n]_{q_n}^2}x\right) \\
&= f'(x)pAx + f''(x)\left((1-A)\frac{x^2}{2} + \frac{1}{2}x\right)
\end{aligned}
$$

uniformly with respect to $x \in [0,c]$, $(c > 0)$. Hence, the proof is completed.

25.3 Concluding Remarks

Remark 25.7. If we choose $p = 0$ in Lemma 25.4, our results will be the same as that of q-Szasz operators.

Remark 25.8. According to chosen q_n, we obtain the following Voronovskaja-type results, respectively:
Let $q_n = 1 - \frac{1}{n}$ when $q_n \to 1$, $\lim_{n\to\infty} q_n^n = A = e^{-1}$, we have

$$\lim_{n\to\infty} [n]_{q_n} \left(S_{n,q_n}(f(t);x;p) - f(x)\right)$$
$$= f'(x)\, pe^{-1}x + f''(x)\left(\left(1-e^{-1}\right)\frac{x^2}{2} + \frac{1}{2}x\right).$$

Let $q_n = 1 - \frac{1}{e^n}$ or $q_n = 1 - \frac{1}{n^2}$ when $q_n \to 1$, $\lim_{n\to\infty} q_n^n = A = 1$, we get

$$\lim_{n\to\infty} [n]_{q_n} \left(S_{n,q_n}(f(t);x;p) - f(x)\right) = f'(x)\, px + f''(x)\,\frac{1}{2}x.$$

References

1. A. Acar, V. Gupta, A. Aral, Rate of convergence for generalized Szsaz operators. Bull. Math. Sci. **1**, 99–113 (2011)
2. P.N. Agrawal, V. Gupta, S.A. Kumar, On a q-analogue of Bernstein-Schurer-Stancu operators. Appl. Math. Comput. **219**, 7754–7764 (2013)
3. F. Altomare, M. Campiti, *Korovkin-Type Approximation Theory and Its Applications*. De Gruyter Studies in Mathematics, vol. 17 (Walter de Gruyter, Berlin, 1994)
4. A. Aral, A generalization of Szasz-Mirakyan operators based on q-integers. Math. Comput. Model. **47**(9–10), 1052–1062 (2008)
5. A. Aral, V. Gupta, The q-derivative and applications to q-Szasz Mirakyan operators. Calcolo **43**(3), 151–170 (2006)
6. M. Becker, Global approximation theorems for Szasz-Mirakyan and Baskakov operators in polynomial weight spaces. Indiana Univ. Math. J. **27**(1), 151–170 (2006)
7. I. Büyükyazıcı, H. Sharma, Approximation properties of two-dimensional q-Bernstein-Chlodowsky-Durrmeyer operators. Numer. Funct. Anal. Optim. **33**(2), 1351–1371 (2012)
8. Q.-B. Cai, X.-M. Zeng, Z. Cui, Approximation properties of the modification of Kantorovich type q-Szasz operators. J. Comput. Anal. Appl. **15**(1), 176–187 (2013)
9. N. Deo, Simultaneous approximation by Lupaş modified operators with weighted function of Szasz operators. J. Inequal. Pure Appl. Math. **5**(4), 5 pp. (2014)
10. R.A. DeVore, G.G. Lorentz, *Constructive Approximation* (Springer, Berlin, 1993)
11. O. Duman, M.A. Özarslan, Szasz-Mirakjan type operators providing a better error estimation. Appl. Math. Lett. **20**, 1184–1188 (2007)
12. S.G. Gal, N.I. Mahmudov, K. Kara, Approximation by complex q-Szasz Kantorovich operators in compact disks, $q > 1$. Complex Anal. Oper. Theory **7**, 1853–1867 (2012)
13. V. Gupta, A note on modified Szasz operators. Bull. Inst. Math. Acad. Sin. **21**(3), 275–278 (1993)
14. V. Gupta, A note on q-Baskakov-Szasz operators. Lobachevskii J. Math. **31**(4), 359–366 (2010)
15. V. Gupta, A Stancu variant of Beta Szasz operators. Georgian Math. J. **21**(1), 75–82 (2014)
16. V. Kac, P. Cheung, *Quantum Calculus* (Springer, Berlin, 2002)
17. A.A. Lupaş, q-Analogue of the Bernstein operators. Semin. Numer. Stat. Calculus, Univ. Cluj-Napoca **9**, 85–92 (1987)
18. N.I. Mahmudov, Approximation by q-Szasz operators (2010). arXiv:1005.3934 [math.FA]

19. N.I. Mahmudov, On q-parametric Szasz-Mirakyan operators. Mediterr. J. Math. **7**(3), 297–311 (2010)
20. N.I. Mahmudov, q-Szasz operators which preserve x^2. Math. Slovaca **63**(5), 1059–1072 (2013)
21. N.I. Mahmudov, V. Gupta, On certain q-analogue of Szasz Kantorovich operators. J. Appl. Math. Comput. **37**(1–2), 407–419 (2011)
22. C.V. Muraru, Note on q-Bernstein-Schurer operators. Babeş-Bolyaj Math. **56**, 489–495 (2011)
23. M. Örkcü, Approximation properties of bivariate extension of q-Szasz-Mirakjan-Kantorovich operators. J. Inequal. Appl. **324**, 10 pp. (2013)
24. M. Örkcü, O. Doğru, q-Szasz-Mirakyan-Kantorovich type operators preserving some test functions. Appl. Math. Lett. **24**(9), 1588–1593 (2011)
25. M. Örkcü, O. Doğru, Statistical approximation of a kind of Kantorovich type q-Szasz-Mirakjan operators. Nonlinear Anal. Theory Methods Appl. **75**(5), 2874–2882 (2012)
26. M.A. Özarslan, q-Szasz Schurer operators. Miscolc Math. Notes **12**, 225–235 (2011)
27. M.A. Özarslan, T. Vedi, q-Bernstein-Schurer-Kantorovich Operators. J. Inequal. Appl. **444**, 15 pp. (2013)
28. G.M. Phillips, On generalized Bernstein polynomials, in *Numerical Analysis*, vol. 98 (World Scientific, River Edge, 1996), pp. 263–269
29. T. Vedi, M.A. Özarslan, Some properties of q-Bernstein-Schurer operators. J. Appl. Funct. Anal. **8**(1), 45–53 (2013)
30. İ. Yüksel, Ü. Dinlemez, Voronovskaja type approximation theorem for q-Szasz-beta operators. Appl. Math. Comput. **235**, 555–559 (2014)

Chapter 26
Approximation to Derivatives of Functions by Linear Operators Acting on Weighted Spaces by Power Series Method

Emre Taş and Tuğba Yurdakadim

Abstract In this chapter, using power series method we study some Korovkin type approximation theorems which deal with the problem of approximating a function by means of a sequence of linear operators acting on weighted spaces.

26.1 Introduction

Much of the literature on approximation theory is focused on the classical approximation operators which tend to converge to value of functions being approximated and ordinary test functions. Efendiev has studied some approximations to derivatives of functions by means of a class of linear operators defined on various weighted spaces. Kucuk and Duman [10] verify the same results using A-summation process. Recent studies demonstrate that summability theory provides an important contribution to improvement of the classical analysis. For example, many authors have studied Korovkin type approximation theorems by using summability theory [1, 6, 11]. In this chapter we study the Korovkin theory, using the power series method, which deals with the problem of approximating derivatives of a function by means of a sequence of linear operators acting on weighted spaces without using the ordinary test functions. This chapter consists of three sections. The first section is devoted to basic definitions and notations used in the chapter. In the second section, we give some approximations to derivatives of function by means of a class of linear operators defined on various weighted spaces by using the power series method. Our motivation to this chapter is [3] and [8]. In the final section we give some concluding remarks.

We first recall some notation and basic definitions used in this chapter.

E. Taş
Department of Mathematics, Ahi Evran University, Kırşehir, Turkey
e-mail: emretas86@hotmail.com

T. Yurdakadim (✉)
Department of Mathematics, Hitit University, Çorum, Turkey
e-mail: tugbayurdakadim@hotmail.com

G.A. Anastassiou, O. Duman (eds.), *Computational Analysis*, Springer Proceedings in Mathematics & Statistics 155, DOI 10.1007/978-3-319-28443-9_26

Throughout the chapter we use the following weighted spaces introduced by Efendiev [8]. Let k be a nonnegative integer. By $C^{(k)}(\mathbb{R})$ we denote the space of all functions having k-th continuous derivatives on $\mathbb{R}$. Now, let $M^{(k)}(\mathbb{R})$ denote the class of all linear operators mapping the set of functions f that are convex of order $(k-1)$ on $\mathbb{R}$, i.e., $f^{(k)}(x) \geq 0$ holds for all $x \in \mathbb{R}$, into the set of all positive functions on $\mathbb{R}$. More precisely, for a fixed nonnegative integer k and a linear operator L,

$$L \in M^{(k)}(\mathbb{R}) \Leftrightarrow L(f) \geq 0 \text{ for every function } f \text{ satisfying } f^{(k)} \geq 0. \tag{26.1}$$

If $k = 0$, then $M^{(0)}(\mathbb{R})$ stands for the class of all positive linear operators. Assume that ρ is a weight function, i.e., $\rho : \mathbb{R} \to \mathbb{R}^+ = (0, +\infty)$ is a function such that $\rho(0) = 1$; ρ is increasing on $\mathbb{R}^+$ and decreasing on $\mathbb{R}^-$; and $\lim_{x\to\pm\infty} \rho(x) = +\infty$. In this case, we consider the following weighted spaces:

$$\begin{aligned}
C_\rho^{(k)}(\mathbb{R}) &= \left\{f \in C^{(k)}(\mathbb{R}) : \text{for some positive } m_f,\ \left|f^{(k)}(x)\right| \leq m_f\rho(x),\ x \in \mathbb{R}\right\}, \\
\tilde{C}_\rho^{(k)}(\mathbb{R}) &= \left\{f \in C_\rho^{(k)}(\mathbb{R}) : \text{for some } l_f,\ \lim_{x\to\pm\infty} \frac{f^{(k)}(x)}{\rho(x)} = l_f\right\}, \\
\hat{C}_\rho^{(k)}(\mathbb{R}) &= \left\{f \in \tilde{C}_\rho^{(k)}(\mathbb{R}) : \lim_{x\to\pm\infty} \frac{f^{(k)}(x)}{\rho(x)} = 0\right\}, \\
B_\rho(\mathbb{R}) &= \left\{g : \mathbb{R} \to \mathbb{R} : \text{for some positive } m_g,\ |g(x)| \leq m_g\rho(x),\ x \in \mathbb{R}\right\}.
\end{aligned}$$

As usual, the weighted space $B_\rho(\mathbb{R})$ is endowed with the norm

$$\|g\|_\rho := \sup_{x\in\mathbb{R}} \frac{|g(x)|}{\rho(x)} \text{ for } g \in B_\rho(\mathbb{R}).$$

If $k = 0$, then we write $M(\mathbb{R})$, $C_\rho(\mathbb{R})$, $\tilde{C}_\rho(\mathbb{R})$ and $\hat{C}_\rho(\mathbb{R})$ instead of $M^{(0)}(\mathbb{R})$, $C_\rho^{(0)}(\mathbb{R})$, $\tilde{C}_\rho^{(0)}(\mathbb{R})$ and $\hat{C}_\rho^{(0)}(\mathbb{R})$, respectively.

26.2 Approximation Theorems by Power Series Methods

We first recall that the system of functions $f_0, f_1, f_2, \ldots, f_m$ continuous on an interval $[a, b]$ is called a Tschebyshev system of order m, or T-system, if any polynomial

$$P(x) = a_0f_0(x) + a_1f_1(x) + \cdots + a_mf_m(x)$$

has not more than m zeros in this interval with the condition that the numbers a_0, $a_1, \ldots, a_m$ are not all equal to zero.

Let (p_j) be real sequence with $p_0 > 0$ and $p_j \geq 0$ ($j \in \mathbb{N}$), and such that the corresponding power series $p(t) := \sum_{j=0}^{\infty} p_j t^j$ has radius of convergence R with $0 < R \leq \infty$. If, for all $t \in (0, R)$,

$$\lim_{t \to R^-} \frac{1}{p(t)} \sum_{j=0}^{\infty} x_j p_j t^j = L \tag{26.2}$$

then we say that $x = (x_j)$ is convergent in the sense of power series method [9, 13]. Note that the power series method is regular if and only if

$$\lim_{t \to R^-} \frac{p_j t^j}{p(t)} = 0, \quad \text{for each } j \in \mathbb{N} \tag{26.3}$$

holds [5].

Let $\{L_j\}$ be a sequence of positive linear operators from C_ρ into B_ρ such that for every $f \in C_\rho$

$$\sup_{t \in (0,R)} \frac{1}{p(t)} \sum_{j=0}^{\infty} \|L_j\|_{C_\rho \to B_\rho} p_j t^j < \infty \tag{26.4}$$

holds. Also V_t given by

$$V_t\{(f(y); x)\} := \frac{1}{p(t)} \sum_{j=0}^{\infty} L_j(f(y); x) p_j t^j$$

which is a positive linear operator from C_ρ into B_ρ is well defined by (26.4).

Throughout the chapter, the operators fulfill conditions (26.3). The classical test functions have been changed in the Korovkin theory [2, 4, 7, 12]. Now, following Theorem 1 of [12], we obtain the following result at once.

Theorem 26.1. *Let* $\{L_j\}$ *be a sequence of positive linear operators from* $C(X, \mathbb{R})$ *into* $C(X, \mathbb{R})$ *for which* (26.4) *holds. If* $f_i \in C(X, \mathbb{R})$

$$\lim_{t \to R^-} \|V_t(f_i) - f_i\| = 0, \; i = 0, 1, 2$$

then for all $f \in C(X, \mathbb{R})$,

$$\lim_{t \to R^-} \|V_t(f) - f\| = 0$$

where X *is a compact Hausdorff space.*

We first consider the case of $k = 0$.

Theorem 26.2. *Let* $\{L_j\}$ *be a sequence of operators from* $C_\rho(\mathbb{R})$ *into* $B_\rho(\mathbb{R})$ *satisfying* (26.4)*, belong to the class* $M(\mathbb{R})$*. Assume that the following conditions hold:*

(*i*) $\{f_0, f_1\}$ *and* $\{f_0, f_1, f_2\}$ *are* T*-systems on* $\mathbb{R}$,

(*ii*) $\lim_{x\to\pm\infty} \dfrac{f_i(x)}{1+|f_2(x)|} = 0,\ i = 0, 1,$

(*iii*) $\lim_{x\to\pm\infty} \dfrac{f_2(x)}{\rho(x)} = m_{f_2} \neq 0,$

(*iv*) $\lim_{t\to R^-} \|V_t(f_i) - f_i\|_\rho = 0,\ i = 0, 1, 2.$

Then, for all $f \in \tilde{C}_\rho(\mathbb{R})$*, we have*

$$\lim_{t\to R^-} \|V_t(f) - f\|_\rho = 0.$$

Proof. Let $f \in \tilde{C}_\rho(\mathbb{R})$ and define a function g on $\mathbb{R}$ as follows:

$$g(y) = m_{f_2} f(y) - l_f f_2(y), \tag{26.5}$$

where m_{f_2} and l_f are certain constants as in the definitions of the weighted spaces. Then, we easily observe that $g \in \hat{C}_\rho(\mathbb{R})$. Now we first prove that

$$\lim_{t\to R^-} \|V_t(g) - g\|_\rho = 0.$$

Since $\{f_0, f_1\}$ is T-system on $\mathbb{R}$, we know from Lemma 2 of [8] that, for each $a \in \mathbb{R}$ satisfying $f_i(a) \neq 0$, i=0,1, there exists a function $\Phi_a(y)$ such that

$$\Phi_a(a) = 0 \text{ and } \Phi_a(y) > 0 \text{ for } y < a,$$

and the function Φ_a has the following form

$$\Phi_a(y) = \gamma_0(a) f_0(y) + \gamma_1(a) f_1(y),$$

where $|\gamma_0(a)| = \left|\dfrac{f_1(a)}{f_0(a)}\right|$ and $|\gamma_1(a)| = 1$. In fact here we define

$$\Phi_a(y) = \begin{cases} F(y), & \text{if } F(y) > 0 \text{ for } y < a \\ -F(y), & \text{if } F(y) < 0 \text{ for } y < a \end{cases},$$

where

$$F(y) = \frac{f_1(a)}{f_0(a)} f_0(y) - f_1(y).$$

Clearly here $F(a) = 0$ and F has no other root by $\{f_0, f_1\}$ being a T-system. On the other hand, by (*ii*) and (*iii*), we see for each $i = 0, 1$, that

$$\frac{f_i(y)}{\rho(y)} = \frac{f_i(y)}{1 + |f_2(y)|}\left(\frac{1}{\rho(y)} + \frac{|f_2(y)|}{\rho(y)}\right) \to 0 \text{ as } y \to \pm\infty. \tag{26.6}$$

Now using the fact that $g \in \hat{C}_\rho(\mathbb{R})$ and also considering (26.6) and (*iii*), for every $\varepsilon > 0$, there exists a positive number u_0 such that the conditions

$$|g(y)| < \varepsilon\rho(y) \tag{26.7}$$

$$|f_i(y)| < \varepsilon\rho(y),\ i = 0, 1 \tag{26.8}$$

$$\rho(y) < s_0 f_2(y) \text{ for a certain constant } s_0, \tag{26.9}$$

hold for all y with $|y| > u_0$. By (26.7) and (26.9), we can write that

$$|g(y)| < \varepsilon s_0 f_2(y), \text{ whenever } |y| > u_0 \tag{26.10}$$

and for a fixed $a > u_0$ such that $f_i(a) \neq 0$, $i = 0, 1$,

$$|g(y)| \leq \frac{M}{m_a}\Phi_a(y) \text{ whenever } |y| \leq u_0 \tag{26.11}$$

where

$$M := \max_{|y|\leq u_0} |g(y)| \text{ and } m_a := \min_{|y|\leq u_0} \Phi_a(y). \tag{26.12}$$

So, combining (26.10) and (26.11), we get

$$|g(y)| < \frac{M}{m_a}\Phi_a(y) + \varepsilon s_0 f_2(y) \text{ for all } y \in \mathbb{R}. \tag{26.13}$$

Now, by using the linearity and monotonicity of the operators L_j, also considering (26.13) and $|\gamma_1(a)| = 1$, we obtain

$$\begin{aligned}
|V_t(g(y);x)| &\leq V_t(|g(y)|\,;x) \\
&\leq \frac{M}{m_a}V_t(\Phi_a(y);x) + \varepsilon s_0 V_t(f_2(y);x) \\
&= \frac{M}{m_a}\left[\gamma_0(a)V_t(f_0(y);x) + \gamma_1(a)V_t(f_1(y);x)\right] \\
&\quad + \varepsilon s_0 V_t(f_2(y);x)
\end{aligned}$$

$$\leq \frac{M}{m_a}\left\{|\gamma_0(a)|\,|V_t(f_0(y);x)-f_0(x)|+|V_t(f_1(y);x)-f_1(x)|\right\}$$
$$+\frac{M}{m_a}\left\{|\gamma_0(a)f_0(x)|+|\gamma_1(a)f_1(x)|\right\}$$
$$+\varepsilon\,|s_0|\,|V_t(f_2(y);x)-f_2(x)|+\varepsilon\,|s_0f_2(x)|\,.$$

Since $|\gamma_1(a)|=1$, then we have

$$\sup_{|x|>u_0}\frac{|V_t(g(y);x)|}{\rho(x)}$$
$$\leq \frac{M}{m_a}\left\{|\gamma_0(a)|\sup_{|x|>u_0}\frac{|V_t(f_0(y);x)-f_0(x)|}{\rho(x)}+\sup_{|x|>u_0}\frac{|V_t(f_1(y);x)-f_1(x)|}{\rho(x)}\right\}$$
$$+\frac{M}{m_a}\left\{|\gamma_0(a)|\sup_{|x|>u_0}\frac{|f_0(x)|}{\rho(x)}+\sup_{|x|>u_0}\frac{|f_1(x)|}{\rho(x)}\right\}$$
$$+\varepsilon\,|s_0|\sup_{|x|>u_0}\frac{|V_t(f_2(y);x)-f_2(x)|}{\rho(x)}+\varepsilon\,|s_0|\sup_{|x|>u_0}\frac{|f_2(x)|}{\rho(x)}.$$

But by (26.6) and *(iii)*, we get that

$$A(\varepsilon):=\frac{M}{m_a}\left\{|\gamma_0(a)|\sup_{|x|>u_0}\frac{|f_0(x)|}{\rho(x)}+\sup_{|x|>u_0}\frac{|f_1(x)|}{\rho(x)}\right\}+\varepsilon\,|s_0|\sup_{|x|>u_0}\frac{|f_2(x)|}{\rho(x)}$$

is finite for every $\varepsilon>0$. Now let

$$B(\varepsilon):=\max\left\{\frac{M\,|\gamma_0(a)|}{m_a},\frac{M}{m_a},\varepsilon\,|s_0|\right\},$$

which is also finite for every $\varepsilon>0$. Then we obtain

$$\sup_{|x|>u_0}\frac{|V_t(g(y);x)|}{\rho(x)}\leq A(\varepsilon)+B(\varepsilon)\sum_{i=0}^{2}\sup_{|x|>u_0}\frac{|V_t(f_i(y);x)-f_i(x)|}{\rho(x)}$$

which implies that

$$\sup_{|x|>u_0}\frac{|V_t(g(y);x)|}{\rho(x)}\leq A(\varepsilon)+B(\varepsilon)\sum_{i=0}^{2}\|V_t(f_i)-f_i\|_{\rho}\,. \tag{26.14}$$

On the other hand, since

$$\|V_t(g) - g\|_\rho \leq \sup_{|x| \leq u_0} \frac{|V_t(g(y);x) - g(x)|}{\rho(x)} + \sup_{|x|>u_0} \frac{|V_t(g(y);x)|}{\rho(x)} + \sup_{|x|>u_0} \frac{|g(x)|}{\rho(x)},$$

it follows from (26.7) and (26.14) that

$$\|V_t(g) - g\|_\rho \leq \varepsilon + A(\varepsilon) + B_1 \|V_t(g) - g\|_{C[-u_0,u_0]} + B(\varepsilon) \sum_{i=0}^{2} \|V_t(f_i) - f_i\|_\rho$$

holds for every $\varepsilon > 0$ where $B_1 = \max_{x \in [-u_0,u_0]} \frac{1}{\rho(x)}$. By (iv), we write immediately that

$$\lim_{t \to R^-} \|V_t(f_i) - f_i\|_{C[-u_0,u_0]} = 0, \; i = 0, 1, 2. \tag{26.15}$$

Since $\{f_0, f_1, f_2\}$ is T-system and $g \in C[-u_0, u_0]$, we get from (26.15) and Theorem 26.1

$$\lim_{t \to R^-} \|V_t(g) - g\|_{C[-u_0,u_0]} = 0.$$

By (26.5), we have that

$$\|V_t(f) - f\|_\rho = \left\| V_t(\frac{1}{m_{f_2}} g + \frac{l_f}{m_{f_2}} f_2) - \frac{1}{m_{f_2}} g + \frac{l_f}{m_{f_2}} f_2 \right\|_\rho \leq \frac{1}{m_{f_2}} \|V_t(g) - g\|_\rho + \frac{l_f}{m_{f_2}} \|V_t(f_2) - f_2\|_\rho,$$

which completes the proof.

Now, we consider the case $k \geq 1$. Let $\{L_j\}$ be a sequence of positive linear operators from $C_\rho^{(k)}$ into B_ρ such that for every $f \in C_\rho^{(k)}$

$$\sup_{t \in (0,R)} \frac{1}{p(t)} \sum_{j=0}^{\infty} \|L_j\|_{C_\rho^{(k)} \to B_\rho} p_j t^j < \infty \tag{26.16}$$

holds. Then V_t which is constructed before, is a positive linear operator from $C_\rho^{(k)}$ into B_ρ and is well defined by (26.16).

Theorem 26.3. *Let $\{L_j\}$ be a sequence of linear operators from $C_\rho^{(k)}(\mathbb{R})$ into $B_\rho(\mathbb{R})$ satisfying (26.16), belong to the class $M^{(k)}(\mathbb{R})$ and f_0, f_1, f_2 be in $C_\rho^{(k)}$. Assume further that the following conditions hold:*

(a) $\left\{f_0^{(k)}, f_1^{(k)}\right\}$ *and* $\left\{f_0^{(k)}, f_1^{(k)}, f_2^{(k)}\right\}$ *are T-systems on* $\mathbb{R}$,

(b) $\lim\limits_{x\to\pm\infty} \dfrac{f_i^{(k)}(x)}{1+\left|f_2^{(k)}(x)\right|} = 0,\ i = 0, 1,$

(c) $\lim\limits_{x\to\pm\infty} \dfrac{f_2^{(k)}(x)}{\rho(x)} = m_{f_2} \neq 0,$

(d) $\lim\limits_{t\to R^-} \left\| V_t(f_i) - f_i^{(k)} \right\|_\rho = 0,\ i = 0, 1, 2.$

Then, for all $f \in \tilde{C}_\rho^{(k)}(\mathbb{R})$, we have

$$\lim_{t\to R^-} \left\| V_t(f) - f^{(k)} \right\|_\rho = 0.$$

Proof. We say that $f, g \in \tilde{C}_\rho^{(k)}(\mathbb{R})$ are equivalent provided that $f^{(k)}(x) = g^{(k)}(x)$ for all $x \in \mathbb{R}$. We denote the equivalent classes of $f \in \tilde{C}_\rho^{(k)}(\mathbb{R})$ by $[f]$. This means that

$$[f] = d^{-k}d^k f,$$

where d^k denotes the k-th derivative operator and d^{-k} denotes the k-th inverse derivative operator. Thus by $\left[\tilde{C}_\rho^{(k)}(\mathbb{R})\right]$ we denote the equivalent weighted spaces of $\tilde{C}_\rho^{(k)}(\mathbb{R})$. Then for $f \in \tilde{C}_\rho^{(k)}(\mathbb{R})$, consider

$$V_t([f]) = V_t(d^{-k}d^k f) =: V_t^*(\Psi),$$

where $f^{(k)} = \Psi \in \tilde{C}_\rho(\mathbb{R})$, and V_t^* is an operator such that $V_t^* = V_t d^{-k}$. Observe that V_t^* is a positive linear operator from $\tilde{C}_\rho(\mathbb{R})$ to $B_\rho(\mathbb{R})$. Indeed, if $\Psi \geq 0$, i.e., $f^{(k)} \geq 0$, then since each V_t belongs to the class $M^{(k)}(\mathbb{R})$, it follows from (26.1) that $V_t([f]) \geq 0$, i.e., $V_t^*(\Psi) \geq 0$. Now, for every $x \in \mathbb{R}$, defining

$$\Psi_i(x) := f_i^{(k)}(x),\ i = 0, 1, 2,$$

it follows from (a)–(d) that $\{\Psi_0, \Psi_1\}$ and $\{\Psi_0, \Psi_1, \Psi_2\}$ are T-systems on $\mathbb{R}$,

$$\lim_{x\to\pm\infty} \frac{\Psi_i(x)}{1+|\Psi_2(x)|} = 0 \text{ for each } i = 0, 1,$$

$$\lim_{x\to\pm\infty} \frac{\Psi_2(x)}{\rho(x)} = m_{\Psi_2} \neq 0,$$

$$\lim_{t\to R^-} \left\| V_t([f_i]) - f_i^{(k)} \right\|_\rho = \lim_{t\to R^-} \left\| V_t^*(\Psi_i) - \Psi_i \right\|_\rho = 0,\ i = 0, 1, 2.$$

Hence all the conditions of Theorem 26.2 are satisfied for the functions Ψ_0, Ψ_1, Ψ_2 and the positive linear operator V_t^*. Therefore, we immediately get

$$\lim_{t\to R^-} \left\| V_t^*(\Psi) - \Psi \right\|_\rho = 0$$

or equivalently

$$\lim_{t\to R^-} \left\| V_t(f) - f^{(k)} \right\|_\rho = 0$$

which completes the proof.

Finally, we have the following result.

Theorem 26.4. *Assume that conditions* (a), (b), *and* (d) *of the above theorem hold. Let* ρ_1 *be a weight function. If*

$$\lim_{x\to\pm\infty} \frac{\rho(x)}{\rho_1(x)} = 0, \tag{26.17}$$

and

$$\lim_{x\to\pm\infty} \frac{f_2^{(k)}(x)}{\rho_1(x)} = m_{f_2} \neq 0 \tag{26.18}$$

then for all $f \in C_\rho^{(k)}(\mathbb{R})$, *we have*

$$\lim_{t\to R^-} \left\| V_t(f) - f^{(k)} \right\|_{\rho_1} = 0.$$

Proof. Let $f \in C_\rho^{(k)}(\mathbb{R})$. Since $\dfrac{|f^{(k)}(x)|}{\rho(x)} \leq m_f$ for every $x \in \mathbb{R}$, we get

$$\lim_{x\to\pm\infty} \frac{|f^{(k)}(x)|}{\rho_1(x)} \leq \lim_{x\to\pm\infty} \frac{|f^{(k)}(x)|}{\rho(x)} \frac{\rho(x)}{\rho_1(x)} \leq m_f \lim_{x\to\pm\infty} \frac{\rho(x)}{\rho_1(x)}.$$

Then by (26.17), we easily obtain that

$$\lim_{x\to\pm\infty} \frac{f^{(k)}(x)}{\rho_1(x)} = 0,$$

which yields

$$f \in \hat{C}_{\rho_1}^{(k)}(\mathbb{R}) \subset \tilde{C}_{\rho_1}^{(k)}(\mathbb{R}).$$

Also observe that, by (26.17), condition (d) is satisfied for the weight function ρ_1. Hence the proof follows from Theorem 26.3 and condition (26.18) at once.

26.3 Concluding Remarks

- In the case of $R = 1, p(t) = \dfrac{1}{1-t}$ and for $j \geq 0, p_j = 1$ the power series method coincides with Abel method which is a sequence-to-function transformation.
- In the case of $R = \infty, p(t) = e^t$ and for $j \geq 0, p_j = \dfrac{1}{j!}$ the power series method coincides with Borel method.

We can therefore give all of the theorems of this chapter for Abel and Borel methods.

References

1. O. Agratini, Statistical convergence of a non-positive approximation process. Chaos Solitons Fractals **44**, 977–981 (2011)
2. F. Altomare, Korovkin-type theorems and approximation by positive linear operators. Surv. Approx. Theory **5**, 92–164 (2010)
3. G.A. Anastassiou, O. Duman, Statistical weighted approximation to derivatives of functions by linear operators. J. Comput. Anal. Appl. **11**, 20–30 (2009)
4. Ö.G. Atlihan, E. Taş, An abstract version of the Korovkin theorem via A-summation process. Acta Math. Hung. **145**, 360–368 (2015)
5. J. Boos, *Classical and Modern Methods in Summability* (Oxford University Press, Oxford, 2000)
6. K. Demirci, F. Dirik, Statistical σ-convergence of positive linear operators. Appl. Math. Lett. **24**, 375–380 (2011)
7. O. Duman, C. Orhan, An abstract version of the Korovkin approximation theorem. Publ. Math. Debr. **69**, 33–46 (2006)
8. R.O. Efendiev, Conditions for convergence of linear operators to derivatives (Russian). Akad. Nauk. Azerb. SSR Dokl. **40**, 3–6 (1984)
9. W. Kratz, U. Stadtmüller, Tauberian theorems for J_p-summability. J. Math. Anal. Appl. **139**, 362–371 (1989)
10. N. Küçük, O. Duman, Summability methods in weighted approximation to derivatives of functions. Serdica Math. J. **41**, 335–368 (2015)
11. T. Nishishiraho, Quantitative equi-uniform approximation processes of integral operators in Banach spaces. Taiwan. J. Math. **10**, 441–465 (2006)
12. I. Özgüç, E. Taş, A Korovkin-type approximation theorem and power series method. Results Math. doi:10.1007/s00025_016_0538_7 (2016)
13. U. Stadtmüller, A. Tali, On certain families of generalized Nörlund methods and power series methods. J. Math. Anal. Appl. **238**, 44–66 (1999)

Chapter 27
Generalized Iterated Fractional Representation Formulae and Inequalities

George A. Anastassiou

Abstract Here we derive very general iterated fractional representation formulae. Based on these we obtain a fractional Ostrowski type inequality, fractional Poincaré type inequalities, fractional Opial type inequalities, and fractional Hilbert–Pachpatte inequalities. All these inequalities are very general.

27.1 Background

We need

Definition 27.1 (See Also [10, p. 99]). The left and right fractional integrals, respectively, of a function f with respect to given function g are defined as follows:

Let $a, b \in \mathbb{R}$, $a < b$, $\alpha > 0$. Here $g \in AC([a,b])$ (absolutely continuous functions) and is strictly increasing, $f \in L_{\infty}([a,b])$. We set

$$\left(I_{a+;g}^{\alpha} f\right)(x) = \frac{1}{\Gamma(\alpha)} \int_a^x (g(x) - g(t))^{\alpha-1} g'(t) f(t)\, dt, \quad x \geq a, \tag{27.1}$$

where Γ is the gamma function, clearly $\left(I_{a+;g}^{\alpha} f\right)(a) = 0$, $I_{a+;g}^{0} f := f$ and

$$\left(I_{b-;g}^{\alpha} f\right)(x) = \frac{1}{\Gamma(\alpha)} \int_x^b (g(t) - g(x))^{\alpha-1} g'(t) f(t)\, dt, \quad x \leq b, \tag{27.2}$$

clearly $\left(I_{b-;g}^{\alpha} f\right)(b) = 0$, $I_{b-;g}^{0} f := f$. When g is the identity function id, we get that $I_{a+;id}^{\alpha} = I_{a+}^{\alpha}$, and $I_{b-;id}^{\alpha} = I_{b-}^{\alpha}$, the ordinary left and right Riemann–Liouville fractional integrals, where

---

G.A. Anastassiou (✉)
Department of Mathematical Sciences, University of Memphis, TN 38152, USA
e-mail: ganastss@memphis.edu

G.A. Anastassiou, O. Duman (eds.), *Computational Analysis*, Springer Proceedings in Mathematics & Statistics 155, DOI 10.1007/978-3-319-28443-9_27

$$\left(I_{a+}^{\alpha}f\right)(x) = \frac{1}{\Gamma(\alpha)}\int_{a}^{x}(x-t)^{\alpha-1}f(t)\,dt, \quad x \geq a, \tag{27.3}$$

$\left(I_{a+}^{\alpha}f\right)(a) = 0$ and

$$\left(I_{b-}^{\alpha}f\right)(x) = \frac{1}{\Gamma(\alpha)}\int_{x}^{b}(t-x)^{\alpha-1}f(t)\,dt, \quad x \leq b, \tag{27.4}$$

$\left(I_{b-}^{\alpha}f\right)(b) = 0.$

Definition 27.2 (See [7]). Let $\alpha > 0$, $\lceil\alpha\rceil = n$, $\lceil\cdot\rceil$ the ceiling of the number. Again here $g \in AC([a,b])$ and strictly increasing. We assume that $\left(f \circ g^{-1}\right)^{(n)} \circ g \in L_{\infty}([a,b])$. We define the left generalized g-fractional derivative of f of order α as follows:

$$\left(D_{a+;g}^{\alpha}f\right)(x) := \frac{1}{\Gamma(n-\alpha)}\int_{a}^{x}(g(x)-g(t))^{n-\alpha-1}g'(t)\left(f \circ g^{-1}\right)^{(n)}(g(t))\,dt, \tag{27.5}$$

$x \geq a$. If $\alpha \notin \mathbb{N}$, by Anastassiou [5], we have that $D_{a+;g}^{\alpha}f \in C([a,b])$. We see that

$$\left(I_{a+;g}^{n-\alpha}\left(\left(f \circ g^{-1}\right)^{(n)} \circ g\right)\right)(x) = \left(D_{a+;g}^{\alpha}f\right)(x), \ x \geq a. \tag{27.6}$$

We set

$$D_{a+;g}^{n}f(x) := \left(\left(f \circ g^{-1}\right)^{(n)} \circ g\right)(x), \tag{27.7}$$

$$D_{a+;g}^{0}f(x) = f(x), \ \forall\, x \in [a,b]. \tag{27.8}$$

When $g = id$, then

$$D_{a+;g}^{\alpha}f = D_{a+;id}^{\alpha}f = D_{*a}^{\alpha}f, \tag{27.9}$$

the usual left Caputo fractional derivative.

Definition 27.3 (See [7]). Here we assume that $\left(f \circ g^{-1}\right)^{(n)} \circ g \in L_{\infty}([a,b])$, where $N \ni n = \lceil\alpha\rceil$, $\alpha > 0$, $g \in AC([a,b])$ and strictly increasing. We define the right generalized g-fractional derivative of f of order α as follows:

$$\left(D_{b-;g}^{\alpha}f\right)(x) := \frac{(-1)^{n}}{\Gamma(n-\alpha)}\int_{x}^{b}(g(t)-g(x))^{n-\alpha-1}g'(t)\left(f \circ g^{-1}\right)^{(n)}(g(t))\,dt, \tag{27.10}$$

all $x \in [a,b]$. If $\alpha \notin \mathbb{N}$, by Anastassiou [6], we get that $\left(D_{b-;g}^{\alpha} f\right) \in C([a,b])$. We see that

$$I_{b-;g}^{n-\alpha}\left((-1)^n \left(f \circ g^{-1}\right)^{(n)} \circ g\right)(x) = \left(D_{b-;g}^{\alpha} f\right)(x), \ a \leq x \leq b. \tag{27.11}$$

We set

$$D_{b-;g}^{n} f(x) = (-1)^n \left(\left(f \circ g^{-1}\right)^{(n)} \circ g\right)(x), \tag{27.12}$$

$$D_{b-;g}^{0} f(x) = f(x), \ \forall x \in [a,b].$$

When $g = id$, then

$$D_{b-;g}^{\alpha} f(x) = D_{b-;id}^{\alpha} f(x) = D_{b-}^{\alpha} f, \tag{27.13}$$

the usual right Caputo fractional derivative.

Set $g([a,b]) = [c,d]$, where $c, d \in \mathbb{R}$, i.e. $g(a) = c$, $g(b) = d$.
Denote by

$$D_{b-;g}^{n\alpha} := D_{b-;g}^{\alpha} D_{b-;g}^{\alpha} \ldots D_{b-;g}^{\alpha} \ (n \text{ times}), n \in \mathbb{N}. \tag{27.14}$$

Also denote by

$$I_{b-;g}^{n\alpha} := I_{b-;g}^{\alpha} I_{b-;g}^{\alpha} \ldots I_{b-;g}^{\alpha} \ (n \text{ times}). \tag{27.15}$$

We proved the following g-right generalized modified Taylor's formula:

Theorem 27.4 (See [7]). *Suppose that $F_k := D_{b-;g}^{k\alpha} f$, for $k = 0, 1, \ldots, n+1$, fulfill: $F_k \circ g^{-1} \in AC([c,d])$ and $\left(F_k \circ g^{-1}\right)' \circ g \in L_\infty([a,b])$, where $0 < \alpha \leq 1$. Then*

$$f(x) = \sum_{i=0}^{n} \frac{(g(b) - g(x))^{i\alpha}}{\Gamma(i\alpha + 1)} \left(D_{b-;g}^{i\alpha} f\right)(b) \tag{27.16}$$

$$+ \frac{1}{\Gamma((n+1)\alpha)} \int_x^b (g(t) - g(x))^{(n+1)\alpha - 1} g'(t) \left(D_{b-;g}^{(n+1)\alpha} f\right)(t) \, dt$$

$$= \sum_{i=0}^{n} \frac{(g(b) - g(x))^{i\alpha}}{\Gamma(i\alpha + 1)} \left(D_{b-;g}^{i\alpha} f\right)(b) \tag{27.17}$$

$$+ \frac{\left(D_{b-;g}^{(n+1)\alpha}\right)(\psi_x)}{\Gamma((n+1)\alpha + 1)} (g(b) - g(x))^{(n+1)\alpha},$$

where $\psi_x \in [x,b]$, any $x \in [a,b]$.

Denote by

$$D_{a+;g}^{n\alpha} := D_{a+;g}^{\alpha} D_{a+;g}^{\alpha} \dots D_{a+;g}^{\alpha} \ (n \text{ times}),\ n \in \mathbb{N}. \tag{27.18}$$

Also denote by

$$I_{a+;g}^{n\alpha} := I_{a+;g}^{\alpha} I_{a+;g}^{\alpha} \dots I_{a+;g}^{\alpha} \ (n \text{ times}). \tag{27.19}$$

We proved the following g-left generalized modified Taylor's formula:

Theorem 27.5 (See [7]). *Suppose that $F_k := D_{a+;g}^{k\alpha} f$, for $k = 0, 1, \dots, n+1$, fulfill: $F_k \circ g^{-1} \in AC([c,d])$ and $\left(F_k \circ g^{-1}\right)' \circ g \in L_\infty([a,b])$, where $0 < \alpha \leq 1$. Then*

$$f(x) = \sum_{i=0}^{n} \frac{(g(x) - g(a))^{i\alpha}}{\Gamma(i\alpha + 1)} \left(D_{a+;g}^{i\alpha} f\right)(a)$$

$$+ \frac{1}{\Gamma((n+1)\alpha)} \int_a^x (g(x) - g(t))^{(n+1)\alpha - 1} g'(t) \left(D_{a+;g}^{(n+1)\alpha} f\right)(t)\, dt \tag{27.20}$$

$$= \sum_{i=0}^{n} \frac{(g(x) - g(a))^{i\alpha}}{\Gamma(i\alpha + 1)} \left(D_{a+;g}^{i\alpha} f\right)(a) \tag{27.21}$$

$$+ \frac{\left(D_{a+;g}^{(n+1)\alpha}\right)(\psi_x)}{\Gamma((n+1)\alpha + 1)} (g(x) - g(a))^{(n+1)\alpha},$$

where $\psi_x \in [a,x]$, any $x \in [a,b]$.

We give the useful

Corollary 27.6. *Here all as in Theorem 27.4. Additionally we assume that*

$$\left(D_{b-;g}^{i\alpha} f\right)(b) = 0, \quad i = 0, 1, \dots, n. \tag{27.22}$$

Then

$$f(x) = \frac{1}{\Gamma((n+1)\alpha)} \int_x^b (g(t) - g(x))^{(n+1)\alpha - 1} g'(t) \left(D_{b-;g}^{(n+1)\alpha} f\right)(t)\, dt, \tag{27.23}$$

$\forall\, x \in [a,b]$.

Corollary 27.7. *Here all as in Theorem 27.5. Additionally we assume that*

$$\left(D_{a+;g}^{i\alpha} f\right)(a) = 0, \quad i = 0, 1, \dots, n. \tag{27.24}$$

Then

$$f(x) = \frac{1}{\Gamma((n+1)\alpha)} \int_a^x (g(x) - g(t))^{(n+1)\alpha-1} g'(t) \left(D_{a+;g}^{(n+1)\alpha} f\right)(t)\, dt, \tag{27.25}$$

$\forall\, x \in [a,b]$.

27.2 Main Results

We make

Remark 27.8. All here as in Corollary 27.7. We observe that

$$\begin{aligned}
&\left| \int_{g(a)}^{g(x)} (g(x) - z)^{(n+1)\alpha-1} \left(\left(D_{a+;g}^{(n+1)\alpha} f\right) \circ g^{-1}\right)(z)\, dz \right| \\
&\leq \int_{g(a)}^{g(x)} (g(x) - z)^{(n+1)\alpha-1} \left| \left(\left(D_{a+;g}^{(n+1)\alpha} f\right) \circ g^{-1}\right)(z) \right| dz \\
&\leq \left\| \left(D_{a+;g}^{(n+1)\alpha} f\right) \circ g^{-1} \right\|_\infty \frac{(g(x) - g(a))^{(n+1)\alpha}}{(n+1)\alpha} < \infty.
\end{aligned} \tag{27.26}$$

That is, $(g(x) - z)^{(n+1)\alpha-1} \left(\left(D_{a+;g}^{(n+1)\alpha} f\right) \circ g^{-1}\right)(z)$ is integrable over $[g(a), g(x)]$. Since $g(t)$ is monotone we get that (see [9])

$$(g(x) - g(t))^{(n+1)\alpha-1} g'(t) \left(\left(D_{a+;g}^{(n+1)\alpha} f\right)(t)\right)$$

is integrable on $[a,x]$. Therefore by Jia [9], it holds that [see (27.25)]

$$f(x) = \frac{1}{\Gamma((n+1)\alpha)} \int_{g(a)}^{g(x)} (g(x) - z)^{(n+1)\alpha-1} \left(D_{a+;g}^{(n+1)\alpha} f\right)\left(g^{-1}(z)\right) dz. \tag{27.27}$$

In particular, we have

$$\left(f \circ g^{-1}\right)(y) = \frac{1}{\Gamma((n+1)\alpha)} \int_{g(a)}^{y} (y - z)^{(n+1)\alpha-1} \left(\left(D_{a+;g}^{(n+1)\alpha} f\right) \circ g^{-1}\right)(z)\, dz, \tag{27.28}$$

where $y = g(x)$.

By assuming $(n+1)\alpha - 1 > 0$ (equivalently $\alpha > \frac{1}{n+1}$) and using Theorem 7.7, p. 117 of [1], we obtain

$$\left(f \circ g^{-1}\right)'(y) = \frac{((n+1)\alpha - 1)}{\Gamma((n+1)\alpha)} \int_{g(a)}^{y} (y-z)^{(n+1)\alpha - 2} \left(\left(D_{a+;g}^{(n+1)\alpha} f\right) \circ g^{-1}\right)(z)\, dz. \tag{27.29}$$

If $(n+1)\alpha - 2 > 0$ (equivalently $\alpha > \frac{2}{n+1}$), we get

$$\left(f \circ g^{-1}\right)''(y) = \frac{((n+1)\alpha - 1)((n+1)\alpha - 2)}{\Gamma((n+1)\alpha)} \cdot \int_{g(a)}^{y} (y-z)^{(n+1)\alpha - 3} \left(\left(D_{a+;g}^{(n+1)\alpha} f\right) \circ g^{-1}\right)(z)\, dz. \tag{27.30}$$

In general, if $(n+1)\alpha - m > 0$ (equivalently $\alpha > \frac{m}{n+1}$), we get that there exists

$$\left(f \circ g^{-1}\right)^{(m)}(y) = \frac{\prod_{j=1}^{m} ((n+1)\alpha - j)}{\Gamma((n+1)\alpha)} \cdot \int_{g(a)}^{y} (y-z)^{(n+1)\alpha - m - 1} \left(\left(D_{a+;g}^{(n+1)\alpha} f\right) \circ g^{-1}\right)(z)\, dz, \tag{27.31}$$

$\forall\, y \in [g(a), g(b)]$. By Anastassiou [1, p. 388], we get that $\left(f \circ g^{-1}\right)^{(m)} \in C([g(a), g(b)])$. By (27.3) we have that

$$\left(f \circ g^{-1}\right)^{(m)}(y) = \frac{\prod_{j=1}^{m} ((n+1)\alpha - j)\, \Gamma((n+1)\alpha - m)}{\Gamma((n+1)\alpha)} \cdot \left(I_{g(a)+}^{(n+1)\alpha - m} \left(\left(D_{a+;g}^{(n+1)\alpha} f\right) \circ g^{-1}\right)\right)(y)$$
$$= \left(I_{g(a)+}^{(n+1)\alpha - m} \left(\left(D_{a+;g}^{(n+1)\alpha} f\right) \circ g^{-1}\right)\right)(y). \tag{27.32}$$

That is

$$\left(f \circ g^{-1}\right)^{(m)}(y) = \left(I_{g(a)+}^{(n+1)\alpha - m} \left(\left(D_{a+;g}^{(n+1)\alpha} f\right) \circ g^{-1}\right)\right)(y), \tag{27.33}$$

$\forall\, y \in [g(a), g(b)]$, and

$$\left(f \circ g^{-1}\right)^{(m)} (g(x)) = \left(I_{g(a)+}^{(n+1)\alpha - m} \left(\left(D_{a+;g}^{(n+1)\alpha} f\right) \circ g^{-1}\right)\right) (g(x)), \tag{27.34}$$

$\forall\, x \in [a, b]$. By Anastassiou [4], now we derive that

$$\left(f \circ g^{-1}\right)^{(m)} (g(x)) = \left(I_{a+;g}^{(n+1)\alpha - m} \left(D_{a+;g}^{(n+1)\alpha} f\right)\right) (x), \tag{27.35}$$

$\forall\, x \in [a, b]$.

Let $\gamma > 0$ with $\lceil \gamma \rceil = m < n + 1$, such that $m < (n + 1)\alpha$ (equivalently, $\alpha > \frac{m}{n+1}$). We have that (case of $\gamma < m$)

$$\left(D_{a+;g}^{\gamma} f\right)(x) \overset{(27.6)}{=} \left(I_{a+;g}^{m-\gamma} \left(\left(f \circ g^{-1}\right)^{(m)} \circ g\right)\right)(x)$$

$$\overset{(27.35)}{=} \left(I_{a+;g}^{m-\gamma} I_{a+;g}^{(n+1)\alpha - m} \left(D_{a+;g}^{(n+1)\alpha} f\right)\right)(x) \tag{27.36}$$

(by the semigroup property of operators $I_{a+;g}^{\varepsilon}$, $\varepsilon > 0$, see [7])

$$= \left(I_{a+;g}^{(n+1)\alpha - \gamma} \left(D_{a+;g}^{(n+1)\alpha} f\right)\right)(x).$$

We have proved that

$$\left(D_{a+;g}^{\gamma} f\right)(x) = \left(I_{a+;g}^{(n+1)\alpha - \gamma} \left(D_{a+;g}^{(n+1)\alpha} f\right)\right)(x), \tag{27.37}$$

$\forall\, x \in [a, b]$, which is continuous by Anastassiou [5].

We have established the following representation formula:

Theorem 27.9. *All as in Corollary 27.7. Let $\gamma > 0$ with $\lceil \gamma \rceil = m < n + 1$, such that $m < (n + 1)\alpha$ (i.e. $1 \geq \alpha > \frac{m}{n+1}$). Then*

$$\left(D_{a+;g}^{\gamma} f\right)(x) = \frac{1}{\Gamma((n+1)\alpha - \gamma)} \cdot \int_a^x (g(x) - g(t))^{(n+1)\alpha - \gamma - 1} g'(t) \left(D_{a+;g}^{(n+1)\alpha} f\right)(t)\, dt, \tag{27.38}$$

$\forall\, x \in [a, b]$, and $\left(D_{a+;g}^{\gamma} f\right) \in C([a, b])$.

Similarly, we obtain the next representation formulae:

Theorem 27.10. *All as in Corollary 27.7. Let* $\gamma > 0$ *with* $\lceil \gamma \rceil = m$*, such that* $\frac{\gamma+m}{n+1} < \alpha \leq 1$*. Then*

$$\left(D_{a+;g}^{2\gamma} f\right)(x) = \frac{1}{\Gamma\left((n+1)\alpha - 2\gamma\right)} \cdot \int_a^x \left(g(x) - g(t)\right)^{(n+1)\alpha - 2\gamma - 1} g'(t) \left(D_{a+;g}^{(n+1)\alpha} f\right)(t)\, dt, \quad (27.39)$$

$\forall\, x \in [a,b]$*, and* $\left(D_{a+;g}^{2\gamma} f\right) \in C([a,b])$.

Theorem 27.11. *All as in Corollary 27.7. Let* $\gamma > 0$ *with* $\lceil \gamma \rceil = m$*, such that* $\frac{m+2\gamma}{n+1} < \alpha \leq 1$*. Then*

$$\left(D_{a+;g}^{3\gamma} f\right)(x) = \frac{1}{\Gamma\left((n+1)\alpha - 3\gamma\right)} \cdot \int_a^x \left(g(x) - g(t)\right)^{(n+1)\alpha - 3\gamma - 1} g'(t) \left(D_{a+;g}^{(n+1)\alpha} f\right)(t)\, dt, \quad (27.40)$$

$\forall\, x \in [a,b]$*, and* $\left(D_{a+;g}^{3\gamma} f\right) \in C([a,b])$.

Theorem 27.12. *All here as in Corollary 27.7. Let* $\gamma > 0$ *with* $\lceil \gamma \rceil = m$*, such that* $\frac{m+(k-1)\gamma}{n+1} < \alpha \leq 1$*,* $k \in \mathbb{N}$*. Then*

$$\left(D_{a+;g}^{k\gamma} f\right)(x) = \frac{1}{\Gamma\left((n+1)\alpha - k\gamma\right)} \cdot \int_a^x \left(g(x) - g(t)\right)^{(n+1)\alpha - k\gamma - 1} g'(t) \left(D_{a+;g}^{(n+1)\alpha} f\right)(t)\, dt, \quad (27.41)$$

$\forall\, x \in [a,b]$*, and* $\left(D_{a+;g}^{k\gamma} f\right) \in C([a,b])$.

We make

Remark 27.13. All here as in Corollary 27.6. We observe that

$$\left| \int_{g(x)}^{g(b)} (z - g(x))^{(n+1)\alpha - 1} \left(\left(D_{b-;g}^{(n+1)\alpha} f\right) \circ g^{-1} \right)(z)\, dz \right|$$

$$\leq \int_{g(x)}^{g(b)} (z - g(x))^{(n+1)\alpha-1} \left| \left(\left(D_{b-;g}^{(n+1)\alpha} f \right) \circ g^{-1} \right)(z) \right| dz \tag{27.42}$$

$$\leq \left\| \left(D_{b-;g}^{(n+1)\alpha} f \right) \circ g^{-1} \right\|_{\infty} \frac{(g(b) - g(x))^{(n+1)\alpha}}{(n+1)\alpha} < \infty.$$

That is, $(z - g(x))^{(n+1)\alpha-1} \left(\left(D_{b-;g}^{(n+1)\alpha} f \right) \circ g^{-1} \right)(z)$ is integrable over $[g(x), g(b)]$. Since $g(t)$ is monotone we get that (see [9])

$$(g(t) - g(x))^{(n+1)\alpha-1} g'(t) \left(\left(D_{b-;g}^{(n+1)\alpha} f \right)(t) \right)$$

is integrable on $[x, b]$. Therefore by Jia [9], it holds that [see (27.23)]

$$f(x) = \frac{1}{\Gamma((n+1)\alpha)} \int_{g(x)}^{g(b)} (z - g(x))^{(n+1)\alpha-1} \left(\left(D_{b-;g}^{(n+1)\alpha} f \right) \circ g^{-1} \right)(z) \, dz. \tag{27.43}$$

In particular, we have

$$\left(f \circ g^{-1}\right)(y) = \frac{1}{\Gamma((n+1)\alpha)} \int_{y}^{g(b)} (z - y)^{(n+1)\alpha-1} \left(\left(D_{b-;g}^{(n+1)\alpha} f \right) \circ g^{-1} \right)(z) \, dz, \tag{27.44}$$

where $y = g(x)$.

By assuming $(n+1)\alpha - 1 > 0$ (equivalently $\alpha > \frac{1}{n+1}$) and using [3], we obtain

$$\left(f \circ g^{-1}\right)'(y) = \frac{(-1)((n+1)\alpha - 1)}{\Gamma((n+1)\alpha)} \cdot \int_{y}^{g(b)} (z - y)^{(n+1)\alpha-2} \left(\left(D_{b-;g}^{(n+1)\alpha} f \right) \circ g^{-1} \right)(z) \, dz. \tag{27.45}$$

If $(n+1)\alpha - 2 > 0$ (equivalently $\alpha > \frac{2}{n+1}$), we get

$$\left(f \circ g^{-1}\right)''(y) = \frac{(-1)^2 ((n+1)\alpha - 1)((n+1)\alpha - 2)}{\Gamma((n+1)\alpha)} \cdot \int_{y}^{g(b)} (z - y)^{(n+1)\alpha-3} \left(\left(D_{b-;g}^{(n+1)\alpha} f \right) \circ g^{-1} \right)(z) \, dz. \tag{27.46}$$

In general, if $(n+1)\alpha - m > 0$ (equivalently $\alpha > \frac{m}{n+1}$), we get that there exists

$$\left(f \circ g^{-1}\right)^{(m)}(y) = \frac{(-1)^m \prod_{j=1}^{m} ((n+1)\alpha - j)}{\Gamma((n+1)\alpha)} \cdot \int_{y}^{g(b)} (z-y)^{(n+1)\alpha - m - 1} \left(\left(D_{b-;g}^{(n+1)\alpha} f\right) \circ g^{-1}\right)(z)\, dz, \quad (27.47)$$

$\forall\, y \in [g(a), g(b)]$.

By Anastassiou [2], we get that $\left(f \circ g^{-1}\right)^{(m)} \in C([g(a), g(b)])$. By (27.4) we have that

$$\left(f \circ g^{-1}\right)^{(m)}(y) = \frac{(-1)^m \prod_{j=1}^{m} ((n+1)\alpha - j)}{\Gamma((n+1)\alpha)} \cdot \Gamma((n+1)\alpha - m) \left(I_{g(b)-}^{(n+1)\alpha - m} \left(\left(D_{b-;g}^{(n+1)\alpha} f\right) \circ g^{-1}\right)\right)(y)$$
$$= (-1)^m \left(I_{g(b)-}^{(n+1)\alpha - m} \left(\left(D_{b-;g}^{(n+1)\alpha} f\right) \circ g^{-1}\right)\right)(y). \quad (27.48)$$

We have proved that

$$\left(f \circ g^{-1}\right)^{(m)}(y) = (-1)^m \left(I_{g(b)-}^{(n+1)\alpha - m} \left(\left(D_{b-;g}^{(n+1)\alpha} f\right) \circ g^{-1}\right)\right)(y), \quad (27.49)$$

$\forall\, y \in [g(a), g(b)]$, and

$$\left(f \circ g^{-1}\right)^{(m)}(g(x)) = (-1)^m \left(I_{g(b)-}^{(n+1)\alpha - m} \left(\left(D_{b-;g}^{(n+1)\alpha} f\right) \circ g^{-1}\right)\right)(g(x)), \quad (27.50)$$

$\forall\, x \in [a, b]$.

By Anastassiou [4], now we derive that

$$\left(f \circ g^{-1}\right)^{(m)}(g(x)) = (-1)^m \left(I_{b-;g}^{(n+1)\alpha - m} \left(D_{b-;g}^{(n+1)\alpha} f\right)\right)(x), \quad (27.51)$$

$\forall\, x \in [a, b]$.

Let $\gamma > 0$ with $\lceil \gamma \rceil = m < n+1$, such that $m < (n+1)\alpha$ (equivalently, $\alpha > \frac{m}{n+1}$). We have that (case of $\gamma < m$)

$$\left(D_{b-;g}^{\gamma} f\right)(x) \overset{(27.11)}{=} \left(I_{b-;g}^{m-\gamma} \left((-1)^m \left(f \circ g^{-1}\right)^{(m)} \circ g\right)\right)(x) \quad (27.52)$$
$$\overset{(27.51)}{=} \left(I_{b-;g}^{m-\gamma} \left((-1)^{2m} I_{b-;g}^{(n+1)\alpha - m} \left(D_{b-;g}^{(n+1)\alpha} f\right)\right)\right)(x)$$

(by semigroup property of operators $I_{b-;g}^{\varepsilon}$, $\varepsilon > 0$, see [7])

$$= \left(I_{b-;g}^{(n+1)\alpha-\gamma} \left(D_{b-;g}^{(n+1)\alpha} f \right) \right) (x) . \tag{27.53}$$

We have proved that

$$\left(D_{b-;g}^{\gamma} f \right) (x) = \left(I_{b-;g}^{(n+1)\alpha-\gamma} \left(D_{b-;g}^{(n+1)\alpha} f \right) \right) (x) , \tag{27.54}$$

$\forall\, x \in [a, b]$, which is continuous by Anastassiou [6].

We have established the following representation formula:

Theorem 27.14. *All as in Corollary 27.6. Let $\gamma > 0$ with $\lceil \gamma \rceil = m < n + 1$, such that $m < (n + 1)\alpha$ (i.e. $1 \geq \alpha > \frac{m}{n+1}$). Then*

$$\left(D_{b-;g}^{\gamma} f \right) (x) = \frac{1}{\Gamma\left((n + 1)\alpha - \gamma \right)} \cdot \int_x^b \left(g(t) - g(x) \right)^{(n+1)\alpha-\gamma-1} g'(t) \left(D_{b-;g}^{(n+1)\alpha} f \right) (t)\, dt, \tag{27.55}$$

$\forall\, x \in [a, b]$, and $\left(D_{b-;g}^{\gamma} f \right) \in C([a, b])$.

We have the next very general representation formulae:

Theorem 27.15. *All here as in Corollary 27.6. Let $\gamma > 0$ with $\lceil \gamma \rceil = m$, such that $\frac{m+(k-1)\gamma}{n+1} < \alpha \leq 1$, $k \in \mathbb{N}$. Then*

$$\left(D_{b-;g}^{k\gamma} f \right) (x) = \frac{1}{\Gamma\left((n + 1)\alpha - k\gamma \right)} \cdot \int_x^b \left(g(t) - g(x) \right)^{(n+1)\alpha-k\gamma-1} g'(t) \left(D_{b-;g}^{(n+1)\alpha} f \right) (t)\, dt, \tag{27.56}$$

$\forall\, x \in [a, b]$, and $\left(D_{b-;g}^{k\gamma} f \right) \in C([a, b])$.

Proof. Similar to Theorem 27.14.

Next, we give a related fractional Ostrowski type inequality:

Theorem 27.16. *Let $g \in AC([a, b])$ and strictly increasing, and $0 < \alpha \leq 1$, $x_0 \in [a, b]$ be fixed. Assume that $F_k^{x_0} := D_{x_0-;g}^{k\alpha} f$, for $k = 0, 1, \ldots, n+1$, fulfill: $F_k^{x_0} \circ g^{-1} \in AC([g(a), g(x_0)])$, and $\left(F_k^{x_0} \circ g^{-1} \right)' \circ g \in L_{\infty}([a, x_0])$, and $\left(D_{x_0-;g}^{i\alpha} f \right) (x_0) = 0$, $i = 1, \ldots, n$.*

Similarly, we assume that $G_k^{x_0} := D_{x_0+;g}^{k\alpha} f$, *for* $k = 0, 1, \ldots, n+1$, *fulfill:* $G_k^{x_0} \circ g^{-1} \in AC([g(x_0), g(b)])$, *and* $\left(G_k^{x_0} \circ g^{-1}\right)' \circ g \in L_\infty([x_0, b])$, *and* $\left(D_{x_0+;g}^{i\alpha} f\right)(x_0) = 0$, $i = 1, \ldots, n$.

Then

$$\left|\frac{1}{b-a}\int_a^b f(x)\,dx - f(x_0)\right|$$

$$\leq \frac{1}{(b-a)\,\Gamma((n+1)\alpha+1)}$$

$$\cdot\left\{(g(b)-g(x_0))^{(n+1)\alpha}(b-x_0)\left\|D_{x_0+;g}^{(n+1)\alpha} f\right\|_{\infty,[x_0,b]}\right.$$

$$\left.+(g(x_0)-g(a))^{(n+1)\alpha}(x_0-a)\left\|D_{x_0-;g}^{(n+1)\alpha} f\right\|_{\infty,[a,x_0]}\right\}. \quad (27.57)$$

Proof. By (27.16), we obtain

$$f(x) - f(x_0)$$

$$= \frac{1}{\Gamma((n+1)\alpha)}\int_x^{x_0} (g(t)-g(x))^{(n+1)\alpha-1} g'(t)\left(D_{x_0-;g}^{(n+1)\alpha} f\right)(t)\,dt, \quad (27.58)$$

$\forall\, x \in [a, x_0]$. Hence it holds

$$|f(x) - f(x_0)|$$

$$\leq \frac{1}{\Gamma((n+1)\alpha)}\int_x^{x_0} (g(t)-g(x))^{(n+1)\alpha-1} g'(t)\left|\left(D_{x_0-;g}^{(n+1)\alpha} f\right)(t)\right| dt$$

$$\leq \frac{\left\|D_{x_0-;g}^{(n+1)\alpha} f\right\|_{\infty,[a,x_0]}}{\Gamma((n+1)\alpha)} \frac{(g(x_0)-g(x))^{(n+1)\alpha}}{(n+1)\alpha}. \quad (27.59)$$

We have proved that

$$|f(x) - f(x_0)| \leq \frac{(g(x_0)-g(x))^{(n+1)\alpha}}{\Gamma((n+1)\alpha+1)}\left\|D_{x_0-;g}^{(n+1)\alpha} f\right\|_{\infty,[a,x_0]}, \quad (27.60)$$

$\forall\, x \in [a, x_0]$. Also, by (27.20), we obtain

$$f(x) - f(x_0)$$

$$= \frac{1}{\Gamma((n+1)\alpha)}\int_{x_0}^{x} (g(x)-g(t))^{(n+1)\alpha-1} g'(t)\left(D_{x_0+;g}^{(n+1)\alpha} f\right)(t)\,dt, \quad (27.61)$$

$\forall\, x \in [x_0, b]$. Hence

$$|f(x) - f(x_0)| \leq \frac{(g(x) - g(x_0))^{(n+1)\alpha}}{\Gamma((n+1)\alpha + 1)} \left\| D_{x_0+;g}^{(n+1)\alpha} f \right\|_{\infty,[x_0,b]}, \tag{27.62}$$

$\forall\, x \in [x_0, b]$. Next we see that

$$\begin{aligned}
&\left| \frac{1}{b-a} \int_a^b f(x)\,dx - f(x_0) \right| \\
&\leq \frac{1}{b-a} \left| \int_a^b (f(x) - f(x_0))\,dx \right| \\
&\leq \frac{1}{b-a} \int_a^b |f(x) - f(x_0)|\,dx \\
&= \frac{1}{b-a} \left\{ \int_a^{x_0} |f(x) - f(x_0)|\,dx + \int_{x_0}^b |f(x) - f(x_0)|\,dx \right\},
\end{aligned} \tag{27.63}$$

which gives

$$\begin{aligned}
&\leq \frac{1}{(b-a)\,\Gamma((n+1)\alpha+1)} \left\{ \left(\int_a^{x_0} (g(x_0) - g(x))^{(n+1)\alpha}\,dx \right) \left\| D_{x_0-;g}^{(n+1)\alpha} f \right\|_{\infty,[a,x_0]} \right. \\
&\quad \left. + \left(\int_{x_0}^b (g(x) - g(x_0))^{(n+1)\alpha}\,dx \right) \left\| D_{x_0+;g}^{(n+1)\alpha} f \right\|_{\infty,[x_0,b]} \right\}
\end{aligned} \tag{27.64}$$

$$\begin{aligned}
&\leq \frac{1}{(b-a)\,\Gamma((n+1)\alpha+1)} \left\{ (g(x_0) - g(a))^{(n+1)\alpha} (x_0 - a) \left\| D_{x_0-;g}^{(n+1)\alpha} f \right\|_{\infty,[a,x_0]} \right. \\
&\quad \left. + (g(b) - g(x_0))^{(n+1)\alpha} (b - x_0) \left\| D_{x_0+;g}^{(n+1)\alpha} f \right\|_{\infty,[x_0,b]} \right\},
\end{aligned} \tag{27.65}$$

proving the claim.

It follows a left fractional Poincaré type inequality:

Theorem 27.17. *Here all as in Theorem 27.12. Let* $p, q > 1 : \frac{1}{p} + \frac{1}{q} = 1$. *We further assume that* $(k \in \mathbb{N})$

$$1 \geq \alpha > \max\left(\frac{m + (k-1)\gamma}{n+1}, \frac{k\gamma q + 1}{(n+1)\,q} \right). \tag{27.66}$$

Then

$$\left\| D^{k\gamma}_{a+;g} f \right\|_{q,[a,b]} \leq \frac{(g(b)-g(a))^{((n+1)\alpha-k\gamma-1)+\frac{1}{p}} (b-a)^{\frac{1}{q}}}{\Gamma((n+1)\alpha-k\gamma)(p((n+1)\alpha-k\gamma-1)+1)^{\frac{1}{p}}} \cdot \left\| \left(D^{(n+1)\alpha}_{a+;g} f \right) \circ g^{-1} \right\|_{q,[g(a),g(b)]}. \tag{27.67}$$

Proof. We use (27.41). We observe that [12]

$$\left| \left(D^{k\gamma}_{a+;g} f \right)(x) \right| \leq \frac{1}{\Gamma((n+1)\alpha-k\gamma)} \cdot \int_a^x (g(x)-g(t))^{(n+1)\alpha-k\gamma-1} g'(t) \left| \left(D^{(n+1)\alpha}_{a+;g} f \right)(t) \right| dt \tag{27.68}$$

(same reasoning as in Remark 27.8)

$$= \frac{1}{\Gamma((n+1)\alpha-k\gamma)} \int_{g(a)}^{g(x)} (g(x)-z)^{(n+1)\alpha-k\gamma-1} \left| \left(D^{(n+1)\alpha}_{a+;g} f \right)\left(g^{-1}(z) \right) \right| dz$$

$$\leq \frac{1}{\Gamma((n+1)\alpha-k\gamma)} \left(\int_{g(a)}^{g(x)} (g(x)-z)^{p((n+1)\alpha-k\gamma-1)} dz \right)^{\frac{1}{p}} \tag{27.69}$$

$$\cdot \left(\int_{g(a)}^{g(b)} \left| \left(D^{(n+1)\alpha}_{a+;g} f \right)\left(g^{-1}(z) \right) \right|^q dz \right)^{\frac{1}{q}}$$

$$= \frac{(g(x)-g(a))^{\frac{p((n+1)\alpha-k\gamma-1)+1}{p}}}{\Gamma((n+1)\alpha-k\gamma)(p((n+1)\alpha-k\gamma-1)+1)^{\frac{1}{p}}} \cdot \left\| \left(D^{(n+1)\alpha}_{a+;g} f \right) \circ g^{-1} \right\|_{q,[g(a),g(b)]}.$$

Thus we have

$$\left| \left(D^{k\gamma}_{a+;g} f \right)(x) \right|^q \leq \frac{(g(x)-g(a))^{(p((n+1)\alpha-k\gamma-1)+1)\frac{q}{p}}}{\Gamma((n+1)\alpha-k\gamma)^q (p((n+1)\alpha-k\gamma-1)+1)^{\frac{q}{p}}} \cdot \left\| \left(D^{(n+1)\alpha}_{a+;g} f \right) \circ g^{-1} \right\|^q_{q,[g(a),g(b)]}. \tag{27.70}$$

Therefore it holds

$$\int_a^b \left| \left(D_{a+;g}^{k\gamma} f \right)(x) \right|^q dx \leq \frac{(g(x) - g(a))^{(p((n+1)\alpha - k\gamma - 1)+1)\frac{q}{p}} (b-a)}{\Gamma((n+1)\alpha - k\gamma)^q (p((n+1)\alpha - k\gamma - 1) + 1)^{\frac{q}{p}}} \cdot \left\| \left(D_{a+;g}^{(n+1)\alpha} f \right) \circ g^{-1} \right\|_{q,[g(a),g(b)]}^q, \tag{27.71}$$

proving the claim.

It follows a right fractional Poincaré type inequality:

Theorem 27.18. *Here all as in Theorem 27.15. Let $p, q > 1 : \frac{1}{p} + \frac{1}{q} = 1$. We further assume that ($k \in \mathbb{N}$)*

$$1 \geq \alpha > \max\left(\frac{m + (k-1)\gamma}{n+1}, \frac{k\gamma q + 1}{(n+1)q} \right). \tag{27.72}$$

Then

$$\left\| D_{b-;g}^{k\gamma} f \right\|_{q,[a,b]} \leq \frac{(g(b) - g(a))^{((n+1)\alpha - k\gamma - 1) + \frac{1}{p}} (b-a)^{\frac{1}{q}}}{\Gamma((n+1)\alpha - k\gamma)(p((n+1)\alpha - k\gamma - 1) + 1)^{\frac{1}{p}}} \cdot \left\| \left(D_{b-;g}^{(n+1)\alpha} f \right) \circ g^{-1} \right\|_{q,[g(a),g(b)]}. \tag{27.73}$$

Proof. As similar to Theorem 27.17 is omitted.

Next comes a left fractional Opial type inequality:

Theorem 27.19. *All here as in Theorem 27.12. Let $p, q > 1 : \frac{1}{p} + \frac{1}{q} = 1$. We further assume that ($k \in \mathbb{N}$)*

$$1 \geq \alpha > \max\left(\frac{m + (k-1)\gamma}{n+1}, \frac{k\gamma q + 1}{(n+1)q} \right). \tag{27.74}$$

Then

$$\int_{g(a)}^{y} \left| \left(\left(D_{a+;g}^{k\gamma} f \right) \circ g^{-1} \right)(w) \right| \left| \left(\left(D_{a+;g}^{(n+1)\alpha} f \right) \circ g^{-1} \right)(w) \right| dw$$

$$\leq \frac{(y - g(a))^{((n+1)\alpha - k\gamma - 1) + \frac{2}{p}}}{2^{\frac{1}{q}} \Gamma((n+1)\alpha - k\gamma) \left[(p((n+1)\alpha - k\gamma - 1) + 1)(p((n+1)\alpha - k\gamma - 1) + 2) \right]^{\frac{1}{p}}}$$

$$\cdot \left(\int_{g(a)}^{y} \left| \left(\left(D_{a+;g}^{(n+1)\alpha} f \right) \circ g^{-1} \right)(w) \right|^q dw \right)^{\frac{2}{q}}, \tag{27.75}$$

$\forall\, y \in [g(a), g(b)]$.

Proof. We use (27.41). We observe that

$$\left|\left(D_{a+;g}^{k\gamma}f\right)(x)\right| \leq \frac{1}{\Gamma((n+1)\alpha-k\gamma)} \cdot \int_a^x (g(x)-g(t))^{(n+1)\alpha-k\gamma-1} g'(t) \left|\left(D_{a+;g}^{(n+1)\alpha}f\right)(t)\right| dt \quad (27.76)$$

(same reasoning as in Remark 27.8)

$$= \frac{1}{\Gamma((n+1)\alpha-k\gamma)} \int_{g(a)}^{g(x)} (g(x)-z)^{(n+1)\alpha-k\gamma-1} \left|\left(D_{a+;g}^{(n+1)\alpha}f\right)\left(g^{-1}(z)\right)\right| dz$$

$$\leq \frac{1}{\Gamma((n+1)\alpha-k\gamma)} \left(\int_{g(a)}^{g(x)} (g(x)-z)^{p((n+1)\alpha-k\gamma-1)} dz\right)^{\frac{1}{p}} \cdot \left(\int_{g(a)}^{g(x)} \left|\left(D_{a+;g}^{(n+1)\alpha}f\right)\left(g^{-1}(z)\right)\right|^q dz\right)^{\frac{1}{q}} \quad (27.77)$$

$$= \frac{1}{\Gamma((n+1)\alpha-k\gamma)} \frac{(g(x)-g(a))^{\frac{p((n+1)\alpha-k\gamma-1)+1}{p}}}{(p((n+1)\alpha-k\gamma-1)+1)^{\frac{1}{p}}} \cdot \left(\int_{g(a)}^{g(x)} \left|\left(D_{a+;g}^{(n+1)\alpha}f\right)\left(g^{-1}(z)\right)\right|^q dz\right)^{\frac{1}{q}}. \quad (27.78)$$

That is,

$$\left|\left(D_{a+;g}^{k\gamma}f\right)(x)\right| \leq \frac{(g(x)-g(a))^{\frac{p((n+1)\alpha-k\gamma-1)+1}{p}}}{\Gamma((n+1)\alpha-k\gamma)(p((n+1)\alpha-k\gamma-1)+1)^{\frac{1}{p}}} \cdot \left(\int_{g(a)}^{g(x)} \left|\left(D_{a+;g}^{(n+1)\alpha}f\right)\left(g^{-1}(z)\right)\right|^q dz\right)^{\frac{1}{q}}, \quad (27.79)$$

$\forall\, x \in [a,b]$. Call $y = g(x)$, then

$$x = g^{-1}(y). \quad (27.80)$$

Hence it holds

$$\left|\left(D_{a+;g}^{k\gamma}f\right)\left(g^{-1}(y)\right)\right| \leq \frac{(y-g(a))^{\frac{p((n+1)\alpha-k\gamma-1)+1}{p}}}{\Gamma((n+1)\alpha-k\gamma)(p((n+1)\alpha-k\gamma-1)+1)^{\frac{1}{p}}} \cdot \left(\int_{g(a)}^{y}\left|\left(D_{a+;g}^{(n+1)\alpha}f\right)\left(g^{-1}(z)\right)\right|^{q}dz\right)^{\frac{1}{q}}, \tag{27.81}$$

$\forall\, y \in [g(a), g(b)]$. Call

$$\eta(y) := \int_{g(a)}^{y}\left|\left(D_{a+;g}^{(n+1)\alpha}f\right)\left(g^{-1}(z)\right)\right|^{q}dz, \tag{27.82}$$

and

$$\eta(g(a)) = 0.$$

Thus

$$\eta'(y) = \left|\left(D_{a+;g}^{(n+1)\alpha}f\right)\left(g^{-1}(y)\right)\right|^{q} \geq 0, \tag{27.83}$$

and

$$\left(\eta'(y)\right)^{\frac{1}{q}} = \left|\left(D_{a+;g}^{(n+1)\alpha}f\right)\left(g^{-1}(y)\right)\right| \geq 0, \tag{27.84}$$

$\forall\, y \in [g(a), g(b)]$. Consequently, we get

$$\left|\left(D_{a+;g}^{k\gamma}f\right)\left(g^{-1}(w)\right)\right|\left|\left(D_{a+;g}^{(n+1)\alpha}f\right)\left(g^{-1}(w)\right)\right| \tag{27.85}$$

$$\leq \frac{(w-g(a))^{\frac{p((n+1)\alpha-k\gamma-1)+1}{p}}}{\Gamma((n+1)\alpha-k\gamma)(p((n+1)\alpha-k\gamma-1)+1)^{\frac{1}{p}}}\left(\eta(w)\,\eta'(w)\right)^{\frac{1}{q}},$$

$\forall\, w \in [g(a), g(b)]$. Then it holds

$$\int_{g(a)}^{y}\left|\left(D_{a+;g}^{k\gamma}f\right)\left(g^{-1}(w)\right)\right|\left|\left(D_{a+;g}^{(n+1)\alpha}f\right)\left(g^{-1}(w)\right)\right|dw$$

$$\leq \frac{1}{\Gamma((n+1)\alpha-k\gamma)(p((n+1)\alpha-k\gamma-1)+1)^{\frac{1}{p}}} \tag{27.86}$$

$$\cdot \int_{g(a)}^{y} (w - g(a))^{\frac{p((n+1)\alpha - k\gamma - 1)+1}{p}} \left(\eta(w)\eta'(w)\right)^{\frac{1}{q}} dw$$

$$\leq \frac{1}{\Gamma((n+1)\alpha - k\gamma)(p((n+1)\alpha - k\gamma - 1)+1)^{\frac{1}{p}}} \tag{27.87}$$

$$\cdot \left(\int_{g(a)}^{y} (w - g(a))^{(p((n+1)\alpha - k\gamma - 1)+1)} dw \right)^{\frac{1}{p}} \left(\int_{g(a)}^{y} \eta(w)\eta'(w)\, dw \right)^{\frac{1}{q}},$$

which gives

$$= \frac{(y - g(a))^{\frac{p((n+1)\alpha - k\gamma - 1)+2}{p}}}{\Gamma((n+1)\alpha - k\gamma)\left[(p((n+1)\alpha - k\gamma - 1)+1)(p((n+1)\alpha - k\gamma - 1)+2)\right]^{\frac{1}{p}}}$$

$$\cdot \left(\frac{\eta^2(y)}{2} \right)^{\frac{1}{q}} \tag{27.88}$$

$$= \frac{(y - g(a))^{((n+1)\alpha - k\gamma - 1)+\frac{2}{p}}}{2^{\frac{1}{q}}\Gamma((n+1)\alpha - k\gamma)\left[(p((n+1)\alpha - k\gamma - 1)+1)(p((n+1)\alpha - k\gamma - 1)+2)\right]^{\frac{1}{p}}}$$

$$\cdot \left(\int_{g(a)}^{y} \left| \left(\left(D_{a+;g}^{(n+1)\alpha} f \right) \circ g^{-1} \right)(w) \right|^{q} dw \right)^{\frac{2}{q}}, \tag{27.89}$$

$\forall\, y \in [g(a), g(b)]$, proving the claim.

Also we give a right fractional Opial type inequality:

Theorem 27.20. *All here as in Theorem 27.15. Let $p, q > 1 : \frac{1}{p} + \frac{1}{q} = 1$. We further assume that $(k \in \mathbb{N})$*

$$1 \geq \alpha > \max\left(\frac{m + (k-1)\gamma}{n+1}, \frac{k\gamma q + 1}{(n+1)q} \right). \tag{27.90}$$

Then

$$\int_{y}^{g(b)} \left| \left(\left(D_{b-;g}^{k\gamma} f \right) \circ g^{-1} \right)(w) \right| \left| \left(\left(D_{b-;g}^{(n+1)\alpha} f \right) \circ g^{-1} \right)(w) \right| dw$$

$$\leq \frac{(g(b) - y)^{((n+1)\alpha - k\gamma - 1)+\frac{2}{p}}}{2^{\frac{1}{q}}\Gamma((n+1)\alpha - k\gamma)\left[(p((n+1)\alpha - k\gamma - 1)+1)(p((n+1)\alpha - k\gamma - 1)+2)\right]^{\frac{1}{p}}} \tag{27.91}$$

$$\cdot \left(\int_{y}^{g(b)} \left| \left(\left(D_{b-;g}^{(n+1)\alpha} f \right) \circ g^{-1} \right) (w) \right|^{q} dw \right)^{\frac{2}{q}},$$

$\forall\, y \in [g(a), g(b)]$.

Proof. It is omitted as similar to Theorem 27.19.

Next we present a left fractional Hilbert–Pachpatte type inequality:

Theorem 27.21. *Here $i = 1, 2$. Let $a_i, b_i \in \mathbb{R}$, $a_i < b_i$, $0 < \alpha_i \leq 1$, and $g_i \in C([a_i, b_i])$ that are strictly increasing, $f_i : [a_i, b_i] \to \mathbb{R}$. Suppose that $F_{\overline{k_i}} := D_{a_i+;g_i}^{\overline{k_i}\alpha_i} f_i$, for $\overline{k_i} = 0, 1, \ldots, n_i + 1$, fulfill $\left(F_{\overline{k_i}} \circ g_i^{-1}\right) \in AC([g_i(a_i), g_i(b_i)])$ and $\left(F_{\overline{k_i}} \circ g_i^{-1}\right)' \circ g_i \in L_\infty([a_i, b_i])$, and $\left(D_{a_i+;g_i}^{j_i\alpha_i} f_i\right)(a_i) = 0$, $j_i = 0, 1, \ldots, n_i$. Let $\gamma_i > 0$ with $\lceil \gamma_i \rceil = m_i$, $k_i \in \mathbb{N}$, $p, q > 1 : \frac{1}{p} + \frac{1}{q} = 1$. We further assume*

$$1 \geq \alpha_1 > \max \left(\frac{m_1 + (k_1 - 1)\gamma_1}{n_1 + 1}, \frac{k_1\gamma_1 q + 1}{(n_1 + 1) q} \right), \tag{27.92}$$

and

$$1 \geq \alpha_2 > \max \left(\frac{m_2 + (k_2 - 1)\gamma_2}{n_2 + 1}, \frac{k_2\gamma_2 p + 1}{(n_2 + 1) p} \right). \tag{27.93}$$

Then

$$\int_{g_1(a_1)}^{g_1(b_1)} \int_{g_2(a_2)}^{g_2(b_2)} \frac{\left| \left(D_{a_1+;g_1}^{k_1\gamma_1} f_1 \right) \left(g_1^{-1}(y_1) \right) \right| \left| \left(D_{a_2+;g_2}^{k_2\gamma_2} f_2 \right) \left(g_2^{-1}(y_2) \right) \right|}{\left[\frac{(y_1 - g_1(a_1))^{p((n_1+1)\alpha_1 - k_1\gamma_1 - 1) + 1}}{p(p((n_1+1)\alpha_1 - k_1\gamma_1 - 1) + 1)} + \frac{(y_2 - g_2(a_2))^{q((n_2+1)\alpha_2 - k_2\gamma_2 - 1) + 1}}{q(q((n_2+1)\alpha_2 - k_2\gamma_2 - 1) + 1)} \right]} dy_1 dy_2$$

$$\leq \frac{(g_1(b_1) - g_1(a_1))(g_2(b_2) - g_2(a_2))}{\Gamma((n_1 + 1)\alpha_1 - k_1\gamma_1)\, \Gamma((n_2 + 1)\alpha_2 - k_2\gamma_2)}$$

$$\cdot \left(\int_{g_1(a_1)}^{g_1(b_1)} \left| \left(D_{a_1+;g_1}^{(n_1+1)\alpha_1} f_1 \right) \left(g_1^{-1}(z_1) \right) \right|^{q} dz_1 \right)^{\frac{1}{q}}$$

$$\cdot \left(\int_{g_2(a_2)}^{g_2(b_2)} \left| \left(D_{a_2+;g_2}^{(n_2+1)\alpha_2} f_2 \right) \left(g_2^{-1}(z_2) \right) \right|^{p} dz_2 \right)^{\frac{1}{p}}. \tag{27.94}$$

Proof. As in the proof of Theorem 27.19, we obtain

$$\left| \left(D_{a_1+;g_1}^{k_1\gamma_1} f_1 \right) \left(g_1^{-1}(y_1) \right) \right|$$

$$\leq \frac{(y_1 - g_1(a_1))^{\frac{p((n_1+1)\alpha_1 - k_1\gamma_1 - 1)+1}{p}}}{\Gamma((n_1+1)\alpha_1 - k_1\gamma_1)(p((n_1+1)\alpha_1 - k_1\gamma_1 - 1)+1)^{\frac{1}{p}}}$$

$$\cdot \left(\int_{g_1(a_1)}^{y_1} \left| \left(D_{a_1+;g_1}^{(n_1+1)\alpha_1} f_1 \right) \left(g_1^{-1}(z_1) \right) \right|^q dz_1 \right)^{\frac{1}{q}}, \tag{27.95}$$

$\forall\, y_1 \in [g_1(a_1), g_1(b_1)]$. Also we obtain

$$\left| \left(D_{a_2+;g_2}^{k_2\gamma_2} f_2 \right) \left(g_2^{-1}(y_2) \right) \right|$$

$$\leq \frac{(y_2 - g_2(a_2))^{\frac{q((n_2+1)\alpha_2 - k_2\gamma_2 - 1)+1}{q}}}{\Gamma((n_2+1)\alpha_2 - k_2\gamma_2)(q((n_2+1)\alpha_2 - k_2\gamma_2 - 1)+1)^{\frac{1}{q}}}$$

$$\cdot \left(\int_{g_2(a_2)}^{y_2} \left| \left(D_{a_2+;g_2}^{(n_2+1)\alpha_2} f_2 \right) \left(g_2^{-1}(z_2) \right) \right|^p dz_2 \right)^{\frac{1}{p}}, \tag{27.96}$$

$\forall\, y_2 \in [g_2(a_2), g_2(b_2)]$. Multiplying (27.95) and (27.96), we get

$$\left| \left(D_{a_1+;g_1}^{k_1\gamma_1} f_1 \right) \left(g_1^{-1}(y_1) \right) \right| \left| \left(D_{a_2+;g_2}^{k_2\gamma_2} f_2 \right) \left(g_2^{-1}(y_2) \right) \right|$$

$$\leq \frac{1}{\Gamma((n_1+1)\alpha_1 - k_1\gamma_1)\,\Gamma((n_2+1)\alpha_2 - k_2\gamma_2)}$$

$$\cdot \left(\frac{(y_1 - g_1(a_1))^{p((n_1+1)\alpha_1 - k_1\gamma_1 - 1)+1}}{(p((n_1+1)\alpha_1 - k_1\gamma_1 - 1)+1)} \right)^{\frac{1}{p}}$$

$$\cdot \left(\frac{(y_2 - g_2(a_2))^{q((n_2+1)\alpha_2 - k_2\gamma_2 - 1)+1}}{(q((n_2+1)\alpha_2 - k_2\gamma_2 - 1)+1)} \right)^{\frac{1}{q}}$$

$$\cdot \left(\int_{g_1(a_1)}^{y_1} \left| \left(D_{a_1+;g_1}^{(n_1+1)\alpha_1} f_1 \right) \left(g_1^{-1}(z_1) \right) \right|^q dz_1 \right)^{\frac{1}{q}}$$

$$\cdot \left(\int_{g_2(a_2)}^{y_2} \left| \left(D_{a_2+;g_2}^{(n_2+1)\alpha_2} f_2 \right) \left(g_2^{-1}(z_2) \right) \right|^p dz_2 \right)^{\frac{1}{p}} \tag{27.97}$$

(using Young's inequality for $a, b \geq 0$, $a^{\frac{1}{p}} b^{\frac{1}{q}} \leq \frac{a}{p} + \frac{b}{q}$)

$$\leq \frac{1}{\Gamma\left(\left(n_1+1\right)\alpha_1-k_1\gamma_1\right)\Gamma\left(\left(n_2+1\right)\alpha_2-k_2\gamma_2\right)}$$
$$\cdot\left[\left(\frac{\left(y_1-g_1\left(a_1\right)\right)^{p\left(\left(n_1+1\right)\alpha_1-k_1\gamma_1-1\right)+1}}{p\left(p\left(\left(n_1+1\right)\alpha_1-k_1\gamma_1-1\right)+1\right)}\right)\right.$$
$$\left.+\left(\frac{\left(y_2-g_2\left(a_2\right)\right)^{q\left(\left(n_2+1\right)\alpha_2-k_2\gamma_2-1\right)+1}}{q\left(q\left(\left(n_2+1\right)\alpha_2-k_2\gamma_2-1\right)+1\right)}\right)\right]$$
$$\cdot\left(\int_{g_1(a_1)}^{y_1}\left|\left(D_{a_1+;g_1}^{(n_1+1)\alpha_1}f_1\right)\left(g_1^{-1}\left(z_1\right)\right)\right|^q dz_1\right)^{\frac{1}{q}}$$
$$\cdot\left(\int_{g_2(a_2)}^{y_2}\left|\left(D_{a_2+;g_2}^{(n_2+1)\alpha_2}f_2\right)\left(g_2^{-1}\left(z_2\right)\right)\right|^p dz_2\right)^{\frac{1}{p}}. \tag{27.98}$$

So far we have

$$\frac{\left|\left(D_{a_1+;g_1}^{k_1\gamma_1}f_1\right)\left(g_1^{-1}\left(y_1\right)\right)\right|\left|\left(D_{a_2+;g_2}^{k_2\gamma_2}f_2\right)\left(g_2^{-1}\left(y_2\right)\right)\right|}{\left[\left(\frac{\left(y_1-g_1\left(a_1\right)\right)^{p\left(\left(n_1+1\right)\alpha_1-k_1\gamma_1-1\right)+1}}{p\left(p\left(\left(n_1+1\right)\alpha_1-k_1\gamma_1-1\right)+1\right)}\right)+\left(\frac{\left(y_2-g_2\left(a_2\right)\right)^{q\left(\left(n_2+1\right)\alpha_2-k_2\gamma_2-1\right)+1}}{q\left(q\left(\left(n_2+1\right)\alpha_2-k_2\gamma_2-1\right)+1\right)}\right)\right]}$$
$$\leq \frac{1}{\Gamma\left(\left(n_1+1\right)\alpha_1-k_1\gamma_1\right)\Gamma\left(\left(n_2+1\right)\alpha_2-k_2\gamma_2\right)}$$
$$\cdot\left(\int_{g_1(a_1)}^{y_1}\left|\left(D_{a_1+;g_1}^{(n_1+1)\alpha_1}f_1\right)\left(g_1^{-1}\left(z_1\right)\right)\right|^q dz_1\right)^{\frac{1}{q}}$$
$$\cdot\left(\int_{g_2(a_2)}^{y_2}\left|\left(D_{a_2+;g_2}^{(n_2+1)\alpha_2}f_2\right)\left(g_2^{-1}\left(z_2\right)\right)\right|^p dz_2\right)^{\frac{1}{p}}. \tag{27.99}$$

The denominator in (27.99) can be zero only when $y_1 = g_1(a_1)$ and $y_2 = g_2(a_2)$. Therefore we obtain

$$\int_{g_1(a_1)}^{g_1(b_1)}\int_{g_2(a_2)}^{g_2(b_2)}\frac{\left|\left(D_{a_1+;g_1}^{k_1\gamma_1}f_1\right)\left(g_1^{-1}\left(y_1\right)\right)\right|\left|\left(D_{a_2+;g_2}^{k_2\gamma_2}f_2\right)\left(g_2^{-1}\left(y_2\right)\right)\right|dy_1dy_2}{\left[\left(\frac{\left(y_1-g_1\left(a_1\right)\right)^{p\left(\left(n_1+1\right)\alpha_1-k_1\gamma_1-1\right)+1}}{p\left(p\left(\left(n_1+1\right)\alpha_1-k_1\gamma_1-1\right)+1\right)}\right)+\left(\frac{\left(y_2-g_2\left(a_2\right)\right)^{q\left(\left(n_2+1\right)\alpha_2-k_2\gamma_2-1\right)+1}}{q\left(q\left(\left(n_2+1\right)\alpha_2-k_2\gamma_2-1\right)+1\right)}\right)\right]}$$
$$\leq \frac{1}{\Gamma\left(\left(n_1+1\right)\alpha_1-k_1\gamma_1\right)\Gamma\left(\left(n_2+1\right)\alpha_2-k_2\gamma_2\right)} \tag{27.100}$$
$$\cdot\left(\int_{g_1(a_1)}^{g_1(b_1)}\left(\int_{g_1(a_1)}^{y_1}\left|\left(D_{a_1+;g_1}^{(n_1+1)\alpha_1}f_1\right)\left(g_1^{-1}\left(z_1\right)\right)\right|^q dz_1\right)^{\frac{1}{q}}dy_1\right)$$

$$\cdot\left(\int_{g_2(a_2)}^{g_2(b_2)}\left(\int_{g_2(a_2)}^{y_2}\left|\left(D_{a_2+;g_2}^{(n_2+1)\alpha_2}f_2\right)\left(g_2^{-1}(z_2)\right)\right|^p dz_2\right)^{\frac{1}{p}} dy_2\right)$$

which gives

$$\leq \frac{1}{\Gamma\left(\left(n_1+1\right)\alpha_1-k_1\gamma_1\right)\Gamma\left(\left(n_2+1\right)\alpha_2-k_2\gamma_2\right)}$$

$$\cdot\left(\int_{g_1(a_1)}^{g_1(b_1)}\left(\int_{g_1(a_1)}^{g_1(b_1)}\left|\left(D_{a_1+;g_1}^{(n_1+1)\alpha_1}f_1\right)\left(g_1^{-1}(z_1)\right)\right|^q dz_1\right)^{\frac{1}{q}} dy_1\right)$$

$$\cdot\left(\int_{g_2(a_2)}^{g_2(b_2)}\left(\int_{g_2(a_2)}^{g_2(b_2)}\left|\left(D_{a_2+;g_2}^{(n_2+1)\alpha_2}f_2\right)\left(g_2^{-1}(z_2)\right)\right|^p dz_2\right)^{\frac{1}{p}} dy_2\right) \quad (27.101)$$

$$= \frac{\left(g_1(b_1)-g_1(a_1)\right)\left(g_2(b_2)-g_2(a_2)\right)}{\Gamma\left(\left(n_1+1\right)\alpha_1-k_1\gamma_1\right)\Gamma\left(\left(n_2+1\right)\alpha_2-k_2\gamma_2\right)}$$

$$\cdot\left(\int_{g_1(a_1)}^{g_1(b_1)}\left|\left(D_{a_1+;g_1}^{(n_1+1)\alpha_1}f_1\right)\left(g_1^{-1}(z_1)\right)\right|^q dz_1\right)^{\frac{1}{q}}$$

$$\cdot\left(\int_{g_2(a_2)}^{g_2(b_2)}\left|\left(D_{a_2+;g_2}^{(n_2+1)\alpha_2}f_2\right)\left(g_2^{-1}(z_2)\right)\right|^p dz_2\right)^{\frac{1}{p}}. \quad (27.102)$$

The theorem is proved.

Finally we present a right fractional Hilbert–Pachpatte type inequality:

Theorem 27.22. *Here $i=1,2$. Let $a_i,b_i\in\mathbb{R}$, $a_i<b_i$, $0<\alpha_i\leq 1$, and $g_i\in C([a_i,b_i])$ that are strictly increasing, $f_i:[a_i,b_i]\to\mathbb{R}$. Suppose that $F_{\overline{k_i}}:=D_{b_i-;g_i}^{\overline{k_i}\alpha_i}f_i$, for $\overline{k_i}=0,1,\ldots,n_i+1$, fulfill $\left(F_{\overline{k_i}}\circ g_i^{-1}\right)\in AC\left(\left[g_i(a_i),g_i(b_i)\right]\right)$ and $\left(F_{\overline{k_i}}\circ g_i^{-1}\right)'\circ g_i\in L_\infty([a_i,b_i])$, and $\left(D_{b_i-;g_i}^{j_i\alpha_i}f_i\right)(b_i)=0$, $j_i=0,1,\ldots,n_i$. Let $\gamma_i>0$ with $\lceil\gamma_i\rceil=m_i$, $k_i\in\mathbb{N}$, $p,q>1:\frac{1}{p}+\frac{1}{q}=1$. We further assume*

$$1\geq\alpha_1>\max\left(\frac{m_1+(k_1-1)\gamma_1}{n_1+1},\frac{k_1\gamma_1 q+1}{(n_1+1)q}\right). \quad (27.103)$$

and

$$1\geq\alpha_2>\max\left(\frac{m_2+(k_2-1)\gamma_2}{n_2+1},\frac{k_2\gamma_2 p+1}{(n_2+1)p}\right). \quad (27.104)$$

Then

$$\int_{g_1(a_1)}^{g_1(b_1)} \int_{g_2(a_2)}^{g_2(b_2)} \frac{\left|\left(D_{b_1-;g_1}^{k_1\gamma_1} f_1\right)\left(g_1^{-1}(y_1)\right)\right| \left|\left(D_{b_2-;g_2}^{k_2\gamma_2} f_2\right)\left(g_2^{-1}(y_2)\right)\right|}{\left[\frac{(g_1(b_1)-y_1)^{p((n_1+1)\alpha_1-k_1\gamma_1-1)+1}}{p(p((n_1+1)\alpha_1-k_1\gamma_1-1)+1)} + \frac{(g_2(b_2)-y_2)^{q((n_2+1)\alpha_2-k_2\gamma_2-1)+1}}{q(q((n_2+1)\alpha_2-k_2\gamma_2-1)+1)}\right]} dy_1 dy_2$$

$$\leq \frac{(g_1(b_1)-g_1(a_1))(g_2(b_2)-g_2(a_2))}{\Gamma((n_1+1)\alpha_1-k_1\gamma_1)\,\Gamma((n_2+1)\alpha_2-k_2\gamma_2)}$$

$$\cdot \left(\int_{g_1(a_1)}^{g_1(b_1)} \left|\left(D_{b_1-;g_1}^{(n_1+1)\alpha_1} f_1\right)\left(g_1^{-1}(z_1)\right)\right|^q dz_1\right)^{\frac{1}{q}}$$

$$\cdot \left(\int_{g_2(a_2)}^{g_2(b_2)} \left|\left(D_{b_2-;g_2}^{(n_2+1)\alpha_2} f_2\right)\left(g_2^{-1}(z_2)\right)\right|^p dz_2\right)^{\frac{1}{p}}. \tag{27.105}$$

Proof. It is similar to Theorem 27.21, thus it is omitted.

Remark 27.23. Some examples for g follow:

$$g(x) = e^x,\ x \in [a,b] \subset \mathbb{R},$$

$$g(x) = \sin x,$$

$$g(x) = \tan x,$$

$$\text{where } x \in \left[-\frac{\pi}{2}+\varepsilon, \frac{\pi}{2}-\varepsilon\right],\ \varepsilon > 0 \text{ small.}$$

Indeed, the above examples of g are strictly increasing and absolutely continuous functions.

One can apply all of our results here for the above specific choices of g. We choose to omit this lengthy job.

References

1. G. Anastassiou, *Fractional Differentiation Inequalities* (Springer, New York, 2009)
2. G. Anastassiou, Fractional representation formulae and right fractional inequalities. Math. Comput. Model. **54**(11–12), 3098–3115 (2011)
3. G. Anastassiou, Opial type inequalities for functions and their ordinary and balanced fractional derivatives. J. Comput. Anal. Appl. **14**(5), 862–879 (2012)
4. G. Anastassiou, The reduction method in fractional calculus and fractional Ostrowski type inequalities. Indian J. Math. **56**(3), 333–357 (2014)
5. G. Anastassiou, Univariate left general high order fractional monotone approximation. Acta Math. Univ. Comenianae (2015, accepted)
6. G. Anastassiou, Univariate right general high order fractional monotone approximation theory. Panam. Math. J. (2015, accepted)

7. G. Anastassiou, Advanced fractional Taylor's formulae. J. Comput. Anal. Appl. (2015, accepted)
8. G. Anastassiou, I. Argyros, A convergence analysis for a certain family of extended iterative methods: part II. Applications to fractional calculus. Ann. Univ. Sci. Budapest Sect. Comp. (2015, accepted)
9. R.-Q. Jia, *Absolutely Continuous Functions*, Chap. 3. https://www.ualberta.ca/126rjia/Math418/Notes/Chap.3.pdf
10. A.A. Kilbas, H.M. Srivastava, J.J. Tujillo, *Theory and Applications of Fractional Differential Equations*. North-Holland Mathematics Studies, vol. 204 (Elsevier, New York, 2006)
11. Z.M. Odibat, N.J. Shawagleh, Generalized Taylor's formula. Appl. Math. Comput. **186**, 286–293 (2007)
12. A.R. Schep, Differentiation of monotone functions (2012). people.math.sc.edu/schep/diffmonotone.pdf

MIX
Papier aus verantwortungsvollen Quellen
Paper from responsible sources
FSC® C105338

If you have any concerns about our products,
you can contact us on
ProductSafety@springernature.com

In case Publisher is established outside the EU,
the EU authorized representative is:
Springer Nature Customer Service Center GmbH
Europaplatz 3, 69115 Heidelberg, Germany

Printed by Libri Plureos GmbH
in Hamburg, Germany